Advances in Experimental Medicine and Biology

Volume 1274

Advances in Experimental Medicine and Biology provides a platform for scientific contributions in the main disciplines of the biomedicine and the life sciences. This series publishes thematic volumes on contemporary research in the areas of microbiology, immunology, neurosciences, biochemistry, biomedical engineering, genetics, physiology, and cancer research. Covering emerging topics and techniques in basic and clinical science, it brings together clinicians and researchers from various fields.

Advances in Experimental Medicine and Biology has been publishing exceptional works in the field for over 40 years, and is indexed in SCOPUS, Medline (PubMed), Journal Citation Reports/Science Edition, Science Citation Index Expanded (SciSearch, Web of Science), EMBASE, BIOSIS, Reaxys, EMBiology, the Chemical Abstracts Service (CAS), and Pathway Studio.

2019 Impact Factor: 2.450 5 Year Impact Factor: 2.324.

More information about this series at http://www.springer.com/series/5584

Yasuyuki Kihara
Editor

Druggable Lipid Signaling Pathways

Springer

Editor
Yasuyuki Kihara
Sanford Burnham Prebys Medical
Discovery Institute
La Jolla, CA, USA

ISSN 0065-2598 ISSN 2214-8019 (electronic)
Advances in Experimental Medicine and Biology
ISBN 978-3-030-50620-9 ISBN 978-3-030-50621-6 (eBook)
https://doi.org/10.1007/978-3-030-50621-6

This Springer imprint is published by the registered company Springer Nature Switzerland AG
The registered company address is: Gewerbestrasse 11, 6330 Cham, Switzerland

Contents

Introduction: Druggable Lipid Signaling Pathways

1

Yasuyuki Kihara

Abstract

Lipids are essential for life. They store energy, constitute cellular membranes, serve as signaling molecules, and modify proteins. In the long history of lipid research, many drugs targeting lipid receptors and enzymes that are responsible for lipid metabolism and function have been developed and applied to a variety of diseases. For example, non-steroidal anti-inflammatory drugs (NSAIDs) are commonly prescribed medications for fever, pain, and inflammation. The NSAIDs block prostaglandin production by inhibiting cyclooxygenases. A recent innovative breakthrough in drug discovery for the lipid biology field was the development of the sphingosine 1-phosphate receptor modulators (fingolimod, siponimod and ozanimod) for the treatment of multiple sclerosis, which were approved by the United States Food and Drug Administration in 2010, 2019 and 2020, respectively. This review series of "Druggable Lipid Signaling Pathways" provides 9 outstanding reviews that summarize the currently available drugs that target lipid signaling pathways and also outlines future directions for drug discovery. The review chapters include lipid signaling pathways (prostanoids, leukotrienes, epoxy fatty acids, sphingolipids, lysophospholipids, endocannabinoids, and phosphoinositides) and lipid signaling proteins (lysophospholipid acyltransferases, phosphoinositide 3-kinase, and G protein-coupled receptors (GPCRs)). Drugs targeting lipid signaling pathways promise to be life changing magic for the future of human health and well-being.

Keywords

Lipid mediator · Drug discovery · Pharmacology · Biochemistry · Molecular biology

The lipid bilayer membrane that protects DNA from external stress is essential for DNA inheritance and cell survival [1]. Membrane lipids contain glycerophospholipids (phosphatidylcholine, phosphatidylethanolamine, phosphatidylserine, phosphatidylinositol), glycerolipids (mono-, di-tri-acyl-glycerols), sphingolipids, and sterol lipids [2–4]. Membrane glycerophospholipids are *de novo* synthesized from glycerol-3-phosphate (known as the Kennedy pathway), which produces phosphatidic acid (PA) via lysophosphatidic acid (LPA) by the sequential actions of glycerol-3-phosphate acyltransferase and LPA acyltransferase (LPAAT). Phospholipases liberate fatty acyls from the glycerophospholipids,

Y. Kihara (✉)
Sanford Burnham Prebys Medical Discovery Institute, La Jolla, CA, USA
e-mail: kihara-yasuyuki@umin.net

Y. Kihara (ed.), *Druggable Lipid Signaling Pathways*, Advances in Experimental Medicine and Biology 1274, https://doi.org/10.1007/978-3-030-50621-6_1

whose diversity is generated by lysophospholipid acyltransferases (known as Land's cycle) [5–8]. Liberated fatty acyls are further metabolized by cyclooxygenases (COXs), lipoxygenases (LOs), and cytochrome P450, resulting in the generation of prostanoids, leukotrienes, and epoxy fatty acids, respectively [9–11]. Drug discoveries for lysophospholipid acyltransferases are summarized by Dr. Hideo Shindou (Chap. 2).

The first NSAID, acetylsalicylic acid (Aspirin®), was commercialized in 1897 before the finding of prostanoids including the prostaglandins (PGs: PGD_2, PGE_2, $PGF_{2\alpha}$, PGI_2) and thromboxane (TXA_2) [12]. Prostanoids are derived from arachidonic acid and play important roles in pain, fever, inflammation, cardiovascular diseases, the reproductive system, and others [9, 13, 14]. Dr. Emanuela Ricciotti summarizes the prostanoid pathways and drugs targeting this pathway (Chap. 3). Leukotrienes are also derived from arachidonic acid, whose receptor antagonists such as pranlukast, montelukast, and zafirulukast are prescribed for treating asthma and rhinitis [9, 15–19]. Dr. C. Henrique Serezani provided a review of leukotriene pathways, particularly focusing on metabolic and cardiovascular diseases (Chap. 4). Polyunsaturated fatty acids including arachidonic acid, docosahexaenoic acid, and eicosapentaenoic acid are metabolized by CYP P450 to produce epoxy fatty acids that are further metabolized to hydroxy fatty acids by soluble epoxide hydrolases (sEHs) [20]. Cindy McReynolds et al. introduced novel drugs targeting sEH for the treatment of neuropathic pain and cardiovascular diseases (Chap. 5).

Sphingolipids (sphingoid bases, ceramides, sphingomyelins, cerebrosides) are also essential components of membranes that have a sphingosine backbone with N-acyl chains and/or head groups [21, 22]. Sphingosine 1-phosphate (S1P) is a lipid mediator that controls lymphocyte trafficking and is responsible for immune diseases [23, 24]. Fingolimod is a pro-drug that is metabolized to fingolimod-phosphate by endogenous sphingosine kinases [25]. Fingolimod is the first FDA-approved orally available drug for the treatment of relapsing-remitting multiple sclerosis, which targets S1P receptors [26–28]. Dr. Victoria A. Blaho reviewed drug discoveries in the sphingolipid pathways (Chap. 6).

Lysophospholipid biology is expanding rapidly by finding receptors for each lysophospholipid (LPA, lysophosphatidic acid; LPI, lysophosphatidylinositol; lysoPS, lysophosphatidylserine; LPGlc, lysophosphatidylglucoside) [26, 27]. LPA receptor antagonists and the LPA metabolic enzyme, autotaxin, inhibitors are expected to treat pulmonary fibrosis, pain, cardiovascular, and neurological diseases [29–32], which was summarized by Dr. Keisuke Yanagida (Chap. 7).

Endocannabinoids (anandamide and 2-arachidonoylglycerol) are essential lipid mediators in the central nervous system and immune system [33]. The endocannabinoid receptors were originally discovered as receptors for cannabis components before endocannabinoids were found. Cannabinoids are clinically and recreationally used worldwide. Dr. Matthew W. Buczynski provides an overview of endocannabinoid biology and drug discovery (Chap. 8).

Phosphoinositides (phosphatidylinositol 3-phosphate (PI3P), phosphatidylinositol 4-phosphate (PI4P), phosphatidylinositol 5-phosphate (PI5P), phosphatidylinositol 3,4-biphosphate ($PI(3,4)P_2$), phosphatidylinositol 3,5-biphosphate ($PI(3,5)P_2$), phosphatidylinositol 3,4-biphosphate ($PI(3,4)P_2$), and phosphatidylinositol 3,4,5-triphosphate (PIP_3)) are essential intracellular signaling molecules downstream of a variety of cell surface receptors [34, 35]. These are derived from phosphatidylinositol by the actions of phosphoinositide kinases. Dr. John E. Burke provided a comprehensive summary of this pathway with particular focus on the phosphoinositide 3-kinase (PI3K) and phosphatidylinositol 4-kinase (PI4K) family (Chap. 9).

All the lipid mediators introduced above bind to their cognate GPCRs that are the most attractive targets for drug discovery. This book's editor, Dr. Yasuyuki Kihara, provided a review of druggable lipid GPCRs that summarizes the histories of lipid GPCR identification, drugs targeting lipid GPCRs, and striking drug designs for future GPCR drug discovery (Chap. 10).

Taken together, the book contains 10 chapters that provide a historical overview as well as the recent advances in studies of lipid signaling pathways with particular emphasis on "druggable" targets. The book is aimed at a broad audience from academic to industry researchers and was authored by the next generation of lipid researchers who were trained and mentored by prestigious lipid scientists including Drs. Frank K. Austen, Charles R. Brown, Jerold Chun, Edward A. Dennis, Garret A. FitzGerald, Timothy Hla, Bruce D. Hammock, Loren H. Parsons, Marc Peters-Golden, Takao Shimizu, Gabor Tigyi, Roger L. Williams, and more.

Acknowledgements Thanks to all the authors who contributed to this book, Dr. Inês Alves (Springer Nature) for inviting me to edit this book, Ms. Ilse Kooijman (Springer Nature), and Ms. Danielle Jones (SBP) for editorial assistance. This work was supported by a grant from NIH/ NINDS R01NS103940 (Y.K.). The content is solely the responsibility of the authors and does not necessarily represent the official views of the National Institutes of Health.

Conflict of Interest Y.K. declares no competing financial interests.

References

1. Shimizu T (2009) Lipid mediators in health and disease: enzymes and receptors as therapeutic targets for the regulation of immunity and inflammation. Annu Rev Pharmacol Toxicol 49:123–150
2. Dennis EA (2016) Liberating chiral lipid mediators, inflammatory enzymes, and LIPID MAPS from biological grease. J Biol Chem 291:24431–24448
3. Fahy E, Subramaniam S, Murphy RC, Nishijima M, Raetz CR, Shimizu T, Spener F, van Meer G, Wakelam MJ, Dennis EA (2009) Update of the LIPID MAPS comprehensive classification system for lipids. J Lipid Res 50(Suppl):S9–S14
4. O'Donnell VB, Dennis EA, Wakelam MJO, Subramaniam S (2019) LIPID MAPS: serving the next generation of lipid researchers with tools, resources, data, and training. Sci Signal 12(563):eaaw2964
5. Hishikawa D, Hashidate T, Shimizu T, Shindou H (2014) Diversity and function of membrane glycerophospholipids generated by the remodeling pathway in mammalian cells. J Lipid Res 55:799–807
6. Kita Y, Shindou H, Shimizu T (2019) Cytosolic phospholipase A2 and lysophospholipid acyltransferases. Biochim Biophys Acta Mol Cell Biol Lipids 1864:838–845
7. Shindou H, Hishikawa D, Harayama T, Eto M, Shimizu T (2013) Generation of membrane diversity by lysophospholipid acyltransferases. J Biochem 154:21–28
8. Shindou H, Shimizu T (2009) Acyl-CoA:lysophospholipid acyltransferases. J Biol Chem 284:1–5
9. Funk CD (2001) Prostaglandins and leukotrienes: advances in eicosanoid biology. Science 294:1871–1875
10. Inceoglu B, Schmelzer KR, Morisseau C, Jinks SL, Hammock BD (2007) Soluble epoxide hydrolase inhibition reveals novel biological functions of epoxyeicosatrienoic acids (EETs). Prostaglandins Other Lipid Mediat 82:42–49
11. Smith WL, Urade Y, Jakobsson PJ (2011) Enzymes of the cyclooxygenase pathways of prostanoid biosynthesis. Chem Rev 111:5821–5865
12. Vane JR, Botting RM (2003) The mechanism of action of aspirin. Thromb Res 110:255–258
13. Ricciotti E, FitzGerald GA (2011) Prostaglandins and inflammation. Arterioscler Thromb Vasc Biol 31:986–1000
14. Smyth EM, Grosser T, Wang M, Yu Y, FitzGerald GA (2009) Prostanoids in health and disease. J Lipid Res 50(Suppl):S423–S428
15. Austen KF (2005) The mast cell and the cysteinyl leukotrienes. Novartis Found Symp 271:166–175; discussion 176-168, 198-169
16. Austen KF, Maekawa A, Kanaoka Y, Boyce JA (2009) The leukotriene E4 puzzle: finding the missing pieces and revealing the pathobiologic implications. J Allergy Clin Immunol 124:406–414. quiz 415-406
17. Haeggstrom JZ, Funk CD (2011) Lipoxygenase and leukotriene pathways: biochemistry, biology, and roles in disease. Chem Rev 111:5866–5898
18. Kanaoka Y, Boyce JA (2004) Cysteinyl leukotrienes and their receptors: cellular distribution and function in immune and inflammatory responses. J Immunol 173:1503–1510
19. Saeki K, Yokomizo T (2017) Identification, signaling, and functions of LTB4 receptors. Semin Immunol 33:30–36
20. Atone J, Wagner K, Hashimoto K, Hammock BD (2019) Cytochrome P450 derived epoxidized fatty acids as a therapeutic tool against neuroinflammatory diseases. Prostaglandins Other Lipid Mediat 147:106385
21. Trayssac M, Hannun YA, Obeid LM (2018) Role of sphingolipids in senescence: implication in aging and age-related diseases. J Clin Invest 128:2702–2712
22. Hannun YA, Obeid LM (2018) Sphingolipids and their metabolism in physiology and disease. Nat Rev Mol Cell Biol 19:175–191

23. Blaho VA, Chun J (2018) 'Crystal' clear? Lysophospholipid receptor structure insights and controversies. Trends Pharmacol Sci 39:953–966
24. Proia RL, Hla T (2015) Emerging biology of sphingosine-1-phosphate: its role in pathogenesis and therapy. J Clin Invest 125:1379–1387
25. Chun J, Kihara Y, Jonnalagadda D, Blaho VA (2019) Fingolimod: lessons learned and new opportunities for treating multiple sclerosis and other disorders. Annu Rev Pharmacol Toxicol 59:149–170
26. Kihara Y, Mizuno H, Chun J (2015) Lysophospholipid receptors in drug discovery. Exp Cell Res 333:171–177
27. Kihara Y, Maceyka M, Spiegel S, Chun J (2014) Lysophospholipid receptor nomenclature review: IUPHAR review 8. Br J Pharmacol 171:3575–3594
28. Kihara Y (2019) Systematic understanding of bioactive lipids in neuro-immune interactions: lessons from an animal model of multiple sclerosis. Adv Exp Med Biol 1161:133–148
29. Yung YC, Stoddard NC, Mirendil H, Chun J (2015) Lysophosphatidic acid signaling in the nervous system. Neuron 85:669–682
30. Yung YC, Stoddard NC, Chun J (2014) LPA receptor signaling: pharmacology, physiology, and pathophysiology. J Lipid Res 55:1192–1214
31. Yanagida K, Ishii S (2011) Non-Edg family LPA receptors: the cutting edge of LPA research. J Biochem 150:223–232
32. Nakanaga K, Hama K, Aoki J (2010) Autotaxin – an LPA producing enzyme with diverse functions. J Biochem 148:13–24
33. Parsons LH, Hurd YL (2015) Endocannabinoid signalling in reward and addiction. Nat Rev Neurosci 16:579–594
34. Baretic D, Williams RL (2014) PIKKs – the solenoid nest where partners and kinases meet. Curr Opin Struct Biol 29:134–142
35. Burke JE (2019) Dynamic structural biology at the protein membrane interface. J Biol Chem 294:3872–3880

Biosynthetic Enzymes of Membrane Glycerophospholipid Diversity as Therapeutic Targets for Drug Development

2

William J. Valentine, Tomomi Hashidate-Yoshida, Shota Yamamoto, and Hideo Shindou

Abstract

Biophysical properties of membranes are dependent on their glycerophospholipid compositions. Lysophospholipid acyltransferases (LPLATs) selectively incorporate fatty chains into lysophospholipids to affect the fatty acid composition of membrane glycerophospholipids. Lysophosphatidic acid acyltransferases (LPAATs) of the 1-acylglycerol-3-phosphate *O*-acyltransferase (AGPAT) family incorporate fatty chains into phosphatidic acid during the *de novo* glycerophospholipid synthesis in the Kennedy pathway. Other LPLATs of both the AGPAT and the membrane bound *O*-acyltransferase (MBOAT) families further modify the fatty chain compositions of membrane glycerophospholipids in the remodeling pathway known as the Lands' cycle. The LPLATs functioning in these pathways possess unique characteristics in terms of their biochemical activities, regulation of expressions, and functions in various biological contexts. Essential physiological functions for LPLATs have been revealed in studies using gene-deficient mice, and important roles for several enzymes are also indicated in human diseases where their mutation or dysregulation causes or contributes to the pathological condition. Now several LPLATs are emerging as attractive therapeutic targets, and further understanding of the mechanisms underlying their physiological and pathological roles will aid in the development of novel therapies to treat several diseases that involve altered glycerophospholipid metabolism.

Keywords

Phospholipid · LPCAT · LPAAT · Lysophospholipid acyltransferase · AGPAT

W. J. Valentine (✉)
Department of Lipid Signaling, National Center for Global Health and Medicine, Tokyo, Japan

Department of Molecular Therapy, National Center of Neurology and Psychiatry, Tokyo, Japan
e-mail: wvalentine@ncnp.go.jp

T. Hashidate-Yoshida · S. Yamamoto
Department of Lipid Signaling, National Center for Global Health and Medicine, Tokyo, Japan

H. Shindou (✉)
Department of Lipid Signaling, National Center for Global Health and Medicine,
Tokyo, Japan

Department of Lipid Science, Graduate School of Medicine, The University of Tokyo, Tokyo, Japan

Japan Agency for Medical Research and Development, Tokyo, Japan
e-mail: hshindou@ri.ncgm.go.jp

Y. Kihara (ed.), *Druggable Lipid Signaling Pathways*, Advances in Experimental Medicine and Biology 1274, https://doi.org/10.1007/978-3-030-50621-6_2

2.1 Function and Diversity of Cellular Membranes

Cellular membranes are comprised mainly of protein and lipids; and glycerophospholipids, along with sphingolipids and cholesterol, are the major lipid components. The glycerophospholipids share a common glycerol-based structure with a phosphate-containing headgroup at the *sn*-3 position of the glycerol backbone, and two hydrophobic fatty chains attached to the *sn*-1 and *sn*-2 positions (Fig. 2.1). This basic structural unit supports formation of lipid bilayers, which serve various cellular functions: to separate the interior of cells from their physical environments; to form compartments within cells that define organelles and structures; to form domains of molecular interactions; and as storage of lipid mediator precursors. Some glycerophospholipids also form monolayers such as outer monolayers of HDL (high-density lipoprotein) particles, LDL (low-density lipoprotein) particles, and lipid droplets; as well as specialized surfactant monolayers such as pulmonary surfactant. A great diversity of glycerophospholipid species exists due to the many possible combinations of headgroups, fatty chains, and linkages of the fatty chains to the glycerol backbone. This diversity varies among cell types and tissues and imparts membranes with unique biophysical and biological properties.

2.1.1 Glycerophospholipid Classes

Glycerophospholipids possess a phosphate-containing headgroup esterified to *sn*-3 of their glycerol backbones. In the simplest glycerophospholipid, phosphatidic acid (PA), the phosphate is the headgroup and no further addition is made. PA is not very abundant in membranes but is a key intermediate in biosynthesis of all other classes of glycerophospholipids. The other classes are formed by esterification of the phosphate group to any of several alcohols (choline, ethanolamine, serine, inositol, or glycerol) forming the glycerophospholipid classes phosphatidylcholine (PC), phosphatidylethanolamine (PE), phosphatidylserine (PS), phosphatidylinositol (PI), and phosphatidylglycerol (PG). Cardiolipin (CL) is structurally unique in having a modified dimeric PG structure, with a glycerol backbone esterified to two PA moieties.

The relative amounts of the different glycerophospholipid classes in cellular membranes varies. PC followed by PE are the most abundant glycerophospholipids in membranes, and together typically account for over half of the glycerophospholipids. Other classes of glycerophospholipids are typically less abundant, PI and PS classes each typically account for ~5–10% of all glycerophospholipids, and lesser amounts of CL, PG, and PA (<5% each) are generally present

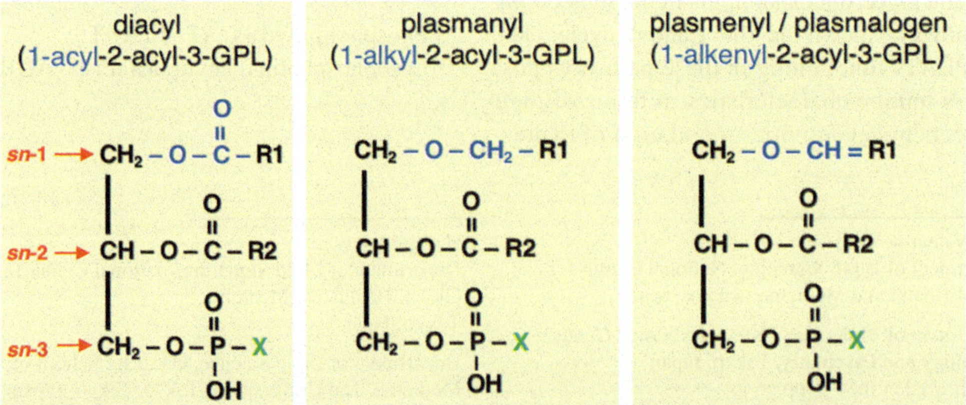

Fig. 2.1 Glycerophospholipid structure
Basic structures of diacyl-, alkyl-acyl-, and alkenyl-acyl (plasmalogen) -glycerophospholipids are shown. X represents -OH, choline, ethanolamine, serine, inositol, or glycerol for PA, PC, PE, PS, PI, and PG, respectively. Stereospecific numbering (*sn*) positions of the glycerol backbone are indicated by red arrows. R1 and R2 represent the fatty chains at *sn*-1 and *sn*-2, respectively. Acyl, alkyl, and alkenyl linkages at *sn*-1 are highlighted in blue. GPL, glycerophospholipid

[1]. However, the less abundant classes of glycerophospholipids may have specialized functions and important roles in various biological contexts, such as roles for PS in cell fusion, phagocytosis, and apoptosis [2] and roles for PI in vesicle trafficking [3]. These less abundant classes may also be concentrated within cells in certain structures, domains, or organelles; for example, CL is a major component of the inner mitochondrial membrane, where it has important roles in mitochondrial structure and function [4].

2.1.2 Fatty Chain Diversity

Fatty chains found in glycerophospholipids vary in terms of carbon chain length and double-bond number. The fatty chain compositions of membrane glycerophospholipids varies among organelles, cell types, tissues, and physiological states; and these compositions can strongly affect biophysical properties of the membranes in terms of fluidity, flexibility, fusion, fission, and curvature [5, 6]. Carbon chain lengths affect membrane thickness and microdomain formation, which can strongly affect functions of transmembrane proteins and membrane-associated proteins. Double-bond number affects membrane thickness, viscosity, flexibility, and strength. Compared to unsaturated and monounsaturated fatty chains, polyunsaturated fatty chains in membrane phospholipids impart flexibility and curvature to the membranes. Polyunsaturated fatty acids in membrane glycerophospholipids also represent important sources of precursor molecules that can be de-esterified by phospholipase A (PLA) 1/2 s and converted to lipid mediators [7, 8].

2.1.3 Glycerophospholipid Linkages

In addition to chain length and saturation, the linkages of the fatty chains to the glycerol backbone further increase the diversity of glycerophospholipid species. Acyl linkages are the most common at both *sn*-1 and *sn*-2. Alkyl and alkenyl linkages may also occur at *sn*-1, forming what are known as plasmanyl- or plasmenyl-phospholipids, respectively (Fig. 2.1). Plasmanyl- and plasmenyl-phospholipids have a different biosynthetic route than diacyl glycerophospholipids. Rather than being synthesized from a PA precursor, as are diacyl species, their synthesis begins in peroxisomes with acylation of dihydroxyacetone phosphate (DHAP) by the enzyme DHAP-acyltransferase (DHAPAT). This reaction produces 1-acyl-DHAP, which is enzymatically converted to 1-alkyl-DHAP and then 1-alkylglycerol-3-phosphate. 1-alkylglycerol-3-phosphate is acylated at *sn*-2 and further metabolized in the endoplasmic reticulum to form the plasmanyl- and plasmenyl-phospholipids [9].

The plasmanyl-phospholipids, with alkyl linkages at *sn*-1, usually are of the PC class of glycerophospholipids. A notable example is platelet-activating factor (PAF; 1-*O*-alkyl-2-acetyl-*sn*-glycero-3-phosphocholine). PAF is a potent inflammatory phospholipid mediator, and produced by lyso-PAF acetyltransferase activities of lysophosphatidylcholine acyltransferase (LPCAT)1 or LPCAT2 from lyso-PAF and acetyl-CoA [10, 11].

The plasmenyl glycerophospholipids, commonly known as plasmalogens, have alkenyl linkages at *sn*-1 and are enriched with polyunsaturated fatty acids at *sn*-2. This unique structure may impart them with special properties, however determining the biological functions of plasmalogens has often been challenging. Overall, plasmalogens account for 15–20% of all phospholipids in cell membranes and are highly enriched in certain tissues. Plasmalogens of the PE class are generally the most abundant, as well as PC plasmalogens in certain tissues. For instance, in adult humans over half of the PE glycerophospholipids in heart and brain may be plasmalogen; and, while PC in brain is low in plasmalogen, ~25% of PC in heart may be plasmalogen. Other tissues rich in plasmalogens include kidney, lung, skeletal muscle, and ocular lens; and plasmalogens may also have unique roles during development and in immune cells [9].

2.1.4 *De novo* Glycerophospholipid Biosynthesis (Kennedy Pathway)

De novo glycerophospholipid production proceeds through pathways that involve production of PA as a precursor in production of all mammalian diacyl glycerophospholipids. These enzymatic reactions were first described by Kennedy and Weiss in 1956 and are now termed the Kennedy pathway [12]. First, glycerol-3-phosphate acyltransferases (GPATs) esterify an acyl chain to glycerol-3-phosphate (G3P) at the *sn*-1 position to produce lyso-PA (LPA; 1-acylglycerol-3-phosphate). Next, LPA-acyltransferases (LPAATs) esterify a second acyl chain to the *sn*-2 position to produce PA. Newly synthesized PA may be further converted by either of two metabolic routes active in glycerophospholipid biosynthesis. The PA may be dephosphorylated by PA phosphatases (PAPs; also known as lipins) to produce diacylglycerol (DAG), which may be further metabolized to produce PC, PE, PS, and triglyceride (TG). Alternatively, PA may be acted upon by cytidine diphosphate (CDP) -DAG synthase to produce CDP-DAG, which may be further metabolized to produce CL, PI and PG [13] (Fig. 2.2).

2.1.5 PC and PE Are Produced from Diacylglycerol

PC and PE are the most abundant glycerophospholipids, and in the Kennedy pathway two parallel pathways convert DAG to either PC or PE. The Kennedy pathway to produce PC proceeds in three sequential reactions: (a) choline kinase (CK) phosphorylates choline to form phosphocholine; (b) CTP:phosphocholine cytidyltransferase (PCYT1) forms CDP-choline from phosphocholine and cytidine triphosphate (CTP); and (c) CDP choline:1,2-DAG cholinephosphotransferase (CPT) utilizes CDP-choline and DAG as substrates to produce PC. In the analogous Kennedy pathway to produce PE, (a) ethanolamine kinase (EK) phosphorylates ethanolamine to form phosphoethanolamine; (b) CTP:phosphoethanolamine cytidylyltransfer-

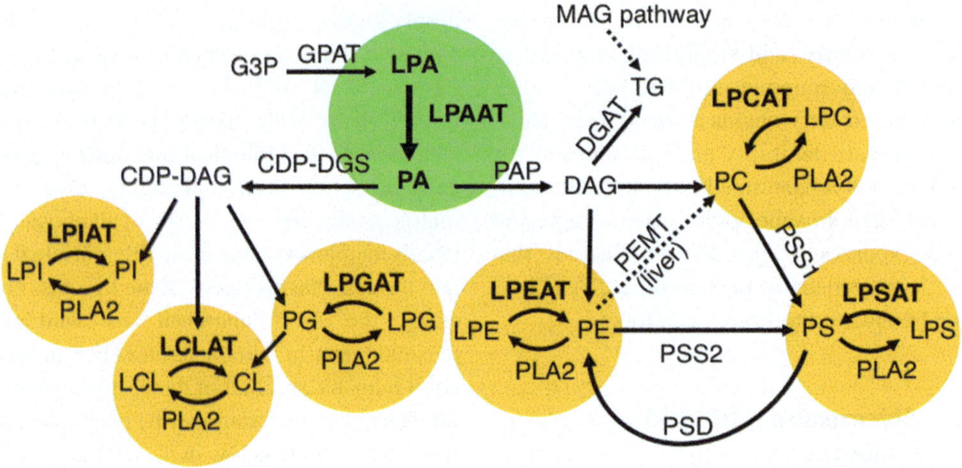

Fig. 2.2 Glycerophospholipid *de novo* synthesis and remodeling pathways
Basic mammalian pathways to produce diacyl-glycerophospholipids is shown (see text for details). Acyl chains at *sn*-2 are specified during production of PA from LPA in Kennedy pathway (green circle), and may be extensively remodeled during Lands' cycle reactions (orange circles). Abbreviations: *CDP-DAG* cytidine diphosphate-diacylglycerol, *CDP-DGS* cytidine diphosphate-diacylglycerol synthase, *DAG* diacylglycerol, *DGAT* diacylglycerol acyltransferase, *G3P* glycerol-3-phosphate, *GPAT* glycerol-3-phosphate acyltransferase, *LPC* lyso-PC, *LPE* lyso-PE, *LPG* lyso-PG, *LPI* lyso-PI, *LPS* lyso-PS, *MAG* monoacylglycerol, *PAP* phosphatidic acid phosphatase, *PEMT* PE methyltransferase, *PSD* PS decarboxylase, *PSS1* PS synthase 1, *PSS2* PS synthase 2, *TG* triglyceride

ase (PCYT2) forms CDP-ethanolamine from CTP and phosphoethanolamine; and (c) CDP-ethanolamine:1,2-diacylglycerol ethanolamine-phosphotransferase (EPT) utilizes CDP-ethanolamine and DAG to produce PE [1].

2.1.6 Additional Reactions to Generate Glycerophospholipid Classes

Other enzymatic reactions occur that mediate conversion of glycerophospholipids between classes. An alternative pathway to produce PC from PE exists that is called the PE methyltransferase (PEMT) pathway. In this pathway, PC production occurs via three sequential methylations of PE, all catalyzed by the enzyme PEMT. This pathway to PC production is only active in liver, where it accounts for ~30% of PC production [14].

PS is produced from PC and PE via base exchange reactions catalyzed by either of two enzymes, PS synthase (PSS) 1 or PSS2. Both enzymes produce PS in regions of the endoplasmic reticulum; PSS1 converts PC to PS, and PSS2 converts PE to PS. PS that has been transported to the mitochondria may be converted back to PE by the actions of another enzyme, PS decarboxylase (PSD) [1].

These pathways for production of PC, PE, and PS all proceed from DAG as a common intermediary. The pathways to produce the other classes of glycerophospholipids PI, CL, and PG all involve a series of enzymatic reactions that utilize CDP-DAG, produced from PA, as a precursor. Thus, PA is a common precursor molecule in *de novo* synthesis of all diacyl glycerophospholipids. PA is usually maintained at relatively low levels in cells, and newly synthesized PA may be rapidly converted to other glycerophospholipid classes. Therefore, the enzymes that regulate fatty chain compositions of PA also regulate the fatty chain compositions of all glycerophospholipids produced in Kennedy pathway [12].

2.2 Regulation of Acyl Chain Compositions

During Kennedy pathway of glycerophospholipid synthesis, acyl chains are determined first when GPAT incorporates an acyl chains into *sn*-1 position of G3P to produce LPA, and again when LPAAT incorporates an acyl chain into *sn*-2 to produce PA. GPAT enzymes preferentially utilize saturated or mono-unsaturated acyl-CoA substrates for incorporation into *sn*-1, and different LPAAT enzymes vary in their acyl-CoA selectivities to utilize saturated, mono-unsaturated, or poly-unsaturated acyl-CoA substrates for incorporation into *sn*-2. Thus, GPAT and LPAAT enzyme selectivities are major determinants of fatty chain compositions at *sn*-1 and -2 of glycerophospholipids formed during the *de novo* synthesis. However, these selectivities do not fully account for the fatty chain diversity found in cellular membranes. The fatty chain compositions at *sn*-2 are further modified as part of a glycerophospholipid acyl chain remodeling pathway known as the Lands' cycle.

Rapid turnover of acyl chains in glycerophospholipids was originally described by William Lands in 1958 [15]. In radioactive tracer experiments of PC synthesis in lung tissue, Lands noticed the ratio of fatty acid incorporation to glycerol was several-fold higher in phospholipids compared to TGs. This suggested that in addition to the biosynthetic route leading to production of both TG and PC from a common DAG precursor, another mechanism to allow exchange of fatty acids on PC in absence of *de novo* synthesis may exist. Lands postulated existence of an enzyme system which catalyzes exchange of fatty acids via formation of a lysophospholipid intermediate followed by reacylation. This enzyme system is now known to catalyze the rapid turnover at *sn*-2 via cycles of PLA2-mediated deacylations and lysophospholipid acyltransferase (LPLAT) -mediated reacylations in a pathway distinct from *de novo* synthesis. PA production as part of the Kennedy pathway and glycerophospholipid remodeling during Lands' cycle are two distinct mechanisms by which acyl chains of glycerophospholipids are specified. Different sets of LPLATs function in these two pathways, and together these pathways account for the acyl chain diversity that exists in membrane glycerophospholipids.

2.2.1 Distinct Sets of LPLATs Function During *de novo* and Remodeling Pathways

The first identified mammalian LPAATs, LPAAT1 and LPAAT2, were cloned by several independent groups in 1997 based on motifs conserved in homologous sequences from plants, bacteria, yeast, nematodes and viruses [16–20]. These motifs included regions of high homology, now known as 1-acylglycerol-3-phosphate *O*-acyltransferase (AGPAT) motifs I-IV, that are important for substrate binding and enzymatic activity [19, 21, 22]. Originally, eleven mammalian AGPAT family members were identified [23]; however, biochemical characterization of the recombinant enzymes indicated many of these AGPAT family members primarily produce glycerolipids other than PA. Now five AGPAT family members are thought to function primarily as LPAATs by acylating LPA at *sn*-2 to produce PA [18, 24–27]. The other AGPAT family members have been reclassified according to their biochemical activities as GPATs, DHAPAT, or Lands' cycle LPLATs (such as LPCATs, LPEATs, LPIATs, LCLATs, and LPGATs) [21, 28]. Subsequently, non-AGPAT family LPLATs were also identified that are membrane bound *O*-acyltransferase (MBOAT) family members [29]. MBOAT family enzymes have activation motifs that are different from AGPAT motifs [30]. Currently four MBOAT family members are known to function as LPLATs in the Lands' cycle. The LPLATs of AGPAT and MBOAT families which function in the Kennedy pathway and Lands' cycle are summarized in Fig. 2.3.

	Family	LPLAT name	Product	HUGO symbol	Other Symbols
Kennedy pathway	AGPAT	LPAAT1	PA	AGPAT1	LPAATα
		LPAAT2	PA	AGPAT2	LPAATβ
		LPAAT3	PA, PI	AGPAT3	LPAATγ
		LPAAT4	PA	AGPAT4	LPAATδ
		AGPAT5		AGPAT5	LPAATε
Lands' cycle	AGPAT	LPEAT2	PE	LPCAT4	AYTL3, AGPAT7
		LCLAT1	CL, PG, PI	LCLAT1	ALCAT1, AGPAT8
		LPCAT1	PC, PE, PAF	LPCAT1	AYTL2, AGPAT9/10
		LPCAT2	PC, PE, PAF	LPCAT2	AYTL1, lysoPAFAT
		LPGAT1	LPG	LPGAT1	NET8, KIAA0205
	MBOAT	LPCAT3	PC, PE, PS	LPCAT3	OACT5, MBOAT5
		LPCAT4	PC, PE, PA	MBOAT2	OACT2
		LPEAT1	PE, PS	MBOAT1	OACT1
		LPIAT1	PI	MBOAT7	LENG4, MBOA7

Fig. 2.3 LPLATs of the AGPAT and MBOAT families LPLATs of the AGPAT family function in the Kennedy pathway (green box), and LPLATs of the AGPAT and MBOAT families function during Lands' cycle remodeling (orange box). Both LPCAT4/MBOAT2 and LPEAT2/LPCAT4 (in red letters) are referred to as "LPCAT4" and sometimes induce confusion in the literature

2.3 LPAAT Enzymes Determine Acyl Chains During *de novo* Synthesis

In *de novo* synthesis of PA, following esterification of a fatty chain to G3P by GPATs, the next step involving esterification of a second fatty chain, to the *sn*-2 position, is catalyzed by LPAAT enzymes of the AGPAT family (Fig. 2.2). The first identified mammalian isoforms, LPAAT1 and LPAAT2, were cloned nearly simultaneously by several groups [16–19]. Eberhardt et al. identified motifs in human LPAAT2 that were conserved in homologous sequences from plants, bacteria, yeast, nematodes and viruses [19], and Aguado et al. reported similarly conserved sequences in human LPAAT1 the following year [20]. These regions included the regions now referred to as AGPAT motifs that are conserved in all AGPAT family members and important in substrate binding and enzymatic activity [21]. Mammalian LPAAT1 and LPAAT2 are the most evolutionarily conserved isoforms and suggested to have evolutionarily arisen as part of the earliest TG synthesis pathway, indicating the importance of these isoforms in TG as well as glycerophospholipid production [31].

2.3.1 LPAAT1 Has Essential Functions in Multiple Tissues

LPAAT1 and LPAAT2 share high homology and have similar but not identical biochemical activities to utilize a variety of fatty chain-CoAs in producing PA [32]. LPAAT1 is broadly expressed, while LPAAT2 shows more restricted expression. Both LPAAT1 and LPAAT2 are expressed in adipocytes [32], however genetic loss of LPAAT2 activity causes congenital generalized lipodystrophy (CGL) [33], indicating that LPAAT2 has essential functions in adipocytes that are not compensated for by LPAAT1 [34].

To understand the biological functions of LPAAT1, LPAAT1 knockout (KO) mice were generated and phenotypically analyzed [35]. LPAAT1 KO mice had widespread disturbances in metabolism and glycerophospholipid homeostasis, causing pathological effects in multiple organ system. The LPAAT1 KO pups were born at less than Mendelian frequency, and those born had reduced body weights that almost did not increase after day 12. The LPAAT1 KO pups had decreased leptin and decreased plasma glucose, and about half died by 4 weeks of age. KO mice had markedly reduced epididymal fat pads, and cervical brown adipose tissue was also reduced but maintained in proportion to the reduced body weights. The LPAAT1 KO mice that survived to adult had reproductive abnormalities including impaired sperm development in males, and ovulation defects in females especially in late follicular maturation. LPAAT1 KO mice had features of seizures/epilepsy, and in hippocampus, where LPAAT1 is normally expressed, LPAAT1 KO mice tended to have reduced CA-region thickness, reduced neuron number, and impaired neurological function. Overall these results indicate LPAAT1 has essential functions in brain and testes, and is required for normal lipid homeostasis [35].

2.3.2 LPAAT2 and Lipodystrophy

LPAAT2 is expressed in several tissues, showing highest expression in adipose tissue, pancreas and liver, and has broad specificity to utilize a variety of fatty acid-CoAs as substrates; including C14:0, C16:0, C18:1, and C18:2 -CoAs to produce PA [32]. LPAAT2 mutations are associated with congenital generalized lipodystrophy (CGL), also known as Berardinelli-Seip lipodystrophy [33]. Four subtypes of the disease are recognized based upon the causative gene mutation (LPAAT2, BSCL2, CAV1, or CAVIN1/PTRF), and LPAAT2 was identified as the causative gene of type 1 CGL through positional cloning [33, 36]. Disease-causing homozygous or compound heterozygous LPAAT2 mutations have been identified that include deletions, nonsense, missense, splice-site and those in the 3′-UTR mutations [34, 37].

CGL is a rare autosomal recessive disorder and the most striking feature is a total lack of subcutaneous body fat from birth. Children with this disorder have increased appetites, undergo accelerated growth, and develop metabolic complications including severe insulin resistance, hypertriglyceridemia, hepatic steatosis and early onset of diabetes. LPAAT2 KO mice have been used to elucidate the pathological mechanisms involved [38]. LPAAT2 KO mice had almost a complete lack of both white and brown adipose tissue, and just 2% body fat compared to 24–29% in wild-type mice. The LPAAT2 KO mice developed extreme insulin resistance, diabetes, and hepatic steatosis. Lipogenic gene expression was increased and fatty acid biosynthesis was accelerated in LPAAT2 KO mouse livers, accompanied by increased monoacylglycerol (MAG) acyltransferase isoform 1, suggesting the MAG pathway for TG production may be hyper-activated under conditions of LPAAT2 deficiency, possibly causing or contributing to the hepatic steatosis. Both LPAAT1 and LPAAT2 are normally expressed in adipose tissue, however LPAAT1 cannot compensate for the observed phenotypic abnormalities in LPAAT2 KO mice, underscoring the essential function of LPAAT2 in this tissue [38].

LPAAT2 functions in TG as well as glycerophospholipid synthesis, and it is possible that LPAAT2 mutations cause CGL primarily by inhibiting TG biosynthesis and storage in adipocytes. The lack of functional adipose tissue in CGL results in TG accumulations in other tissues like skeletal muscle and liver and contributes to the disrupted metabolic homeostasis [39]. It has been suggested that the lipodystrophy due to mutations in LPAAT2 and other CGL-causing genes might be mechanistically caused by defective lipid droplet formation in adipocytes. However, LPAAT2-generated phospholipids may also impact adipocyte function and TG storage, and further studies are required to understand the relative contribution of the LPAAT2-generated TGs and glycerophospholipids to metabolic homeostasis that is disrupted in CGL [34].

2.3.3 LPAAT3 Incorporates DHA into Membranes

Consumption of fish oil, which is rich in very long-chain omega-3 fatty acids such as docosahexaenoic acid (DHA) and eicosapentaenoic acid (EPA), is well-known to promote cardiovascular health and have beneficial effects in other tissues [40]. Incorporation of very long-chain, highly unsaturated chains like DHA and EPA into membrane glycerophospholipids may impart fluidity and flexibility to the membranes, as well as impact functions of membrane proteins [41]. DHA and EPA also have other biological functions as ligands for lipid sensing molecules such as PPARs [42] and as precursor molecules for specialized pro-resolving mediators [43]. Thus, incorporation of DHA and EPA into membrane glycerophospholipids may both alter membrane properties as well regulate their availability as ligands or precursors of bioactive molecules. We have focused on the functions of DHA in biological membranes, because DHA-CoA is selectively utilized as a substrate by LPAAT3 during glycerophospholipid synthesis [44, 45].

LPAAT3 was originally identified as an enzyme broadly expressed in tissues, regulated by PPAR-alpha in mouse heart [24], and involved in golgi dynamics [46]. Substrate specificity of LPAAT3 was shown to incorporate very long-chain polyunsaturated fatty acids into LPA [25, 44, 45] and to selectively incorporate DHA to increase production of DHA-containing glycerophospholipids [45, 47–49]. LPAAT3 KO mice were generated, and they show marked and specific reductions in DHA-containing glycerophospholipids in several tissues, demonstrating a central role for LPAAT3 in generation of DHA-containing glycerophospholipids [47, 48].

DHA-containing glycerophospholipids are abundant in several tissues including retina, testes, brain, heart, and skeletal muscle where they may have important physiological functions. DHA deficiency is associated with several human diseases including hyperlipidemia, cardiovascular disease, cognitive dysfunction, retinal degeneration, and male infertility [50]. We have investigated regulation and biological functions of LPAAT3-generated DHA-containing glycerophospholipids in sperm, retina, and skeletal muscle cells [47–49].

2.3.4 LPAAT3 Incorporates DHA into Glycerophospholipids During Sperm Cell Development

Male infertility frequently occurs due to declines in sperm function. Mammalian sperm contains a high amount of very long-chain polyunsaturated fatty acids including DHA, and defective human sperm populations show declines in DHA in both esterified and unesterified fatty acid fractions [51]. Thus, high DHA content in sperm may be required to impart sperm cell plasma membranes with necessary properties for specialized functions involved in sperm cell maturation, morphology, motility, or fusion with the egg [52].

LPAAT3 was investigated as a candidate enzyme to incorporate DHA into sperm cell membranes. LPAAT3 was highly expressed in mouse testicles, and expression increased during development, coinciding with increased enzymatic activities to incorporate DHA into PC and PE [25, 45]. Immunohistochemical analyses indicated LPAAT3 was expressed both in Leydig cells as well as in spermatids, where LPAAT3 levels increased during spermatogenesis. DHA levels and metabolism in the testes were investigated in LPAAT3 KO mice. Microsome fractions from testes of LPAAT3 KO mice had moderately decreased LPAAT activities to utilize arachidonic acid-CoA but greatly decreased activities to utilize DHA-CoA. Moreover, DHA-containing PC, PE, and PS were highly and selectively decreased in testes, establishing the physiological role of LPAAT3 to incorporate DHA into membrane glycerophospholipids [47].

LPAAT3 KO mice showed male infertility due to defective spermiogenesis. LPAAT3 KO sperm had morphological defects, including sperm heads that bent backwards. Scanning electron microscopy revealed abnormal membranous structures wrapped around the sperm heads which should have been removed when the sperm were released into the seminiferous tubules. This suggests DHA-containing glycerophospholipids generated by LPAAT3 impart flexibility to sperm membranes that allows rapid endocytosis required for normal spermiation [47] and may partly explain why omega-3 supplementation promotes male fertility [53].

2.3.5 LPAAT3 and DHA in Visual Function

Retinal tissue is abundant in DHA-containing glycerophospholipids, and DHA may have protective roles to prevent or delay the progression of several retinal disorders [54, 55]. The role of LPAAT3 in visual function was examined in wild-type and LPAAT3 KO mice [48]. In wild-type mice, LPAAT3 was highly expressed in retinas, and protein expression increased during development from age 11 days through 8 weeks. Compared to wild-type tissues, retinal tissues from LPAAT3 KO mice had decreased levels of DHA in several glycerophospholipids (PA, PC, PE, and PS) accompanied by corresponding increases in several arachidonic acid-containing species. Imaging mass spectrometry showed DHA-containing PC was present in outer segments of photoreceptors in wild-type mice but nearly absent in KO mice. Histologically, LPAAT3 KO retinas appeared normal until 2 weeks of age, but by 3–8 weeks of age they showed abnormal retinal layer morphologies including incomplete elongation of the outer segment, decreased thickness of the outer nuclear layers, and disordered disc morphology in photoreceptor cells. LPAAT3 KO mice showed attenuation of visual function as assessed by electroretinography, with 50–80% decreases in a- and b-wave amplitudes at 8 weeks of age. These results established that LPAAT3 has an essential role in visual function by producing DHA-containing glycerophospholipids, which are required to form, organize, and maintain proper disc shape in photoreceptor cells [48].

2.3.6 LPAAT3 and DHA in Skeletal Muscle

Skeletal muscle is adaptive tissue, and adaptation of skeletal muscle to endurance exercise training is associated with metabolic benefits including reduced obesity and improved glucose handling. Although the biological function and mechanism is unknown, several studies indicate that DHA-containing glycerophospholipids are also increased in endurance-trained muscle and correlates with oxidative status of the muscles [56–60]. In mice, exercise training increased phospholipid-DHA in glycolytic extensor digitorum longus muscle, while phospholipid-DHA was constitutively high in oxidative soleus muscle even without training [56]. In rats, PE-DHA content was enriched in the oxidative compared to glycolytic vastus lateralis muscles [60]. In human volunteers, a 4 weeks regime of one-leg exercise training resulted in increased DHA-containing GPLs that was accompanied by increased citrate synthase activity compared to the untrained leg [58].

Endurance training activates transcriptional responses that mediate adaptive responses in skeletal muscle. PPAR-delta and AMPK are activated by exercise and promote increased oxidative metabolic capacity and transitioning of glycolytic myofibers to more oxidative fiber types [61, 62]. PGC1-alpha is activated downstream of AMPK and is a coactivator of PPAR-delta. In one study, mice overexpressing PGC1-alpha in skeletal muscle had enhanced levels of several DHA-containing glycerophospholipid species that were also increased by endurance training, indicating exercise-induced transcriptional responses may also enhance DHA incorporation into glycerophospholipids of muscle [56].

The role of LPAAT3 to produce DHA-containing glycerophospholipids in skeletal muscle was investigated using primary skeletal muscle myoblasts. LPAAT3 expression was transiently upregulated during differentiation into myotubes and functioned to increase DHA-containing PC and PE [49]. Treatment of differentiated myofibers with PPAR-delta and AMPK agonists also enhanced LPAAT3 expression and increased levels of DHA-containing PC and PE. These results showed LPAAT3 expression is regulated by two mechanisms in myoblasts, transient upregulation as part of the myogenic transcriptional program during myogenesis, and in differentiated myotubes in response to PPAR-delta and AMPK agonists. The PPAR-delta agonist GW1516 and AMPK agonist AICAR are exercise mimicking drugs, and when administered to mice promote endurance and transcriptionally upregulate genes that promote oxidative metabolism [63, 64]. LPAAT3 expression and DHA incorporation was upregulated by these same compounds in cultured myotubes, suggesting LPAAT3 may also be upregulated by exercise-induced pathways to increase DHA in endurance-trained muscle [49].

2.4 Acyl Chain Remodeling by Lands' Cycle Enzymes

In glycerolipids, an asymmetric distribution of fatty acids at *sn*-1 and *sn*-2 positions imparts diversity to cellular membranes that cannot be fully explained enzymatic reactions of *de novo* glycerophospholipid synthesis, and further diversity at *sn*-2 especially is generated by cycles of deacylation and reacylation reactions known as Lands' cycle (remodeling pathway). PC, the most abundant glycerophospholipid in membranes, may undergo extensive Lands' cycle remodeling mediated by the actions of PLA2s and LPCAT enzymes. Several LPCAT enzymes of both the AGPAT and MBOAT families have been identified, and a growing list of functional roles pertinent to physiological and/or pathological conditions is being assigned to these enzymes, garnering much interest in their biological activities and therapeutic potentials. Similar remodeling of other classes of glycerophospholipids is mediated by other LPLATs classified as LPEATs, LCLATs, LPIATs, or LPGATs, and as their biological activities are being increasingly uncovered, interest in their potential as therapeutic targets is also emerging (Table 2.1).

2.4.1 LPCAT1 Produces Lung Surfactant Essential for Respiratory Function

Pulmonary is surfactant essential for breathing and comprised mainly of lipids (~90%) and protein (~10%), and nearly half of the lipid is PC(16:0/16:0), dipalmitoyl-PC (DPPC) [65]. DPPC along with surfactant proteins form a specialized monolayer in alveolar spaces which

Table 2.1 LPLATs as possible therapeutic targets

Enzyme	Major biological functions	Human disease/*in vivo* studies
LPAAT1	Brain, testes, lipid homeostasis	[35]
LPAAT2	Lipid homeostasis, lipodystrophy	[33, 38]
LPAAT3	Sperm maturation, visual function, muscle	[47, 48]
LPCAT1	Lung surfactant, visual function, cancer, non-inflammatory PAF production	[76–78, 81]
LPCAT2	Inflammatory PAF production, neuropathic pain	[92]
LPCAT3	TG transport, VLDL secretion and lipogenesis in liver, lipid absorption in small intestine	[115–119, 126, 127]
LPIAT1	Brain development, liver disease modifier	[134–136]

reduces surface tension, facilitates gas exchange, and prevents alveolar collapse [66, 67]. Pulmonary surfactant is produced by alveolar type II cells, and deficiency of pulmonary surfactant is associated with several pulmonary diseases. Respiratory distress syndrome, the leading cause of death in pre-term infants, is caused by lack of surfactant in premature lung structures [68, 69].

Identification of the biosynthetic routes of DPPC production in lung had been long sought, and both the Kennedy pathway and Lands' cycle were thought to have significant roles. Involvement of Lands' cycle enzymes was indicated by biochemical activities present in alveolar type II cells and rat lung tissue to incorporate labelled palmitate (C16:0) preferentially into the *sn*-2 position of DPPC rather than TGs, presumably through acylation of LPC [70, 71]. LPCAT activities had been known to exist in several tissues, and in 2006 the first mammalian LPCAT enzyme was independently identified by two groups [72, 73]. Both studies identified LPCAT1 as the candidate enzyme present in alveolar type II cells that produces DPPC.

LPCAT1 has AGPAT motifs, and Nakanishi et al. cloned mouse LPCAT1 based on homology to other AGPAT family members [72]. Mouse LPCAT1 was highly expressed in lung and had activity to produce PC using saturated acyl-CoA as donors. Rat LPCAT1 was also highly expressed in lung and enriched in alveolar type II cells rather than macrophages. Later, they also characterized the human LPCAT1 homolog, which was suggested to also produce DPPC in lung [74]. Chen et al. also identified LPCAT1 as an enzyme that was upregulated in mouse and rat lungs shortly before birth. They cloned both the mouse and rat enzymes, which were 99% identical at amino acid level. The enzymes both possessed a $H(X)_4D$ activation domain sequence that was known to be present in other AGPAT family members [75]. Rat LPCAT1 was overexpressed in mammalian cells and produced DPPC [73].

These studies indicated that LPCAT1 may be responsible for production of DPPC in pulmonary surfactant [72–74]. Additional studies utilizing LPCAT1-deficient mice generated by two different groups have confirmed that LPCAT1 is a major producer of DPPC in pulmonary surfactant [76, 77]. The two LPCAT1-deficient mouse strains showed varying degrees of respiratory dysfunction; this variation may be due to the genetic backgrounds or different gene deletion strategies. Bridges et al. generated mice bearing a hypomorphic allele of LPCAT1. These mice had difficulty to transition to air breathing and showed varying degrees of perinatal lethality due to respiratory distress. Perinatal lethality correlated with decreased LPCAT1 mRNA, decreased disaturated PC in lungs, and functionally poor surfactant [76]. Harayama et al. generated LPCAT1 KO mice lacking exon 3 which contains the $H(X)_4D$ AGPAT activation domain. These mice had dramatically reduced LPCAT activity to utilize C16:0-CoA. The KO mice had ~40% reduced DPPC in lung but did not show major signs of neonatal lung distress. However, the mice had enhanced susceptibility to a ventilator-induced lung injury model, which was partially rescued by intratracheal administration of DPPC [77]. These studies establish a critical function for LPCAT1 in producing sufficient levels of DPPC in pulmonary surfactant to prevent respiratory distress or failure under different physiological challenges.

2.4.2 LPCAT1 and Cancer

Besides producing DPPC in pulmonary surfactant, LPCAT1 has additional important roles in physiological and disease states. LPCAT1 is important for visual function, and a natural mutation in LPCAT1 in rd11 mice causes rod and cone cell loss, followed by retinal degeneration [78]. LPCAT1 also has a role to produce PAF under non-inflammatory conditions [22]. Another reported function of LPCAT1 is to protect against polyunsaturated fatty acid-mediated cytotoxicity [79].

In addition, LPCAT1 has a role in cancer. LPCAT1 expression was reported to be high or involved in progression of a variety of cancers including colorectal adenocarcinomas, prostate cancer, lung adenocarcinoma, hepatocellular carcinoma, oral squamous cell carcinoma, breast cancer, esophageal cancer, gastric cancer, clear cell renal cell carcinoma, and glioblastoma [80–82]. Recently Bi et al. reported LPCAT1 is highly upregulated in a wide array of cancers, and LPCAT1-produced disaturated PCs may be required for proliferative signaling via oncogenic growth factor signaling [81]. Initially they examined the impact of EGFRvIII, an activating mutation of EGFR and common driver in glioblastoma, on membrane compositions in glioblastoma cells. In EGFRvIII-transduced glioblastoma cells, disaturated PC species and LPCAT1 mRNA were increased, and shRNA-knockdown of LPCAT1 reduced disaturated PC species, particularly PC(28:0), PC(30:0), and PC(32:0). Knockdown of LPCAT1 in glioblastoma cancer cells harboring EGFRvIII mutations inhibited EGFR phosphorylation and signaling, and this could be reversed by addition of either DPPC liposomes or saturated C14:0/C16:0 fatty acids. Knockdown of LPCAT1 also inhibited glioblastoma cell viability and colony-forming ability, and this was reversed by supplementation of DPPC or a constitutively-active AKT allele. Their results suggest LPCAT1 may promote glioblastoma growth by maintaining high saturated PC levels that support EGFR oncogenic signaling.

The authors went on to show that LPCAT1 amplification is a widespread occurrence across all human cancers. They examined LPCAT1 expression in publicly available databases, and LPCAT1 was found to have increased copy number in 30% of all cancer patients, and inverse correlations between LPCAT1 expression and overall survival was detected in multiple cancer types. LPCAT1 knockdown reduced cancer cell viability *in vitro*, and inducible genetic depletion of LPCAT1 inhibited tumor growth in mouse xenograft models. Their study suggests LPCAT1 inhibition may be a promising therapeutic strategy for a variety of oncogenic growth factor driven cancers [81].

2.4.3 LPCAT2 Produces PAF

LPCAT2 was identified in 2007 as a PAF-producing enzyme in the AGPAT family [83]. In the following year, LPCAT1, which had already been identified as the first LPCAT enzyme [72], was also reported to also have PAF-producing activity [22, 77]. The enzymes having lyso-PAF acetyltransferase activity had been long-sought following the first identification of PAF in 1972 as the substance responsible for platelet aggregation [84]. LPCAT2 can synthesize not only PAF but also polyunsaturated fatty acid-containing PC [77, 83]. Unlike most other LPLATs, the activity of LPCAT2 is calcium ion-dependent. LPCAT2 is expressed in inflammatory cells, such as macrophages and neutrophils, and the enhancement of its PAF biosynthesis activity by lipopolysaccharide (LPS) stimulation is another characteristic of LPCAT2 [83].

Recent studies demonstrated three distinct mechanisms that regulate LPCAT2 activity and expression level in macrophages (Fig. 2.4). First, LPCAT2 is rapidly activated (within 30 s) after either PAF or adenosine triphosphate (ATP) stimulation via intracellular calcium-stimulated activation of PLCβ-PKCα signaling. These pathways induce phosphorylation of LPCAT2 at Ser-34, which enhances LPCAT2's PAF biosynthesis activity [85]. Another mechanism of phosphorylation is via toll-like receptor 4 (TLR4) activation. Thirty minutes of stimulation by the TLR4 ligand LPS activates the downstream pathway,

MyD88-p38 MAPK-MK2, leading to phosphorylation of LPCAT2 at the same site Ser-34 as Ca^{2+}-PLCβ-PKCα signaling [86]. Finally, LPS stimulation for 16 h increases mRNA expression of LPCAT2, which is attenuated by co-treatment with the steroidal anti-inflammatory drug dexamethasone [83]. These studies indicate LPCAT2 is upregulated and shows enhanced PAF-producing activity as part of inflammatory responses, but this is not the case for LPCAT1.

Although it has been clearly demonstrated that LPCAT1 produces PAF in *in vitro* assays, LPCAT1's activity was not enhanced by inflammatory stimuli [85, 86]. It has been suggested that LPCAT1 may have a role to limit PAF production during the onset of diabetic retinopathy by catalyzing synthesis of alkyl-PC from lyso-PAF and acyl-CoA, thereby limiting lyso-PAF availability for LPCAT2-mediated PAF production [87]. However, further investigations are needed to clarify the roles of LPCAT1 to produce or regulate production of PAF *in vivo*.

2.4.4 LPCAT2 and PAF Production in Neuropathic Pain

Several reports have demonstrated LPCAT2 is involved in various pathological conditions such as allergic asthma [88], colorectal cancer resistance [89], nonalcoholic fatty liver disease [90], and multiple sclerosis [91]. It was recently reported that LPCAT2 is involved in neuropathic pain, a highly debilitating chronic pain syndrome [92]. Neuropathic pain occurs from damage to the nervous system; such as by trauma, cancer, diabetes, infection or chemotherapy; which does not resolve even after normal tissue healing. Unfortunately, although clinical neuropathic pain affects 7–10% of the human population [93], the cardinal symptom allodynia, pain caused by innocuous stimuli, is refractory to currently available treatments [94, 95].

In the 1980s, the involvement of PAF in pain processes was indicated by studies that observed enhancement of pain sensitivity following local injection of PAF into skin of animals [96, 97] and humans [98]. In addition, intrathecal PAF injection into the spinal cord also produced allodynia, but injection of its precursor lyso-PAF did not [99, 100]. The biological activities of PAF are elicited via its cognate G protein-coupled receptor, PAF receptor (PAFR) [101]. In rodent models of nerve injury-induced neuropathic pain, it has been reported that treatment with PAFR antagonists, such as WEB2086 or CV-3988, alleviate allodynia symptoms [100, 102]. In accord with this, neuropathic allodynia was attenuated in PAFR KO mice [100]. Both PAFR and LPCAT2 are highly expressed in macrophages (in the periphery) and microglia (in the spinal cord). Peripheral nerve injury increases the number of macrophages in the dorsal root ganglia, where cell bodies of primary sensory neurons are present, and also increases microglia in the spinal dorsal horn. Along with these increases, the protein expressions of PAFR and LPCAT2 also rise [92, 100]. Recently, LPCAT2 KO mice were established, and while the abundance of other PCs were similar to wild-type mice, PAF levels in the spinal cord of this mouse were decreased to below detectable levels under normal physiological conditions. Moreover, nerve injury-induced neuropathic allodynia was attenuated in LPCAT2 KO mice, similar to the PAFR KO mice [92]. These results indicate the importance of LPCAT2-PAFR axis in the pathology of neuropathic pain.

Several studies on the pathogenesis of neuropathic pain have reported that peripheral nerve injury significantly increases the expression levels of several ATP receptors (i.e. P2X4R, P2X7R, P2Y12R) and induces ATP release, via vesicular nucleotide transporter, in the spinal dorsal horn [103–106]. Furthermore, increased TLR4 expression and phosphorylation of p38 MAPK in microglia were also observed in response to nerve injury [107, 108]. Nerve injury also induces LPCAT2 PAF-producing activity by activating ATP receptors and TLR4 signaling, and PAF stimulation induces activation of LPCAT2 via PAFR to further enhance PAF production. Therefore, a positive feedback "PAF-pain loop" is suggested, as PAF-induced increases in PAF production resulted in exacerbation and chronicity of neuropathic pain [92] (Fig. 2.4).

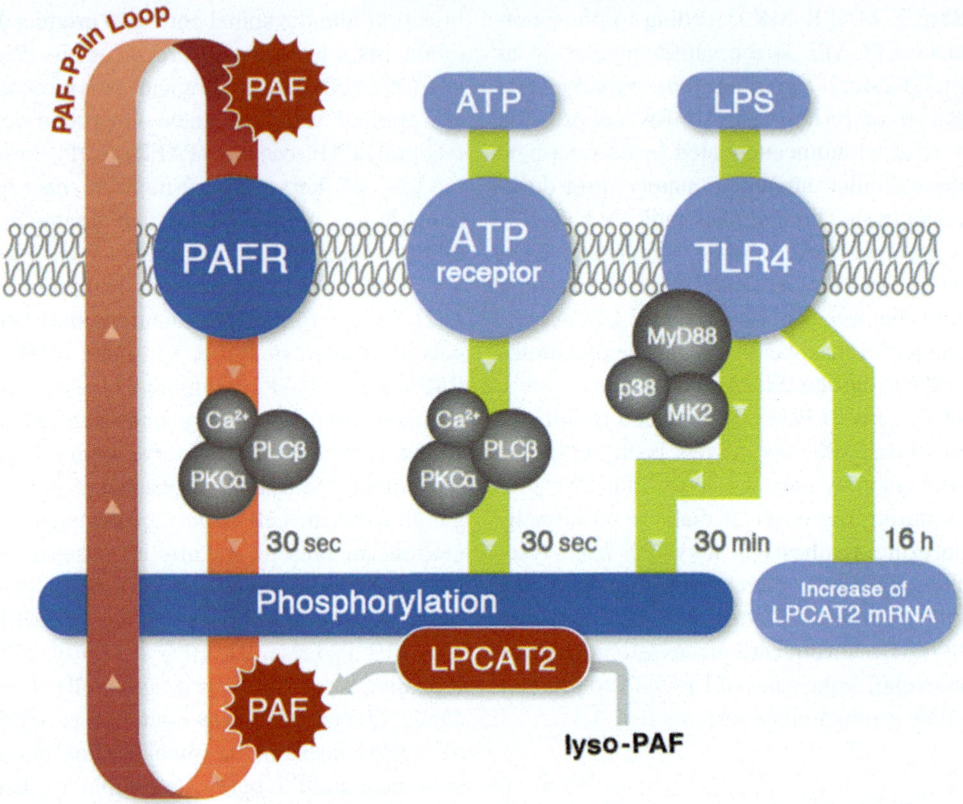

Fig. 2.4 Regulatory mechanisms of LPCAT2-produced PAF levels and the PAF-Pain Loop hypothesis
(i) PAFR- and/or ATP receptors-mediated second-scale phosphorylation of LPCAT2, (ii) TLR4 and its downstream pathways-mediated minute-scale phosphorylation, and (iii) hour-scale LPCAT2 mRNA upregulation via TLR4. Nerve injury may activate these regulatory systems and increase PAF levels. The increased PAF further enhances PAFR-mediated PAF production, and may result in exacerbation of neuropathic pain (PAF-Pain Loop)

Many pharmaceutical companies have already developed PAFR antagonists as possible drug candidates to treat inflammatory conditions such as asthma, but in most cases these drugs have so far shown limited efficacy [10, 109]. However, in light of LPCAT2's role in the "PAF-pain loop", it may be warranted to re-evaluate the possible clinical utility of these PAFR antagonists as analgesics. Moreover, LPCAT2 may also be a novel therapeutic target for analgesic drugs, suggesting that PAF loop blockers such as PAFR antagonists or LPCAT2 inhibitors might represent a new class of analgesic drugs different from the current classes of nonsteroidal anti-inflammatory drugs (NSAIDs) and opioids. Also, because PAF feedback loop may operate not only in neuropathic pain but also in various other pathological conditions with PAF involvement, a recently identified selective LPCAT2 inhibitor, TSI-01 [110], and its derivatives also have potential to contribute to the development of novel drugs for PAF-related diseases. Importantly, the biological significance of the activities of LPCAT2 to produce PCs other than PAF requires further investigation, which may expand our understanding of LPCAT2 as a therapeutic target [92].

2.4.5 LPCAT3 Functions in Small Intestine, Liver and Several Cells

LPCAT3 is a member of the MBOAT family that catalyzes the transfer reaction of acyl group to proteins and lipids [30]. LPCAT3 synthesizes linoleic acid (C18:2) and arachidonic acid (C20:4) containing PC, PE and PS from LPC, LPE and LPS [29, 111–113]. Recently, several reports suggested that LPCAT3 has several important roles in lipid homeostasis [82, 114]. LPCAT3 global KO mice showed neonatal lethality and died between postnatal day 2 and 3 weeks [115–117]. It was found to be caused by nutritional failure in small intestine because the pups were partially rescued by injections of 10% glucose solution or oral gavage with PCs/olive oil [115, 117]. However, the different studies reported different effects of LPCAT3 deficiency on TG accumulation in small intestine. We reported LPCAT3 deletion caused dramatic accumulation of lipid droplets in enterocytes of global LPCAT3 KO pups [116], and Rong et al. observed a similar accumulation in intestine-specific KO pups [115]. In contrast, Li et al. reported no TG accumulation in intestines of LPCAT3 global KO pups [117], nor following intestine-specific LPCAT3 depletion in adult mice [118]. It is unclear why intestinal lipid accumulation was observed in some studies but not others; the differences might reflect the different gene deletion strategies, gene-targeted-regions, or other factors.

The expression of proteins related to lipid absorption in enterocytes, such as NPC1L1, CD36 and FATP4, were decreased in LPCAT3 global KO mice [117] as well as following inducible intestine-specific depletion [118]. Body weight was decreased following intestine-specific LPCAT3 depletion, and total TG and cholesterol levels were also decreased in plasma and serum [115, 119]. In addition, small intestines were longer and intestinal stem cells showed hyperproliferation in the intestine-specific LPCAT3-KO mice [117, 119]. These results suggest that arachidonic acid rich cell membranes synthesized by LPCAT3 have an essential role in intestinal absorption of dietary lipid and may also have a function to limit proliferation of intestinal stem cells. The role in lipid absorption suggests that intestine-specific inhibition of LPCAT3 might be a useful therapeutic strategy to treat hyperlipidemia.

LPCAT3 was also highly expressed in liver [29, 111]. Its expression was regulated by LXR and PPAR-delta in hepatocytes [120–123]. LXR activates lipogenic genes such as SREBP-1c, SCD-1 and FAS [124, 125]; and plasma TG and phospholipids were changed in LXR agonist-treated mouse as well as LXR-alpha/beta double-KO mice [125]. These results suggest that LPCAT3 could contribute to fatty acid homeostasis in liver. In fact, plasma VLDL secretion was increased by hepatic LPCAT3 knockdown in mice [126]; and glucose tolerance was improved and plasma TG levels were decreased by transient overexpression of LPCAT3 in liver [127].

Compared to the changes observed following transient knockdown and overexpression, liver TG was increased and plasma TG, VLDL secretion levels, and VLDL particle size were decreased in LPCAT3 KO mice [115, 116]. These differences between transient LPCAT3 knockdown and LPCAT3 KO mice may reflect residual LPCAT3-generated arachidonic acid- and linoleic acid-containing phospholipids and lysophospholipids that are present under knockdown conditions, due to limited duration or incomplete efficiency of knockdown by the LPCAT3-targeting shRNAs. Both the effective transport of TG by MTP (microsomal triglyceride transfer protein) as well as SREBP-1c maturation in the endoplasmic reticulum by protein cleavage can be promoted by arachidonic acid-containing phospholipids synthesized by LPCAT3 [116, 128], further supporting that LPCAT3 inhibition might improve dyslipidemia.

LPCAT3 has also been shown to function or have roles in other pathological conditions. LPCAT3 expression levels inversely correlated with the disease stage of atherosclerosis, and atherosclerosis was facilitated by LPCAT3 deletion in macrophages [123, 129, 130]. LPCAT3 may also function in cancer because intestinal tumorigenesis was promoted by LPCAT3 deletion [119]. LPCAT3 also has important roles in adipocyte differentiation and macrophage polarization [131, 132], indicating LPCAT3 has diverse physiological roles and may be a promising therapeutic target in several diseases.

2.4.6 LPIAT1 Functions in Brain Development and Modifies Liver Disease Risk

LPIAT1 is a MBOAT family member and incorporates arachidonic acid-CoA into LPI to produce arachidonic acid-containing PI [133]. Two independent groups generated global LPIAT1 KO mice and reported the KO mice are small, born at low frequency, and have poor survival. Both groups also reported abnormal brain development and decreased arachidonic acid-containing PI and PI phosphates in brain [134, 135]. Lee et al. measured LPLAT activities in the KO tissues, and activity to incorporate arachidonic acid-CoA into LPI was almost abolished in several tissues including brain, whereas the activity to incorporate arachidonic acid-CoA into other lysophospholipids in brain was almost unchanged, establishing that LPIAT1 is a predominant LPIAT involved in production of arachidonic acid-containing PI in mice [134]. Lee et al. also performed detailed analyses of the defective brain development and reported delayed neuronal migration, decreased sizes of cerebral cortex and hippocampus, and disordered cortical lamination and neuronal processes [134]. Anderson et al. reported abnormal brain development as well, and arachidonic acid-containing PI and PI phosphates were decreased in both brain and liver. They also reported that in brain, but not in liver, arachidonic acid-containing PC(38:4) and PE(38:4) were both decreased by about half, which might also contribute to the defective brain development [135]. In humans, homozygous inactivating mutations of LPIAT1 are associated with intellectual disability, autism-like symptoms, and epilepsy; indicating a critical role for LPIAT1-generated arachidonic acid-containing PI in brain development in humans as well as mice [136].

LPIAT1 is also suggested to be a genetic modifier of liver disease risk. Liver cirrhosis is a major disease burden worldwide in terms of disability and death and may develop due to a number of different factors that cause chronic liver injury such as hepatitis C, hepatitis B, alcoholism, non-alcoholic steatohepatitis, and autoimmune hepatitis. High variabilities in disease progression exist due partly to genetic modifiers, and a common genetic variant of LPIAT1 (rs641738) has been reported to be associated with an array of alcohol and non-alcoholic fatty liver diseases. The role of LPIAT1 rs641738 variant in modifying fatty liver disease is not well understood but apparently involves altering LPIAT1 expression and the acyl chain compositions of membrane glycerophospholipids, which is proposed to affect inflammation-driven hepatic fibrosis and other symptoms in a variety of contexts including alcohol-related cirrhosis [137], non-alcoholic fatty liver disease [138], and hepatitis C [139]. However, results among different studies with different cohorts of patients have been inconsistent with some studies not supporting LPIAT1 rs641738 as a disease-modifying variant, suggesting that liver disease may have different genetic factors in different populations [140]. Additional mechanistic studies are required to understand the roles of LPIAT1 in liver disease and possible function as a genetic modifier.

2.5 Summary and Future Directions

It has become increasingly clear that the glycerophospholipid compositions of cell membranes, including their fatty chain compositions, affects a variety of physiological and disease states. Identification and biochemical characterization of the key enzymes involved, including LPLATs of the AGPAT and MBOAT families, has shed light on how the membrane compositions may be regulated in different biological contexts. Genetic deletion of these enzymes in mice has revealed unique physiological roles of many of these enzymes, and is expected to translate into new understanding of related human physiological and pathological conditions. Development of potent and selective inhibitors/activators of LPLATs will allow further discovery and elucidation of their biological functions, and pave the way to realize their promise as therapeutic targets.

Acknowledgements The Lipid Signaling Project, National Center for Global Health and Medicine receives funding from ONO Pharmaceuticals. This work was supported by AMED-CREST 19gm0910011 (to HS), AMED-P-CREATE 19 cm0106116 (to HS), AMED Program for Basic and Clinical Research on Hepatitis JP19fk0210041 (to HS).

Conflicts of Interest The authors have no conflicts of interest affecting the objectivity of this review.

References

1. Vance JE (2015) Phospholipid synthesis and transport in mammalian cells. Traffic 16(1):1–18. https://doi.org/10.1111/tra.12230
2. Schlegel RA, Williamson P (2001) Phosphatidylserine, a death knell. Cell Death Differ 8(6):551–563. https://doi.org/10.1038/sj.cdd.4400817
3. Ohashi M, Jan de Vries K, Frank R, Snoek G, Bankaitis V, Wirtz K, Huttner WB (1995) A role for phosphatidylinositol transfer protein in secretory vesicle formation. Nature 377(6549):544–547. https://doi.org/10.1038/377544a0
4. Schlame M, Greenberg ML (2017) Biosynthesis, remodeling and turnover of mitochondrial cardiolipin. Biochim Biophys Acta Mol Cell Biol Lipids 1862(1):3–7. https://doi.org/10.1016/j.bbalip.2016.08.010
5. Barelli H, Antonny B (2016) Lipid unsaturation and organelle dynamics. Curr Opin Cell Biol 41:25–32. https://doi.org/10.1016/j.ceb.2016.03.012
6. Hishikawa D, Hashidate T, Shimizu T, Shindou H (2014) Diversity and function of membrane glycerophospholipids generated by the remodeling pathway in mammalian cells. J Lipid Res 55(5):799–807. https://doi.org/10.1194/jlr.R046094
7. Antonny B, Vanni S, Shindou H, Ferreira T (2015) From zero to six double bonds: phospholipid unsaturation and organelle function. Trends Cell Biol 25(7):427–436. https://doi.org/10.1016/j.tcb.2015.03.004
8. Shimizu T (2009) Lipid mediators in health and disease: enzymes and receptors as therapeutic targets for the regulation of immunity and inflammation. Annu Rev Pharmacol Toxicol 49:123–150. https://doi.org/10.1146/annurev.pharmtox.011008.145616
9. Braverman NE, Moser AB (2012) Functions of plasmalogen lipids in health and disease. Biochim Biophys Acta 1822(9):1442–1452. https://doi.org/10.1016/j.bbadis.2012.05.008
10. Ishii S, Shimizu T (2000) Platelet-activating factor (PAF) receptor and genetically engineered PAF receptor mutant mice. Prog Lipid Res 39(1):41–82
11. Prescott SM, Zimmerman GA, Stafforini DM, McIntyre TM (2000) Platelet-activating factor and related lipid mediators. Annu Rev Biochem 69:419–445. https://doi.org/10.1146/annurev.biochem.69.1.419
12. Kennedy EP, Weiss SB (1956) The function of cytidine coenzymes in the biosynthesis of phospholipides. J Biol Chem 222(1):193–214
13. Reue K, Wang H (2019) Mammalian lipin phosphatidic acid phosphatases in lipid synthesis and beyond: metabolic and inflammatory disorders. J Lipid Res 60(4):728–733. https://doi.org/10.1194/jlr.S091769
14. Vance DE, Ridgway ND (1988) The methylation of phosphatidylethanolamine. Prog Lipid Res 27(1):61–79
15. Lands WE (1958) Metabolism of glycerolipides; a comparison of lecithin and triglyceride synthesis. J Biol Chem 231(2):883–888
16. Kume K, Shimizu T (1997) cDNA cloning and expression of murine 1-acyl-sn-glycerol-3-phosphate acyltransferase. Biochem Biophys Res Commun 237(3):663–666. https://doi.org/10.1006/bbrc.1997.7214
17. Stamps AC, Elmore MA, Hill ME, Kelly K, Makda AA, Finnen MJ (1997) A human cDNA sequence with homology to non-mammalian lysophosphatidic acid acyltransferases. Biochem J 326(Pt 2):455–461. https://doi.org/10.1042/bj3260455
18. West J, Tompkins CK, Balantac N, Nudelman E, Meengs B, White T, Bursten S, Coleman J, Kumar A, Singer JW, Leung DW (1997) Cloning and expression of two human lysophosphatidic acid acyltransferase cDNAs that enhance cytokine-induced signaling responses in cells. DNA Cell Biol 16(6):691–701. https://doi.org/10.1089/dna.1997.16.691
19. Eberhardt C, Gray PW, Tjoelker LW (1997) Human lysophosphatidic acid acyltransferase. cDNA cloning, expression, and localization to chromosome 9q34.3. J Biol Chem 272(32):20299–20305. https://doi.org/10.1074/jbc.272.32.20299
20. Aguado B, Campbell RD (1998) Characterization of a human lysophosphatidic acid acyltransferase that is encoded by a gene located in the class III region of the human major histocompatibility complex. J Biol Chem 273(7):4096–4105. https://doi.org/10.1074/jbc.273.7.4096
21. Yamashita A, Nakanishi H, Suzuki H, Kamata R, Tanaka K, Waku K, Sugiura T (2007) Topology of acyltransferase motifs and substrate specificity and accessibility in 1-acyl-sn-glycero-3-phosphate acyltransferase 1. Biochim Biophys Acta 1771(9):1202–1215. https://doi.org/10.1016/j.bbalip.2007.07.002
22. Harayama T, Shindou H, Ogasawara R, Suwabe A, Shimizu T (2008) Identification of a novel noninflammatory biosynthetic pathway of platelet-activating factor. J Biol Chem 283(17):11097–11106. https://doi.org/10.1074/jbc.M708909200

23. Bradley RM, Duncan RE (2018) The lysophosphatidic acid acyltransferases (acylglycerophosphate acyltransferases) family: one reaction, five enzymes, many roles. Curr Opin Lipidol 29(2):110–115. https://doi.org/10.1097/MOL.0000000000000492
24. Lu B, Jiang YJ, Zhou Y, Xu FY, Hatch GM, Choy PC (2005) Cloning and characterization of murine 1-acyl-sn-glycerol 3-phosphate acyltransferases and their regulation by PPARalpha in murine heart. Biochem J 385(Pt 2):469–477. https://doi.org/10.1042/BJ20041348
25. Yuki K, Shindou H, Hishikawa D, Shimizu T (2009) Characterization of mouse lysophosphatidic acid acyltransferase 3: an enzyme with dual functions in the testis. J Lipid Res 50(5):860–869. https://doi.org/10.1194/jlr.M800468-JLR200
26. Eto M, Shindou H, Shimizu T (2014) A novel lysophosphatidic acid acyltransferase enzyme (LPAAT4) with a possible role for incorporating docosahexaenoic acid into brain glycerophospholipids. Biochem Biophys Res Commun 443(2):718–724. https://doi.org/10.1016/j.bbrc.2013.12.043
27. Shindou H, Shimizu T (2009) Acyl-CoA:lysophospholipid acyltransferases. J Biol Chem 284(1):1–5. https://doi.org/10.1074/jbc.R800046200
28. Yamashita A, Hayashi Y, Matsumoto N, Nemoto-Sasaki Y, Oka S, Tanikawa T, Sugiura T (2014) Glycerophosphate/Acylglycerophosphate acyltransferases. Biology (Basel) 3(4):801–830. https://doi.org/10.3390/biology3040801
29. Hishikawa D, Shindou H, Kobayashi S, Nakanishi H, Taguchi R, Shimizu T (2008) Discovery of a lysophospholipid acyltransferase family essential for membrane asymmetry and diversity. Proc Natl Acad Sci U S A 105(8):2830–2835. https://doi.org/10.1073/pnas.0712245105
30. Shindou H, Eto M, Morimoto R, Shimizu T (2009) Identification of membrane O-acyltransferase family motifs. Biochem Biophys Res Commun 383(3):320–325. https://doi.org/10.1016/j.bbrc.2009.04.013
31. Korbes AP, Kulcheski FR, Margis R, Margis-Pinheiro M, Turchetto-Zolet AC (2016) Molecular evolution of the lysophosphatidic acid acyltransferase (LPAAT) gene family. Mol Phylogenet Evol 96:55–69. https://doi.org/10.1016/j.ympev.2015.12.001
32. Agarwal AK, Sukumaran S, Cortes VA, Tunison K, Mizrachi D, Sankella S, Gerard RD, Horton JD, Garg A (2011) Human 1-acylglycerol-3-phosphate O-acyltransferase isoforms 1 and 2: biochemical characterization and inability to rescue hepatic steatosis in Agpat2(−/−) gene lipodystrophic mice. J Biol Chem 286(43):37676–37691. https://doi.org/10.1074/jbc.M111.250449
33. Agarwal AK, Arioglu E, De Almeida S, Akkoc N, Taylor SI, Bowcock AM, Barnes RI, Garg A (2002) AGPAT2 is mutated in congenital generalized lipodystrophy linked to chromosome 9q34. Nat Genet 31(1):21–23. https://doi.org/10.1038/ng880
34. Garg A, Agarwal AK (2009) Lipodystrophies: disorders of adipose tissue biology. Biochim Biophys Acta 1791(6):507–513. https://doi.org/10.1016/j.bbalip.2008.12.014
35. Agarwal AK, Tunison K, Dalal JS, Nagamma SS, Hamra FK, Sankella S, Shao X, Auchus RJ, Garg A (2017) Metabolic, reproductive, and neurologic abnormalities in Agpat1-null mice. Endocrinology 158(11):3954–3973. https://doi.org/10.1210/en.2017-00511
36. Garg A, Wilson R, Barnes R, Arioglu E, Zaidi Z, Gurakan F, Kocak N, O'Rahilly S, Taylor SI, Patel SB, Bowcock AM (1999) A gene for congenital generalized lipodystrophy maps to human chromosome 9q34. J Clin Endocrinol Metab 84(9):3390–3394. https://doi.org/10.1210/jcem.84.9.6103
37. Broekema MF, Massink MPG, De Ligt J, Stigter ECA, Monajemi H, De Ridder J, Burgering BMT, van Haaften GW, Kalkhoven E (2018) A single complex Agpat2 allele in a patient with partial lipodystrophy. Front Physiol 9:1363. https://doi.org/10.3389/fphys.2018.01363
38. Cortes VA, Curtis DE, Sukumaran S, Shao X, Parameswara V, Rashid S, Smith AR, Ren J, Esser V, Hammer RE, Agarwal AK, Horton JD, Garg A (2009) Molecular mechanisms of hepatic steatosis and insulin resistance in the AGPAT2-deficient mouse model of congenital generalized lipodystrophy. Cell Metab 9(2):165–176. https://doi.org/10.1016/j.cmet.2009.01.002
39. Wang H, Airola MV, Reue K (2017) How lipid droplets "TAG" along: glycerolipid synthetic enzymes and lipid storage. Biochim Biophys Acta Mol Cell Biol Lipids 1862(10 Pt B):1131–1145. https://doi.org/10.1016/j.bbalip.2017.06.010
40. Hishikawa D, Valentine WJ, Iizuka-Hishikawa Y, Shindou H, Shimizu T (2017) Metabolism and functions of docosahexaenoic acid-containing membrane glycerophospholipids. FEBS Lett 591(18):2730–2744. https://doi.org/10.1002/1873-3468.12825
41. Rawicz W, Olbrich KC, McIntosh T, Needham D, Evans E (2000) Effect of chain length and unsaturation on elasticity of lipid bilayers. Biophys J 79(1):328–339. https://doi.org/10.1016/S0006-3495(00)76295-3
42. Forman BM, Chen J, Evans RM (1997) Hypolipidemic drugs, polyunsaturated fatty acids, and eicosanoids are ligands for peroxisome proliferator-activated receptors alpha and delta. Proc Natl Acad Sci U S A 94(9):4312–4317. https://doi.org/10.1073/pnas.94.9.4312
43. Serhan CN, Levy BD (2018) Resolvins in inflammation: emergence of the pro-resolving superfamily of mediators. J Clin Invest 128(7):2657–2669. https://doi.org/10.1172/JCI97943
44. Koeberle A, Shindou H, Harayama T, Shimizu T (2010) Role of lysophosphatidic acid acyltransferase 3 for the supply of highly polyunsaturated fatty acids in TM4 Sertoli cells. FASEB J 24(12):4929–4938. https://doi.org/10.1096/fj.10-162818

45. Koeberle A, Shindou H, Harayama T, Yuki K, Shimizu T (2012) Polyunsaturated fatty acids are incorporated into maturating male mouse germ cells by lysophosphatidic acid acyltransferase 3. FASEB J 26(1):169–180. https://doi.org/10.1096/fj.11-184879
46. Schmidt JA, Brown WJ (2009) Lysophosphatidic acid acyltransferase 3 regulates Golgi complex structure and function. J Cell Biol 186(2):211–218. https://doi.org/10.1083/jcb.200904147
47. Iizuka-Hishikawa Y, Hishikawa D, Sasaki J, Takubo K, Goto M, Nagata K, Nakanishi H, Shindou H, Okamura T, Ito C, Toshimori K, Sasaki T, Shimizu T (2017) Lysophosphatidic acid acyltransferase 3 tunes the membrane status of germ cells by incorporating docosahexaenoic acid during spermatogenesis. J Biol Chem 292(29):12065–12076. https://doi.org/10.1074/jbc.M117.791277
48. Shindou H, Koso H, Sasaki J, Nakanishi H, Sagara H, Nakagawa KM, Takahashi Y, Hishikawa D, Iizuka-Hishikawa Y, Tokumasu F, Noguchi H, Watanabe S, Sasaki T, Shimizu T (2017) Docosahexaenoic acid preserves visual function by maintaining correct disc morphology in retinal photoreceptor cells. J Biol Chem 292(29):12054–12064. https://doi.org/10.1074/jbc.M117.790568
49. Valentine WJ, Tokuoka SM, Hishikawa D, Kita Y, Shindou H, Shimizu T (2018) LPAAT3 incorporates docosahexaenoic acid into skeletal muscle cell membranes and is upregulated by PPARdelta activation. J Lipid Res 59(2):184–194. https://doi.org/10.1194/jlr.M077321
50. Spector AA, Kim HY (2015) Cytochrome P450 epoxygenase pathway of polyunsaturated fatty acid metabolism. Biochim Biophys Acta 1851(4):356–365. https://doi.org/10.1016/j.bbalip.2014.07.020
51. Koppers AJ, Garg ML, Aitken RJ (2010) Stimulation of mitochondrial reactive oxygen species production by unesterified, unsaturated fatty acids in defective human spermatozoa. Free Radic Biol Med 48(1):112–119. https://doi.org/10.1016/j.freeradbiomed.2009.10.033
52. Lenzi A, Gandini L, Maresca V, Rago R, Sgro P, Dondero F, Picardo M (2000) Fatty acid composition of spermatozoa and immature germ cells. Mol Hum Reprod 6(3):226–231. https://doi.org/10.1093/molehr/6.3.226
53. Gonzalez-Ravina C, Aguirre-Lipperheide M, Pinto F, Martin-Lozano D, Fernandez-Sanchez M, Blasco V, Santamaria-Lopez E, Candenas L (2018) Effect of dietary supplementation with a highly pure and concentrated docosahexaenoic acid (DHA) supplement on human sperm function. Reprod Biol 18(3):282–288. https://doi.org/10.1016/j.repbio.2018.06.002
54. Ebert S, Weigelt K, Walczak Y, Drobnik W, Mauerer R, Hume DA, Weber BH, Langmann T (2009) Docosahexaenoic acid attenuates microglial activation and delays early retinal degeneration. J Neurochem 110(6):1863–1875. https://doi.org/10.1111/j.1471-4159.2009.06286.x
55. Querques G, Forte R, Souied EH (2011) Retina and omega-3. J Nutr Metab 2011:748361. https://doi.org/10.1155/2011/748361
56. Senoo N, Miyoshi N, Goto-Inoue N, Minami K, Yoshimura R, Morita A, Sawada N, Matsuda J, Ogawa Y, Setou M, Kamei Y, Miura S (2015) PGC-1alpha-mediated changes in phospholipid profiles of exercise-trained skeletal muscle. J Lipid Res 56(12):2286–2296. https://doi.org/10.1194/jlr.M060533
57. Goto-Inoue N, Yamada K, Inagaki A, Furuichi Y, Ogino S, Manabe Y, Setou M, Fujii NL (2013) Lipidomics analysis revealed the phospholipid compositional changes in muscle by chronic exercise and high-fat diet. Sci Rep 3:3267. https://doi.org/10.1038/srep03267
58. Helge JW, Wu BJ, Willer M, Daugaard JR, Storlien LH, Kiens B (2001) Training affects muscle phospholipid fatty acid composition in humans. J Appl Physiol (1985) 90(2):670–677. https://doi.org/10.1152/jappl.2001.90.2.670
59. Andersson A, Sjodin A, Hedman A, Olsson R, Vessby B (2000) Fatty acid profile of skeletal muscle phospholipids in trained and untrained young men. Am J Physiol Endocrinol Metab 279(4):E744–E751. https://doi.org/10.1152/ajpendo.2000.279.4.E744
60. Mitchell TW, Turner N, Hulbert AJ, Else PL, Hawley JA, Lee JS, Bruce CR, Blanksby SJ (2004) Exercise alters the profile of phospholipid molecular species in rat skeletal muscle. J Appl Physiol (1985) 97(5):1823–1829. https://doi.org/10.1152/japplphysiol.00344.2004
61. Fan W, Evans RM (2017) Exercise mimetics: impact on health and performance. Cell Metab 25(2):242–247. https://doi.org/10.1016/j.cmet.2016.10.022
62. Wang YX, Zhang CL, Yu RT, Cho HK, Nelson MC, Bayuga-Ocampo CR, Ham J, Kang H, Evans RM (2004) Regulation of muscle fiber type and running endurance by PPARdelta. PLoS Biol 2(10):e294. https://doi.org/10.1371/journal.pbio.0020294
63. Manio MC, Inoue K, Fujitani M, Matsumura S, Fushiki T (2016) Combined pharmacological activation of AMPK and PPARdelta potentiates the effects of exercise in trained mice. Physiol Rep 4(5). https://doi.org/10.14814/phy2.12625
64. Narkar VA, Downes M, Yu RT, Embler E, Wang YX, Banayo E, Mihaylova MM, Nelson MC, Zou Y, Juguilon H, Kang H, Shaw RJ, Evans RM (2008) AMPK and PPARdelta agonists are exercise mimetics. Cell 134(3):405–415. https://doi.org/10.1016/j.cell.2008.06.051
65. Batenburg JJ (1992) Surfactant phospholipids: synthesis and storage. Am J Phys 262(4 Pt 1):L367–L385. https://doi.org/10.1152/ajplung.1992.262.4.L367
66. Veldhuizen EJ, Haagsman HP (2000) Role of pulmonary surfactant components in surface film formation and dynamics. Biochim Biophys Acta 1467(2):255–270. https://doi.org/10.1016/s0005-2736(00)00256-x

67. Creuwels LA, van Golde LM, Haagsman HP (1997) The pulmonary surfactant system: biochemical and clinical aspects. Lung 175(1):1–39
68. Whitsett JA, Wert SE, Weaver TE (2010) Alveolar surfactant homeostasis and the pathogenesis of pulmonary disease. Annu Rev Med 61:105–119. https://doi.org/10.1146/annurev.med.60.041807.123500
69. Matthay MA, Zemans RL (2011) The acute respiratory distress syndrome: pathogenesis and treatment. Annu Rev Pathol 6:147–163. https://doi.org/10.1146/annurev-pathol-011110-130158
70. Post M, Schuurmans EA, Batenburg JJ, Van Golde LM (1983) Mechanisms involved in the synthesis of disaturated phosphatidylcholine by alveolar type II cells isolated from adult rat lung. Biochim Biophys Acta 750(1):68–77. https://doi.org/10.1016/0005-2760(83)90205-9
71. Vereyken JM, Montfoort A, van Golde LM (1972) Some studies on the biosynthesis of the molecular species of phosphatidylcholine from rat lung and phosphatidylcholine and phosphatidylethanolamine from rat liver. Biochim Biophys Acta 260(1):70–81. https://doi.org/10.1016/0005-2760(72)90075-6
72. Nakanishi H, Shindou H, Hishikawa D, Harayama T, Ogasawara R, Suwabe A, Taguchi R, Shimizu T (2006) Cloning and characterization of mouse lung-type acyl-CoA:lysophosphatidylcholine acyltransferase 1 (LPCAT1). Expression in alveolar type II cells and possible involvement in surfactant production. J Biol Chem 281(29):20140–20147. https://doi.org/10.1074/jbc.M600225200
73. Chen X, Hyatt BA, Mucenski ML, Mason RJ, Shannon JM (2006) Identification and characterization of a lysophosphatidylcholine acyltransferase in alveolar type II cells. Proc Natl Acad Sci U S A 103(31):11724–11729. https://doi.org/10.1073/pnas.0604946103
74. Harayama T, Shindou H, Shimizu T (2009) Biosynthesis of phosphatidylcholine by human lysophosphatidylcholine acyltransferase 1. J Lipid Res 50(9):1824–1831. https://doi.org/10.1194/jlr.M800500-JLR200
75. Heath RJ, Rock CO (1998) A conserved histidine is essential for glycerolipid acyltransferase catalysis. J Bacteriol 180(6):1425–1430
76. Bridges JP, Ikegami M, Brilli LL, Chen X, Mason RJ, Shannon JM (2010) LPCAT1 regulates surfactant phospholipid synthesis and is required for transitioning to air breathing in mice. J Clin Invest 120(5):1736–1748. https://doi.org/10.1172/JCI38061
77. Harayama T, Eto M, Shindou H, Kita Y, Otsubo E, Hishikawa D, Ishii S, Sakimura K, Mishina M, Shimizu T (2014) Lysophospholipid acyltransferases mediate phosphatidylcholine diversification to achieve the physical properties required in vivo. Cell Metab 20(2):295–305. https://doi.org/10.1016/j.cmet.2014.05.019
78. Friedman JS, Chang B, Krauth DS et al (2010) Loss of lysophosphatidylcholine acyltransferase 1 leads to photoreceptor degeneration in rd11 mice. Proc Natl Acad Sci U S A 107(35):15523–15528. https://doi.org/10.1073/pnas.1002897107
79. Akagi S, Kono N, Ariyama H, Shindou H, Shimizu T, Arai H (2016) Lysophosphatidylcholine acyltransferase 1 protects against cytotoxicity induced by polyunsaturated fatty acids. FASEB J 30(5):2027–2039. https://doi.org/10.1096/fj.201500149
80. Wei C, Dong X, Lu H, Tong F, Chen L, Zhang R, Dong J, Hu Y, Wu G, Dong X (2019) LPCAT1 promotes brain metastasis of lung adenocarcinoma by up-regulating PI3K/AKT/MYC pathway. J Exp Clin Cancer Res 38(1):95. https://doi.org/10.1186/s13046-019-1092-4
81. Bi J, Ichu TA, Zanca C et al (2019) Oncogene amplification in growth factor signaling pathways renders cancers dependent on membrane lipid remodeling. Cell Metab 30(3):525–538 e528. https://doi.org/10.1016/j.cmet.2019.06.014
82. Wang B, Tontonoz P (2019) Phospholipid remodeling in physiology and disease. Annu Rev Physiol 81:165–188. https://doi.org/10.1146/annurev-physiol-020518-114444
83. Shindou H, Hishikawa D, Nakanishi H, Harayama T, Ishii S, Taguchi R, Shimizu T (2007) A single enzyme catalyzes both platelet-activating factor production and membrane biogenesis of inflammatory cells. Cloning and characterization of acetyl-CoA:LYSO-PAF acetyltransferase. J Biol Chem 282(9):6532–6539. https://doi.org/10.1074/jbc.M609641200
84. Benveniste J, Henson PM, Cochrane CG (1972) Leukocyte-dependent histamine release from rabbit platelets. The role of IgE, basophils, and a platelet-activating factor. J Exp Med 136(6):1356–1377. https://doi.org/10.1084/jem.136.6.1356
85. Morimoto R, Shindou H, Tarui M, Shimizu T (2014) Rapid production of platelet-activating factor is induced by protein kinase Calpha-mediated phosphorylation of lysophosphatidylcholine acyltransferase 2 protein. J Biol Chem 289(22):15566–15576. https://doi.org/10.1074/jbc.M114.558874
86. Morimoto R, Shindou H, Oda Y, Shimizu T (2010) Phosphorylation of lysophosphatidylcholine acyltransferase 2 at Ser34 enhances platelet-activating factor production in endotoxin-stimulated macrophages. J Biol Chem 285(39):29857–29862. https://doi.org/10.1074/jbc.M110.147025
87. Cheng L, Han X, Shi Y (2009) A regulatory role of LPCAT1 in the synthesis of inflammatory lipids, PAF and LPC, in the retina of diabetic mice. Am J Physiol Endocrinol Metab 297(6):E1276–E1282. https://doi.org/10.1152/ajpendo.00475.2009
88. Liang L, Willis-Owen SAG, Laprise C et al (2015) An epigenome-wide association study of total serum immunoglobulin E concentration. Nature 520(7549):670–674. https://doi.org/10.1038/nature14125
89. Cotte AK, Aires V, Fredon M, Limagne E, Derangere V, Thibaudin M, Humblin E, Scagliarini A, de

Barros JP, Hillon P, Ghiringhelli F, Delmas D (2018) Lysophosphatidylcholine acyltransferase 2-mediated lipid droplet production supports colorectal cancer chemoresistance. Nat Commun 9(1):322. https://doi.org/10.1038/s41467-017-02732-5
90. Hall Z, Bond NJ, Ashmore T, Sanders F, Ament Z, Wang X, Murray AJ, Bellafante E, Virtue S, Vidal-Puig A, Allison M, Davies SE, Koulman A, Vacca M, Griffin JL (2017) Lipid zonation and phospholipid remodeling in nonalcoholic fatty liver disease. Hepatology 65(4):1165–1180. https://doi.org/10.1002/hep.28953
91. Kihara Y, Yanagida K, Masago K, Kita Y, Hishikawa D, Shindou H, Ishii S, Shimizu T (2008) Platelet-activating factor production in the spinal cord of experimental allergic encephalomyelitis mice via the group IVA cytosolic phospholipase A2-lyso-PAFAT axis. J Immunol 181(7):5008–5014. https://doi.org/10.4049/jimmunol.181.7.5008
92. Shindou H, Shiraishi S, Tokuoka SM, Takahashi Y, Harayama T, Abe T, Bando K, Miyano K, Kita Y, Uezono Y, Shimizu T (2017) Relief from neuropathic pain by blocking of the platelet-activating factor-pain loop. FASEB J 31(7):2973–2980. https://doi.org/10.1096/fj.201601183R
93. van Hecke O, Austin SK, Khan RA, Smith BH, Torrance N (2014) Neuropathic pain in the general population: a systematic review of epidemiological studies. Pain 155(4):654–662. https://doi.org/10.1016/j.pain.2013.11.013
94. Woolf CJ, Salter MW (2000) Neuronal plasticity: increasing the gain in pain. Science 288(5472):1765–1769. https://doi.org/10.1126/science.288.5472.1765
95. Braz J, Solorzano C, Wang X, Basbaum AI (2014) Transmitting pain and itch messages: a contemporary view of the spinal cord circuits that generate gate control. Neuron 82(3):522–536. https://doi.org/10.1016/j.neuron.2014.01.018
96. Bonnet J, Loiseau AM, Orvoen M, Bessin P (1981) Platelet-activating factor acether (PAF-acether) involvement in acute inflammatory and pain processes. Agents Actions 11(6–7):559–562. https://doi.org/10.1007/bf01978740
97. Dallob A, Guindon Y, Goldenberg MM (1987) Pharmacological evidence for a role of lipoxygenase products in platelet-activating factor (PAF)-induced hyperalgesia. Biochem Pharmacol 36(19):3201–3204. https://doi.org/10.1016/0006-2952(87)90633-2
98. Basran GS, Page CP, Paul W, Morley J (1984) Platelet-activating factor: a possible mediator of the dual response to allergen? Clin Allergy 14(1):75–79
99. Morita K, Morioka N, Abdin J, Kitayama S, Nakata Y, Dohi T (2004) Development of tactile allodynia and thermal hyperalgesia by intrathecally administered platelet-activating factor in mice. Pain 111(3):351–359. https://doi.org/10.1016/j.pain.2004.07.016
100. Hasegawa S, Kohro Y, Shiratori M, Ishii S, Shimizu T, Tsuda M, Inoue K (2010) Role of PAF receptor in proinflammatory cytokine expression in the dorsal root ganglion and tactile allodynia in a rodent model of neuropathic pain. PLoS One 5(5):e10467. https://doi.org/10.1371/journal.pone.0010467
101. Honda Z, Nakamura M, Miki I, Minami M, Watanabe T, Seyama Y, Okado H, Toh H, Ito K, Miyamoto T et al (1991) Cloning by functional expression of platelet-activating factor receptor from Guinea-pig lung. Nature 349(6307):342–346. https://doi.org/10.1038/349342a0
102. Okubo M, Yamanaka H, Kobayashi K, Kanda H, Dai Y, Noguchi K (2012) Up-regulation of platelet-activating factor synthases and its receptor in spinal cord contribute to development of neuropathic pain following peripheral nerve injury. Mol Pain 8:8. https://doi.org/10.1186/1744-8069-8-8
103. Tsuda M, Shigemoto-Mogami Y, Koizumi S, Mizokoshi A, Kohsaka S, Salter MW, Inoue K (2003) P2X4 receptors induced in spinal microglia gate tactile allodynia after nerve injury. Nature 424(6950):778–783. https://doi.org/10.1038/nature01786
104. Tozaki-Saitoh H, Tsuda M, Miyata H, Ueda K, Kohsaka S, Inoue K (2008) P2Y12 receptors in spinal microglia are required for neuropathic pain after peripheral nerve injury. J Neurosci 28(19):4949–4956. https://doi.org/10.1523/JNEUROSCI.0323-08.2008
105. Kobayashi K, Takahashi E, Miyagawa Y, Yamanaka H, Noguchi K (2011) Induction of the P2X7 receptor in spinal microglia in a neuropathic pain model. Neurosci Lett 504(1):57–61. https://doi.org/10.1016/j.neulet.2011.08.058
106. Masuda T, Ozono Y, Mikuriya S, Kohro Y, Tozaki-Saitoh H, Iwatsuki K, Uneyama H, Ichikawa R, Salter MW, Tsuda M, Inoue K (2016) Dorsal horn neurons release extracellular ATP in a VNUT-dependent manner that underlies neuropathic pain. Nat Commun 7:12529. https://doi.org/10.1038/ncomms12529
107. Jin SX, Zhuang ZY, Woolf CJ, Ji RR (2003) p38 mitogen-activated protein kinase is activated after a spinal nerve ligation in spinal cord microglia and dorsal root ganglion neurons and contributes to the generation of neuropathic pain. J Neurosci 23(10):4017–4022
108. Tanga FY, Nutile-McMenemy N, DeLeo JA (2005) The CNS role of Toll-like receptor 4 in innate neuroimmunity and painful neuropathy. Proc Natl Acad Sci U S A 102(16):5856–5861. https://doi.org/10.1073/pnas.0501634102
109. Kasperska-Zajac A, Brzoza Z, Rogala B (2008) Platelet-activating factor (PAF): a review of its role in asthma and clinical efficacy of PAF antagonists in the disease therapy. Recent Patents Inflamm Allergy Drug Discov 2(1):72–76
110. Tarui M, Shindou H, Kumagai K, Morimoto R, Harayama T, Hashidate T, Kojima H, Okabe T,

Nagano T, Nagase T, Shimizu T (2014) Selective inhibitors of a PAF biosynthetic enzyme lysophosphatidylcholine acyltransferase 2. J Lipid Res 55(7):1386–1396. https://doi.org/10.1194/jlr.M049205
111. Zhao Y, Chen YQ, Bonacci TM, Bredt DS, Li S, Bensch WR, Moller DE, Kowala M, Konrad RJ, Cao G (2008) Identification and characterization of a major liver lysophosphatidylcholine acyltransferase. J Biol Chem 283(13):8258–8265. https://doi.org/10.1074/jbc.M710422200
112. Gijon MA, Riekhof WR, Zarini S, Murphy RC, Voelker DR (2008) Lysophospholipid acyltransferases and arachidonate recycling in human neutrophils. J Biol Chem 283(44):30235–30245. https://doi.org/10.1074/jbc.M806194200
113. Matsuda S, Inoue T, Lee HC, Kono N, Tanaka F, Gengyo-Ando K, Mitani S, Arai H (2008) Member of the membrane-bound O-acyltransferase (MBOAT) family encodes a lysophospholipid acyltransferase with broad substrate specificity. Genes Cells 13(8):879–888. https://doi.org/10.1111/j.1365-2443.2008.01212.x
114. Kita Y, Shindou H, Shimizu T (2019) Cytosolic phospholipase A2 and lysophospholipid acyltransferases. Biochim Biophys Acta Mol Cell Biol Lipids 1864(6):838–845. https://doi.org/10.1016/j.bbalip.2018.08.006
115. Rong X, Wang B, Dunham MM, Hedde PN, Wong JS, Gratton E, Young SG, Ford DA, Tontonoz P (2015) Lpcat3-dependent production of arachidonoyl phospholipids is a key determinant of triglyceride secretion. elife 4. https://doi.org/10.7554/eLife.06557
116. Hashidate-Yoshida T, Harayama T, Hishikawa D et al (2015) Fatty acid remodeling by LPCAT3 enriches arachidonate in phospholipid membranes and regulates triglyceride transport. elife 4. https://doi.org/10.7554/eLife.06328
117. Li Z, Jiang H, Ding T, Lou C, Bui HH, Kuo MS, Jiang XC (2015) Deficiency in lysophosphatidylcholine acyltransferase 3 reduces plasma levels of lipids by reducing lipid absorption in mice. Gastroenterology 149(6):1519–1529. https://doi.org/10.1053/j.gastro.2015.07.012
118. Kabir I, Li Z, Bui HH, Kuo MS, Gao G, Jiang XC (2016) Small intestine but not liver Lysophosphatidylcholine acyltransferase 3 (Lpcat3) deficiency has a dominant effect on plasma lipid metabolism. J Biol Chem 291(14):7651–7660. https://doi.org/10.1074/jbc.M115.697011
119. Wang B, Rong X, Palladino END, Wang J, Fogelman AM, Martin MG, Alrefai WA, Ford DA, Tontonoz P (2018) Phospholipid remodeling and cholesterol availability regulate intestinal stemness and tumorigenesis. Cell Stem Cell 22(2):206–220. e204. https://doi.org/10.1016/j.stem.2017.12.017
120. Zelcer N, Khanlou N, Clare R, Jiang Q, Reed-Geaghan EG, Landreth GE, Vinters HV, Tontonoz P (2007) Attenuation of neuroinflammation and Alzheimer's disease pathology by liver x receptors. Proc Natl Acad Sci U S A 104(25):10601–10606. https://doi.org/10.1073/pnas.0701096104
121. Demeure O, Lecerf F, Duby C, Desert C, Ducheix S, Guillou H, Lagarrigue S (2011) Regulation of LPCAT3 by LXR. Gene 470(1–2):7–11. https://doi.org/10.1016/j.gene.2010.09.002
122. Rong X, Albert CJ, Hong C, Duerr MA, Chamberlain BT, Tarling EJ, Ito A, Gao J, Wang B, Edwards PA, Jung ME, Ford DA, Tontonoz P (2013) LXRs regulate ER stress and inflammation through dynamic modulation of membrane phospholipid composition. Cell Metab 18(5):685–697. https://doi.org/10.1016/j.cmet.2013.10.002
123. Ishibashi M, Varin A, Filomenko R, Lopez T, Athias A, Gambert P, Blache D, Thomas C, Gautier T, Lagrost L, Masson D (2013) Liver x receptor regulates arachidonic acid distribution and eicosanoid release in human macrophages: a key role for lysophosphatidylcholine acyltransferase 3. Arterioscler Thromb Vasc Biol 33(6):1171–1179. https://doi.org/10.1161/ATVBAHA.112.300812
124. Repa JJ, Liang G, Ou J, Bashmakov Y, Lobaccaro JM, Shimomura I, Shan B, Brown MS, Goldstein JL, Mangelsdorf DJ (2000) Regulation of mouse sterol regulatory element-binding protein-1c gene (SREBP-1c) by oxysterol receptors, LXRalpha and LXRbeta. Genes Dev 14(22):2819–2830. https://doi.org/10.1101/gad.844900
125. Schultz JR, Tu H, Luk A, Repa JJ, Medina JC, Li L, Schwendner S, Wang S, Thoolen M, Mangelsdorf DJ, Lustig KD, Shan B (2000) Role of LXRs in control of lipogenesis. Genes Dev 14(22):2831–2838. https://doi.org/10.1101/gad.850400
126. Li Z, Ding T, Pan X, Li Y, Li R, Sanders PE, Kuo MS, Hussain MM, Cao G, Jiang XC (2012) Lysophosphatidylcholine acyltransferase 3 knockdown-mediated liver lysophosphatidylcholine accumulation promotes very low density lipoprotein production by enhancing microsomal triglyceride transfer protein expression. J Biol Chem 287(24):20122–20131. https://doi.org/10.1074/jbc.M111.334664
127. Cash JG, Hui DY (2016) Liver-specific overexpression of LPCAT3 reduces postprandial hyperglycemia and improves lipoprotein metabolic profile in mice. Nutr Diabetes 6:e206. https://doi.org/10.1038/nutd.2016.12
128. Rong X, Wang B, Palladino EN, de Aguiar Vallim TQ, Ford DA, Tontonoz P (2017) ER phospholipid composition modulates lipogenesis during feeding and in obesity. J Clin Invest 127(10):3640–3651. https://doi.org/10.1172/JCI93616
129. Tanaka H, Zaima N, Sasaki T, Yamamoto N, Inuzuka K, Yata T, Iwaki T, Umemura K, Sano H, Suzuki Y, Urano T, Setou M, Unno N (2017) Lysophosphatidylcholine Acyltransferase-3 expression is associated with atherosclerosis progression. J Vasc Res 54(4):200–208. https://doi.org/10.1159/000473879

130. Thomas C, Jalil A, Magnani C, Ishibashi M, Quere R, Bourgeois T, Bergas V, Menegaut L, Patoli D, Le Guern N, Labbe J, Gautier T, de Barros JPP, Lagrost L, Masson D (2018) LPCAT3 deficiency in hematopoietic cells alters cholesterol and phospholipid homeostasis and promotes atherosclerosis. Atherosclerosis 275:409–418. https://doi.org/10.1016/j.atherosclerosis.2018.05.023
131. Taniguchi K, Hikiji H, Okinaga T, Hashidate-Yoshida T, Shindou H, Ariyoshi W, Shimizu T, Tominaga K, Nishihara T (2015) Essential role of Lysophosphatidylcholine acyltransferase 3 in the induction of macrophage polarization in PMA-treated U937 cells. J Cell Biochem 116(12):2840–2848. https://doi.org/10.1002/jcb.25230
132. Eto M, Shindou H, Koeberle A, Harayama T, Yanagida K, Shimizu T (2012) Lysophosphatidylcholine acyltransferase 3 is the key enzyme for incorporating arachidonic acid into glycerophospholipids during adipocyte differentiation. Int J Mol Sci 13(12):16267–16280. https://doi.org/10.3390/ijms131216267
133. Lee HC, Inoue T, Imae R, Kono N, Shirae S, Matsuda S, Gengyo-Ando K, Mitani S, Arai H (2008) Caenorhabditis elegans mboa-7, a member of the MBOAT family, is required for selective incorporation of polyunsaturated fatty acids into phosphatidylinositol. Mol Biol Cell 19(3):1174–1184. https://doi.org/10.1091/mbc.e07-09-0893
134. Lee HC, Inoue T, Sasaki J, Kubo T, Matsuda S, Nakasaki Y, Hattori M, Tanaka F, Udagawa O, Kono N, Itoh T, Ogiso H, Taguchi R, Arita M, Sasaki T, Arai H (2012) LPIAT1 regulates arachidonic acid content in phosphatidylinositol and is required for cortical lamination in mice. Mol Biol Cell 23(24):4689–4700. https://doi.org/10.1091/mbc.E12-09-0673
135. Anderson KE, Kielkowska A, Durrant TN, Juvin V, Clark J, Stephens LR, Hawkins PT (2013) Lysophosphatidylinositol-acyltransferase-1 (LPIAT1) is required to maintain physiological levels of PtdIns and PtdInsP(2) in the mouse. PLoS One 8(3):e58425. https://doi.org/10.1371/journal.pone.0058425
136. Johansen A, Rosti RO, Musaev D et al (2016) Mutations in MBOAT7, encoding Lysophosphatidylinositol acyltransferase I, Lead to intellectual disability accompanied by epilepsy and autistic features. Am J Hum Genet 99(4):912–916. https://doi.org/10.1016/j.ajhg.2016.07.019
137. Buch S, Stickel F, Trepo E et al (2015) A genome-wide association study confirms PNPLA3 and identifies TM6SF2 and MBOAT7 as risk loci for alcohol-related cirrhosis. Nat Genet 47(12):1443–1448. https://doi.org/10.1038/ng.3417
138. Mancina RM, Dongiovanni P, Petta S et al (2016) The MBOAT7-TMC4 variant rs641738 increases risk of nonalcoholic fatty liver disease in individuals of European descent. Gastroenterology 150(5):1219–1230 e1216. https://doi.org/10.1053/j.gastro.2016.01.032
139. Thabet K, Asimakopoulos A, Shojaei M et al (2016) MBOAT7 rs641738 increases risk of liver inflammation and transition to fibrosis in chronic hepatitis C. Nat Commun 7:12757. https://doi.org/10.1038/ncomms12757
140. Sookoian S, Flichman D, Garaycoechea ME, Gazzi C, Martino JS, Castano GO, Pirola CJ (2018) Lack of evidence supporting a role of TMC4-rs641738 missense variant-MBOAT7- intergenic downstream variant-in the susceptibility to nonalcoholic fatty liver disease. Sci Rep 8(1):5097. https://doi.org/10.1038/s41598-018-23453-9

Druggable Prostanoid Pathway

3

Liudmila L. Mazaleuskaya and Emanuela Ricciotti

Abstract

Prostanoids (prostaglandins, prostacyclin and thromboxane) belong to the oxylipin family of biologically active lipids generated from arachidonic acid (AA). Protanoids control numerous physiological and pathological processes. Cyclooxygenase (COX) is a rate-limiting enzyme involved in the conversion of AA into prostanoids. There are two COX isozymes: the constitutive COX-1 and the inducible COX-2. COX-1 and COX-2 have similar structures, catalytic activities, and subcellular localizations but differ in patterns of expression and biological functions. Non-selective COX-1/2 or traditional, non-steroidal anti-inflammatory drugs (tNSAIDs) target both COX isoforms and are widely used to relieve pain, fever and inflammation. However, the use of NSAIDs is associated with various side effects, particularly in the gastrointestinal tract. NSAIDs selective for COX-2 inhibition (coxibs) were purposefully designed to spare gastrointestinal toxicity, but predisposed patients to increased cardiovascular risks. These health complications from NSAIDs prompted interest in the downstream effectors of the COX enzymes as novel drug targets. This chapter describes various safety issues with tNSAIDs and coxibs, and discusses the current development of novel classes of drugs targeting the prostanoid pathway, including nitrogen oxide- and hydrogen sulfide-releasing NSAIDs, inhibitors of prostanoid synthases, dual inhibitors, and prostanoid receptor agonists and antagonists.

Keywords

Prostanoids · Prostaglandin · Cyclooxygenase · NSAID · mPGES-1 inhibitor · Prostanoid receptor agonist/antagonist

3.1 Overview/History

Prostanoids are bioactive oxylipins formed from arachidonic acid (AA, 5,8,11,14-eicosatetraenoic acid), a polyunsaturated omega-6 fatty acid (FA) 20:4(ω-6) esterified to the membrane phospholipids. Prostanoids include prostaglandin (PG) E_2 (PGE_2), $F_{2\alpha}$ ($PGF_{2\alpha}$), D_2 (PGD_2), prostaglandin I_2 (PGI_2), commonly known as prostacyclin, and thromboxane A_2 (TxA_2) [1]. The era of prostanoid

L. L. Mazaleuskaya · E. Ricciotti (✉)
Department of Systems Pharmacology and Translational Therapeutics, University of Pennsylvania Perelman School of Medicine, Philadelphia, PA, USA

Institute for Translational Medicine and Therapeutics, University of Pennsylvania Perelman School of Medicine, Philadelphia, PA, USA
e-mail: emanuela@pennmedicine.upenn.edu

Y. Kihara (ed.), *Druggable Lipid Signaling Pathways*, Advances in Experimental Medicine and Biology 1274, https://doi.org/10.1007/978-3-030-50621-6_3

research started in the early 1930 with the observation by Kurzrok and Leib that human semen altered the contractile activity of human uterine strips [2]. It followed by the discovery by Goldblatt and von Euler that identified lipid-soluble acids present in the seminal vesicles, named prostaglandins, because erroneously believed to be produced in the prostate gland, that caused smooth muscle-contracting and vasodepressor activities [3, 4]. Bergström isolated the first prostaglandins (PGE_1 and $PGF_{2\alpha}$) in 1957 [5], and Bergström and Samuelsson showed in detail how prostaglandins form from AAand determined their structures in 1963 [6]. In 1971, Vane showed that acetylsalicylic acid, the active compound present in aspirin-like drugs, worked by inhibiting the formation of prostaglandins [7]. In 1976, Vane isolated vascular wall-synthesized PGI_2, which caused blood vessel relaxation and inhibited platelet function by counteracting thromboxane action [8]. Vane, Bergström and Samuelsson received the 1982 Nobel Prize in Physiology or Medicine for their pioneering work on prostaglandins (www.nobel.se/medicine/laureates/1982/index.html).

The synthesis of prostanoids is the result of the coordinated actions of the activity of several enzymes (Fig. 3.1). AA is first released from membrane phospholipids by phospholipase A_2 [9]. Free AA then can be metabolized by prostaglandin endoperoxide G/H synthase (PTGS), also known as cyclooxygenase (COX), into PGH_2. PGH_2 is an unstable cyclic endoperoxide that is subjected to further metabolism into TxA_2, PGI_2, PGD_2, PGE_2, and $PGF_{2\alpha}$ by tissue-specific isomerases, synthases and hydrolases, located on the cytosolic surface of the endoplasmic reticulum or in the cytosol. TxA synthase (TxS), PGI synthase (PGIS) and PGF synthase (PGFS) are involved in the biosynthesis of TxA_2, PGI_2, $PGF_{2\alpha}$, respectively. PGE_2 can be generated by the action of cytosolic PGE synthase (cPGES) and microsomal PGES (mPGES)-1 and 2 [10]. PGD_2 is generated mainly by the action of hematopoietic PGDS (H-PGDS) and to a lesser extent by lipocalin-like PGDS (L-PGDS) [11]. There is a preferential coupling of individual synthases to either COX-1 or COX-2: COX-1 has a preference for coupling with TXAS, PGFS and cPGES; COX-2 has a preference for PGIS and mPGES-1 [12]. Thus, through a combination of regulated expression, tissue selective expression and specific COX-synthase coupling, COX products are formed in a cell-type or disease-state specific manner, such that most cells make one or two dominant prostanoids.

Prostanoids are autacoids, which act locally on surrounding cells, in an autocrine and paracrine fashion, to mediate a diverse range of physiological and pathological processes. Prostanoids are synthesized de-novo in a cell, have a short half-life, on average less than 1 min, and are inactivated in two steps; prostanoid metabolites formed in the liver are eliminated mainly in the urine. The measure of these inert but chemically stable urinary metabolites by mass-spectrometry provides insights into their *in vivo* biology and pharmacology [13, 14].

Prostanoids act through G-protein coupled receptors localized at the plasma membrane or at the nuclear envelope and characterized by different downstream signal transduction pathways. There are four receptors to bind PGE_2 ($EP_{1,2,3,4}$), two receptors to bind PGD_2 (DP_1 and DP_2) and one receptor each to bind PGF_{2a} (FP), PGI_2 (IP) and TxA_2 (TP; with alpha and beta subtypes found only in humans), with additional subtypes generated through alternative splicing on the cell membrane (Fig. 3.1) [15]. Prostanoid receptors are linked to specific G-proteins, which in turn facilitate changes in the concentrations of intracellular signaling [15].

Prostanoid receptors can be categorized in 3 functional clusters that reflect their biological action, cellular signaling and phylogenetic origin:

- DP_1, EP_2, EP_4 and IP, the "relaxant" receptors, which signal through $G\alpha_s$-mediated increases in intracellular cyclic adenosine monophosphate (cAMP);
- TP, FP and EP_1, the "contractile" receptors, which signal through $G\alpha_q$-mediated increases in inositol phosphate and intracellular Ca^{2+};
- EP_3 and DP_2, the "inhibitory" receptors, which signal through $G\alpha_i$-mediated decreases cAMP formation [15].

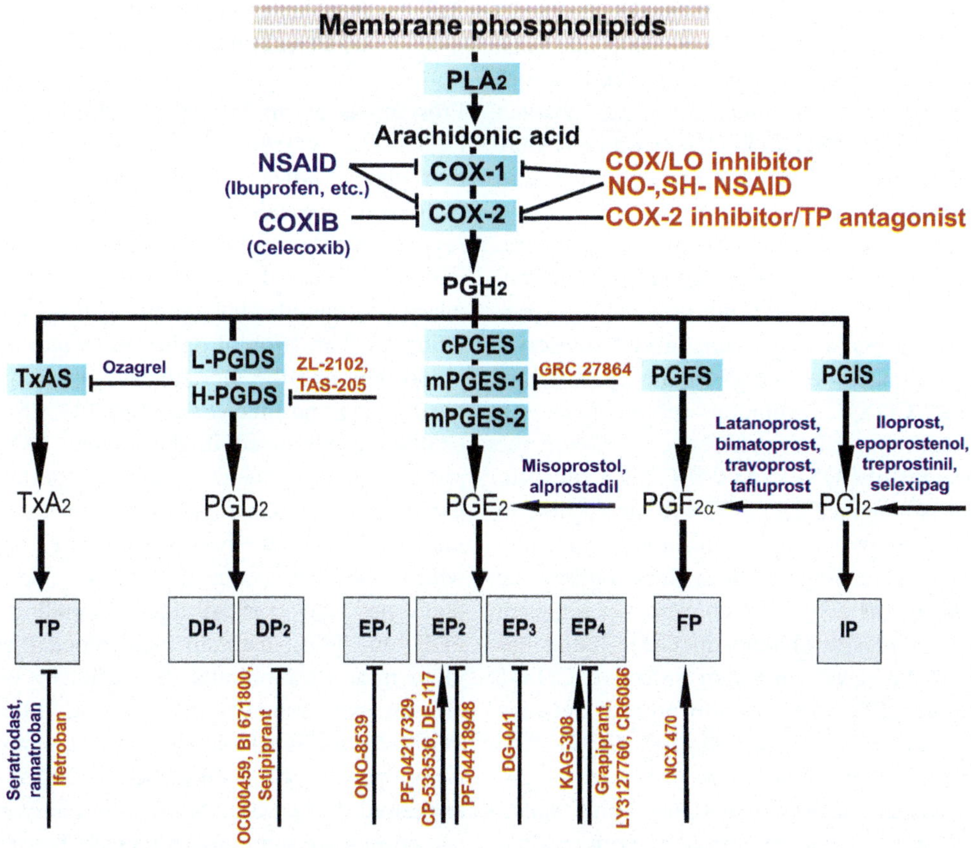

Fig. 3.1 Biosynthetic pathway of prostanoids and druggable targets. Drugs marketed for clinical use are shown in blue. Compounds under clinical development are shown in red. Agonists are indicated with the symbol ←. Antagonists/Inhibitors are indicated with the symbol ⊢

3.2 Cyclooxygenases

COX exists in two isoforms, referred to as COX-1 (the constitutive isoform) and COX-2 (the inducible isoform), that arose from gene duplication early in evolution [1]. COXs are heme-containing glycoproteins, members of the myeloperoxidase superfamily, which possess both cyclooxygenase and peroxidase activities [16]. COXs are encoded by genes localized on different chromosomes. The intron-exon junctions in the two genes are highly conserved and few significant differences are seen in the dimerization, membrane-binding and catalytic domains between COXs from different species. COXs are glycosylated on asparagine in all organisms and the N-glycosylation appears to be necessary for proper protein folding [17]. There is 61% amino acid homology in human COXs.

Both COX isoforms are homodimers that function as conformational heterodimers having allosteric and catalytic subunits. Each COX monomer consists of three structural domains: an N-terminal epidermal growth factor-like domain that functions as a dimerization domain, a membrane binding domain, and a large C-terminal globular catalytic domain, which has the COX and peroxidase active sites [18]. Both COX-1 and COX-2 work as a dimer located in the luminal surface of the endoplasmic reticulum and associated with the inner membrane of the nuclear envelope. COX-2 is also located in the Golgi apparatus and associated with lipid droplets. COX-1 and COX-2 dimers exhibit half-of-sites activity, where each monomer binds distinct ligands [19]. COX-1/COX-2 heterodimers may also exist, but their role in biology remains to be established [20].

Despite the close similarities in structure, catalytic function, and subcellular localization, COX-1 and COX-2 differ markedly in their profiles of expression. There are several transcriptional regulatory elements in the COX-2, but not the COX-1, promoter region [21]. An important structural difference between the active sites of COX-1 and COX-2 is a substitution of isoleucine 523 in COX-1 for a valine in COX-2. The switch of a voluminous amino acid residue (isoleucine) with a less bulky one (valine) opens a hydrophobic pocket adjacent to the active site channel in COX-2 that has allowed the development of selective COX-2 inhibitors [22]. Another difference between COX-1 and COX-2 is the ability of COX-2 to efficiently oxygenate neutral derivatives of AA, such as 2-arachidonoylglycerol and anandamide, two endogenous ligands for the cannabinoid receptors [19]. Moreover, COX-2 has broader substrate specificity than COX-1 and can oxygenate other FAs [19].

In simplistic terms, COX-1 is ubiquitously and constitutively expressed in mammalian tissues and cells, and it is the dominant (but not exclusive) source of prostanoids for physiological functions. COX-2, in contrast, is generally present in mammalian tissues at very low levels, but it is readily induced by inflammatory mediators and mitogenic stimuli, and thus, contributes to the formation of prostanoids in pathophysiological conditions. COX-2 is constitutively expressed in some tissues like the brain, kidney, testis and vasculature [15]. However, both COXs contribute to the generation of autoregulatory and homeostatic prostanoids, and both enzymes can contribute to prostanoid formation during pathophysiological conditions.

3.3 Biological Functions of Prostanoids

Prostanoids are implicated in controlling physiological (i.e., gastric cytoprotection, renal function, platelet aggregation and vascular tone) and pathophysiological (i.e., inflammation, pain, fever, asthma) processes [23]. COX-1 is the main isoform expressed in the gastrointestinal tract, where prostanoids play a protective role by regulating the formation of gastric acid and mucosal blood flow and the secretion of gastrin, pepsin, bicarbonate and mucus secretion [24, 25]. Both COXs are expressed in the kidney: COX-1 is present mainly in the cortical and medullary collecting ducts, mesangial cells, arteriolar endothelium, and the epithelium of Bowman's capsule, while COX-2 is localized in the macula densa, podocytes and cortical thick ascending limb. PGE_2 and PGI_2 maintain renal blood flow and glomerular filtration rate, and regulate water and sodium excretion. On the opposite, TxA_2 and $PGF_{2\alpha}$ reduce renal blood flow and glomerular filtration rate [26]. Both COX-1 and COX-2 are expressed in the vasculature, where prostanoids can promote or limit the interactions between the vessel walls and blood cells and control the local vascular tone. Studies in animal models have revealed that COX-2 is implicated in the development of vascular remodeling, atherothrombotic events and aneurysm formation [23]. In the blood vessels, the main products of both endothelial and smooth muscle cell layers are PGI_2 and PGE_2. Normal blood flow shear forces make COX-2 responsible for most PGI_2 production in the blood vessels [27, 28]. PGI_2 mediates vasodilatation, inhibits leukocyte adhesion, prevents smooth muscle cell proliferation, and opposes platelet activation. PGE_2 has vasodilatory and antiplatelet properties by activating EP_2 and EP_4, but it also can favor vascular hyperplasia and vasoconstriction from activation of EP_1 and EP_3 [29]. PGD_2 mediates vasodilation and increases vascular permeability [30], relaxation of vascular and nonvascular smooth muscle [31, 32], inhibition of platelet aggregation [33]. L-PGDS-derived PGD_2 can control blood pressure and thrombosis through an autocrine effect on the vasculature or through L-PGDS functioning as a lipophilic carrier [11]. $PGF_{2\alpha}$ is a potent constrictor of both pulmonary arteries and veins in humans. TxA_2 mediates vasoconstrictions and platelet aggregation. TxA_2 is the main product synthetized by platelet COX-1, the only COX isoform expressed in mature platelets. TxA_2 causes vasoconstriction and platelet aggregation. It amplifies further platelet activation by other platelet agonists and the recruitment of other platelets [29]. Consequently, for effective inhibition of platelet activation, platelet TxA_2 blockade must be almost complete (~95%) to limit thrombotic events [34].

Prostanoids are involved in multiple stages of the inflammatory response and they mediate the cardinal signs of acute inflammation: *rubor* (redness), *calor* (fever), *tumor* (swelling) and *dolor* (pain) [23]. They regulate vascular permeability, expression of cellular adhesion molecules, extravasation and release of cytokines and chemokines, and peripheral and central sensitization. Moreover, PGE_2 mediates the resetting of the hypothalamic temperature set-point.

Prostanoid biosynthesis is significantly increased in inflamed tissues. COX-2 expression is markedly upregulated during inflammation and it represents the major source of pro-inflammatory prostanoids, although COX-1 also plays a role in the initial phase of the inflammatory response [35].

PGE_2 and PGI_2 are the predominant pro-inflammatory prostanoids. Both markedly enhance vascular permeability and leukocyte infiltration by promoting blood flow in the inflamed region. In addition, PGE_2 can cross the blood-brain barrier and mediate the febrile response [36]. $PGF_{2\alpha}$ plays a role in both acute and chronic inflammation [37, 38].

TxA_2 induces lymphocyte activation and adhesion molecule expression, and similarly to PGE_2 and PGI_2, it can also act as an anti-inflammatory mediator [15].

The chemotactic function of PGD_2, a major product of mast cells, contributes to inflammation in allergic responses, particularly in the lung. PGD_2 also increases vascular permeability and vasodilation. However, PGD_2 has been shown to have anti-inflammatory proprieties as well [39–41].

Prostanoids play pivotal roles in pain. Prostanoids increase the sensitivity of nociceptors in inflamed tissue and augment pain signal transmission in the spinal cord and in the brain [42]. In particular, PGI_2 and PGE_2 cause peripheral and central hyperalgesia when bound to IP, EP_1, EP_3, and EP_4 receptors by reducing the threshold of nociceptor sensory neurons to stimulation. Perhaps PGD_2 and $PGF_{2\alpha}$ also contribute to central sensitization [23].

In the lung, TxA_2, PGD_2 and PGE_2 can induce bronchoconstriction, while PGI_2, PGE_2 and PGD_2 can mediate bronchodilation. They play a major role in allergic responses such as asthma. PGE_2 can have anti-inflammatory and anti-asthmatic effects via activation of EP_3.

PGD_2 contributes to the characteristic pattern of vascular and cellular changes associated with both the early and late phases of the allergic response via DP_2 [43].

3.4 Approved Drugs Targeting the Prostanoid Pathway

Since the major role of prostanoids, particularly PGE_2, in inflammation, pain and fever, inhibitors of their synthesis have been used successfully as anti-inflammatory agents [44]. The class of drugs, which inhibit COX activity, is known as the non-steroidal anti-inflammatory drugs (NSAIDs). They are among the most used drugs in the world, and the current opioid crisis has placed a new emphasis on their use since they don't cause addiction. The use of prostanoids themselves as therapeutic agents is limited largely due to their short half-life and because systemic administration of these agents is frequently associated with significant adverse side effects. There are, however, a few prostanoid analogs that are currently used in clinic for different therapeutic indications. Moreover, drugs that limit the effect of TXA_2 (TP antagonist and TXAS inhibitors) are currently on the market in Asia but not in the US; a dual TP antagonist and TXAS inhibitor is sold as an antiplatelet drug in Italy.

3.4.1 NSAIDs

Aspirin is a prototypic COX inhibitor. It was synthesized serendipitously from salicylic acid, a natural compound with known analgesic activity, more than 100 years ago [45]. After the discovery of the mechanism of action of aspirin in 1971, traditional NSAIDs (tNSAIDs), which block both COX-1 and COX-2, such as ibuprofen, naproxen, diclofenac, were developed to mimic aspirin pharmacological effects [46]. The purposefully designed selective COX-2 inhibitors (coxibs) were developed after the discovery of COX-2 in 1991 [47] and the structural differences between the two COX isozymes as discussed above.

NSAIDs are weak organic acids, highly protein-bound and lipophilic. They are a chemically heterogeneous group of compounds that can be grouped on the basis of their chemical structure [48]. The kinetics, potency and selectivity of COX inhibitors can be assessed *in vitro* using purified enzyme systems, isolated cellular preparations or whole blood assay [49, 50]. Based on the selectivity for COX isoform blockade, NSAIDs are classified into nonselective COX inhibitors (ibuprofen, naproxen, indomethacin) and selective COX-2 inhibitors (celecoxib, rofecoxib, etoricoxib) [45]. NSAIDs can also be classified based on the kinetics of the interaction with COX-1 and/or COX-2. They can be distinguished into:

- Inhibitors purely competitive with respect to AA (e.g., ibuprofen, piroxicam) [19, 51].
- Time-dependent, slowly reversible non-covalent inhibitors (e.g., indomethacin, flurbiprofen, naproxen, celecoxib, rofecoxib) [19, 51].
- Irreversible inhibitors (e.g., aspirin). Aspirin irreversibly inhibits COX activity by acetylation of a serine residue in the active site for COX (Ser^{529} in COX-1 and Ser^{516} in COX-2). This produces a steric blockage that prevents AA access to the active site [51].

Low-dose aspirin (75–100 mg/day) has an effective anti-platelet action. Since platelets do not have a nucleus, they cannot regenerate new COX-1 protein, after it is acetylated by aspirin. Thus, aspirin unique pharmacokinetics (short half-life, about 39 min) and pharmacodynamics (irreversible inhibition of COX isozyme) allow a selective, cumulative inhibition of COX-1–mediated TxA_2 production and platelet function throughout the dosing interval. Since acetylation of platelet COX-1 by aspirin requires initial low affinity anchoring to the Arg^{120} residue of the COX channel, a common docking site for all NSAIDs, concomitant treatment with ibuprofen or naproxen can interfere with the antiplatelet effect of low dose aspirin [52]. NSAIDs can cause serious adverse events, primarily gastrointestinal and cardiovascular complications in some individuals, related to the relative selectivity of a drug for COX-1 and COX-2 and the duration of drug exposure [25, 53].

Coxibs were designed to retain the analgesic and anti-inflammatory effects of tNSAIDs, while sparing COX-1 in the gastric mucosa and platelets, and thus reducing gastrointestinal toxicity [54]. The first coxibs, celecoxib and rofecoxib, were approved by the Food and Drug Administration (FDA) in the late 1990s. Three randomized, double-blind clinical trials, based mainly on short-term use with gastrointestinal monitoring, showed that coxibs present less gastrointestinal toxicity than tNSAIDs [55–57]. Subsequently, novel COX-2 inhibitors with improved biochemical selectivity were developed, i.e., etoricoxib, valdecoxib, parecoxib (the prodrug of valdecoxib, the first injectable coxib), and lumiracoxib (structural analog of diclofenac) [58]. Unfortunately, long-term chemoprevention studies, launched to study the beneficial effect of

COX-2 inhibitors in the sporadic colorectal adenoma recurrence, revealed that celecoxib and rofecoxib were associated with increased risk for cardiovascular mortality [59, 60]. This prompted a voluntarily withdrawal of rofecoxib from the market in late 2004. In early 2005, the FDA requested that all sponsors of tNSAIDs and coxibs included a black box warning on the drug labels highlighting the potential for increased cardiovascular risk and the well-described gastrointestinal bleeding associated with the drug use [61]. In addition, the FDA asked the maker of valdecoxib to withdraw the drug from the market, and initiated analysis of all available at that time clinical data on celecoxib to determine if additional regulatory action was needed. In the meantime, the European Medicines Agency (EMA) decided that coxibs should be contraindicated in patients with coronary heart disease (CHD) or stroke [54].

Later, placebo-controlled trials showed that all FDA-approved COX-2-selective NSAIDs cause cardiovascular complications, including atherosclerosis, thrombosis, myocardial infarction, stroke, systemic and pulmonary hypertension, heart failure and sudden cardiac death in approximately 1 to 2% of people exposed, but the absolute risk increases with higher drug doses, frequency of use, and established cardiovascular disease [62]. Observational studies consistent with a cardiovascular risk have also been reported for several tNSAIDs, which are highly selective for COX-2 inhibition, such as diclofenac and meloxicam [53, 62].

In 2013, the Coxib and traditional NSAID Trialists' (CNT) Collaboration published a meta-analysis of 754 randomized controlled trials comparing traditional NSAIDs and coxibs for their cardiovascular and upper gastrointestinal effects [63]. Coxibs and tNSAIDs, excluding naproxen, were reported to increase the risk of major vascular events (myocardial infarction, stroke or vascular death) by roughly 40% [63]. Elevated risk from diclofenac was comparable to that from coxibs, and is consistent with diclofenac acting as a COX-2-selective inhibitor [64, 65]. This indicates that any NSAID blocking COX-2 with insufficient inhibition of platelet COX-1 has elevated risk of major coronary events. In contrast, naproxen did not increase the risk of major vascular or coronary events consistent with its aspirin-like COX-1 inhibition in platelets described above [63]. All studied coxibs and tNSAIDs, including naproxen, doubled the risk of hospitalization due to heart failure [63]. Moreover, all studied NSAIDs elevated the risk of upper gastrointestinal complications (perforation, obstruction, or bleeding), with high-dose diclofenac causing a gastrointestinal risk similar to high-dose coxibs [53].

In 2015, after the review of the CNT meta-analyses and other clinical data, the FDA strengthened the warning that non-aspirin NSAIDs can cause heart attacks and stroke [53]. In turn, the EMA extended restrictive labeling from coxibs, implemented in 2005, to diclofenac and high-dose ibuprofen [66]. Two recent attempts to compare the cardiovascular safety of celecoxib with tNSAIDs, Standard Care vs. Celecoxib Outcome Trial (SCOT) and Prospective Randomized Evaluation of Celecoxib Integrated Safety versus Ibuprofen or Naproxen (PRECISION), were undermined by multiple limitations of the trial design [53, 66, 67]. At present, celecoxib is the only coxib remaining on the market in the United States.

Coxibs and most tNSAIDs inhibit COX-2-dependent formation of cardioprotective PGs, PGI_2 and PGE_2, with inconsistent and ineffective platelet TxA_2 suppression [68], except for naproxen. High-dose naproxen affords sustained inhibition of platelet COX-1/TxA_2, particularly in those individuals who experience a long half-life of this drug [69]. These effects may increase the risk for cardiovascular events through a direct effect on the myocardium and the vasculature and/or by altering hemodynamic stability and renal function [34, 53]. The clinically manifested elements of this hazard (a predisposition to thrombosis, hypertension, cardiac failure, and arrhythmogenesis) have been recapitulated in rodent models in which the COX-2–dependent formation of PGI_2 or its action is disrupted in the vascular or in cardiomyocytes [70, 71].

3.4.2 Prostanoid Analogs

A PGE_1 analog, misoprostol, provides gastric cytoprotection during long term NSAID treatment [53]. It is also used as an effective abortifacient in combination with the anti-progesterone mifepristone for medical abortion [72]. Another PGE_1 analog, alprostadil, is used to treat impotence or for maintenance of the ductus arteriosus in neonates awaiting cardiac surgery [73, 74].

PGI_2 analogs, iloprost, epoprostenol, treprostinil, selexipag (the first approved oral IP receptor agonist with a non-prostanoid structure [75, 76]), are on the market for the treatment of pulmonary arterial hypertension (PAH) [77].

Several $PGF_{2\alpha}$ analogs, latanoprost, bimatoprost, travoprost and tafluprost, have been marketed for the treatment of ocular hypertension and glaucoma [78–85]. In recent years, NO-donating $PGF_{2\alpha}$ analogs have been developed to treat glaucoma in patients that do not respond to the treatment with $PGF_{2\alpha}$ analogs [86, 87]. Latanoprostene bunod, LBN, was the first drug in this class approved by the FDA in November 2017 [78]. Upon exposure to corneal esterases, LBN is rapidly cleaved into latanoprost and butanediol mononitrate, an NO-donating moiety [88]. NO-mediated vasodilation, anti-inflammatory and antiplatelet effects make this class of drugs especially attractive as impaired ocular blood flow and inflammation contribute to glaucoma progression [89, 90]. Another compound in this class is NCX 470, composed of a bimatoprost molecule and the NO-donating moiety 6-(nitrooxy)hexanoic acid [91]. In animal models of glaucoma and ocular hypertension, NCX 470 was more efficacious in reducing intraocular pressure (IOP) than the equimolar doses of bimatoprost, the most efficacious PG analog approved to date [91]. Currently, a clinical trial is ongoing comparing NCX 470 with latanoprost for the safety and efficacy in patients with open-angle glaucoma or ocular hypertension (ClinicalTrials.gov Identifier NCT03657797).

3.5 Prostanoid Pathway-Targeting Drugs in Clinical Development

The future of prostanoid-based therapies appears to reside in selective prostanoid synthase inhibitors, mainly mPGES-1 and H-PGDS inhibitors, prostanoid receptor modulators and dual inhibitors (COX/5-lipoxygenase (LO), COX-2/TP, mPGES-1/5-LO). Moreover, a novel approach to develop NSAIDs with improved safety profiles has focused on linking existing drugs to a molecule of nitric oxide (NO), hydrogen sulfide (H_2S) or nitric oxide-hydrogen sulfide, and generation of NO, H_2S, nitric oxide-hydrogen sulfide-(NOSH) NSAIDs [92–94]. NO and H_2S are gaseous mediators that have protective effects in the gastrointestinal tract and counteract cardiovascular side effects of the NSAIDs on blood pressure [95, 96].

3.5.1 Inhibitors of mPGES-1

Cardiovascular side effects from COX-2-selective NSAIDs [53] prompted search for safer but equally effective drugs for pain and inflammation. Scientists focused on the new drug targets in the arachidonic acid cascade, specifically on the downstream effectors of the COX isoforms to block the synthesis or action of PGE_2. The development of two new classes of analgesics began, the inhibitors of mPGES-1 and antagonists of the EP_4 receptor.

Numerous animal studies showed that NSAID efficacy might be preserved, while the cardiovascular and gastrointestinal risks are reduced, by blocking mPGES-1 downstream of COX-2 [14, 97–104]. Genetic deletion of mPGES-1 was as effective as NSAID treatment in rodent models of inflammatory pain [97, 98]. Similar to interference with COX-2 expression or function, deletion of mPGES-1 in mice suppresses pro-inflammatory PGE_2; but unlike with COX-2, deficiency in mPGES-1 elevates cardio-protective

PGI_2, reflected by increased urinary metabolite excretion [100]. Both, suppression of PGE_2 and augmentation of PGI_2 in mPGES-1-lacking mice, result from the shunting of the elevated PGH_2 substrate away from mPGES-1 to PGIS [100]. The beneficial cardiovascular outcomes from mPGES-1 deletion in mice include a decrease in the proliferative response to vascular injury [103], reduced aortic aneurysm formation [101], slowed development of atherosclerosis [100], and protection from thrombosis and hypertension [14]; although contrasting effects on the blood pressure were reported in mice with different genetic backgrounds [105]. Besides that global deletion of mPGES-1 in mice reduces the post-ischemic brain infarction and neurological dysfunction in a cerebral ischemia/reperfusion model [99]. Finally, mPGES-1 deficiency in mice does not affect ozone-induced airway inflammation or airway hyper-responsiveness suggesting that pharmacological inhibition of mPGES-1 may not predispose patients at risk to airway dysfunction [102]. Taken together, these studies suggest that pharmacological inhibition of mPGES-1 may retain anti-inflammatory effects from PGE_2 suppression, but due to PGI_2 elevation, targeting of mPGES-1 might avoid the cardiovascular risks associated with selective COX-2 inhibitors.

On the other hand, several preclinical studies suggest that mPGES-1-mediated inflammatory response might be beneficial in certain settings, thus, complicating the development of drugs blocking mPGES-1. The whole-body or myeloid cell-specific deletion of mPGES-1 was reported to adversely interfere with cardiac remodeling after myocardial infarction [106, 107]. In a mouse model of myocardial ischemia/reperfusion injury, deletion of mPGES-1 impairs cardiac microvascular perfusion and increases inflammatory cell infiltration [108]. Furthermore, although PGE_2 is the main mediator of pain [109], PGI_2 may contribute to pain responses and even predominate in some rodent models [42, 110]. So even if substrate shunting to PGI_2 offers cardio-protective effects, it might also compromise the analgesic efficacy of mPGES-1 inhibitors in some settings. Finally, the predominant product of substrate shunting amongst the prostanoids varies by cell type [111], which further complicates development of drugs for systemic inhibition of mPGES-1. Recent preclinical studies suggest that cell-specific targeting of mPGES-1 in macrophages might be advantageous over systemic blockade of the enzyme, and might provide comparable analgesic and anti-inflammatory efficacy with improved toxicity. Deletion of mPGES-1 selectively in macrophages recapitulates the analgesic effects of global mPGES-1 depletion in an inflammatory pain model, restrains the proliferative response to vascular injury [111] and attenuates development of atherosclerosis [112]. Therefore, macrophage mPGES-1 provides a promising therapeutic target alternative to COXs, and requires further investigation.

In 2005, the development of a high-throughput screening assay for novel mPGES-1 blockers at Merck Frosst led to the discovery of a series of phenanthrene imidazole inhibitors, which were among the first synthetic inhibitors of mPGES-1 [104, 113, 114]. Compound MF63 was the most promising lead in this class [113]. MF63 was highly potent and selective for mPGES-1i inhibition in cell-free and cell-based assays; most importantly, MF63 was the first reported mPGES-1 inhibitor to exhibit potency in human whole blood [113]. Due to structural differences in mPGES-1 between human and rat enzymes, MF63 could not be screened in well-established rat models of inflammatory pain. Indeed, despite being highly potent at the human and guinea pig enzymes, MF63 was inactive against rat and mouse mPGES-1. It was not until 2010, when scientists from the Karolinska Institute reported that the entrance to the active site of mPGES-1 in the mouse and rat enzyme contained more bulky and sterically hindered amino acids than the human enzyme [115]. To screen MF63 in vivo, a humanized knock-in (KI) mouse model was generated, where the murine mPGES-1 gene was replaced with a human mPGES-1 minigene

[104]. MF63 dose-dependently inhibited pain sensitivity in a hyperalgesia mouse model with comparable potency to a selective COX-2 inhibitor. MF63 was well tolerated by KI mice and squirrel monkeys, and did not cause gastrointestinal lesions in contrast to a nonselective NSAID indomethacin. When dosed orally in guinea pigs, MF63 was efficacious in the models of pyresis, chronic and acute pain with the analgesic efficacy comparable to diclofenac [104].

Pharmacokinetic analysis of MF63 revealed a short plasma half-life in rats and the rhesus monkeys [114]. Based on MF63 structure, a series of derivatives were synthesized with an improved pharmacokinetic profile [114]. Extensive structure-activity relationship (SAR) studies resulted in compound MK-7285, the most active mPGES-1 inhibitor reported to date [114, 116]. MK-7285 [116], was highly selective and potent at the human mPGES-1 enzyme and in the whole blood assay. MK-7285 exhibited improved pharmacokinetics in rats, and enhanced analgesic efficacy over MF63 in a guinea pig pain model [114]. Despite excellent activity in animals and suitability for clinical development, no further studies on the mPGES-1 inhibitor MK-7285 have been reported.

In 2011 and 2012, two potent and selective inhibitors of mPGES-1, LY3023703 and LY3031207 from Eli Lilly advanced in clinical trials for the treatment of osteoarthritis (OA) pain [117–120]. LY3023703 and LY3031207 were the first mPGES-1 inhibitors to reach clinical development. Both compounds were 2-aminoimidazole-based inhibitors that demonstrated analgesic efficacy and robust safety margins in preclinical studies [117]. LY3023703 exhibited high potency in human blood ex vivo, 10,000-fold selectivity against COX-1 and COX-2 enzymes, and analgesic efficacy in a rodent arthritis model [118]. Based on LY3023703 structure, a new mPGES-1 inhibitor, LY3031207, was made with improved physicochemical properties and a superior pharmacokinetic profile. LY3031207 demonstrated high oral bioavailability, quick absorption, sustained plasma levels over a long period, although it was not as potent as LY3023703 [117]. In a single ascending dose study, LY3023703 was well tolerated at the highest dose tested in healthy subjects, but when screened in a multiple ascending dose study, one subject experienced elevated serum aminotransferase levels indicating liver damage [118]. Hepatotoxicity in this subject met criteria for a severe adverse event and precluded further development of LY3023703 [118]. After single doses in healthy subjects, LY3031207 exhibited good tolerability up to 900 mg, but triggered acute liver injury in six subjects in a multiple ascending dose study. Notably, the hepatotoxicity identified in humans contrasted the preclinical toxicology findings in rats and dogs, and thus, could not be predicted from the animal studies. Thus, further development of LY3031207 was terminated due to a rare form of dose-dependent liver injury with hypersensitivity features [120]. Oxidation of the 2-aminoimidazole ring to epoxide reactive metabolites was reported as a cause of hepatotoxicity in both compounds [119]. Taken together, LY3023703 and LY3031207 studies emphasize the limitations of the animal toxicology studies and the need for careful safety monitoring throughout Phase I trials.

The most successful mPGES-1 inhibitor to date that has recently advanced in a Phase IIb clinical trial is a substituted pyrimidine GRC 27864 from Glenmark Pharmaceuticals [121]. GRC 27864 exhibited high potency in the enzymatic assay (IC_{50} = 5 nM) and in human whole blood (IC_{50} = 376 nM), and inhibited PGE_2 biosynthesis in chondrocytes and synovial fibroblasts taken from patients with rheumatoid arthritis (RA) and OA [121]. GRC 27864 demonstrated significant analgesic efficacy in preclinical models of OA and inflammatory arthritis. When tested in Phase I clinical trial, GRC 27864 was well tolerated at single doses up to 1000 mg in healthy adults (NCT02179645, NCT02361034). Similarly to celecoxib, GRC 27864 inhibited *ex vivo* LPS-induced PGE_2 in whole blood of treated subjects in a dose-dependent manner. This was accompanied by a reduction of PGE-M by both drugs. However, unlike celecoxib, GRC 27864 did not inhibit PGI-M and Tx-M, indicating that GRC 27864 spared PGI_2 with highly selective PGE_2 inhibition. This suggests that systemic inhibition of PGE_2 by GRC 27864 might avoid cardiovascular toxicity associated with NSAIDs selective for COX-2 inhibition [122]. In two additional studies, multiple oral doses of GRC 27864 were well tolerated in healthy adults and elderly subjects without any dose limiting adverse events. GRC 27864 has recently advanced in a Phase IIb dose range finding study to evaluate safety and efficacy in patients with moderate osteoarthritic pain [123].

3.5.2 Inhibitors of H-PGDS

PGD_2 is a well-established pro-inflammatory mediator in allergy-associated disorders and muscle dystrophy [124–127]. Beneficial effects of blocking PGD_2 production or action by targeting H-PGDS enzyme or DP_2/CRTH2 receptors, respectively, have been reported for many allergic diseases including asthma, rhinitis, atopic dermatitis [128–132], as well as Duchenne Muscular Dystrophy (DMD) [133, 134]. However, in contrast to DP antagonists (see below), clinical research on H-PGDS inhibitors is still in its infancy. From a vast plethora of compounds blocking H-PGDS [135] only two H-PGDS inhibitors, ZL-2102 and TAS-205, have recently advanced into clinical development [136].

ZL-2102 is a selective and reversible H-PGDS inhibitor. ZL-2102 was designed to treat chronic obstructive pulmonary disease (COPD), asthma and idiopathic pulmonary fibrosis. Currently, ZL-2102 is in Phase I clinical trial to assess safety, tolerability, and preliminary pharmacokinetics in healthy male subjects (NCT02397005).

Patients with DMD exhibit elevated levels of H-PGDS in necrotic muscle fibers and increased urinary tetranor-PGD metabolite (t-PGDM) with advancing age, suggesting the involvement of H-PGDS/PGD_2 axis in the etiology and progression of this disease [137, 138]. TAS-205 is a highly selective inhibitor of H-PGDS, which reduced muscle necrosis, promoted recovery of locomotor activity, and suppressed urinary t-PGDM in a mouse model of DMD [139]. TAS-205 was successfully tested over a range of doses in Phase I trial and was considered safe and tolerable in DMD patients [136]. Recently, a Phase IIa study was completed but the results are still to be released on the efficacy of TAS-205 after 24-week repeated oral dosing in patients with DMD (NCT02752048).

3.5.3 Inhibitors of TXAS

Several inhibitors of TXAS were created to treat bronchoconstriction by suppressing TxA_2-mediated activation of the airway smooth muscle in asthma [140]. Early animal studies reported efficacy of TXAS inhibitors [140] and one of them, ozagrel hydrochloride (OKY-046), was marketed in Japan for bronchial asthma in adults [141, 142]. However, in the Western studies, TXAS inhibitors dazmegrel (UK-38485) and pirmagrel (CGS-13080) were ineffective in improving the lung function in asthmatics [143, 144]. Furthermore, dazmegrel was under clinical development in Europe for the treatment of thrombosis, ischemic heart disease and arrhythmias, but failed to advance into clinic [145].

3.5.4 Dual Inhibitors (COX/LO, COX-2/TP, mPGES-1/LO)

Selective inhibition of one of the pathways in the AA cascade may cause overproduction of undesirable metabolites from other pathways. The shunting of the arachidonate-derived lipids towards leukotrienes, the products of 5-LO pathway, is a well-recognized contributing factor to NSAID-induced asthma and allergic reactions in sensitive patients and gastrointestinal side effects [146]. Therefore, development of multi-target agents with dual inhibitory activity at COXs and 5-LO (dual COX/LO inhibitors) has been ongoing since the 1970s [147].

The design of ligands to inhibit COX-2/5-LO enzymes has been based on the modification of already existing NSAIDs by adding LO pharmacophores or with the development of new scaffolds by computational techniques. Licofelone or

ML-3000 was the most successful dual COX/LO inhibitor that showed anti-inflammatory, analgesic and anti-asthmatic effects in several animal models. It reached Phase III clinical trials for the treatment of OA [148], but results were mixed and the drug has never been submitted for regulatory approval.

Flavocoxid is an FDA-regulated prescription medical food composed of a concentrated proprietary mixture of the flavonoid molecules baicalin and catechin extracted from Scutellaria baicalensis and Acacia catechu, respectively, and was shown to inhibit both COX-2 and 5-LO and show antioxidant activity [149, 150]. Flavocoxid successfully completed Phase II clinical trials for the treatment of knee OA [151, 152]. When tested in healthy volunteers, flavocoxid neither affected hemostasis nor inhibited the anticoagulation effects of aspirin, and thus might have utility in cardiovascular patients with chronic inflammation [153]. Lastly, flavocoxid demonstrated favorable gastrointestinal tolerability and a > 30% reduction or cessation in the use of gastroprotecive medications by the OA patients [154].

Another strategy that is under consideration is to develop a multitarget drug, COX-2 inhibitor/TP antagonist, that combines the anti-inflammatory activity of a coxib with the anti-aggregating and anti-atherotrombotic activities of a TP antagonist, and exploits recognized therapeutic advantages of the coxib as a GI-safer anti-inflammatory drug. Since lumiracoxib possesses a competitive antagonism at the TP receptor at high micromolar concentration [155], a series of lumiracoxib derivatives are under evaluation for their COX inhibition and TP antagonism activities [156, 157]. A possible pitfall of this approach could be an increase in bleeding episodes.

Dual inhibition of mPGES-1 and 5-LO or mPGES-1 and Leukotriene C4 Synthase, an enzyme downstream of 5-LO, represent a promising strategy in the development of novel anti-inflammatory drugs [158–160].

3.5.5 Prostanoid Receptor Modulators

Another approach to modulate prostanoid action in a more targeted fashion is at the level of the prostanoid receptors. Global inhibition of all the prostanoids, many of which have important homeostatic functions, lies at the basis of NSAID-associated toxicity. Taking into account cell-selective localization of the prostanoid receptors and their diverse functions, pharmacological targeting of a specific prostanoid receptor may ensure better therapeutic efficacy and lower toxicity. For the past decade, major strides have been made in the development of prostanoid receptor-based therapeutics. This section discusses the most clinically advanced modulators of the prostanoid receptors with the compounds currently in clinical development summarized in Table 3.1.

3.5.6 DP Antagonists

PGD_2 binds DP_2 receptor that plays a role in the pathogenesis of allergic disorders, such as asthma. DP_2 receptor is highly expressed on various immune cells, while its natural ligand PGD_2 is elevated in the bronchoalveolar lavage fluid of asthmatics and triggers a characteristic cough [126]. Patients with severe asthma patients showed elevated airway remodeling, goblet metaplasia, and cellular migration after DP_2 activation [127]. Thus, blockade of PGD_2/DP_2 signaling might be a plausible therapeutic strategy to alleviate asthma symptoms.

OC000459 is a potent and highly selective DP_2 antagonist that has been successfully tested in asthmatic patients. In human cell-based studies, OC000459 inhibited mast cell-dependent activation of both Th2 lymphocytes and eosinophils. In clinical trials, OC000459 inhibited allergic airway inflammation by blocking PGD_2-mdiated elevation in serum and airway

Table 3.1 Prostanoid receptor modulators in current clinical development

Target	Compound	Therapeutic indication/clinical status	References
DP_2 antagonist	OC000459	Asthma (Phase II completed)	NCT02560610
DP_2 antagonist	BI 671800	Asthma	[162–164]
		Seasonal allergic rhinitis	
DP_2 antagonist	Setipiprant	Androgenetic alopecia (Phase II)	NCT02781311
EP_1 antagonist	ONO-8539	Heartburn, overactive bladder	[210, 211]
EP_2 agonist	CP-533536	Bone healing (Phase II)	NCT00533377
EP_2 agonist	Taprenepag isopropyl	Ocular hypertension, glaucoma (Phase II)	[170]
EP_2 agonist	Omidenepag isopropyl	Ocular hypertension, glaucoma (Approved in Japan in 2018; Phase III, USA)	[173, 174]
EP_2 antagonist	PF-04418948	Safety and tolerability (Phase I)	NCT01002963
EP_3 antagonist	DG-041	Antithrombotic, antiulcer agent	[212]
EP_4 agonist	KAG-308	Ulcerative colitis (Phase II)	[180]
EP_4 antagonist	LY3127760	Safety and tolerability (Phase I completed)	[197]
EP_4 antagonist/ PD-1 blocker	Grapiprant/ Pembrolizumab	NSCLC adenocarcinoma (Phase I/II); MSS colorectal cancer (Phases Ib)	NCT03696212 NCT03658772
EP_4 antagonist/ PD-1 blocker	ONO-4578/ Nivolumab	Advanced or metastatic solid tumors (Phase I)	NCT03155061
EP_4 antagonist/ antifolate	CR6086/ Methotrexate	RA (Phase II)	[188] NCT03163966
FP agonist	NCX 470	Ocular hypertension and open-angle glaucoma (Phase II)	[91] NCT03657797
TP antagonist	Ifetroban	Systemic scleroderma or scleroderma-associated PAH; aspirin-exacerbated respiratory disease, solid tumors, DMD, vascular dilation (all in Phase II);	NCT02682511 NCT03028350 NCT03326063 NCT03694249
		Hepatorenal syndrome, portal hypertension in cirrhotic patients (the last two are completed)	NCT03340675 NCT03962855 NCT01436500 NCT02802228

eosinophils, improved the quality of life, and overall lung function [128–130]. Furthermore, OC000459 reduced nasal and ocular symptoms in patients with allergic rhinitis [131].

BI 671800, a highly specific and potent DP_2 antagonist, was effective in a number of animal models of asthma and allergic dermatitis [161]. When tested in clinical trials, BI 671800 inhibited DP_2-dependent eosinophil activation by PGD_2 and demonstrated efficacy in treating seasonal allergic rhinitis [162]. Moreover, BI 671800 led to a moderate but significant improvement in the lung function in poorly controlled asthma [163, 164].

Setipiprant, ACT-129968, is another highly potent, selective and orally bioavailable antagonist of DP_2 receptor [165]. In clinical trials, setipiprant was evaluated as a therapy for asthma and seasonal allergic rhinitis. When tested in asthmatic patients, setipiprant decreased allergen-induced airway responses and the associated airway hyperresponsiveness [166]. However, when screened in seasonal allergic rhinitis, setipiprant dose-dependent efficacy observed in Phase II was not confirmed in a Phase III trial [167]. Currently, setipiprant is in a clinical trial for the treatment of androgenetic alopecia in males (NCT02781311).

3.5.7 Targeting EP Receptors

EP_2 Agonists Activation of EP_2 receptor was reported to reduce intra-ocular pressure (IOP) by affecting both uveoscleral and conventional pathways of aqueous humor outflow [168, 169]. Thus, a series of EP_2 agonists have been successfully tested in animal models of glaucoma and ocular hypertension, and have advanced into clinical development [169–173]. Taprenepag isopropyl, PF-04217329, is a prodrug of a highly selective and potent EP_2 receptor agonist CP-544326 [169]. Taprenepag isopropyl exerted IOP-lowering effects in dog, rabbit and non-human primate models of glaucoma and ocular hypertension [169]. In Phase II clinical trial, taprenepag isopropyl monotherapy significantly lowered IOP and was comparable in its effect to latanoprost; co-administration of taprenepag and latanoprost reduced IOP to a greater degree than latanoptost alone suggesting an additive effect on IOP and a possibility of combination therapy [170]. Another EP_2 agonist in this class is omidenepag isopropyl, DE-117, which stimulates drainage through both pathways of aqueous humor outflow [172]. When tested in the Japanese patients, omidenepag isopropyl demonstrated sustained IOP-lowering effects and was recommended as an alternative treatment for non-responders to latanoprost [173, 174]. In September 2018, omidenepag isopropyl ophthalmic solution was approved in Japan for the treatment of glaucoma and ocular hypertension; Phase III clinical trials are currently ongoing in the United States [174].

EP_2 agonist CP-533536 was shown effective in bone healing in animals and clinical studies [183, 184]. CP-533536 increased bone formation in rat tibia, and healed long bone segmental and fracture defects in dogs [183, 184]. Subsequently, bone-healing abilities of CP-533536 were evaluated in patients with closed fracture of the tibial shaft (NCT00533377).

EP_4 Agonists EP_4 receptor is constitutively expressed in the colonic epithelium and its levels are increased during inflammatory bowel disease (IBD) [175]. Mice deficient in EP_4 receptor are susceptible to colitis, while administration of an EP_4 agonist protects the animals through regeneration of the intestinal epithelium and suppression of the immune response [176]. Based on these findings, it was suggested that an EP_4 agonist could be used for the treatment of patients with IBD. Rivenprost, ONO-4819, a potent and selective EP_4 receptor agonist [177], was tested in patients with mild to moderate ulcerative colitis (UC) for its effects on disease activity and inflammatory response [178]. An improvement in the ulcerative colitis symptoms was observed in three out of four patients, and one patient achieved complete remission. While more studies with larger patient cohorts would be necessary to conclude on the UC efficacy of rivenprost, its clinical development was terminated in 2009 (NCT00296556). A novel EP_4 agonist KAG-308 was reported to be advantageous over rivenprost due to favorable oral bioavailability [179]. KAG-308 showed a profound anti-inflammatory effect in a mouse model of UC, suppressed the onset of colitis and promoted mucosal healing [179]. KAG-308 was successfully tested in healthy volunteers, and the compound advanced in Phase II clinical trials in patients with ulcerative colitis [180].

EP_4 Antagonists Blockade of PGE_2 action through antagonism of its receptor EP_4 presents another alternative to COX-1/2 inhibition in pain management. The EP_4 receptor is the primary mediator of the PGE_2-triggered sensitization of sensory neurons and inflammation [181–184]. PGE_2/EP_4 signaling is not the only mechanism involved in pain and inflammation, but animal studies with EP_4 genetic deletion or pharmacological inhibition implicate the role of this receptor in central sensitization, thus, confirming EP_4 involvement in nociceptive responses [181, 182, 185].

Over the past decade, several pharmaceutical companies have been reporting on EP_4 antagonists with analgesic and anti-inflammatory properties [110, 186–189]. Grapiprant (CJ-023,423), a potent and highly selective EP_4 antagonist, is among the most advanced compounds in its class

[182]. Grapiprant showed analgesic efficacy in animal models of acute and chronic inflammatory pain in various species including rats, rabbits, cats, dogs and horses [182, 190–193], and has been recently approved by the FDA for OA in dogs [194–196]. Grapiprant is currently in clinical development as a co-therapy for cancer indications (see below).

LY3127760, a highly selective and potent EP_4 antagonist, has been successfully tested in first-in-human studies and advanced in further clinical development [197]. LY3127760 exhibited a superior pharmacological profile than grapiprant with a > ten-fold increase in potency in a human whole blood assay, and comparable analgesic efficacy to diclofenac in rat models of joint pain and arthritis [189]. In Phase I trial, LY3127760 demonstrated a comparable safety profile and similar pharmacodynamic effects on sodium and potassium excretion observed with celecoxib; this supported LY3127760 advancement into Phase II trials [197].

PGE_2/EP_4 signaling can amplify pro-inflammatory cytokine production and expand inflammatory T lymphocytes, thus, contributing to the onset and progression of RA [185, 198]. Blocking EP_4 receptor with CR6086, a novel EP_4 antagonist with immunomodulatory properties, presents the rational for the use of this class of agents as disease-modifying anti-rheumatic drugs [188]. CR6086 was effective in the rodent models of arthritis, and when used in combination with methotrexate, CR6086 improved the efficacy of a fully immunosuppressive dose of the anti-rheumatic drug [188]. CR6086 is currently tested in a Phase II clinical trial in patients with early RA as a combination therapy with methotrexate (NCT03163966).

Numerous studies reported on the role of PGE_2 signaling via EP_2 and EP_4 receptors in the elevation of exhausted cytotoxic $CD8^+$ T cells, which cannot mount effective immune responses in chronic viral infections and cancer [199–201]. T cell exhaustion is mediated by multiple inhibitory receptors, such as programmed cell death protein 1 (PD-1), whose major role is to limit T cell activity during inflammation in the peripheral tissues and to prevent autoimmunity. As a mechanism of immune resistance, tumor cells upregulate PD-1 ligands, which bind to PD-1 on lymphocytes and block T cell-mediated anti-tumor immune responses. Thus, suppression of PGE_2/EP_4 signaling may be a complimentary strategy to PD-1 blockade, and may potentiate the effectiveness of the cancer immunotherapy. Currently, EP_4 antagonist grapiprant, ARY-007, is being tested in combination with anti-PD-1 immunotherapy pembrolizumab, an IgG_4 isotype PD-1 antibody, in patients with advanced or metastatic non-small cell lung cancer (NSCLC) adenocarcinoma (NCT03696212). Moreover, the safety and tolerability of the grapiprant and pembrolizumab combination is being assessed in patients with advanced or progressive microsatellite stable (MSS) colorectal cancer (NCT03658772). Finally, a novel EP_4 antagonist ONO-4578 is being assessed as a monotherapy or in combination with nivolumab, a fully human IgG_4 PD-1 antibody, in subjects with advanced or metastatic solid tumors (NCT03155061).

3.5.8 TP Antagonists

Ifetroban is an orally active, potent and selectiveTP receptor antagonist [202]. In pre-clinical studies, ifetroban exhibited high receptor affinity for TP on different cell-types, including platelets, vascular and airway smooth muscle cells [203]. Ifetroban showed to inhibit platelet aggregation and relax contracted isolated vascular tissues [204] *ex vivo*. It was studied in animal models of thrombosis, stroke, myocardial ischemia and hypertension [204]. In humans, ifetroban showed to be a long-acting inhibitor of TxA_2 agonist-induced platelet aggregation and platelet shape change in healthy volunteers [205]. Ifetroban is under clinical investigation and phase II studies are undergoing for several therapeutic indications (Table 3.1). The safety and pharmacokinetics of 3 days of intravenous ifetroban was accessed in patients with hepatorenal syndrome (NCT01436500). Moreover, the safety of ifetroban was evaluated in Aspirin Exacerbated Respiratory Disease (AERD) patients (NCT02216357). Recently, the assessment of the safety and efficacy of 3 months of treatment with ifetroban for portal hypertension in cirrhotic patients was completed (NCT02802228). Currently, the efficacy of oral treatment with ifetroban in AERD patients is under evaluation (NCT03326063 and NCT03028350). Moreover, the safety and efficacy of oral ifetroban in patients with diffuse cutaneous systemic sclerosis or systemic sclerosis-associated PAH are being evaluated (NCT02682511). Furthermore, the assessment of the safety and feasibility of ifetroban administration in patients with malignant solid tumors at high risk of metastatic recurrence is under evaluation (NCT03694249). Two new clinical studies are about to start recruiting patients to evaluate the safety, pharmacokinetics and efficacy of two doses of oral ifetroban in subjects with DMD (NCT03340675) and to study the safety of oral and in patients with established cardiovascular disease (NCT03962855). TP antagonist seratrodast is marketed in Japan and China for the treatment of asthma [206]. However, this drug is not yet licensed for use in other countries. TP antagonist ramatroban, that is also a DP_2 antagonist [207], was initially approved in Japan for the treatment of allergic rhinitis and asthma [208]. Successively, it was approved for the treatment of coronary artery disease [209].

3.6 Concluding Remarks

NSAIDs, including both non-selective COX-1/2 and selective COX-2 inhibitors, are currently the major class of drugs targeting the prostanoid pathway. NSAIDs continue to be widely used for pain and inflammation, despite associated cardiovascular and gastrointestinal toxicity. Considering a high demand for pain management medicine, an urgent need for non-addictive analgesics due to the opioid crisis, great health care costs and loss of productivity due to chronic pain conditions, development of new strategies and drugs alternative to existing NSAIDs, is of paramount importance. The ongoing strategies are summarized as followed:

- A new class of drugs, targeting enzymes downstream of COX-1/2, like mPGES-1, promises to retain the same anti-inflammatory efficacy but assures a more favorable cardiovascular safety profile. However, the hallmark of mPGES-1 targeting is substrate shunting towards other terminal synthases with the predominant prostanoid product varying by cell type. What remains to be determined is whether blockade of mPGES-1 affects the signaling networks beyond the prostanoid pathway, and whether substrate shunting leads to functional consequences in a disease state, like chronic pain conditions or cardiovascular syndromes of inflammation. The same concern exists for the drugs targeting H-PGDS, when inhibition of PGD_2 may lead to a concurrent elevation of other eicosanoids. Alternative to systemic inhibition, cell-specific targeting of mPGES-1 in macrophages provides a promising therapeutic target for pain and inflammation with potential cardiovascular benefits. Preclinical studies are currently ongoing to address the therapeutic potential of targeted blockade of macrophage mPGES-1 in inflammatory pain models.
- Modulation of prostanoid receptors is another active area of therapeutic research in the prostanoid field. Targeting of prostanoid receptors may not have widespread anti-inflammatory effects of NSAIDs, since individual prostanoids and their receptors play compensatory roles, and act sequentially to initiate and sustain the inflammatory response; however, prostanoid receptor modulators may be useful in pathological states where a single prostanoid is the main causative agent. The most clinically advanced therapeutics in this class are DP antagonists for allergic indications, EP_2/FP agonists for glaucoma and ocular hypertension, and a TP antagonist for various inflammatory conditions. Combinatorial approaches of EP_4 antagonism with other drugs are being investigated in oncology and RA. A promising strategy for various cancer indications is a combination of an EP_4 antagonist with PD-1 blocking immunotherapy.

- In addition to novel drug development, alternative strategies are emerging to improve the safety and tolerability of existing NSAIDs. These strategies include a personalized selection of the appropriate NSAID based on each patient's clinical background and genetic makeup, a better selection of NSAID doses and schedules (the lowest effective drug dose used for the shortest period of time), and development of biomarkers that may predict the response to NSAID treatment in individual patients.

References

1. Smith WL, DeWitt DL, Garavito RM (2000) Cyclooxygenases: structural, cellular, and molecular biology. Annu Rev Biochem 69:145–182
2. Kurzrok R, Lieb CC (1930) Biochemical studies of human semen. II. The action of semen on the human uterus. Proc Soc Exp Biol Med 28:268–272
3. Goldblatt MW (1935) Properties of human seminal plasma. J Physiol 84:208–218
4. Von Euler US (1935) A depressor substance in the vesicular gland. J Physiolol (London) 84:21
5. Bergström S, Sjövall J (1957) The isolation of prostaglandin. Acta Chem Scand 11:1086
6. Bergstroem S, Ryhage R, Samuelsson B, Sjoevall J (1963) Prostaglandins and related factors. 15. The structures of prostaglandin E1, F1-alpha, and F1-Beta. J Biol Chem 238:3555–3564
7. Vane JR (1971) Inhibition of prostaglandin synthesis as a mechanism of action for aspirin-like drugs. Nat New Biol 231:232–235
8. Moncada S, Gryglewski R, Bunting S, Vane JR (1976) An enzyme isolated from arteries transforms prostaglandin endoperoxides to an unstable substance that inhibits platelet aggregation. Nature 263:663–665
9. Leslie CC (2015) Cytosolic phospholipase A(2): physiological function and role in disease. J Lipid Res 56:1386–1402
10. Samuelsson B, Morgenstern R, Jakobsson PJ (2007) Membrane prostaglandin E synthase-1: a novel therapeutic target. Pharmacol Rev 59:207–224
11. Song WL, Ricciotti E, Liang X, Grosser T, Grant GR, FitzGerald GA (2018) Lipocalin-like prostaglandin D synthase but not hemopoietic prostaglandin D synthase deletion causes hypertension and accelerates thrombogenesis in mice. J Pharmacol Exp Ther 367:425–432
12. Ueno N, Takegoshi Y, Kamei D, Kudo I, Murakami M (2005) Coupling between cyclooxygenases and terminal prostanoid synthases. Biochem Biophys Res Commun 338:70–76
13. Song WL, Wang M, Ricciotti E, Fries S, Yu Y, Grosser T, Reilly M, Lawson JA, FitzGerald GA (2008) Tetranor PGDM, an abundant urinary metabolite reflects biosynthesis of prostaglandin D2 in mice and humans. J Biol Chem 283:1179–1188
14. Cheng Y, Wang M, Yu Y, Lawson J, Funk CD, Fitzgerald GA (2006) Cyclooxygenases, microsomal prostaglandin E synthase-1, and cardiovascular function. J Clin Invest 116:1391–1399
15. Ricciotti E, FitzGerald GA (2011) Prostaglandins and inflammation. Arterioscler Thromb Vasc Biol 31:986–1000
16. Daiyasu H, Toh H (2000) Molecular evolution of the myeloperoxidase family. J Mol Evol 51:433–445
17. Otto JC, DeWitt DL, Smith WL (1993) N-glycosylation of prostaglandin endoperoxide synthases-1 and -2 and their orientations in the endoplasmic reticulum. J Biol Chem 268:18234–18242
18. Chandrasekharan NV, Simmons DL (2004) The cyclooxygenases. Genome Biol 5:241
19. Smith WL, Malkowski MG (2019) Interactions of fatty acids, nonsteroidal anti-inflammatory drugs, and coxibs with the catalytic and allosteric subunits of cyclooxygenases-1 and -2. J Biol Chem 294:1697–1705
20. Yu Y, Fan J, Chen XS, Wang D, Klein-Szanto AJ, Campbell RL, FitzGerald GA, Funk CD (2006) Genetic model of selective COX2 inhibition reveals novel heterodimer signaling. Nat Med 12:699–704
21. Tanabe T, Tohnai N (2002) Cyclooxygenase isozymes and their gene structures and expression. Prostaglandins Other Lipid Mediat 68–69:95–114
22. Kurumbail RG, Stevens AM, Gierse JK, McDonald JJ, Stegeman RA, Pak JY, Gildehaus D, Miyashiro JM, Penning TD, Seibert K, Isakson PC, Stallings WC (1996) Structural basis for selective inhibition of cyclooxygenase-2 by anti-inflammatory agents. Nature 384:644–648
23. Smyth EM, Grosser T, Wang M, Yu Y, FitzGerald GA (2009) Prostanoids in health and disease. J Lipid Res 50(Suppl):S423–S428
24. Patrignani P, Tacconelli S, Bruno A, Sostres C, Lanas A (2011) Managing the adverse effects of nonsteroidal anti-inflammatory drugs. Expert Rev Clin Pharmacol 4:605–621
25. Bjarnason I, Scarpignato C, Holmgren E, Olszewski M, Rainsford KD, Lanas A (2018) Mechanisms of damage to the gastrointestinal tract from nonsteroidal anti-inflammatory drugs. Gastroenterology 154:500–514
26. Hao CM, Breyer MD (2008) Physiological regulation of prostaglandins in the kidney. Annu Rev Physiol 70:357–377
27. Topper JN, Cai J, Falb D, Gimbrone MA Jr (1996) Identification of vascular endothelial genes differentially responsive to fluid mechanical stimuli: cyclooxygenase-2, manganese superoxide dismutase, and endothelial cell nitric oxide synthase are selectively

up-regulated by steady laminar shear stress. Proc Natl Acad Sci U S A 93:10417–10422
28. Ricciotti E, Yu Y, Grosser T, Fitzgerald GA (2013) COX-2, the dominant source of prostacyclin. Proc Natl Acad Sci U S A 110:E183
29. Weksler BB (2015) Prostanoids and NSAIDs in cardiovascular biology and disease. Curr Atheroscler Rep 17:41
30. Braun M, Schror K (1992) Prostaglandin D2 relaxes bovine coronary arteries by endothelium-dependent nitric oxide-mediated cGMP formation. Circ Res 71:1305–1313
31. Narumiya S, Toda N (1985) Different responsiveness of prostaglandin D2-sensitive systems to prostaglandin D2 and its analogues. Br J Pharmacol 85:367–375
32. Hall IP (2000) Second messengers, ion channels and pharmacology of airway smooth muscle. Eur Respir J 15:1120–1127
33. Song WL, Stubbe J, Ricciotti E, Alamuddin N, Ibrahim S, Crichton I, Prempeh M, Lawson JA, Wilensky RL, Rasmussen LM, Pure E, FitzGerald GA (2012) Niacin and biosynthesis of PGD(2)by platelet COX-1 in mice and humans. J Clin Invest 122:1459–1468
34. Patrono C (2016) Cardiovascular effects of cyclooxygenase-2 inhibitors: a mechanistic and clinical perspective. Br J Clin Pharmacol 82:957–964
35. McAdam BF, Mardini IA, Habib A, Burke A, Lawson JA, Kapoor S, FitzGerald GA (2000) Effect of regulated expression of human cyclooxygenase isoforms on eicosanoid and isoeicosanoid production in inflammation. J Clin Invest 105:1473–1482
36. Engblom D, Saha S, Engstrom L, Westman M, Audoly LP, Jakobsson PJ, Blomqvist A (2003) Microsomal prostaglandin E synthase-1 is the central switch during immune-induced pyresis. Nat Neurosci 6:1137–1138
37. Basu S (2007) Novel cyclooxygenase-catalyzed bioactive prostaglandin F2alpha from physiology to new principles in inflammation. Med Res Rev 27:435–468
38. Sugimoto Y, Yamasaki A, Segi E, Tsuboi K, Aze Y, Nishimura T, Oida H, Yoshida N, Tanaka T, Katsuyama M, Hasumoto K, Murata T, Hirata M, Ushikubi F, Negishi M, Ichikawa A, Narumiya S (1997) Failure of parturition in mice lacking the prostaglandin F receptor. Science 277:681–683
39. Murakami Y, Akahoshi T, Hayashi I, Endo H, Hashimoto A, Kono S, Kondo H, Kawai S, Inoue M, Kitasato H (2003) Inhibition of monosodium urate monohydrate crystal-induced acute inflammation by retrovirally transfected prostaglandin D synthase. Arthritis Rheum 48:2931–2941
40. Zayed N, Afif H, Chabane N, Mfuna-Endam L, Benderdour M, Martel-Pelletier J, Pelletier JP, Motiani RK, Trebak M, Duval N, Fahmi H (2008) Inhibition of interleukin-1beta-induced matrix metalloproteinases 1 and 13 production in human osteoarthritic chondrocytes by prostaglandin D2. Arthritis Rheum 58:3530–3540
41. Rajakariar R, Hilliard M, Lawrence T, Trivedi S, Colville-Nash P, Bellingan G, Fitzgerald D, Yaqoob MM, Gilroy DW (2007) Hematopoietic prostaglandin D2 synthase controls the onset and resolution of acute inflammation through PGD2 and 15-deoxyDelta12 14 PGJ2. Proc Natl Acad Sci U S A 104:20979–20984
42. Chen L, Yang G, Grosser T (2013) Prostanoids and inflammatory pain. Prostaglandins Other Lipid Mediat 104-105:58–66
43. Pettipher R, Hansel TT, Armer R (2007) Antagonism of the prostaglandin D2 receptors DP1 and CRTH2 as an approach to treat allergic diseases. Nat Rev Drug Discov 6:313–325
44. Grosser T, Theken KN, FitzGerald GA (2017) Cyclooxygenase inhibition: pain, inflammation, and the cardiovascular system. Clin Pharmacol Ther 102:611–622
45. Patrignani P, Patrono C (2015) Cyclooxygenase inhibitors: from pharmacology to clinical read-outs. Biochim Biophys Acta 1851:422–432
46. Brune K, Hinz B (2004) The discovery and development of antiinflammatory drugs. Arthritis Rheum 50:2391–2399
47. Kujubu DA, Fletcher BS, Varnum BC, Lim RW, Herschman HR (1991) TIS10, a phorbol ester tumor promoter-inducible mRNA from Swiss 3T3 cells, encodes a novel prostaglandin synthase/cyclooxygenase homologue. J Biol Chem 266:12866–12872
48. FitzGerald GA, Patrono C (2001) The coxibs, selective inhibitors of cyclooxygenase-2. N Engl J Med 345:433–442
49. Patrono C, Ciabattoni G, Pinca E, Pugliese F, Castrucci G, De Salvo A, Satta MA, Peskar BA (1980) Low dose aspirin and inhibition of thromboxane B2 production in healthy subjects. Thromb Res 17:317–327
50. Patrignani P, Panara MR, Greco A, Fusco O, Natoli C, Iacobelli S, Cipollone F, Ganci A, Creminon C, Maclouf J et al (1994) Biochemical and pharmacological characterization of the cyclooxygenase activity of human blood prostaglandin endoperoxide synthases. J Pharmacol Exp Ther 271:1705–1712
51. Tacconelli S, Bruno A, Grande R, Ballerini P, Patrignani P (2017) Nonsteroidal anti-inflammatory drugs and cardiovascular safety – translating pharmacological data into clinical readouts. Expert Opin Drug Saf 16:791–807
52. Patrono C, Baigent C (2014) Nonsteroidal anti-inflammatory drugs and the heart. Circulation 129:907–916
53. Grosser T, Ricciotti E, FitzGerald GA (2017) The cardiovascular pharmacology of nonsteroidal anti-inflammatory drugs. Trends Pharmacol Sci 38:733–748
54. Grosser T, Fries S, FitzGerald GA (2006) Biological basis for the cardiovascular consequences of COX-2

inhibition: therapeutic challenges and opportunities. J Clin Invest 116:4–15
55. Silverstein FE, Faich G, Goldstein JL, Simon LS, Pincus T, Whelton A, Makuch R, Eisen G, Agrawal NM, Stenson WF, Burr AM, Zhao WW, Kent JD, Lefkowith JB, Verburg KM, Geis GS (2000) Gastrointestinal toxicity with celecoxib vs nonsteroidal anti-inflammatory drugs for osteoarthritis and rheumatoid arthritis: the CLASS study: a randomized controlled trial. Celecoxib Long-term Arthritis Safety Study. JAMA 284:1247–1255
56. Bombardier C, Laine L, Reicin A, Shapiro D, Burgos-Vargas R, Davis B, Day R, Ferraz MB, Hawkey CJ, Hochberg MC, Kvien TK, Schnitzer TJ, V. S. Group (2000) Comparison of upper gastrointestinal toxicity of rofecoxib and naproxen in patients with rheumatoid arthritis. VIGOR Study Group. N Engl J Med 343:1520–1528. 1522 p following 1528
57. Schnitzer TJ, Burmester GR, Mysler E, Hochberg MC, Doherty M, Ehrsam E, Gitton X, Krammer G, Mellein B, Matchaba P, Gimona A, Hawkey CJ, T. S. Group (2004) Comparison of lumiracoxib with naproxen and ibuprofen in the Therapeutic Arthritis Research and Gastrointestinal Event Trial (TARGET), reduction in ulcer complications: randomised controlled trial. Lancet 364:665–674
58. Tacconelli S, Capone ML, Patrignani P (2004) Clinical pharmacology of novel selective COX-2 inhibitors. Curr Pharm Des 10:589–601
59. Baron JA, Sandler RS, Bresalier RS, Quan H, Riddell R, Lanas A, Bolognese JA, Oxenius B, Horgan K, Loftus S, Morton DG, Investigators APT (2006) A randomized trial of rofecoxib for the chemoprevention of colorectal adenomas. Gastroenterology 131:1674–1682
60. Bertagnolli MM, Eagle CJ, Zauber AG, Redston M, Solomon SD, Kim K, Tang J, Rosenstein RB, Wittes J, Corle D, Hess TM, Woloj GM, Boisserie F, Anderson WF, Viner JL, Bagheri D, Burn J, Chung DC, Dewar T, Foley TR, Hoffman N, Macrae F, Pruitt RE, Saltzman JR, Salzberg B, Sylwestrowicz T, Gordon GB, Hawk ET, Investigators APCS (2006) Celecoxib for the prevention of sporadic colorectal adenomas. N Engl J Med 355:873–884
61. Kuehn BM (2005) FDA panel: keep COX-2 drugs on market: black box for COX-2 labels, caution urged for all NSAIDs. JAMA 293:1571–1572
62. Grosser T, Yu Y, Fitzgerald GA (2010) Emotion recollected in tranquility: lessons learned from the COX-2 saga. Annu Rev Med 61:17–33
63. Coxib NTC, Traditional N, Bhala J, Emberson AM, Abramson S, Arber N, Baron JA, Bombardier C, Cannon C, Farkouh ME, FitzGerald GA, Goss P, Halls H, Hawk E, Hawkey C, Hennekens C, Hochberg M, Holland LE, Kearney PM, Laine L, Lanas A, Lance P, Laupacis A, Oates J, Patrono C, Schnitzer TJ, Solomon S, Tugwell P, Wilson K, Wittes J, Baigent C (2013) Vascular and upper gastrointestinal effects of non-steroidal anti-inflammatory drugs: meta-analyses of individual participant data from randomised trials. Lancet 382:769–779
64. Capone ML, Tacconelli S, Di Francesco L, Sacchetti A, Sciulli MG, Patrignani P (2007) Pharmacodynamic of cyclooxygenase inhibitors in humans. Prostaglandins Other Lipid Mediat 82:85–94
65. Patrignani P, Panara MR, Sciulli MG, Santini G, Renda G, Patrono C (1997) Differential inhibition of human prostaglandin endoperoxide synthase-1 and -2 by nonsteroidal anti-inflammatory drugs. J Physiol Pharmacol 48:623–631
66. Patrono C, Baigent C (2017) Coxibs, Traditional NSAIDs, and cardiovascular safety post-PRECISION: what we thought we knew then and what we think we know now. Clin Pharmacol Ther 102:238–245
67. FitzGerald GA (2017) Imprecision: limitations to interpretation of a large randomized clinical trial. Circulation 135:113–115
68. McAdam BF, Catella-Lawson F, Mardini IA, Kapoor S, Lawson JA, FitzGerald GA (1999) Systemic biosynthesis of prostacyclin by cyclooxygenase (COX)-2: the human pharmacology of a selective inhibitor of COX-2. Proc Natl Acad Sci U S A 96:272–277
69. Capone ML, Tacconelli S, Sciulli MG, Anzellotti P, Di Francesco L, Merciaro G, Di Gregorio P, Patrignani P (2007) Human pharmacology of naproxen sodium. J Pharmacol Exp Ther 322:453–460
70. Wang D, Patel VV, Ricciotti E, Zhou R, Levin MD, Gao E, Yu Z, Ferrari VA, Lu MM, Xu J, Zhang H, Hui Y, Cheng Y, Petrenko N, Yu Y, FitzGerald GA (2009) Cardiomyocyte cyclooxygenase-2 influences cardiac rhythm and function. Proc Natl Acad Sci U S A 106:7548–7552
71. Yu Y, Ricciotti E, Scalia R, Tang SY, Grant G, Yu Z, Landesberg G, Crichton I, Wu W, Pure E, Funk CD, FitzGerald GA (2012) Vascular COX-2 modulates blood pressure and thrombosis in mice. Sci Transl Med 4:132ra154
72. Chen MJ, Creinin MD (2015) Mifepristone with buccal misoprostol for medical abortion: a systematic review. Obstet Gynecol 126:12–21
73. Hanchanale V, Eardley I (2014) Alprostadil for the treatment of impotence. Expert Opin Pharmacother 15:421–428
74. Akkinapally S, Hundalani SG, Kulkarni M, Fernandes CJ, Cabrera AG, Shivanna B, Pammi M (2018) Prostaglandin E1 for maintaining ductal patency in neonates with ductal-dependent cardiac lesions. Cochrane Database Syst Rev 2:CD011417
75. Kuwano K, Hashino A, Asaki T, Hamamoto T, Yamada T, Okubo K, Kuwabara K (2007) 2-[4-[(5,6-diphenylpyrazin-2-yl)(isopropyl)amino] butoxy]-N-(methylsulfonyl)acetam ide (NS-304), an orally available and long-acting prostacyclin receptor agonist prodrug. J Pharmacol Exp Ther 322:1181–1188
76. Honorato Perez J (2017) Selexipag, a selective prostacyclin receptor agonist in pulmonary arterial

hypertension: a pharmacology review. Expert Rev Clin Pharmacol 10:753–762
77. Del Pozo R, Hernandez Gonzalez I, Escribano-Subias P (2017) The prostacyclin pathway in pulmonary arterial hypertension: a clinical review. Expert Rev Respir Med 11:491–503
78. Matsou A, Anastasopoulos E (2018) Investigational drugs targeting prostaglandin receptors for the treatment of glaucoma. Expert Opin Investig Drugs 27:777–785
79. Klimko PG, Sharif NA (2019) Discovery, characterization and clinical utility of prostaglandin agonists for the treatment of glaucoma. Br J Pharmacol 176:1051–1058
80. Mukhopadhyay P, Bian L, Yin H, Bhattacherjee P, Paterson C (2001) Localization of EP(1) and FP receptors in human ocular tissues by in situ hybridization. Invest Ophthalmol Vis Sci 42:424–428
81. Takagi Y, Nakajima T, Shimazaki A, Kageyama M, Matsugi T, Matsumura Y, Gabelt BT, Kaufman PL, Hara H (2004) Pharmacological characteristics of AFP-168 (tafluprost), a new prostanoid FP receptor agonist, as an ocular hypotensive drug. Exp Eye Res 78:767–776
82. Ota T, Aihara M, Saeki T, Narumiya S, Araie M (2007) The IOP-lowering effects and mechanism of action of tafluprost in prostanoid receptor-deficient mice. Br J Ophthalmol 91:673–676
83. Faulkner R, Sharif NA, Orr S, Sall K, Dubiner H, Whitson JT, Moster M, Craven ER, Curtis M, Pailliotet C, Martens K, Dahlin D (2010) Aqueous humor concentrations of bimatoprost free acid, bimatoprost and travoprost free acid in cataract surgical patients administered multiple topical ocular doses of LUMIGAN or TRAVATAN. J Ocul Pharmacol Ther 26:147–156
84. Woodward DF, Liang Y, Krauss AH (2008) Prostamides (prostaglandin-ethanolamides) and their pharmacology. Br J Pharmacol 153:410–419
85. Nakajima T, Matsugi T, Goto W, Kageyama M, Mori N, Matsumura Y, Hara H (2003) New fluoroprostaglandin F(2alpha) derivatives with prostanoid FP-receptor agonistic activity as potent ocular-hypotensive agents. Biol Pharm Bull 26:1691–1695
86. Scherer WJ (2002) A retrospective review of non-responders to latanoprost. J Ocul Pharmacol Ther 18:287–291
87. Sakurai M, Higashide T, Ohkubo S, Takeda H, Sugiyama K (2014) Association between genetic polymorphisms of the prostaglandin F2alpha receptor gene, and response to latanoprost in patients with glaucoma and ocular hypertension. Br J Ophthalmol 98:469–473
88. Krauss AH, Impagnatiello F, Toris CB, Gale DC, Prasanna G, Borghi V, Chiroli V, Chong WK, Carreiro ST, Ongini E (2011) Ocular hypotensive activity of BOL-303259-X, a nitric oxide donating prostaglandin F2alpha agonist, in preclinical models. Exp Eye Res 93:250–255
89. Costa VP, Harris A, Anderson D, Stodtmeister R, Cremasco F, Kergoat H, Lovasik J, Stalmans I, Zeitz O, Lanzl I, Gugleta K, Schmetterer L (2014) Ocular perfusion pressure in glaucoma. Acta Ophthalmol 92:e252–e266
90. Resch H, Garhofer G, Fuchsjager-Mayrl G, Hommer A, Schmetterer L (2009) Endothelial dysfunction in glaucoma. Acta Ophthalmol 87:4–12
91. Impagnatiello F, Toris CB, Batugo M, Prasanna G, Borghi V, Bastia E, Ongini E, Krauss AH (2015) Intraocular pressure-lowering activity of NCX 470, a novel nitric oxide-donating bimatoprost in preclinical models. Invest Ophthalmol Vis Sci 56:6558–6564
92. Borhade N, Pathan AR, Halder S, Karwa M, Dhiman M, Pamidiboina V, Gund M, Deshattiwar JJ, Mali SV, Deshmukh NJ, Senthilkumar SP, Gaikwad P, Tipparam SG, Mudgal J, Dutta MC, Burhan AU, Thakre G, Sharma A, Deshpande S, Desai DC, Dubash NP, Jain AK, Sharma S, Nemmani KV, Satyam A (2012) NO-NSAIDs. Part 3: nitric oxide-releasing prodrugs of non-steroidal anti-inflammatory drugs. Chem Pharm Bull (Tokyo) 60:465–481
93. Davies NM, Roseth AG, Appleyard CB, McKnight W, Del Soldato P, Calignano A, Cirino G, Wallace JL (1997) NO-naproxen vs. naproxen: ulcerogenic, analgesic and anti-inflammatory effects. Aliment Pharmacol Ther 11:69–79
94. Fiorucci S, Del Soldato P (2003) NO-aspirin: mechanism of action and gastrointestinal safety. Dig Liver Dis 35(Suppl 2):S9–S19
95. Fiorucci S, Distrutti E, Cirino G, Wallace JL (2006) The emerging roles of hydrogen sulfide in the gastrointestinal tract and liver. Gastroenterology 131:259–271
96. van Goor H, van den Born JC, Hillebrands JL, Joles JA (2016) Hydrogen sulfide in hypertension. Curr Opin Nephrol Hypertens 25:107–113
97. Trebino CE, Stock JL, Gibbons CP, Naiman BM, Wachtmann TS, Umland JP, Pandher K, Lapointe JM, Saha S, Roach ML, Carter D, Thomas NA, Durtschi BA, McNeish JD, Hambor JE, Jakobsson PJ, Carty TJ, Perez JR, Audoly LP (2003) Impaired inflammatory and pain responses in mice lacking an inducible prostaglandin E synthase. Proc Natl Acad Sci U S A 100:9044–9049
98. Kamei D, Yamakawa K, Takegoshi Y, Mikami-Nakanishi M, Nakatani Y, Oh-Ishi S, Yasui H, Azuma Y, Hirasawa N, Ohuchi K, Kawaguchi H, Ishikawa Y, Ishii T, Uematsu S, Akira S, Murakami M, Kudo I (2004) Reduced pain hypersensitivity and inflammation in mice lacking microsomal prostaglandin e synthase-1. J Biol Chem 279:33684–33695
99. Ikeda-Matsuo Y, Ota A, Fukada T, Uematsu S, Akira S, Sasaki Y (2006) Microsomal prostaglandin E synthase-1 is a critical factor of stroke-reperfusion injury. Proc Natl Acad Sci U S A 103:11790–11795
100. Wang M, Zukas AM, Hui Y, Ricciotti E, Pure E, FitzGerald GA (2006) Deletion of microsomal pros-

taglandin E synthase-1 augments prostacyclin and retards atherogenesis. Proc Natl Acad Sci U S A 103:14507–14512

101. Wang M, Lee E, Song W, Ricciotti E, Rader DJ, Lawson JA, Pure E, FitzGerald GA (2008) Microsomal prostaglandin E synthase-1 deletion suppresses oxidative stress and angiotensin II-induced abdominal aortic aneurysm formation. Circulation 117:1302–1309
102. Wang M, Cooper PR, Jiang M, Zhao H, Hui Y, Yao Y, Tate JC, Damera G, Lawson JA, Jester WF Jr, Haczku A, Panettieri RA Jr, FitzGerald GA (2010) Deletion of microsomal prostaglandin E synthase-1 does not alter ozone-induced airway hyper-responsiveness. J Pharmacol Exp Ther 334:63–68
103. Wang M, Ihida-Stansbury K, Kothapalli D, Tamby MC, Yu Z, Chen L, Grant G, Cheng Y, Lawson JA, Assoian RK, Jones PL, Fitzgerald GA (2011) Microsomal prostaglandin e2 synthase-1 modulates the response to vascular injury. Circulation 123:631–639
104. Xu D, Rowland SE, Clark P, Giroux A, Cote B, Guiral S, Salem M, Ducharme Y, Friesen RW, Methot N, Mancini J, Audoly L, Riendeau D (2008) MF63 [2-(6-chloro-1H-phenanthro[9,10-d] imidazol-2-yl)-isophthalonitrile], a selective microsomal prostaglandin E synthase-1 inhibitor, relieves pyresis and pain in preclinical models of inflammation. J Pharmacol Exp Ther 326:754–763
105. Facemire CS, Griffiths R, Audoly LP, Koller BH, Coffman TM (2010) The impact of microsomal prostaglandin e synthase 1 on blood pressure is determined by genetic background. Hypertensivon 55:531–538
106. Degousee N, Fazel S, Angoulvant D, Stefanski E, Pawelzik SC, Korotkova M, Arab S, Liu P, Lindsay TF, Zhuo S, Butany J, Li RK, Audoly L, Schmidt R, Angioni C, Geisslinger G, Jakobsson PJ, Rubin BB (2008) Microsomal prostaglandin E2 synthase-1 deletion leads to adverse left ventricular remodeling after myocardial infarction. Circulation 117:1701–1710
107. Degousee N, Simpson J, Fazel S, Scholich K, Angoulvant D, Angioni C, Schmidt H, Korotkova M, Stefanski E, Wang XH, Lindsay TF, Ofek E, Pierre S, Butany J, Jakobsson PJ, Keating A, Li RK, Nahrendorf M, Geisslinger G, Backx PH, Rubin BB (2012) Lack of microsomal prostaglandin e2 synthase-1 in bone marrow-derived myeloid cells impairs left ventricular function and increases mortality after acute myocardial infarction. Circulation 125:2904–2913
108. Zhu L, Xu C, Huo X, Hao H, Wan Q, Chen H, Zhang X, Breyer RM, Huang Y, Cao X, Liu DP, FitzGerald GA, Wang M (2019) The cyclooxygenase-1/mPGES-1/endothelial prostaglandin EP4 receptor pathway constrains myocardial ischemia-reperfusion injury. Nat Commun 10:1888
109. Kawabata A (2011) Prostaglandin E2 and pain–an update. Biol Pharm Bull 34:1170–1173
110. Chen M, Boilard E, Nigrovic PA, Clark P, Xu D, Fitzgerald GA, Audoly LP, Lee DM (2008) Predominance of cyclooxygenase 1 over cyclooxygenase 2 in the generation of proinflammatory prostaglandins in autoantibody-driven K/BxN serum-transfer arthritis. Arthritis Rheum 58:1354–1365
111. Chen L, Yang G, Xu X, Grant G, Lawson JA, Bohlooly YM, FitzGerald GA (2013) Cell selective cardiovascular biology of microsomal prostaglandin E synthase-1. Circulation 127:233–243
112. Chen L, Yang G, Monslow J, Todd L, Cormode DP, Tang J, Grant GR, DeLong JH, Tang SY, Lawson JA, Pure E, Fitzgerald GA (2014) Myeloid cell microsomal prostaglandin E synthase-1 fosters atherogenesis in mice. Proc Natl Acad Sci U S A 111:6828–6833
113. Cote B, Boulet L, Brideau C, Claveau D, Ethier D, Frenette R, Gagnon M, Giroux A, Guay J, Guiral S, Mancini J, Martins E, Masse F, Methot N, Riendeau D, Rubin J, Xu D, Yu H, Ducharme Y, Friesen RW (2007) Substituted phenanthrene imidazoles as potent, selective, and orally active mPGES-1 inhibitors. Bioorg Med Chem Lett 17:6816–6820
114. Giroux A, Boulet L, Brideau C, Chau A, Claveau D, Cote B, Ethier D, Frenette R, Gagnon M, Guay J, Guiral S, Mancini J, Martins E, Masse F, Methot N, Riendeau D, Rubin J, Xu D, Yu H, Ducharme Y, Friesen RW (2009) Discovery of disubstituted phenanthrene imidazoles as potent, selective and orally active mPGES-1 inhibitors. Bioorg Med Chem Lett 19:5837–5841
115. Pawelzik SC, Uda NR, Spahiu L, Jegerschold C, Stenberg P, Hebert H, Morgenstern R, Jakobsson PJ (2010) Identification of key residues determining species differences in inhibitor binding of microsomal prostaglandin E synthase-1. J Biol Chem 285:29254–29261
116. Gosselin F, Lau S, Nadeau C, Trinh T, O'Shea PD, Davies IW (2009) A practical synthesis of m-prostaglandin E synthase-1 inhibitor MK-7285. J Org Chem 74:7790–7797
117. Schiffler MA, Antonysamy S, Bhattachar SN, Campanale KM, Chandrasekhar S, Condon B, Desai PV, Fisher MJ, Groshong C, Harvey A, Hickey MJ, Hughes NE, Jones SA, Kim EJ, Kuklish SL, Luz JG, Norman BH, Rathmell RE, Rizzo JR, Seng TW, Thibodeaux SJ, Woods TA, York JS, Yu XP (2016) Discovery and characterization of 2-Acylaminoimidazole microsomal prostaglandin E synthase-1 inhibitors. J Med Chem 59:194–205
118. Jin Y, Smith CL, Hu L, Campanale KM, Stoltz R, Huffman LG Jr, McNearney TA, Yang XY, Ackermann BL, Dean R, Regev A, Landschulz W (2016) Pharmacodynamic comparison of LY3023703, a novel microsomal prostaglandin e synthase 1 inhibitor, with celecoxib. Clin Pharmacol Ther 99:274–284
119. Norman BH, Fisher MJ, Schiffler MA, Kuklish SL, Hughes NE, Czeskis BA, Cassidy KC,

Abraham TL, Alberts JJ, Luffer-Atlas D (2018) Identification and mitigation of reactive metabolites of 2-aminoimidazole-containing microsomal prostaglandin E synthase-1 inhibitors terminated due to clinical drug-induced liver injury. J Med Chem 61:2041–2051
120. Jin Y, Regev A, Kam J, Phipps K, Smith C, Henck J, Campanale K, Hu L, Hall DG, Yang XY, Nakano M, McNearney TA, Uetrecht J, Landschulz W (2018) Dose-dependent acute liver injury with hypersensitivity features in humans due to a novel microsomal prostaglandin E synthase 1 inhibitor. Br J Clin Pharmacol 84:179–188
121. Banerjee A, Pawar MY, Patil S, Yadav PS, Kadam PA, Kattige VG, Deshpande DS, Pednekar PV, Pisat MK, Gharat LA (2014) Development of 2-aryl substituted quinazolin-4(3H)-one, pyrido[4,3-d]pyrimidin-4(3H)-one and pyrido[2,3-d]pyrimidin-4(3H)-one derivatives as microsomal prostaglandin E(2) synthase-1 inhibitors. Bioorg Med Chem Lett 24:4838–4844
122. Sant S, Tandon M, Menon V, Gudi G, Kattige V, Joshi NK, Korukonda K, Levine-Dolberg O (2018) GRC 27864, novel, microsomal prostaglandin E synthase-1 enzyme inhibitor: phase 1 study to evaluate safety, PK and biomarkers in healthy, adult subjects. Osteoarthr Cartil 26:S351–S352
123. Glenmark pharmaceuticals provides update on clinical development of GRC 27864. Capital Market. January 15, 2018. www.businessstandard.com/article/news-cm/glenmark-pharmaceuticals-provides-update-on-clinical-development-of-grc-27864-118011500958_1.html Date of access May 24, 2019
124. Arima M, Fukuda T (2011) Prostaglandin D(2) and T(H)2 inflammation in the pathogenesis of bronchial asthma. Korean J Intern Med 26:8–18
125. Peebles RS Jr (2019) Prostaglandins in asthma and allergic diseases. Pharmacol Ther 193:1–19
126. Sulaiman I, Lim JC, Soo HL, Stanslas J (2016) Molecularly targeted therapies for asthma: current development, challenges and potential clinical translation. Pulm Pharmacol Ther 40:52–68
127. Stinson SE, Amrani Y, Brightling CE (2015) D prostanoid receptor 2 (chemoattractant receptor-homologous molecule expressed on TH2 cells) protein expression in asthmatic patients and its effects on bronchial epithelial cells. J Allergy Clin Immunol 135:395–406
128. Barnes N, Pavord I, Chuchalin A, Bell J, Hunter M, Lewis T, Parker D, Payton M, Collins LP, Pettipher R, Steiner J, Perkins CM (2012) A randomized, double-blind, placebo-controlled study of the CRTH2 antagonist OC000459 in moderate persistent asthma. Clin Exp Allergy 42:38–48
129. Singh D, Cadden P, Hunter M, Pearce Collins L, Perkins M, Pettipher R, Townsend E, Vinall S, O'Connor B (2013) Inhibition of the asthmatic allergen challenge response by the CRTH2 antagonist OC000459. Eur Respir J 41:46–52
130. Pettipher R, Hunter MG, Perkins CM, Collins LP, Lewis T, Baillet M, Steiner J, Bell J, Payton MA (2014) Heightened response of eosinophilic asthmatic patients to the CRTH2 antagonist OC000459. Allergy 69:1223–1232
131. Horak F, Zieglmayer P, Zieglmayer R, Lemell P, Collins LP, Hunter MG, Steiner J, Lewis T, Payton MA, Perkins CM, Pettipher R (2012) The CRTH2 antagonist OC000459 reduces nasal and ocular symptoms in allergic subjects exposed to grass pollen, a randomised, placebo-controlled, double-blind trial. Allergy 67:1572–1579
132. Rittchen S, Heinemann A (2019) Therapeutic potential of hematopoietic prostaglandin D2 synthase in allergic inflammation. Cell 8:pii: E619
133. Mohri I, Aritake K, Taniguchi H, Sato Y, Kamauchi S, Nagata N, Maruyama T, Taniike M, Urade Y (2009) Inhibition of prostaglandin D synthase suppresses muscular necrosis. Am J Pathol 174:1735–1744
134. Kamauchi S, Urade Y (2011) Hematopoietic prostaglandin D synthase inhibitors for the treatment of duchenne muscular dystrophy. Brain Nerve 63:1261–1269
135. Thurairatnam S (2012) Hematopoietic prostaglandin D synthase inhibitors. Prog Med Chem 51:97–133
136. Takeshita E, Komaki H, Shimizu-Motohashi Y, Ishiyama A, Sasaki M, Takeda S (2018) A phase I study of TAS-205 in patients with Duchenne muscular dystrophy. Ann Clin Transl Neurol 5:1338–1349
137. Nakagawa T, Takeuchi A, Kakiuchi R, Lee T, Yagi M, Awano H, Iijima K, Takeshima Y, Urade Y, Matsuo M (2013) A prostaglandin D2 metabolite is elevated in the urine of Duchenne muscular dystrophy patients and increases further from 8 years old. Clin Chim Acta 423:10–14
138. Takeshita E, Komaki H, Tachimori H, Miyoshi K, Yamamiya I, Shimizu-Motohashi Y, Ishiyama A, Saito T, Nakagawa E, Sugai K, Sasaki M (2018) Urinary prostaglandin metabolites as Duchenne muscular dystrophy progression markers. Brain and Development 40:918–925
139. Tanaka K, Aritake K, Tayama M, Sasaki E, Utsugi T, Sasaoka T, Urade Y (2014) Novel inhibitor of hematopoietic prostaglandin D synthase improves the muscle disorder in an experimental model of Duchenne muscular dystrophy. Neuromuscul Disord 24:821
140. Hernandez JM, Janssen LJ (2015) Revisiting the usefulness of thromboxane-A2 modulation in the treatment of bronchoconstriction in asthma. Can J Physiol Pharmacol 93:111–117
141. Kunitoh H, Watanabe K, Nagatomo A, Okamoto H, Nakagawa T (1998) A double-blind, placebo-controlled trial of the thromboxane synthetase blocker OKY-046 on bronchial hypersensitivity in bronchial asthma patients. J Asthma 35:355–360
142. Kurosawa M (1995) Role of thromboxane A2 synthetase inhibitors in the treatment of patients with bronchial asthma. Clin Ther 17:2–11; discussion 11

143. Gardiner PV, Young CL, Holmes K, Hendrick DJ, Walters EH (1993) Lack of short-term effect of the thromboxane synthetase inhibitor UK-38,485 on airway reactivity to methacholine in asthmatic subjects. Eur Respir J 6:1027–1030
144. Manning PJ, Stevens WH, Cockcroft DW, O'Byrne PM (1991) The role of thromboxane in allergen-induced asthmatic responses. Eur Respir J 4:667–672
145. Huang R, Southall N, Wang Y, Yasgar A, Shinn P, Jadhav A, Nguyen D-T, Austin CP (2011) The NCGC pharmaceutical collection: a comprehensive resource of clinically approved drugs enabling repurposing and chemical genomics. Sci Transl Med 3:80ps16
146. Charlier C, Michaux C (2003) Dual inhibition of cyclooxygenase-2 (COX-2) and 5-lipoxygenase (5-LOX) as a new strategy to provide safer non-steroidal anti-inflammatory drugs. Eur J Med Chem 38:645–659
147. JJ P, Manju SL, Ethiraj KR, Elias G (2018) Safer anti-inflammatory therapy through dual COX-2/5-LOX inhibitors: a structure-based approach. Eur J Pharm Sci 121:356–381
148. Alvaro-Gracia JM (2004) Licofelone–clinical update on a novel LOX/COX inhibitor for the treatment of osteoarthritis. Rheumatology (Oxford) 43(Suppl 1):i21–i25
149. Bitto A, Squadrito F, Irrera N, Pizzino G, Pallio G, Mecchio A, Galfo F, Altavilla D (2014) Flavocoxid, a nutraceutical approach to blunt inflammatory conditions. Mediat Inflamm 2014:790851
150. Burnett BP, Jia Q, Zhao Y, Levy RM (2007) A medicinal extract of Scutellaria baicalensis and Acacia catechu acts as a dual inhibitor of cyclooxygenase and 5-lipoxygenase to reduce inflammation. J Med Food 10:442–451
151. Levy RM, Khokhlov A, Kopenkin S, Bart B, Ermolova T, Kantemirova R, Mazurov V, Bell M, Caldron P, Pillai L, Burnett BP (2010) Efficacy and safety of flavocoxid, a novel therapeutic, compared with naproxen: a randomized multicenter controlled trial in subjects with osteoarthritis of the knee. Adv Ther 27:731–742
152. Levy R, Khokhlov A, Kopenkin S, Bart B, Ermolova T, Kantemirova R, Mazurov V, Bell M, Caldron P, Pillai L, Burnett B (2010) Efficacy and safety of flavocoxid compared with naproxen in subjects with osteoarthritis of the knee- a subset analysis. Adv Ther 27:953–962
153. Pillai L, Levy RM, Yimam M, Zhao Y, Jia Q, Burnett BP (2010) Flavocoxid, an anti-inflammatory agent of botanical origin, does not affect coagulation or interact with anticoagulation therapies. Adv Ther 27:400–411
154. Pillai L, Burnett BP, Levy RM, G. S. C. Group (2010) GOAL: multicenter, open-label, post-marketing study of flavocoxid, a novel dual pathway inhibitor anti-inflammatory agent of botanical origin. Curr Med Res Opin 26:1055–1063
155. Selg E, Buccellati C, Andersson M, Rovati GE, Ezinga M, Sala A, Larsson AK, Ambrosio M, Lastbom L, Capra V, Dahlen B, Ryrfeldt A, Folco GC, Dahlen SE (2007) Antagonism of thromboxane receptors by diclofenac and lumiracoxib. Br J Pharmacol 152:1185–1195
156. Bertinaria M, Shaikh MA, Buccellati C, Cena C, Rolando B, Lazzarato L, Fruttero R, Gasco A, Hoxha M, Capra V, Sala A, Rovati GE (2012) Designing multitarget anti-inflammatory agents: chemical modulation of the lumiracoxib structure toward dual thromboxane antagonists-COX-2 inhibitors. ChemMedChem 7:1647–1660
157. Carnevali S, Buccellati C, Bolego C, Bertinaria M, Rovati GE, Sala A (2017) Nonsteroidal anti-inflammatory drugs: exploiting bivalent COXIB/TP antagonists for the control of cardiovascular risk. Curr Med Chem 24:3218–3230
158. Elkady M, Niess R, Schaible AM, Bauer J, Luderer S, Ambrosi G, Werz O, Laufer SA (2012) Modified acidic nonsteroidal anti-inflammatory drugs as dual inhibitors of mPGES-1 and 5-LOX. J Med Chem 55:8958–8962
159. Shang E, Wu Y, Liu P, Liu Y, Zhu W, Deng X, He C, He S, Li C, Lai L (2014) Benzo[d]isothiazole 1,1-dioxide derivatives as dual functional inhibitors of 5-lipoxygenase and microsomal prostaglandin E(2) synthase-1. Bioorg Med Chem Lett 24:2764–2767
160. Devi NS, Paragi-Vedanthi P, Bender A, Doble M (2018) Common structural and pharmacophoric features of mPGES-1 and LTC4S. Future Med Chem 10:259–268
161. Kupczyk M, Kuna P (2017) Targeting the PGD2/CRTH2/DP1 signaling pathway in asthma and allergic disease: current status and future perspectives. Drugs 77:1281–1294
162. Krug N, Gupta A, Badorrek P, Koenen R, Mueller M, Pivovarova A, Hilbert J, Wetzel K, Hohlfeld JM, Wood C (2014) Efficacy of the oral chemoattractant receptor homologous molecule on TH2 cells antagonist BI 671800 in patients with seasonal allergic rhinitis. J Allergy Clin Immunol 133:414–419
163. Miller D, Wood C, Bateman E, LaForce C, Blatchford J, Hilbert J, Gupta A, Fowler A (2017) A randomized study of BI 671800, a CRTH2 antagonist, as add-on therapy in poorly controlled asthma. Allergy Asthma Proc 38:157–164
164. Hall IP, Fowler AV, Gupta A, Tetzlaff K, Nivens MC, Sarno M, Finnigan HA, Bateman ED, Rand Sutherland E (2015) Efficacy of BI 671800, an oral CRTH2 antagonist, in poorly controlled asthma as sole controller and in the presence of inhaled corticosteroid treatment. Pulm Pharmacol Ther 32:37–44
165. Fretz H, Valdenaire A, Pothier J, Hilpert K, Gnerre C, Peter O, Leroy X, Riederer MA (2013) Identification of 2-(2-(1-naphthoyl)-8-fluoro-3,4-dihydro-1H-pyrido[4,3-b]indol-5(2H)-yl)acetic acid (setipiprant/ACT-129968), a potent, selective, and orally bioavailable chemoattractant receptor-homologous

molecule expressed on Th2 cells (CRTH2) antagonist. J Med Chem 56:4899–4911
166. Diamant Z, Sidharta PN, Singh D, O'Connor BJ, Zuiker R, Leaker BR, Silkey M, Dingemanse J (2014) Setipiprant, a selective CRTH2 antagonist, reduces allergen-induced airway responses in allergic asthmatics. Clin Exp Allergy 44:1044–1052
167. Ratner P, Andrews CP, Hampel FC, Martin B, Mohar DE, Bourrelly D, Danaietash P, Mangialaio S, Dingemanse J, Hmissi A, van Bavel J (2017) Efficacy and safety of setipiprant in seasonal allergic rhinitis: results from phase 2 and phase 3 randomized, double-blind, placebo- and active-referenced studies. Allergy Asthma Clin Immunol 13:18
168. Nilsson SF, Drecoll E, Lutjen-Drecoll E, Toris CB, Krauss AH, Kharlamb A, Nieves A, Guerra T, Woodward DF (2006) The prostanoid EP2 receptor agonist butaprost increases uveoscleral outflow in the cynomolgus monkey. Invest Ophthalmol Vis Sci 47:4042–4049
169. Prasanna G, Carreiro S, Anderson S, Gukasyan H, Sartnurak S, Younis H, Gale D, Xiang C, Wells P, Dinh D, Almaden C, Fortner J, Toris C, Niesman M, Lafontaine J, Krauss A (2011) Effect of PF-04217329 a prodrug of a selective prostaglandin EP(2) agonist on intraocular pressure in preclinical models of glaucoma. Exp Eye Res 93:256–264
170. Schachar RA, Raber S, Courtney R, Zhang M (2011) A phase 2, randomized, dose-response trial of taprenepag isopropyl (PF-04217329) versus latanoprost 0.005% in open-angle glaucoma and ocular hypertension. Curr Eye Res 36:809–817
171. Yanochko GM, Affolter T, Eighmy JJ, Evans MG, Khoh-Reiter S, Lee D, Miller PE, Shiue MH, Trajkovic D, Jessen BA (2014) Investigation of ocular events associated with Taprenepag isopropyl, a topical EP2 agonist in development for treatment of glaucoma. J Ocul Pharmacol Ther 30:429–439
172. Fuwa M, Toris CB, Fan S, Taniguchi T, Ichikawa M, Odani-Kawabata N, Iwamura R, Yoneda K, Matsugi T, Shams NK, Zhang JZ (2018) Effects of a novel selective EP2 receptor agonist, Omidenepag isopropyl, on aqueous humor dynamics in laser-induced ocular hypertensive monkeys. J Ocul Pharmacol Ther 34:531–537
173. Aihara M, Lu F, Kawata H, Iwata A, Liu K, Odani-Kawabata N, Shams NK (2019) Phase 2, randomized, dose-finding studies of Omidenepag isopropyl, a selective EP2 agonist, in patients with primary open-angle Glaucoma or ocular hypertension. J Glaucoma 28:375–385
174. Duggan S (2018) Omidenepag isopropyl ophthalmic solution 0.002%: first global approval. Drugs 78:1925–1929
175. Nitta M, Hirata I, Toshina K, Murano M, Maemura K, Hamamoto N, Sasaki S, Yamauchi H, Katsu K (2002) Expression of the EP4 prostaglandin E2 receptor subtype with rat dextran sodium sulphate colitis: colitis suppression by a selective agonist, ONO-AE1-329. Scand J Immunol 56:66–75
176. Kabashima K, Saji T, Murata T, Nagamachi M, Matsuoka T, Segi E, Tsuboi K, Sugimoto Y, Kobayashi T, Miyachi Y, Ichikawa A, Narumiya S (2002) The prostaglandin receptor EP4 suppresses colitis, mucosal damage and CD4 cell activation in the gut. J Clin Invest 109:883–893
177. Yoshida K, Oida H, Kobayashi T, Maruyama T, Tanaka M, Katayama T, Yamaguchi K, Segi E, Tsuboyama T, Matsushita M, Ito K, Ito Y, Sugimoto Y, Ushikubi F, Ohuchida S, Kondo K, Nakamura T, Narumiya S (2002) Stimulation of bone formation and prevention of bone loss by prostaglandin E EP4 receptor activation. Proc Natl Acad Sci U S A 99:4580–4585
178. Nakase H, Fujiyama Y, Oshitani N, Oga T, Nonomura K, Matsuoka T, Esaki Y, Murayama T, Teramukai S, Chiba T, Narumiya S (2010) Effect of EP4 agonist (ONO-4819CD) for patients with mild to moderate ulcerative colitis refractory to 5-aminosalicylates: a randomized phase II, placebo-controlled trial. Inflamm Bowel Dis 16:731–733
179. Watanabe Y, Murata T, Amakawa M, Miyake Y, Handa T, Konishi K, Matsumura Y, Tanaka T, Takeuchi K (2015) KAG-308, a newly-identified EP4-selective agonist shows efficacy for treating ulcerative colitis and can bring about lower risk of colorectal carcinogenesis by oral administration. Eur J Pharmacol 754:179–189
180. Markovic T, Jakopin Z, Dolenc MS, Mlinaric-Rascan I (2017) Structural features of subtype-selective EP receptor modulators. Drug Discov Today 22:57–71
181. Lin CR, Amaya F, Barrett L, Wang H, Takada J, Samad TA, Woolf CJ (2006) Prostaglandin E2 receptor EP4 contributes to inflammatory pain hypersensitivity. J Pharmacol Exp Ther 319:1096–1103
182. Nakao K, Murase A, Ohshiro H, Okumura T, Taniguchi K, Murata Y, Masuda M, Kato T, Okumura Y, Takada J (2007) CJ-023,423, a novel, potent and selective prostaglandin EP4 receptor antagonist with antihyperalgesic properties. J Pharmacol Exp Ther 322:686–694
183. Chen Q, Muramoto K, Masaaki N, Ding Y, Yang H, Mackey M, Li W, Inoue Y, Ackermann K, Shirota H, Matsumoto I, Spyvee M, Schiller S, Sumida T, Gusovsky F, Lamphier M (2010) A novel antagonist of the prostaglandin E(2) EP(4) receptor inhibits Th1 differentiation and Th17 expansion and is orally active in arthritis models. Br J Pharmacol 160:292–310
184. Boyd MJ, Berthelette C, Chiasson JF, Clark P, Colucci J, Denis D, Han Y, Levesque JF, Mathieu MC, Stocco R, Therien A, Rowland S, Wrona M, Xu D (2011) A novel series of potent and selective EP(4) receptor ligands: facile modulation of agonism and antagonism. Bioorg Med Chem Lett 21:484–487
185. McCoy JM, Wicks JR, Audoly LP (2002) The role of prostaglandin E2 receptors in the pathogenesis of rheumatoid arthritis. J Clin Invest 110:651–658
186. Colucci J, Boyd M, Berthelette C, Chiasson JF, Wang Z, Ducharme Y, Friesen R, Wrona M, Levesque

JF, Denis D, Mathieu MC, Stocco R, Therien AG, Clarke P, Rowland S, Xu D, Han Y (2010) Discovery of 4-[1-[([1-[4-(trifluoromethyl)benzyl]-1H-indol-7-yl]carbonyl)amino]cyclopropyl]be nzoic acid (MF-766), a highly potent and selective EP4 antagonist for treating inflammatory pain. Bioorg Med Chem Lett 20:3760–3763
187. Chandrasekhar S, Yu XP, Harvey AK, Oskins JL, Lin C, Wang X, Blanco MJ, Fisher MJ, Kuklish SL, Schiffler MA, Vetman T, Warshawsky AM, York JS, Bendele AM, Chambers MG (2017) Analgesic and anti-inflammatory properties of novel, selective, and potent EP4 receptor antagonists. Pharmacol Res Perspect 5:e00316
188. Caselli G, Bonazzi A, Lanza M, Ferrari F, Maggioni D, Ferioli C, Giambelli R, Comi E, Zerbi S, Perrella M, Letari O, Di Luccio E, Colovic M, Persiani S, Zanelli T, Mennuni L, Piepoli T, Rovati LC (2018) Pharmacological characterisation of CR6086, a potent prostaglandin E2 receptor 4 antagonist, as a new potential disease-modifying anti-rheumatic drug. Arthritis Res Ther 20:39
189. Blanco MJ, Vetman T, Chandrasekhar S, Fisher MJ, Harvey A, Kuklish SL, Chambers M, Lin C, Mudra D, Oskins J, Wang XS, Yu XP, Warshawsky AM (2016) Identification and biological activity of 6-alkyl-substituted 3-methyl-pyridine-2-carbonyl amino dimethyl-benzoic acid EP4 antagonists. Bioorg Med Chem Lett 26:2303–2307
190. Lebkowska-Wieruszewska B, De Vito V, Owen H, Poapholatep A, Giorgi M (2017) Pharmacokinetics of grapiprant, a selective EP4 prostaglandin PGE2 receptor antagonist, after 2 mg/kg oral and i.v. administrations in cats. J Vet Pharmacol Ther 40:e11–e15
191. De Vito V, Salvadori M, Poapolathep A, Owen H, Rychshanova R, Giorgi M (2017) Pharmacokinetic/pharmacodynamic evaluation of grapiprant in a carrageenan-induced inflammatory pain model in the rabbit. J Vet Pharmacol Ther 40:468–475
192. Nagahisa A, Okumura T (2017) Pharmacology of grapiprant, a novel EP4 antagonist: receptor binding, efficacy in a rodent postoperative pain model, and a dose estimation for controlling pain in dogs. J Vet Pharmacol Ther 40:285–292
193. Knych HK, Seminoff K, McKemie DS (2018) Detection and pharmacokinetics of grapiprant following oral administration to exercised thoroughbred horses. Drug Test Anal 10(8):1237–1243
194. Rausch-Derra LC, Huebner M, Rhodes L (2015) Evaluation of the safety of long-term, daily oral administration of grapiprant, a novel drug for treatment of osteoarthritic pain and inflammation, in healthy dogs. Am J Vet Res 76:853–859
195. Rausch-Derra L, Huebner M, Wofford J, Rhodes L (2016) A prospective, randomized, masked, placebo-controlled multisite clinical study of Grapiprant, an EP4 prostaglandin receptor antagonist (PRA), in dogs with osteoarthritis. J Vet Intern Med 30:756–763
196. Rausch-Derra LC, Rhodes L, Freshwater L, Hawks R (2016) Pharmacokinetic comparison of oral tablet and suspension formulations of grapiprant, a novel therapeutic for the pain and inflammation of osteoarthritis in dogs. J Vet Pharmacol Ther 39:566–571
197. Jin Y, Smith C, Hu L, Coutant DE, Whitehurst K, Phipps K, McNearney TA, Yang X, Ackermann B, Pottanat T, Landschulz W (2018) LY3127760, a selective prostaglandin E4 (EP4) receptor antagonist, and celecoxib: a comparison of pharmacological profiles. Clin Transl Sci 11:46–53
198. Sakata D, Yao C, Narumiya S (2010) Emerging roles of prostanoids in T cell-mediated immunity. IUBMB Life 62:591–596
199. Wherry EJ, Ha SJ, Kaech SM, Haining WN, Sarkar S, Kalia V, Subramaniam S, Blattman JN, Barber DL, Ahmed R (2007) Molecular signature of CD8+ T cell exhaustion during chronic viral infection. Immunity 27:670–684
200. Chen JH, Perry CJ, Tsui YC, Staron MM, Parish IA, Dominguez CX, Rosenberg DW, Kaech SM (2015) Prostaglandin E2 and programmed cell death 1 signaling coordinately impair CTL function and survival during chronic viral infection. Nat Med 21:327–334
201. Miao J, Lu X, Hu Y, Piao C, Wu X, Liu X, Huang C, Wang Y, Li D, Liu J (2017) Prostaglandin E2 and PD-1 mediated inhibition of antitumor CTL responses in the human tumor microenvironment. Oncotarget 8:89802–89810
202. Dockens RC, Santone KS, Mitroka JG, Morrison RA, Jemal M, Greene DS, Barbhaiya RH (2000) Disposition of radiolabeled ifetroban in rats, dogs, monkeys, and humans. Drug Metab Dispos 28:973–980
203. Ogletree ML, Harris DN, Schumacher WA, Webb ML, Misra RN (1993) Pharmacological profile of BMS 180,291: a potent, long-acting, orally active thromboxane A2/prostaglandin endoperoxide receptor antagonist. J Pharmacol Exp Ther 264:570–578
204. Rosenfeld L, Grover GJ, Stier CT Jr (2001) Ifetroban sodium: an effective TxA2/PGH2 receptor antagonist. Cardiovasc Drug Rev 19:97–115
205. Webb ML, Liu EC, Monshizadegan H, Hedberg A, Misra RN, Goldenberg H, Harris DN (1993) Binding and function of a potent new thromboxane receptor antagonist, BMS 180,291, in human platelets. J Pharmacol Exp Ther 264:1387–1394
206. Samara E, Cao G, Locke C, Granneman GR, Dean R, Killian A (1997) Population analysis of the pharmacokinetics and pharmacodynamics of seratrodast in patients with mild to moderate asthma. Clin Pharmacol Ther 62:426–435
207. Royer JF, Schratl P, Carrillo JJ, Jupp R, Barker J, Weyman-Jones C, Beri R, Sargent C, Schmidt JA, Lang-Loidolt D, Heinemann A (2008) A novel antagonist of prostaglandin D2 blocks the locomotion of eosinophils and basophils. Eur J Clin Investig 38:663–671

208. Endo S, Akiyama K (1996) Thromboxane A2 receptor antagonist in asthma therapy. Nihon Rinsho 54:3045–3048
209. Fiedler VB, Seuter F, Perzborn E (1990) Effects of the novel thromboxane antagonist bay U 3405 on experimental coronary artery disease. Stroke 21:IV149–IV151
210. Chapple CR, Abrams P, Andersson KE, Radziszewski P, Masuda T, Small M, Kuwayama T, Deacon S (2014) Phase II study on the efficacy and safety of the EP1 receptor antagonist ONO-8539 for nonneurogenic overactive bladder syndrome. J Urol 191:253–260
211. Kondo T, Sei H, Yamasaki T, Tomita T, Ohda Y, Oshima T, Fukui H, Watari J, Miwa H (2017) A novel prostanoid EP1 receptor antagonist, ONO-8539, reduces acid-induced heartburn symptoms in healthy male volunteers: a randomized clinical trial. J Gastroenterol 52:1081–1089
212. Tilly P, Charles AL, Ludwig S, Slimani F, Gross S, Meilhac O, Geny B, Stefansson K, Gurney ME, Fabre JE (2014) Blocking the EP3 receptor for PGE2 with DG-041 decreases thrombosis without impairing haemostatic competence. Cardiovasc Res 101:482–491

Targeting Leukotrienes as a Therapeutic Strategy to Prevent Comorbidities Associated with Metabolic Stress

4

Theresa Ramalho, Nayara Pereira, Stephanie L. Brandt, and C. Henrique Serezani

Abstract

Leukotrienes (LTs) are potent lipid mediators that exert a variety of functions, ranging from maintaining the tone of the homeostatic immune response to exerting potent proinflammatory effects. Therefore, LTs are essential elements in the development and maintenance of different chronic diseases, such as asthma, arthritis, and atherosclerosis. Due to the pleiotropic effects of LTs in the pathogenesis of inflammatory diseases, studies are needed to discover potent and specific LT synthesis inhibitors and LT receptor antagonists. Even though most clinical trials using LT inhibitors or antagonists have failed due to low efficacy and/or toxicity, new drug development strategies are driving the discovery for LT inhibitors to prevent inflammatory diseases. A newly important detrimental role for LTs in comorbidities associated with metabolic stress has emerged in the last few years and managing LT production and/or actions could represent an exciting new strategy to prevent or treat inflammatory diseases associated with metabolic disorders. This review is intended to shed light on the synthesis and actions of leukotrienes, the most common drugs used in clinical trials, and discuss the therapeutic potential of preventing LT function in obesity, diabetes, and hyperlipidemia.

Keywords

Leukotrienes · Inflammation · Immune regulation · Clinical trials · Metabolic disorders · Diabetes

T. Ramalho · N. Pereira
Department of Immunology, University of Sao Paulo, Sao Paulo, Brazil

S. L. Brandt
Division of Infectious Diseases, Department of Medicine, Vanderbilt University Medical Center, Nashville, TN, USA

C. H. Serezani (✉)
Division of Infectious Diseases, Department of Medicine, Vanderbilt University Medical Center, Nashville, TN, USA

Department of Pathology, Microbiology, and Immunology, Vanderbilt University Medical Center, Nashville, TN, USA

Vanderbilt Institute for Inflammation, infection, and immunity (VI4), Vanderbilt University Medical Center, Nashville, TN, USA

Vanderbilt Center for Immunobiology (VCI), Vanderbilt University Medical Center, Nashville, TN, USA
e-mail: h.serezan@vanderbilt.edu

Y. Kihara (ed.), *Druggable Lipid Signaling Pathways*, Advances in Experimental Medicine and Biology 1274, https://doi.org/10.1007/978-3-030-50621-6_4

4.1 A Brief History of the Leukotrienes (LTs)

Leukotrienes (LTs) are part of a large family of lipids termed eicosanoids, which also include prostaglandins and hydroxyeicosatetraenoic acids (HETEs). Eicosanoids are polyunsaturated fatty acids composed of twenty carbons, aptly named from the Greek word "eicosa", meaning twenty. These lipids are essential mediators in amplifying or dampening inflammation, and promoting tissue remodeling and wound healing. Eicosanoids are not typically stored within cells, but are produced and secreted rapidly after cell activation.

The history of LTs begins in the 1930s with the identification of an alcohol soluble molecule present in the sputum of asthmatic patients which is capable of stimulating smooth muscle cells [1]. A similar phenomenon was described by a substance generated *de novo* from guinea pig lungs by snake venom or antigen-stimulated after ovalbumin sensitization [2, 3]. Due to the slow and long-lasting contraction caused by this substance in the smooth muscle cells of the guinea pig intestine, this unknown substance has been called "Slow Reacting Substance (SRS)." In 1958, Walter Brocklehurst demonstrated that during anaphylaxis caused by antigen challenge in the guinea pig ileum, another substance distinct from histamine was capable of promoting smooth muscle contraction [4]. This contraction was slow, long-lasting, and resistant to antihistamines, and it was, for the first time, considered an effect of "slow-reacting substance of anaphylaxis (SRS-A)" [5, 6]. Only in the 1970s, SRS was identified as a member of the LT class and named as LTC_4 [7, 8].

LTs were named by Samuelsson in 1977, who recognized that leukocyctes produce oxygenated metabolites with three conjugated double bonds in the chemical structure [9]. Therefore, the term leukotrienes were used to identify structures containing three double bonds "tri-enes" that are generated by "leuko"cytes. An extension of these studies led to the characterization of a product derived from arachidonic acid (AA) metabolism, named LTB (12R-dihydroxy-6,14-cis-8,10-trans-eicosatetraenoic acid) [10, 11]. Later, leukotriene A_4 (LTA_4) (5,6-epoxy-7,9,11,14-eicosatetraenoic acid) was identified as a precursor required for the formation of LTB_4 and LTC_4, which for a long time were known as SRS [7, 8].

4.2 LT Biosynthesis and Their Pharmacological Inhibitors

LT production occurs mainly in immune cells activated by different receptors, such as FcRs, $G\alpha_q$, or $G\alpha_i$-coupled receptors and growth factors on the cell membrane [12].

The synthesis of LTs involves several rate-limiting steps that comprise the activation of phospholipase A_2 (PLA_2) and arachidonic acid (AA) release from phospholipids in the cellular membranes [13, 14] (Fig. 4.1). AA is converted to 5-hydroperoxy-eicosatetraenoic acid (5-HPETE) and subsequently to an unstable intermediate, LTA_4, by the enzyme 5-lipoxygenase (5-LO). When 5-LO is activated, it translocates from its cytosolic or nucleoplasmic location to the perinuclear envelope, where it acts in concert with 5-LO-activating protein (FLAP), which is required for 5-LO to function enzymatically in intact cells in order to generate LTA_4 [14–16]. In neutrophils, LTA_4 hydrolase (LTA_4H) is highly active and preferentially generates LTB_4 and dihydroxy LT [17]. In mast cells, eosinophils, basophils, and some types of macrophages, LTA_4 is conjugated with reduced glutathione by LTC_4 synthase to form LTC_4 [18]. Following specific export, LTC_4 is converted by the extracellular enzymes γ-GT and γ-GL to LTD_4, and LTE_4 by dipeptidase, and are collectively known as cysteinyl leukotrienes (CysLTs) [18].

The primary cellular sources of LTB_4 in both murine and humans are granulocytes, monocytes, and macrophages (Table 4.1) [19], while the main cellular source of CysLTs are eosinophils [20, 21]. However, murine (RAW264.7 and J774) and human (THP1 and U937) macrophage cell lines express low levels of 5-LO and produce barely detectable levels of LTB_4, which make these cell lines unsuitable models for LT research [22]. Some cell types, even non-immune cells,

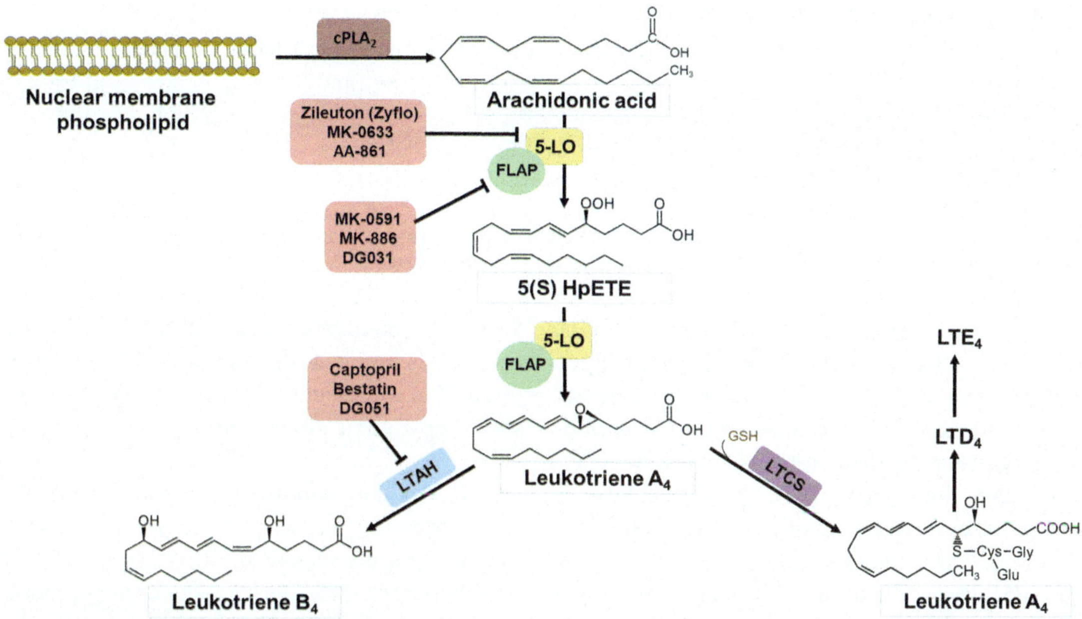

Fig. 4.1 Leukotriene biosynthesis and drug targets. Inflammatory receptor signaling triggers the $cPLA_2$ activation and release of arachidonic acid (AA) that is further metabolized to LTA_4 by 5-LO and FLAP in a twostep reaction involving the formation of 5(S)HpETE and the intermediate LTA_4. The enzyme LTA_4H metabolizes LTA_4 to LTB_4 or LTC_4 by LTC_4S. LT biosynthesis can be inhibited pharmacologically by specific inhibitors at various synthesis steps: LT, leukotriene, $cPLA_2$, cytosolic phospholipase A_2; 5-LO, 5-lipoxygenase; FLAP, 5-LO-activating protein; LTC_4S, LTC_4 synthase; LTA_4H, LTA_4 hydrolase

Table 4.1 Expression of LT synthesis enzymes

Cell types	5-LO	FLAP	LTA_4 hydrolase	LTC_4 synthase
Neutrophil	+++	+	+++	±
Monocyte/macrophage	++	+	+++	+
Dendritic cell	+	+	++	+
Mast cell	+	+	+	++
Eosinophil	+	+	+	++
Endothelial cell	–	–	+	+
Red blood cell	–	–	+	–
Keratinocyte	±	–	+	+

+ indicates relative amounts of enzymes or receptors produced by various cell types. – indicates a lack of expression. ± indicates minimal to no expression

have been reported to express some but not all of the LT-synthesis enzymes, which renders these cells incapable of synthesizing leukotrienes independently (Table 4.2). However, these cells may contribute to the synthesis of LTs in a process known as transcellular biosynthesis [23]. Examples of transcellular biosynthesis of LTB_4 are with neutrophils/erythrocytes and keratinocytes/endothelial cells [24–28]. The capabilities of individual cell types to produce various types of LTs depend on the expression of LT synthesis enzymes (Table 4.2).

5-LO 5-LO activity is reliant on various signals, including calcium release and phosphorylation, which control the catalytic site and the translocation of 5-LO within the cell [13, 14]. In a resting cell, 5-LO location varies depending on the cell type [29]. In neutrophils and peritoneal macrophages, 5-LO is located in the cytosol, whereas in

Table 4.2 Relative induction of LTs produced in response to various stimuli

Cell type	Cytokines/Growth factors	Bacteria	Opsonized pathogen	Fungi	Viral
Neutrophil	++	++	+++	++	++
Monocyte/macrophage	++	++	+++	++	++
Dendritic cell	+	++	ND	+	+
Mast cell	+	+	ND	+	+
Endothelial cell	±	±	±	±	±

+ indicates relative amounts of LTB_4 or cysLTs produced in response to various stimuli or receptors produced by various cell types. ± indicates minimal to no production of LTs. *ND* not determined

alveolar macrophages and Langerhans cells, 5-LO is located within the nucleus [30–32]. Upon cell activation, increased intracellular calcium levels induce 5-LO translocation to the perinuclear or plasmatic membrane where it can metabolize AA into LTA_4 [33]. 5-LO activity is triggered by various stimuli such as pathogens, cytokines, and immune complexes (Table 4.2) [34, 35]. Upon infection, pathogens have limited abilities to increase intracellular calcium levels and therefore are weak 5-LO activators alone. However, treating infected cells with a calcium ionophore or opsonized zymosan particles can significantly enhance LT synthesis [36, 37]. The molecular mechanisms that regulate 5-LO activation are reviewed here [38, 39].

The 5-LO pathway has been involved in the development and progression of many pathologic disorders, such as asthma, allergic rhinitis, cardiovascular diseases, and rheumatoid arthritis [15]. Since there is an increasing number of indications that anti-LT drugs can show protective effects, the development of 5-LO inhibitor agents becomes increasingly important. Although many 5-LO inhibitors have been developed, Zileuton (Zyflo, Leutrol – Abbot) is the only 5-LO inhibitor approved to use in humans. However, this inhibitor can induce liver toxicity and has a short half-live (drug needs to be administered up to 4 times a day) [40, 41]. Zileuton is an iron-ligand inhibitor that inhibits enzyme activity through chelation of the central iron atom and/or by stabilizing the ferrous oxidized state [42]. Zileuton inhibits an estimated 26–86% of endogenous LT production [43].

Many other 5-LO inhibitors were also tested for human use to treat asthma, ulcerative colitis, cardiac failure, and irritable bowel diseases, but they all failed in either phase II or III clinical trials. At this moment, the inhibitor MK-0633 is in phase II of a clinical trial to treat asthma, but also show hepatotoxic effects [44].

FLAP Another important alternative to inhibit LT production is to prevent FLAP actions. Merck-Frost discovered the first class of FLAP inhibitors (MK-886) during a screening of indole compounds derived from indomethacin and sulindac. MK-886 works by preventing the binding to arachidonic acid on FLAP and preventing the transfer of the substrate to 5-LO [45]. Further clinical testing was discontinued due to inefficacy since the treatment of asthmatics with MK-886 showed 50% inhibition of LTB_4 in *ex vivo*-stimulated whole blood and urinary LTE_4 levels [46]. A few years later, Merck developed the second generation of FLAP inhibitors. Like MK-886, MK-0591 blocks 5-LO activity by binding to FLAP, preventing 5-LO activation [47]. The further development of MK-0591 was discontinued after results from further clinical testing showed limited therapeutic potential [47]. Recently, the FLAP crystal structure was resolved, which promises to be very useful in drug design [48]. One FLAP inhibitor, DG031 (Veliflapon, deCODE genetics), is being reformulated for use in phase III clinical trials in the prevention of myocardial infarction [49].

LTA_4H LTA_4H is a complex enzyme that exhibits an epoxide hydrolase activity responsible for stereo-specific conversion of LTA_4 to LTB_4, and an aminopeptidase with preference tripeptides [50]. The most specific and potent LTA_4H inhibitors are derived from broad-spectrum aminopeptidase inhibitors, with ubenimex (also known as bestatin) representing the unique LTB_4 pathway inhibitor to gain approval to treat human disease [51]. At least six LTA_4H inhibitors of the current generation have entered clinical trials, with five of these reaching Phase II [51]. Furthermore, the LTA_4H inhibitors with broad aminopeptidase inhibitory activity have moved the farthest in the clinic. Although the efforts to generate several potent and specific LTA_4H inhibitors, these drugs have failed to demonstrate clinical efficacy or have harmful side effects. Recently, Acebilustat (CTX-4430) showed functional selectivity for LTA_4H epoxide hydrolase, and three trials in phase I showed that Acebilustat inhibits *ex-vivo* stimulated production of LTB_4, CYP3A4 induction, and reduces inflammatory mediators in the blood and sputum of cystic fibrosis patients [52–54]. These data resulted in advancing Acebilustat to a phase II clinical trial to investigate expiratory volume in cystic fibrosis patients. Also, Acebilustat is being tested in the lesions of patients with acne vulgaris. Although these phase II studies have been finished, the reports were not published yet [52–54]. LTA_4H expression has been associated as a risk factor for myocardial infarction (MI). It was shown that DG051 (deCODE genetics) have effects on dose-dependent reductions in blood LTB_4 levels in patients with a history of MI or coronary artery disease and that this drug is safe and well-tolerated, with an excellent pharmacokinetic profile (phase I study) [51]. Recently, a phase II clinical trial started to test DG051 ability to prevent MI in 400 patients, but whether this drug is effective remains to be published [51].

4.3 LT Receptors, Signaling and Their Antagonists

LT receptors are G protein-coupled receptors (GPCR) located on the outer plasma membrane of both structural and immune cells. These receptors comprise a family of structurally related GPCRs that elicit intracellular signaling by activating $G\alpha_q$ or $G\alpha_i$ [17] (Fig. 4.2).

CysLTRs The CysLTs/CysLTR axis is initiated by a variety of signals induced by agents that enhance calcium production, TLRs, cytokines, and opsonins and amplified by the inflammatory milieu and therefore could lead to adaptive immune responses [21, 34]. Since cysLTs have brief half-lives *in vivo*, it is expected that LTC_4 and LTD_4 might exert acute/subacute effects in cells, but due to the high stability of LTE_4, it is anticipated that this cysLTs might have chronic effects *in vivo* [21]. CysLTs have been implicated in a myriad of diseases, from cancer to the pathophysiology of bronchial asthma and allergic rhinitis [55]. This is because both a 5-LO inhibitor and $CysLT_1$ antagonist have proof efficacy to treat these diseases [56–58]. These receptor blockers were developed from functional studies before any components of the pathway had been cloned and molecularly characterized [21].

The receptors for cysLTs are members of the GPCR family, and so far, at least five different receptors have been identified: $CysLT_1$, $CysLT_2$, $P2Y_{12}$, GPR99, and GPR17. These receptors show different affinities for cysLTs, being that $CysLT_1$ binds $LTD_4 > LTC_4 = LTE_4$. $CysLT_2$ binds both $LTC_4 = LTD_4 > LTE_4$ [55]. Recently, GPR17, GPR99, and $P2Y_{12}$ were described as a high-affinity receptor for LTE_4 [59] (Fig. 4.2). $CysLT_1$ is widely expressed in spleen, lungs, small intestine, colon, and skeletal muscle, while $CysLT_2$ is exclusively detected in heart, adrenals, spleen, lymph nodes, and brain [60]. $CysLT_1$ is mainly expressed in monocyte/macrophages, neutrophils, and mast cells [21, 61]. Interestingly, $CysLT_2$ is expressed in eosinophil, muscle cells, and interstitial macrophages [21, 61] (Table 4.3).

$CysLT_1$ and $CysLT_2$ are GPCRs that use both pertussis toxin (PTX)-sensitive ($G\alpha_i$) and PTX-insensitive G proteins ($G\alpha_q$) that culminates in decrease cAMP abundance or Ca^{2+} mobilization and phospholipase C activation, respectively [55]. These varieties of downstream signaling effectors lead to differential activation of many downstream kinases, such as JNK/AP1 signaling, PI3K, Akt, MAPK/ERK1/2, and different PKCs [55]. The activation of pleiotropic signaling pro-

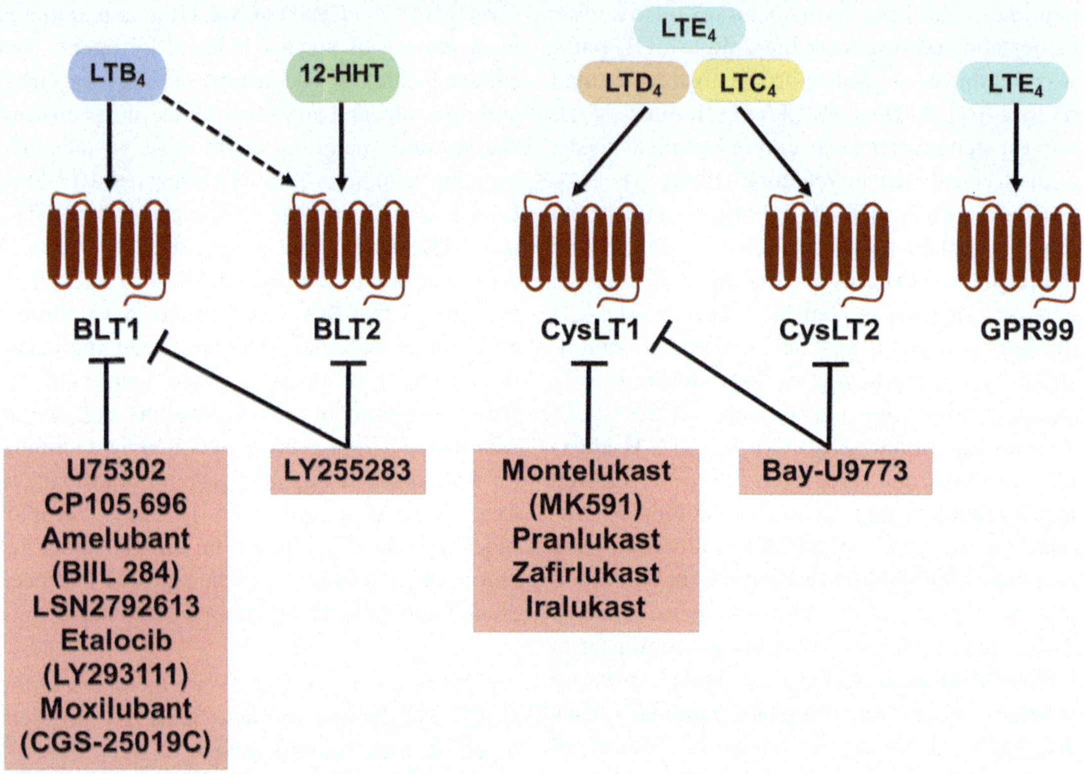

Fig. 4.2 Leukotriene receptors and specific antagonists. Leukotrienes bind to its specific receptors (BLT_1, BLT_2, $CysLT_1$, $CysLT_2$, and GPR99) with different affinities, leading to specific cell changes. The pharmacological agents directed against to prevent receptor signaling. 12-HHT, 12-Hydroxyheptadecatrenoic acid is an eicosanoid that shows a high affinity to BLT_2

Table 4.3 Cellular distribution LT receptors

Cell type	BLT_1	BLT_2	$CysLT_1$	$CysLT_2$
Neutrophil	+	+	±	±
Monocyte/macrophage	+	+	+	+
Dendritic cell	+	+	+	+
Mast cell	+	+	+	+
Eosinophil	+	+	+	+
Endothelial cell	+	+	+	+
Red blood cell	−	−	−	−
Keratinocyte	+	+	+	ND
B Lymphocytes	ND	+	+	ND
CD4 T cells	+	+	+	ND
CD8 T cells	+	+	ND	ND
Hematopoietic stem cells	ND	+	+	ND

+ detection; − not detected, ± minimal expression of the receptors in various cell types. *ND* not determined

grams is responsible for the actions of various transcription factors and, consequently, different pathologies [55]. $CysLT_1$ mediates bronchoconstriction, mucous secretion, and edema in the airways, being an essential therapeutic target in the treatment of asthma [62]. $CysLT_1$ abundance can be influenced by Th2 cytokines and as part of a positive feedback loop [62]. $CysLT_1$ also amplifies Th2 responses [62].

Since LTE_4 can elicit cellular responses in $CysLT_1$ and $CysLT_2$ deficient cells [63], the search for another LTE_4 receptor has been at the center of intense research. Two GPCRs were found to bind to LTE_4. The first LTE_4 receptor identified was $P2Y_{12}$ [64]. Genetic deletion of this receptor prevented LTE_4-mediated production of chemokines and cytokines, as well as ERK1/2 activation in mast cell [21, 63]. GPR99 was initially described as a receptor for alpha-ketoglutarate, and then demonstrated to be imporant for LTE_4 signaling [21, 59]. GPR99 is a $G\alpha_q$-coupled receptor expressed in respiratory epithelial cells and mediates mucin release in animal models of asthma [65]. GPR17 is a member of the P2Y receptor family and can recognize both cysLTs and uracil nucleotides [66]. GPR17 activation decreases cAMP, suggesting that this receptor is coupled to $G\alpha_i$. Interestingly, it has been suggested that GPR17 inhibits $CysLT_1$ in models of allergy [67].

$CysLT_1$ antagonists in the market are montelukast (Singulair), pranlukast (Onon), and zafirlukast (Accolate), and are prescribed to treat asthma and rhinitis. Montelukast, which is administered orally once daily, is the most prescribed antagonist for asthmatic patients [57]. Pranlukast is an orally administered, selective, and competitive $CysLT_1$ antagonist [68]. This drug is effective for chronic bronchial asthma. Furthermore, pranlukast has also shown to be effective in decreasing nasal eosinophil cationic protein and obstruction [69]. Zafirlukast is approved for the treatment of asthma in patients with 7 years of age or older [70].

LTB_4 Receptors LTB_4 induces cellular changes via its interaction with two G protein-coupled receptors, BLT_1, and BLT_2 [71]. LTB_4 was one of the first chemoattractant identified, but the discovery of LTB_4 receptors was challenging. The high-affinity receptor, BLT_1, was cloned in 1997 from the neutrophil-like HL-60 cells and showed a direct association between LTB_4 and BLT_1 in mediating neutrophil chemotaxis [72]. The low-affinity LTB_4 receptor, BLT_2, was also identified by Yokomizo et al. [73]. The genes of both receptors are located close to each other in human as well as mouse genomes [74]. Distribution of BLT_1 and BLT_2 on cells and tissues vary between mouse and human [71]. In human cells, BLT_1 expression is limited to leukocytes, and BLT_2 is ubiquitously found on many cell types (Table 4.3). On mouse cells, BLT_1 expression is detected on leukocytes, and BLT_2 is reported to be limited to the intestinal epithelium and keratinocytes [71, 74]. BLT_1 can be coupled to $G\alpha_i$, $G\alpha_q$, or $G\alpha_{16}$ that results in decreased cyclic AMP (cAMP) levels and increased intracellular calcium levels, respectively [75]. LTB_4 activation of BLT_1 enhances antimicrobial effector function activities in macrophages and neutrophils [76]. BLT_1 activation increases host defense during infections caused by different pathogens, including bacteria, viruses, fungi, and protozoan parasites. It does so by enhancing phagocytosis mediated by different phagocytic receptors, including FcγR1, Dectin1, mannose receptor, and CR3 in macrophages and neutrophils [77]. LTB_4/BLT_1 axis also enhances phagocyte antimicrobial effectors, such as reactive oxygen and nitrogen species, as well as the production of antimicrobial peptides [76]. Also, we and others have shown that topically or aerosolized LTB_4 can enhance *in vivo* host defense in the lungs or skin by enhancing phagocyte antimicrobial arsenal [78]. Furthermore, $LTB_4/BLT_1/BLT_2$ activation is a crucial step in mounting the inflammatory response by activation inflammatory transcription factors, such as NFκB and AP1, and inducing a proinflammatory program characterized by the production of cytokines and chemokines, such as IL-1β, TNFα, IL-6, CXCL1, CXCL2 [79, 80]. In mice, BLT_1 activation is involved in multiple inflammatory diseases as arthritis [81], autoimmune uveitis [82], atherosclerosis [83], and asthma [84]. Also, it has been shown that the LTB_4-BLT_1 axis stimulates macrophages by enhancing the expression of the Toll-like/IL1 receptor (TIR) adaptor MyD88 by amplifying STAT1 phosphorylation. This allows the activation of TIRs and therefore potentiates the inflammatory response [79].

Although there are reports that leukocytes in mice do not express BLT_2, others have detected BLT_2 expression in murine phagocytes [85, 86], but its role in inflammatory response and pathogenesis is poorly understood. Although BLT_2 has been shown to induce pro-inflammatory programs, others have shown that BLT_2 is required to decrease airway eosinophilia and decrease IL-13 abundance in murine asthmatic lungs without influencing IL-4 and IgE production [87]. BLT_2 on mast cells mediates the recruitment and accumulation of these cells in response to LTB_4 production at the sites of inflammation [84]. BLT_2-deficient mice show susceptibility to drug-induced inflammatory colitis [88] suggesting an enhanced epithelial barrier function developed by BLT_2. In humans, bladder cancer cells express high levels of BLT_2, and the treatment with BLT_2 antagonist (LY255283) or BLT_2 siRNA results in apoptosis and inhibits cancer cells proliferation [89], suggesting that BLT_2 antagonists might be candidates for therapeutic agents against cancer.

BLT_2 can also bind to other lipid mediators, indicating that the effects of BLT_2 actions are not limited to LTB_4 actions. BLT_2 plays a beneficial role in skin wound healing through 12-Hydroxyheptadecatrenoic acid (12-HHT) actions that mediate TNF-α and MMP production [71]. 12-HHT is important for keratinocyte wound healing in diabetic mice [90]. Additionally, 12(S)-HETE and BLT_2 effects are important for enhancing VEGF expression and promoting wound healing [91]. Although other lipid mediators promote beneficial wound healing responses through BLT_2 signaling, less is known about specific LTB_4/BLT_2 effects on phagocyte effector functions during infections.

Although there are specific and potent BLT_1-specific antagonists (such as CP105696 [92] and U75302 [93, 94]) none of them are on the market for clinical use. Etalocib (LY293111) have potency for binding to BLT_1 on isolated neutrophils; however, it has off-target agonist activity toward PPARγ [95, 96] [96–98]. Amelubant (BIIL 284) is a synthetic prodrug that can be converted to the active forms BIIL260 and BILL315. These active forms bind BLT_1 with high affinity, while for BIIL 284 binds BLT_1 weakly. Amelubant was tested in at least 18 clinical trials [99]. In the phase I study, Amelubant showed complete inhibition of LTB_4-induced BLT_1 signaling [51]. In a COPD study, Amelubant reduced macrophage frequency in spontaneous or induced sputum [51]. These results opened a new opportunity to start a phase II clinical trial in the treatment of COPD [100], rheumatoid arthritis (RA) [101], asthma [102], and cystic fibrosis [103]. However, these trials showed that Amelubant is not effective in tests of exercise endurance in COPD patients [100], asthma tests, and in adults and children with cystic fibrosis was terminated early due to adverse events [104].

4.4 Leukotrienes in Metabolic and Cardiovascular Diseases

Obesity, Diabetes, and Insulin Resistance In obesity, hypertrophic adipose tissue stores excess lipids, which trigger inflammation and dysregulate the production of inflammatory mediators [105]. For over a decade, studies have identified high expression of 5-LO and FLAP in adipocytes and adipose tissue (AT)-associated macrophages from subjects with obesity [106, 107]. Increased expression of LT synthesizing enzymes correlates with increased LTB_4 in adipose tissue and LTE_4 in the urine of mice and people with obesity [108]. Increased LTE_4 also correlates with high BMI and leptin (a hormone that controls ingestive behavior and energy balance) [109]. These data suggest that obesity is significantly associated with increased urinary LT levels [109]. A study showed that primary mature adipocytes from both healthy humans and mice, secretes high amounts of both LTB_4 and CyLTs, and it is further increased when mice are fed in a high-fat diet (HFD) [110]. Enhanced LT production enhances T cell recruitment and macrophages infiltration, leading to adipose tissue inflammation and obesity-induced insulin resistance [110]. Pharmacologic and genetic deletion of BLT_1 prevents high fat-induced insulin resistance in liver and skeletal muscle, and prevents inflammation in adipose tissue, while protecting mice from systemic insulin resistance and hepatic steatosis [108] [111]. Importantly, LTB_4 also show detrimental effects induced by low insulin levels. We have shown that 5-LO deficiency protects mice from insulin resistance, and the muscle of those mice have improved insulin signaling which corroborates the lower expression of inflammatory cytokines and higher levels of anti-inflammatory markers in this tissue compared to WT mice in a model of diabetes induced by streptozotocin (STZ) [112].

In type 1 diabetes (T1D), shreds of evidence are not enough to affirm that systemic inflammation caused by LTB_4 is the main reason for insulin resistance. In this case, it is speculated that both features caused by LTB_4 can be interlinked between themselves. However, in HFD-induced production of LTB_4 directly inhibits insulin sensitivity in adipose tissue in a manner dependent on T cell and macrophages activation [113]. This data lead us to speculate that in T1D, LTB_4 may be a common link among metabolic alterations, inflammation, and insulin resistance. Therefore, we speculate BLT_1 antagonist or LTA_4H inhibition as a novel strategy to improve insulin resistance in metabolic diseases.

Catabolic Diseases T1D is also associated with adipose tissue remodeling [114]. Nevertheless, low levels of insulin lead to a catabolic state characterized by adiposity loss [115]. This catabolic state affects the accumulation of lipids in the circulation of people with T1D positively, and these circulatory lipids do not follow a standard route to the liver [116]. Although replacement of insulin in subjects with T1D mitigates hyperglycemic complications, lipid metabolism is still impaired in T1D, which associates with comorbidities, such as fat depots in blood vessels [117]. Mice overexpressing FLAP in adipose tissue are leaner and show increased energy expenditure, due to browning of white adipose tissue [118]. We have shown that systemic inhibition of BLT_1 with U75302 protects diabetic mice from adiposity loss and hyperlipidemia [114]. Moreover, the liver of diabetic U75302-treated mice recovers the levels of fatty acid synthase (FAS) and its transcription factor, SREBP1 [114]. This data reinforces that the blockade of LTB_4 signaling would improve catabolic state and impaired lipid metabolism in diabetic subjects. Moreover, we have shown that in macrophages, LTB_4 promotes massive accumulation of lipid droplets, which reflects a pronounced energetic metabolism associated with uncoupled mitochondrial respiration [114]. Interestingly, T1D mice treated with U75302 downregulated all metabolic parameters observed in macrophages [114]. More studies are being conducted by our group to better integrate LTB_4 effects on lipid metabolism and inflammation. Interestingly, our preliminary data suggest that 5-LO deficiency protects mice from adiposity loss in a model of STZ-induced diabetes. We are speculating that poorly controlled diabetes leads to browning of adipose tissue and therefore, to a catabolic state that could negatively affect the quality of life of people with diabetes [119].

Hyperlipidemia Hyperlipidemia is characterized by a chronic state of inflammation and activation of resident macrophages [120]. Our group has demonstrated that LTB_4 dictates systemic inflammatory response in a model of hyperglycemia [121] in a manner dependent on macrophages [122]. In a murine model of atherosclerosis, the inhibition of the 5-LO pathway in apolipoprotein E–deficient ($ApoE^{-/-}$) mice reduces liver injury and enhances cardiovascular protection [123–125]. Hypercholesterolemia is a central risk factor for the initiation and progression of atherosclerotic lesions [126]. LTs enhances the progression of hyperlipidemia-dependent vascular disease and are associated with atherogenesis, cardiovascular disease, myocardial infarction, and stroke [123, 127]. Accordingly, the treatment of apoE-deficient mice with BLT_1 antagonist decreases the development of atherosclerosis [128] and reduces infarct size in a murine model of myocardial ischemia/reperfusion injury [129]. Furthermore, endothelial overexpression of $CysLT_2$ increases vascular permeability, myocardial ischemia/reperfusion damage, and cardiomyocyte apoptosis in peri-infarct areas [130–132]. The treatment of mice with the $CysLT_1$ antagonist montelukast showed cardio-protective capacities in both animal models and clinical trials, suggesting that blocking cysLTs could show beneficial effects in models during cardiac events [133–137].

4.5 Future Directions for Drug Discovery

Since LTs were discovered in the 1970s, and about 20 years since the availability of LT-targeting drugs, several new roles for LTs were discovered. These new insights range from new receptors, new biological functions, and new roles in different disease states. One potential new frontier to be explored is a possible role for LTs in the neuro-immune axis during obesity. There are many potentials for the crosstalk between LTs and sympathetic neuron- associated macrophages (SAMs) and lipid storage. Possibly, LTB_4 may act as a promoter of SAMs NE-uptake activity in adipose tissue of obese subjects. In parallel, in asthma, beta-adrenergic receptors activation is impaired in airway smooth muscle [138], and for this reason, β2-adrenoceptor agonists are often used for the treatment of asthma [139]. β2-adrenoceptor agonists reduce cysLT synthesis by eosinophils, therefore, a potential cross-talk between β2-adrenoceptor agonists and LT production or actions could potentially affect the outcome of allergic diseases.

References

1. Harkavy J (1930) Spam-producing substance in the sputum of patiensts with bronchial asthma. Arch Intern Med 45:641
2. Kellaway C, Trethewie E (1940) The liberation of a slow-reacting smooth muscle-stimulating substance in anaphylaxis. Q J Exp Physiol Cogn Med Sci 30:121–145
3. Feldberg W, Kellaway CH (1938) Liberation of histamine and formation of lysocithin-like substances by cobra venom. J Physiol 94:187–226
4. Brocklehurst W (1960) The release of histamine and formation of a slow-reacting substance (SRS-A) during anaphylactic shock. J Physiol 151:416–435
5. Orange RP, Murphy RC, Karnovsky ML, Austen KF (1973) The physicochemical characteristics and purification of slow-reacting substance of anaphylaxis. J Immunol 110:760–770
6. Morris HR, Taylor GW (1978) Slow-reacting substance of anaphylaxis purification and characterisation. FEBS Lett 87:203–206
7. Murphy RC, Hammarström S, Samuelsson B (1979) Leukotriene C: a slow-reacting substance from murine mastocytoma cells. Proc Natl Acad Sci 76:4275–4279
8. Hammarström S, Murphy RC, Samuelsson B, Clark DA, Mioskowski C, Corey E (1979) Structure of leukotriene C identification of the amino acid part. Biochem Biophys Res Commun 91:1266–1272
9. Samuelsson B, Borgeat P, Hammarström S, Murphy R (1979) Introduction of a nomenclature: leukotrienes. Prostaglandins 17:785–787
10. Borgeat P, Samuelsson B (1979) Arachidonic acid metabolism in polymorphonuclear leukocytes: effects of ionophore A23187. Proc Natl Acad Sci 76:2148–2152
11. Borgeat P, Samuelsson B (1979) Arachidonic acid metabolism in polymorphonuclear leukocytes: unstable intermediate in formation of dihydroxy acids. Proc Natl Acad Sci 76:3213–3217
12. Fitzpatrick FA, Soberman R (2001) Regulated formation of eicosanoids. J Clin Invest 107:1347–1351
13. Leslie CC (2015) Cytosolic phospholipase A_2: physiological function and role in disease. J Lipid Res 56:1386–1402
14. Dennis EA, Cao J, Hsu YH, Magrioti V, Kokotos G (2011) Phospholipase A2 enzymes: physical structure, biological function, disease implication, chemical inhibition, and therapeutic intervention. Chem Rev 111:6130–6185
15. Rådmark O, Werz O, Steinhilber D, Samuelsson B (2015) 5-Lipoxygenase, a key enzyme for leukotriene biosynthesis in health and disease. Biochim Biophys Acta 1851:331–339
16. Evans JF, Ferguson AD, Mosley RT, Hutchinson JH (2008) What's all the FLAP about?: 5-lipoxygenase-activating protein inhibitors for inflammatory diseases. Trends Pharmacol Sci 29:72–78
17. Peters-Golden M, Henderson WR Jr (2007) Leukotrienes. N Engl J Med 357:1841–1854
18. Peters-Golden M, Gleason MM, Togias A (2006) Cysteinyl leukotrienes: multi-functional mediators in allergic rhinitis. Clin Exp Allergy 36:689–703
19. Werz O (2002) 5-lipoxygenase: cellular biology and molecular pharmacology. Curr Drug Targets Inflamm Allergy 1:23–44
20. Liu M, Yokomizo T (2015) The role of leukotrienes in allergic diseases. Allergol Int 64:17–26
21. Kanaoka Y, Austen KF (2019) Roles of cysteinyl leukotrienes and their receptors in immune cell-related functions. Adv Immunol 142:65–84
22. Nolfo R, Rankin JA (1990) U937 and THP-1 cells do not release LTB4, LTC4, or LTD4 in response to A23187. Prostaglandins 39:157–165
23. Sala A, Folco G, Murphy RC (2010) Transcellular biosynthesis of eicosanoids. Pharmacol Rep 62:503–510
24. McGee JE, Fitzpatrick FA (1986) Erythrocyte-neutrophil interactions: formation of leukotriene B4

by transcellular biosynthesis. Proc Natl Acad Sci U S A 83:1349–1353
25. Iversen L, Fogh K, Ziboh VA, Kristensen P, Schmedes A, Kragballe K (1993) Leukotriene B4 formation during human neutrophil keratinocyte interactions: evidence for transformation of leukotriene A4 by putative keratinocyte leukotriene A4 hydrolase. J Invest Dermatol 100:293–298
26. Breton J, Woolf D, Young P, Chabot-Fletcher M (1996) Human keratinocytes lack the components to produce leukotriene B4. J Invest Dermatol 106:162–167
27. Janssen-Timmen U, Vickers PJ, Wittig U, Lehmann WD, Stark HJ, Fusenig NE, Rosenbach T, Radmark O, Samuelsson B, Habenicht AJ (1995) Expression of 5-lipoxygenase in differentiating human skin keratinocytes. Proc Natl Acad Sci U S A 92:6966–6970
28. Feinmark SJ (1992) The role of the endothelial cell in leukotriene biosynthesis. Am Rev Respir Dis 146:S51–S55
29. Peters-Golden M, Brock TG (2001) 5-Lipoxygenase and the nucleus: where, when, how, and why? In: Samuelsson B, Paoletti R, Folco GC, Granström E, Nicosia S (eds) Advances in Prostaglandin and Leukotriene research: basic science and new clinical applications. Springer Netherlands, Dordrecht, pp 9–15
30. Glover S, de Carvalho MS, Bayburt T, Jonas M, Chi E, Leslie CC, Gelb MH (1995) Translocation of the 85-kDa phospholipase A2 from cytosol to the nuclear envelope in rat basophilic leukemia cells stimulated with calcium ionophore or IgE/antigen. J Biol Chem 270:15359–15367
31. Peters-Golden M, McNish RW (1993) Redistribution of 5-lipoxygenase and cytosolic phospholipase A2 to the nuclear fraction upon macrophage activation. Biochem Biophys Res Commun 196:147–153
32. Woods JW, Coffey MJ, Brock TG, Singer II, Peters-Golden M (1995) 5-Lipoxygenase is located in the euchromatin of the nucleus in resting human alveolar macrophages and translocates to the nuclear envelope upon cell activation. J Clin Invest 95:2035–2046
33. Peters-Golden M (1998) Cell biology of the 5-lipoxygenase pathway. Am J Respir Crit Care Med 157:S227–S231; discussion S231-222, S247-228
34. Canetti C, Hu B, Curtis JL, Peters-Golden M (2003) Syk activation is a leukotriene B4-regulated event involved in macrophage phagocytosis of IgG-coated targets but not apoptotic cells. Blood 102:1877–1883
35. Brandt SL, Serezani CH (2017) Too much of a good thing: how modulating LTB4 actions restore host defense in homeostasis or disease. Semin Immunol 33:37–43
36. Gosselin J, Borgeat P (1997) Epstein-Barr virus modulates 5-lipoxygenase product synthesis in human peripheral blood mononuclear cells. Blood 89:2122–2130
37. Grone M, Scheffer J, Konig W (1992) Modulation of leukotriene generation by invasive bacteria. Immunology 77:400–407
38. Brock TG, Peters-Golden M (2007) Activation and regulation of cellular eicosanoid biosynthesis. ScientificWorldJournal 7:1273–1284
39. Radmark O, Samuelsson B (2009) 5-Lipoxygenase: mechanisms of regulation. J Lipid Res 50(Suppl):S40–S45
40. Zaman K, Hanigan MH, Smith A, Vaughan J, Macdonald T, Jones DR, Hunt JF, Gaston B (2006) Endogenous S-nitrosoglutathione modifies 5-lipoxygenase expression in airway epithelial cells. Am J Respir Cell Mol Biol 34:387–393
41. Bruno F, Spaziano G, Liparulo A, Roviezzo F, Nabavi SM, Sureda A, Filosa R, D'Agostino B (2018) Recent advances in the search for novel 5-lipoxygenase inhibitors for the treatment of asthma. Eur J Med Chem 153:65–72
42. McGill KA, Busse WW (1996) Zileuton. Lancet 348:519–524
43. Dubé LM, Swanson LJ, Awni W (1999) Zileuton, a leukotriene synthesis inhibitor in the management of chronic asthma. Clinical pharmacokinetics and safety. Clin Rev Allergy Immunol 17:213–221
44. Ducharme Y, Blouin M, Brideau C, Châteauneuf A, Gareau Y, Grimm EL, Juteau H, Laliberté S, MacKay B, Massé F (2010) The discovery of setileuton, a potent and selective 5-lipoxygenase inhibitor. ACS Med Chem Lett 1:170–174
45. Ford-Hutchinson AW, Gresser M, Young RN (1994) 5-Lipoxygenase. Annu Rev Biochem 63:383–417
46. Friedman BS, Bel EH, Buntinx A, Tanaka W, Han YH, Shingo S, Spector R, Sterk P (1993) Oral leukotriene inhibitor (MK-886) blocks allergen-induced airway responses. Am Rev Respir Dis 147:839–844
47. Steele VE, Holmes CA, Hawk ET, Kopelovich L, Lubet RA, Crowell JA, Sigman CC, Kelloff GJ (1999) Lipoxygenase inhibitors as potential cancer chemopreventives. Cancer Epidemiol Biomark Prev 8:467–483
48. Ferguson AD, McKeever BM, Xu S, Wisniewski D, Miller DK, Yamin TT, Spencer RH, Chu L, Ujjainwalla F, Cunningham BR, Evans JF, Becker JW (2007) Crystal structure of inhibitor-bound human 5-lipoxygenase-activating protein. Science 317:510–512
49. NCT00353067, C. g. Veliflapon (DG-031) o prevent heart attackes or smoke in patients with a history of heart attack or unstable angina. https://clinicaltrials.gov/ct2/show/NCT00353067
50. Stsiapanava A, Samuelsson B, Haeggström JZ (2017) Capturing LTA. Proc Natl Acad Sci U S A 114:9689–9694
51. Bhatt L, Roinestad K, Van T, Springman EB (2017) Recent advances in clinical development of leukotriene B4 pathway drugs. Semin Immunol 33:65–73
52. NCT03964558, C. g. I. Mass balance recovery, metabolite, profile and metabolite identification of Acebilustat. https://clinicaltrials.gov/ct2/show/NCT03964558
53. NCT02443688, C. g. I. EMPIRE CF: A Phase 2 study to evaluate the efficacy, safety and tolerability of CTX-4430 in adult cys-

tic fibrosis patients <p class="MsoNormal" style="margin-bottom:0cm;margin-bottom:.0001pt;text-align: justify; line-height:150%">. https://clinicaltrials.gov/ct2/show/results/NCT02443688#wrapper

54. NCT02385760, C. g. I. CTX-4430 for the treatment of moderate to severe facial Acne vulgaris. https://clinicaltrials.gov/ct2/show/NCT02385760
55. Kanaoka Y, Boyce JA (2004) Cysteinyl leukotrienes and their receptors: cellular distribution and function in immune and inflammatory responses. J Immunol 173:1503–1510
56. Parameswaran K, Liang H, Fanat A, Watson R, Snider DP, O'Byrne PM (2004) Role for cysteinyl leukotrienes in allergen-induced change in circulating dendritic cell number in asthma. J Allergy Clin Immunol 114:73–79
57. Stelmach I, Bobrowska-Korzeniowska M, Majak P, Stelmach W, Kuna P (2005) The effect of montelukast and different doses of budesonide on IgE serum levels and clinical parameters in children with newly diagnosed asthma. Pulm Pharmacol Ther 18:374–380
58. Gounaris E, Heiferman MJ, Heiferman JR, Shrivastav M, Vitello D, Blatner NR, Knab LM, Phillips JD, Cheon EC, Grippo PJ, Khazaie K, Munshi HG, Bentrem DJ (2015) Zileuton, 5-lipoxygenase inhibitor, acts as a chemopreventive agent in intestinal polyposis, by modulating polyp and systemic inflammation. PLoS One 10:e0121402
59. Kanaoka Y, Maekawa A, Austen KF (2013) Identification of GPR99 protein as a potential third cysteinyl leukotriene receptor with a preference for leukotriene E4 ligand. J Biol Chem 288:10967–10972
60. Heise CE, O'Dowd BF, Figueroa DJ, Sawyer N, Nguyen T, Im D-S, Stocco R, Bellefeuille JN, Abramovitz M, Cheng R (2000) Characterization of the human cysteinyl leukotriene 2 receptor. J Biol Chem 275:30531–30536
61. Kanaoka Y, Boyce JA (2014) Cysteinyl leukotrienes and their receptors; emerging concepts. Allergy Asthma Immunol Res 6:288–295
62. Okunishi K, Dohi M, Nakagome K, Tanaka R, Yamamoto K (2004) A novel role of cysteinyl leukotrienes to promote dendritic cell activation in the antigen-induced immune responses in the lung. J Immunol 173:6393–6402
63. Austen KF, Maekawa A, Kanaoka Y, Boyce JA (2009) The leukotriene E4 puzzle: finding the missing pieces and revealing the pathobiologic implications. J Allergy Clin Immunol 124:406–414; quiz 415-406
64. Foster HR, Fuerst E, Lee TH, Cousins DJ, Woszczek G (2013) Characterisation of P2Y(12) receptor responsiveness to cysteinyl leukotrienes. PLoS One 8:e58305
65. Bankova LG, Lai J, Yoshimoto E, Boyce JA, Austen KF, Kanaoka Y, Barrett NA (2016) Leukotriene E4 elicits respiratory epithelial cell mucin release through the G-protein-coupled receptor, GPR99. Proc Natl Acad Sci U S A 113:6242–6247
66. Ciana P, Fumagalli M, Trincavelli ML, Verderio C, Rosa P, Lecca D, Ferrario S, Parravicini C, Capra V, Gelosa P, Guerrini U, Belcredito S, Cimino M, Sironi L, Tremoli E, Rovati GE, Martini C, Abbracchio MP (2006) The orphan receptor GPR17 identified as a new dual uracil nucleotides/cysteinyl-leukotrienes receptor. EMBO J 25:4615–4627
67. Maekawa A, Balestrieri B, Austen KF, Kanaoka Y (2009) GPR17 is a negative regulator of the cysteinyl leukotriene 1 receptor response to leukotriene D4. Proc Natl Acad Sci U S A 106:11685–11690
68. Tohda Y, Nakahara H, Kubo H, Haraguchi R, Fukuoka M, Nakajima S (1999) Effects of ONO-1078 (pranlukast) on cytokine production in peripheral blood mononuclear cells of patients with bronchial asthma. Clin Exp Allergy 29:1532–1536
69. Gotoh M, Okubo K, Hashiguchi K, Wakabayashi K, Kanzaki S, Tanaka N, Fujioka M, Kawashima K, Suematsu K, Sasaki K, Iwasaki M, Yamamotoya H (2012) Noninvasive biological evaluation of response to pranlukast treatment in pediatric patients with Japanese cedar pollinosis. Allergy Asthma Proc 33:459–466
70. Stoloff SW (2000) The role of LTRAs in the management of persistent asthma. Postgrad Med 108:22–31
71. Yokomizo T (2015) Two distinct leukotriene B4 receptors, BLT1 and BLT2. J Biochem 157:65–71
72. Yokomizo T, Izumi T, Chang K, Takuwa Y, Shimizu T (1997) A G-protein-coupled receptor for leukotriene B4 that mediates chemotaxis. Nature 387:620–624
73. Yokomizo T, Kato K, Terawaki K, Izumi T, Shimizu T (2000) A second leukotriene B(4) receptor, BLT2. A new therapeutic target in inflammation and immunological disorders. J Exp Med 192:421–432
74. Tager AM, Luster AD (2003) BLT1 and BLT2: the leukotriene B(4) receptors. Prostaglandins Leukot Essent Fatty Acids 69:123–134
75. Peres CM, Aronoff DM, Serezani CH, Flamand N, Faccioli LH, Peters-Golden M (2007) Specific leukotriene receptors couple to distinct G proteins to effect stimulation of alveolar macrophage host defense functions. J Immunol 179:5454–5461
76. Brandt SL, Serezani CH (2017) Too much of a good thing: how modulating LTB. Semin Immunol 33:37–43
77. Okamoto F, Saeki K, Sumimoto H, Yamasaki S, Yokomizo T (2010) Leukotriene B4 augments and restores Fc gamma Rs-dependent phagocytosis in macrophages. J Biol Chem 285:41113–41121
78. Mancuso P, Lewis C, Serezani CH, Goel D, Peters-Golden M (2010) Intrapulmonary administration of leukotriene B4 enhances pulmonary host defense against pneumococcal pneumonia. Infect Immun 78:2264–2271
79. Serezani CH, Lewis C, Jancar S, Peters-Golden M (2011) Leukotriene B4 amplifies NF-κB activation in mouse macrophages by reducing SOCS1 inhibition of MyD88 expression. J Clin Invest 121:671–682

80. Wang Z, Filgueiras LR, Wang S, Serezani AP, Peters-Golden M, Jancar S, Serezani CH (2014) Leukotriene B4 enhances the generation of proinflammatory microRNAs to promote MyD88-dependent macrophage activation. J Immunol 192:2349–2356
81. Kim ND, Chou RC, Seung E, Tager AM, Luster AD (2006) A unique requirement for the leukotriene B4 receptor BLT1 for neutrophil recruitment in inflammatory arthritis. J Exp Med 203:829–835
82. Liao T, Ke Y, Shao W-H, Haribabu B, Kaplan HJ, Sun D, Shao H (2006) Blockade of the interaction of leukotriene b4 with its receptor prevents development of autoimmune uveitis. Invest Ophthalmol Vis Sci 47:1543–1549
83. Heller EA, Liu E, Tager AM, Sinha S, Roberts JD, Koehn SL, Libby P, Aikawa ER, Chen JQ, Huang P, Freeman MW, Moore KJ, Luster AD, Gerszten RE (2005) Inhibition of atherogenesis in BLT1-deficient mice reveals a role for LTB4 and BLT1 in smooth muscle cell recruitment. Circulation 112:578–586
84. Gelfand EW (2017) Importance of the leukotriene B4-BLT1 and LTB4-BLT2 pathways in asthma. Semin Immunol 33:44–51
85. Lee AJ, Cho KJ, Kim JH (2015) MyD88-BLT2-dependent cascade contributes to LPS-induced interleukin-6 production in mouse macrophage. Exp Mol Med 47:e156
86. Zhang Y, Olson RM, Brown CR (2017) Macrophage LTB4 drives efficient phagocytosis of Borrelia burgdorferi via BLT1 or BLT2. J Lipid Res 58:494–503
87. Matsunaga Y, Fukuyama S, Okuno T, Sasaki F, Matsunobu T, Asai Y, Matsumoto K, Saeki K, Oike M, Sadamura Y, Machida K, Nakanishi Y, Kubo M, Yokomizo T, Inoue H (2013) Leukotriene B4 receptor BLT2 negatively regulates allergic airway eosinophilia. FASEB J 27:3306–3314
88. Iizuka Y, Okuno T, Saeki K, Uozaki H, Okada S, Misaka T, Sato T, Toh H, Fukayama M, Takeda N, Kita Y, Shimizu T, Nakamura M, Yokomizo T (2010) Protective role of the leukotriene B4 receptor BLT2 in murine inflammatory colitis. FASEB J 24:4678–4690
89. Seo J-M, Cho K-J, Kim E-Y, Choi MH, Chung BC, Kim J-H (2011) Up-regulation of BLT2 is critical for the survival of bladder cancer cells. Exp Mol Med 43:129
90. Liu M, Saeki K, Matsunobu T, Okuno T, Koga T, Sugimoto Y, Yokoyama C, Nakamizo S, Kabashima K, Narumiya S, Shimizu T, Yokomizo T (2014) 12-Hydroxyheptadecatrienoic acid promotes epidermal wound healing by accelerating keratinocyte migration via the BLT2 receptor. J Exp Med 211:1063–1078
91. Kim GY, Lee JW, Cho SH, Seo JM, Kim JH (2009) Role of the low-affinity leukotriene B4 receptor BLT2 in VEGF-induced angiogenesis. Arterioscler Thromb Vasc Biol 29:915–920
92. Showell H, Breslow R, Conklyn M, Hingorani G, Koch K (1996) Characterization of the pharmacological profile of the potent LTB4 antagonist CP-105,696 on murine LTB4 receptors in vitro. Br J Pharmacol 117:1127
93. Richards IM, Griffin RL, Oostveen JA, Morris J, Wishka DG, Dunn CJ (1989) Effect of the selective leukotriene B4 antagonist U-75302 on antigen-induced bronchopulmonary eosinophilia in sensitized Guinea pigs. Am Rev Respir Dis 140:1712–1716
94. Lawson C, Wishka D, Morris J, Fitzpatrick F (1989) Receptor antagonism of leukotriene B4 myotropic activity by the 2, 6 disubstituted pyridine analog U-75302: characterization on lung parenchyma strips. J Lipid Mediat 1:3–12
95. Adrian TE, Hennig R, Friess H, Ding X (2008) The role of PPARgamma receptors and leukotriene B(4) receptors in mediating the effects of LY293111 in pancreatic Cancer. PPAR Res 2008:827096
96. Tong WG, Ding XZ, Hennig R, Witt RC, Standop J, Pour PM, Adrian TE (2002) Leukotriene B4 receptor antagonist LY293111 inhibits proliferation and induces apoptosis in human pancreatic cancer cells. Clin Cancer Res 8:3232–3242
97. Hennig R, Ding XZ, Tong WG, Witt RC, Jovanovic BD, Adrian TE (2004) Effect of LY293111 in combination with gemcitabine in colonic cancer. Cancer Lett 210:41–46
98. Zhang W, McQueen T, Schober W, Rassidakis G, Andreeff M, Konopleva M (2005) Leukotriene B4 receptor inhibitor LY293111 induces cell cycle arrest and apoptosis in human anaplastic large-cell lymphoma cells via JNK phosphorylation. Leukemia 19:1977–1984
99. ClinicalTrials.gov, L. f. Amebulant, and U. S. N. I. o. Health. 2017. https://clinicaltrials.gov/ct2/results?term=amelubant&Search=Search
100. . Trial Synopsis 543.10: Effect of 14-Day Treatment with BIIL 284 BS on Patients with COPD (Double-Blind, Placebo-Controlled, Randomised, Parallel Group Study).Boehringer Ingelheim. http://www.trials.boehringer-ingelheim.com/public/trial_results_documents/543/543.10_U01-1368.pdf
101. . Trial Synopsis 543.14: A double-blind, randomized, three parallel group placebocontrolled study to investigate pharmacokinetics, effect on expression of CD11b/CD18 (Mac-1), as well as safety and efficacy of two oral doses of BIIL 284 BS (dosage: 25 mg daily, 150 mg daily) in patients with rheumatoid arthritis over two weeks. Boehringer Ingelheim. http://www.trials.boehringer-ingelheim.com/public/trial_results_documents/543/543.14_U01-1167.pdf.
102. 543.11, T. S. The effect BIIL 284 BS (14 day treatment) on induced-sputum variables in patients with bronchial asthma (a double-blind. randomized, placebo-controlled parallel study). https://trials.boehringer-ingelheim.com/trial_results/clinical_trials_overview/clinical_trial_result.c=n.i=9.html
103. . Trial Synopsis 543.37: A randomized, double-blind within dose, placebo-controlled study to investigate the safety, tolerability and pharmacokinetics of repeated oral doses (15-day dosing) of BIIL 284

BS in adult (150 mg) and pediatric (75 mg) cystic fibrosis patients. Boehringer Ingelheim. http://www.trials.boehringer-ingelheim.com/public/trial_results_documents/543/543.37_U03-3277.pdf.
104. Konstan MW, Döring G, Heltshe SL, Lands LC, Hilliard KA, Koker P, Bhattacharya S, Staab A, Hamilton A, I. a. C. o. B. T. 543.45 (2014) A randomized double blind, placebo controlled phase 2 trial of BIIL 284 BS (an LTB4 receptor antagonist) for the treatment of lung disease in children and adults with cystic fibrosis. J Cyst Fibros 13:148–155
105. Baker RG, Hayden MS, Ghosh S (2011) NF-κB, inflammation, and metabolic disease. Cell Metab 13:11–22
106. Bäck M, Sultan A, Ovchinnikova O, Hansson GK (2007) 5-Lipoxygenase-activating protein: a potential link between innate and adaptive immunity in atherosclerosis and adipose tissue inflammation. Circ Res 100:946–949
107. Kaaman M, Rydén M, Axelsson T, Nordström E, Sicard A, Bouloumié A, Langin D, Arner P, Dahlman I (2006) ALOX5AP expression, but not gene haplotypes, is associated with obesity and insulin resistance. Int J Obes 30:447–452
108. Spite M, Hellmann J, Tang Y, Mathis SP, Kosuri M, Bhatnagar A, Jala VR, Haribabu B (2011) Deficiency of the leukotriene B4 receptor, BLT-1, protects against systemic insulin resistance in diet-induced obesity. J Immunol 187:1942–1949
109. Giouleka P, Papatheodorou G, Lyberopoulos P, Karakatsani A, Alchanatis M, Roussos C, Papiris S, Loukides S (2011) Body mass index is associated with leukotriene inflammation in asthmatics. Eur J Clin Investig 41:30–38
110. Mothe-Satney I, Filloux C, Amghar H, Pons C, Bourlier V, Galitzky J, Grimaldi PA, Féral CC, Bouloumié A, Van Obberghen E, Neels JG (2012) Adipocytes secrete leukotrienes: contribution to obesity-associated inflammation and insulin resistance in mice. Diabetes 61:2311–2319
111. Li P, Oh DY, Bandyopadhyay G, Lagakos WS, Talukdar S, Osborn O, Johnson A, Chung H, Mayoral R, Maris M, Ofrecio JM, Taguchi S, Lu M, Olefsky JM (2015) LTB4 promotes insulin resistance in obese mice by acting on macrophages, hepatocytes and myocytes. Nat Med 21:239–247
112. Guimarães FR, Sales-Campos H, Nardini V, da Costa TA, Fonseca MTC, Júnior VR, Sorgi CA, da Silva JS, Chica JEL, Faccioli LH, de Barros Cardoso CR (2018) The inhibition of 5Lipoxygenase (5-LO) products leukotriene B4 (LTB). Clin Immunol 190:74–83
113. Ying W, Wollam J, Ofrecio JM, Bandyopadhyay G, El Ouarrat D, Lee YS, Oh DY, Li P, Osborn O, Olefsky JM (2017) Adipose tissue B2 cells promote insulin resistance through leukotriene LTB4/LTB4R1 signaling. J Clin Invest 127:1019–1030
114. Ramalho T, Ramalingam L, Filgueiras L, Festuccia W, Jancar S, Moustaid-Moussa N (2019) Leukotriene-B4 modulates macrophage metabolism and fat loss in type 1 diabetic mice. J Leukoc Biol 106(3):665–675
115. Rosenfalck AM, Almdal T, Hilsted J, Madsbad S (2002) Body composition in adults with type 1 diabetes at onset and during the first year of insulin therapy. Diabet Med 19:417–423
116. Wolf P, Fellinger P, Pfleger L, Smajis S, Beiglböck H, Gajdošík M, Anderwald CH, Trattnig S, Luger A, Winhofer Y, Krššák M, Krebs M (2019) Reduced hepatocellular lipid accumulation and energy metabolism in patients with long standing type 1 diabetes mellitus. Sci Rep 9:2576
117. Jones Iv AR, Coleman EL, Husni NR, Deeney JT, Raval F, Steenkamp D, Dooms H, Nikolajczyk BS, Corkey BE (2017) Type 1 diabetes alters lipid handling and metabolism in human fibroblasts and peripheral blood mononuclear cells. PLoS One 12:e0188474
118. Elias I, Ferré T, Vilà L, Muñoz S, Casellas A, Garcia M, Molas M, Agudo J, Roca C, Ruberte J, Bosch F, Franckhauser S (2016) ALOX5AP overexpression in adipose tissue leads to LXA4 production and protection against diet-induced obesity and insulin resistance. Diabetes 65:2139–2150
119. Tamucci KA, Namwanje M, Fan L, Qiang L (2018) The dark side of browning. Protein Cell 9:152–163
120. Honold L, Nahrendorf M (2018) Resident and monocyte-derived macrophages in cardiovascular disease. Circ Res 122:113–127
121. Filgueiras LR, Brandt SL, Wang S, Wang Z, Morris DL, Evans-Molina C, Mirmira RG, Jancar S, Serezani CH (2015) Leukotriene B4-mediated sterile inflammation promotes susceptibility to sepsis in a mouse model of type 1 diabetes. Sci Signal 8:ra10
122. Ramalho T, Filgueiras L, Silva-Jr I, Pessoa AFM, Jancar S (2018) Impaired wound healing in type 1 diabetes is dependent on 5-lipoxygenase products. Sci Rep (Nature) 8(1):14164
123. Zhao L, Moos MP, Gräbner R, Pédrono F, Fan J, Kaiser B, John N, Schmidt S, Spanbroek R, Lötzer K, Huang L, Cui J, Rader DJ, Evans JF, Habenicht AJ, Funk CD (2004) The 5-lipoxygenase pathway promotes pathogenesis of hyperlipidemia-dependent aortic aneurysm. Nat Med 10:966–973
124. Mehrabian M, Allayee H, Wong J, Shi W, Wang XP, Shaposhnik Z, Funk CD, Lusis AJ, Shih W (2002) Identification of 5-lipoxygenase as a major gene contributing to atherosclerosis susceptibility in mice. Circ Res 91:120–126
125. Hlawaty H, Jacob MP, Louedec L, Letourneur D, Brink C, Michel JB, Feldman L, Bäck M (2009) Leukotriene receptor antagonism and the prevention of extracellular matrix degradation during atherosclerosis and in-stent stenosis. Arterioscler Thromb Vasc Biol 29:518–524
126. Clarkson S, Newburgh LH (1926) The relation between atherosclerosis and ingested cholesterol in the rabbit. J Exp Med 43:595–612
127. Bäck M, Hansson GK (2006) Leukotriene receptors in atherosclerosis. Ann Med 38:493–502

128. Ketelhuth DF, Hermansson A, Hlawaty H, Letourneur D, Yan ZQ, Bäck M (2015) The leukotriene B4 receptor (BLT) antagonist BIIL284 decreases atherosclerosis in ApoE−/− mice. Prostaglandins Other Lipid Mediat 121:105–109
129. de Hoog VC, Bovens SM, de Jager SC, van Middelaar BJ, van Duijvenvoorde A, Doevendans PA, Pasterkamp G, de Kleijn DP, Timmers L (2015) BLT1 antagonist LSN2792613 reduces infarct size in a mouse model of myocardial ischaemia-reperfusion injury. Cardiovasc Res 108:367–376
130. Moos MP, Mewburn JD, Kan FW, Ishii S, Abe M, Sakimura K, Noguchi K, Shimizu T, Funk CD (2008) Cysteinyl leukotriene 2 receptor-mediated vascular permeability via transendothelial vesicle transport. FASEB J 22:4352–4362
131. Jiang W, Hall SR, Moos MP, Cao RY, Ishii S, Ogunyankin KO, Melo LG, Funk CD (2008) Endothelial cysteinyl leukotriene 2 receptor expression mediates myocardial ischemia-reperfusion injury. Am J Pathol 172:592–602
132. Hui Y, Cheng Y, Smalera I, Jian W, Goldhahn L, Fitzgerald GA, Funk CD (2004) Directed vascular expression of human cysteinyl leukotriene 2 receptor modulates endothelial permeability and systemic blood pressure. Circulation 110:3360–3366
133. Becher UM, Ghanem A, Tiyerili V, Fürst DO, Nickenig G, Mueller CF (2011) Inhibition of leukotriene C4 action reduces oxidative stress and apoptosis in cardiomyocytes and impedes remodeling after myocardial injury. J Mol Cell Cardiol 50:570–577
134. Nobili E, Salvado MD, Folkersen L, Castiglioni L, Kastrup J, Wetterholm A, Tremoli E, Hansson GK, Sironi L, Haeggström JZ, Gabrielsen A (2012) Cysteinyl leukotriene signaling aggravates myocardial hypoxia in experimental atherosclerotic heart disease. PLoS One 7:e41786
135. Allayee H, Hartiala J, Lee W, Mehrabian M, Irvin CG, Conti DV, Lima JJ (2007) The effect of montelukast and low-dose theophylline on cardiovascular disease risk factors in asthmatics. Chest 132:868–874
136. Ingelsson E, Yin L, Bäck M (2012) Nationwide cohort study of the leukotriene receptor antagonist montelukast and incident or recurrent cardiovascular disease. J Allergy Clin Immunol 129:702–707.e702
137. Hoxha M, Rovati GE, Cavanillas AB (2017) The leukotriene receptor antagonist montelukast and its possible role in the cardiovascular field. Eur J Clin Pharmacol 73:799–809
138. Emala C, Black C, Curry C, Levine MA, Hirshman CA (1993) Impaired beta-adrenergic receptor activation of adenylyl cyclase in airway smooth muscle in the basenji-greyhound dog model of airway hyperresponsiveness. Am J Respir Cell Mol Biol 8:668–675
139. Amrani Y, Bradding P (2017) β2-Adrenoceptor function in asthma. Adv Immunol 136:1–28

Epoxy Fatty Acids Are Promising Targets for Treatment of Pain, Cardiovascular Disease and Other Indications Characterized by Mitochondrial Dysfunction, Endoplasmic Stress and Inflammation

5

Cindy McReynolds, Christophe Morisseau, Karen Wagner, and Bruce Hammock

Abstract

Bioactive lipid mediators resulting from the metabolism of polyunsaturated fatty acids (PUFA) are controlled by many pathways that regulate the levels of these mediators and maintain homeostasis to prevent disease. PUFA metabolism is driven primarily through three pathways. Two pathways, the cyclooxygenase (COX) and lipoxygenase (LO) enzymatic pathways, form metabolites that are mostly inflammatory, while the third route of metabolism results from the oxidation by the cytochrome P450 enzymes to form hydroxylated PUFA and epoxide metabolites. These epoxygenated fatty acids (EpFA) demonstrate largely anti-inflammatory and beneficial properties, in contrast to the other metabolites formed from the degradation of PUFA. Dysregulation of these systems often leads to chronic disease. Pharmaceutical targets of disease focus on preventing the formation of inflammatory metabolites from the COX and LO pathways, while maintaining the EpFA and increasing their concentration in the body is seen as beneficial to treating and preventing disease. The soluble epoxide hydrolase (sEH) is the major route of metabolism of EpFA. Inhibiting its activity increases concentrations of beneficial EpFA, and often disease states correlate to mutations in the sEH enzyme that increase its activity and decrease the concentrations of EpFA in the body. Recent approaches to increasing EpFA include synthetic mimics that replicate biological activity of EpFA while preventing their metabolism, while other approaches focus on developing small molecule inhibitors to the sEH. Increasing EpFA concentrations in the body has demonstrated multiple beneficial effects in treating many diseases, including inflammatory and painful conditions, cardiovascular disease, neurological and disease of the central nervous system. Demonstration of efficacy in so many disease states can be

C. McReynolds · K. Wagner
Department of Entomology and Nematology, and U.C. Davis Comprehensive Cancer Center, University of California Davis, Davis, CA, USA

EicOsis, Davis, CA, USA

C. Morisseau · B. Hammock (✉)
Department of Entomology and Nematology, and U.C. Davis Comprehensive Cancer Center, University of California Davis, Davis, CA, USA
e-mail: bdhammock@ucdavis.edu

Y. Kihara (ed.), *Druggable Lipid Signaling Pathways*, Advances in Experimental Medicine and Biology 1274, https://doi.org/10.1007/978-3-030-50621-6_5

explained by the fundamental mechanism that EpFA have of maintaining healthy microvasculature and preventing mitochondrial and endoplasmic reticulum stress. While there are no FDA approved methods that target the sEH or other enzymes responsible for metabolizing EpFA, current clinical efforts to test for efficacy by increasing EpFA that include inhibiting the sEH or administration of EpFA mimics that block metabolism are in progress.

Keywords

PUFA · sEH · Oxylipin · Epoxide · Lipid metabolism

5.1 Introduction

5.1.1 Production of Hormones and Other Chemical Mediators

Lipids are an important component of human health, providing a source of energy, maintaining cellular integrity and acting as regulators of cell signaling. These bioactive lipids include steroids, diacyl glycerol (DAG), sphingolipids, phosphatidylinositol phosphate (PIP), phosphatidylcholine (PC), and polyunsaturated fatty acids (PUFA). They range in function from energy storage and generation through beta-oxidation (PUFA), cellular proliferation (PIP), insulin regulation (DAG), cellular protection (sphingolipids), pulmonary function (PC), and maintaining cell structure (cholesterol). However, the metabolism of these lipids often produces more potent biological mediators than the parent molecule. For example, PUFAs, which circulate through blood as triglycerides or as free fatty acids, can also be incorporated into adipose tissue or the membranes of cells that are further released upon insult or response to cell signaling, and their metabolism produces potent inflammatory and anti-inflammatory compounds that have profound effects on the body. Due to their potent effects, inhibitors that alter their metabolism represent one of the earliest drug targets of the pharmaceutical industry. For example, aspirin was discovered in the late 1800s although its mechanism of action as both a reversible and irreversible cyclooxygenase (COX) inhibitor that blocked the formation of prostaglandins (PG) was not identified until the early 1970s, and is still debated [1]. Blocking COX activity and the formation of inflammatory PG compounds resulted in one of the largest selling classes of drugs, NSAIDs, on the market today. In addition to COX metabolism, PUFA are also metabolized by the lipoxygenase (LO) and cytochrome P450 (CYP450) enzymes that have more recently attracted attention for their potential in modulating disease. LO inhibitors are targeted for their ability to inhibit the formation of inflammatory leukotrienes, and Zileuton, a 5-LO inhibitor, is used for the treatment of asthma [2] by blocking the formation of inflammatory leukotrienes. In contrast to COX and LO metabolism, the CYP branch of PUFA metabolism results in the formation of both inflammatory hydroxylated metabolites as well as ant-inflammatory fatty-acid epoxides (Fig. 5.1). Disease altering strategies targeting the CYP450 branch of the pathway focuses on increasing these beneficial epoxy fatty acids (EpFA). However, so far, few drugs specifically targeting this pathway have reached the market although several currently approved drugs alter EpFA concentrations through their action on enzymes and their metabolites in the CYP450 branch of the arachidonic acid cascade. This chapter will focus on the biological activity of these largely beneficial lipid epoxides, as well as strategies for developing pharmaceutical interventions to increase their concentrations in the body.

5.1.2 Polyunsaturated Fatty Acids: The Essential Fatty Acids

PUFA are named for the presence of two or more double bonds in the mid-long chain carbon backbone that ranges in length from 16 to 24 carbons or longer. They are considered essential because the body cannot synthesize them naturally and must consume them in order to maintain health;

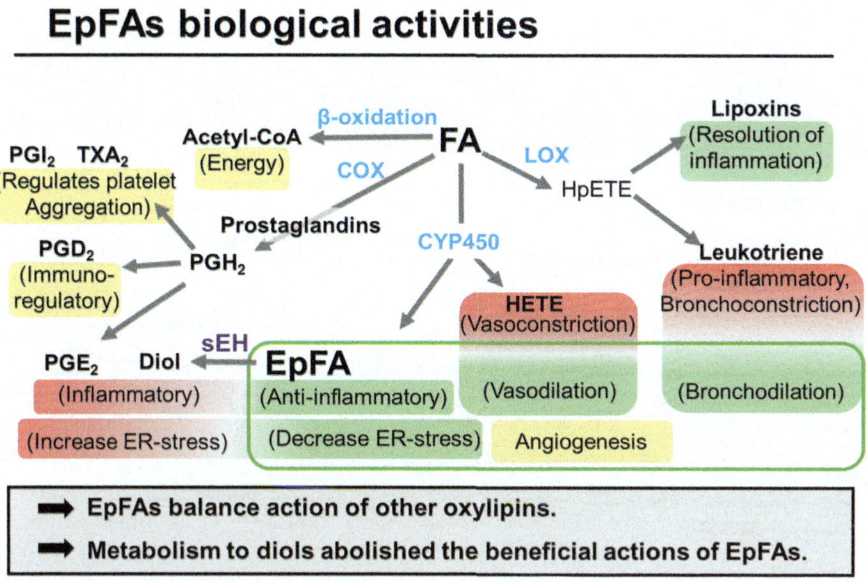

Fig. 5.1 Metabolic fate of polyunsaturated fatty acids
Free fatty acids are primarily metabolized through β-oxidation, Cyclooxygenase (COX), Lipoxygenase (LOX or LO), and CytochromeP450 (CYP450). A simplistic overview of the biological action of the resulting metabolites is shown in parenthesis.
COX 1 and COX 2 metabolize PUFA to PGH2 which is the precursor to other inflammatory prostanoids and thromboxanes that regulate the immune system (PGD_2), increase pain and inflammation (PGE_2) and control platelet aggregation (PHI_2 and TXA_2). [108]
5-LO metabolizes PUFA to 5-HpETE or **15-LO** to 15-HpETE, both are precursors to inflammatory leukotrienes and cytotoxic leukotoxins. These metabolites are important in exacerbating asthma by acting as powerful bronchoconstrictors, and they can also sustain inflammatory reactions through chemotaxis of inflammatory mediators. LO metabolism of omega-3 fatty acids result in pro-resolving lipid mediators called Resolvins, Protectins and Maresins. [139]
CYP450 metabolizes PUFA to anti-inflammatory epoxy fatty acids (EpFAs), n-terminal hydroxylated n-HETE, ω-1 oxidation, or allylic hydroxylations. 20-HETE regulates blood pressure by acting as a potent vasoconstrictor in kidneys and preventing sodium reabsorption in nephrons [140, 141]. The biological significance of mid-chain hydoxylations is less understood; however, biological significance has been observed with 12-HETE in corneal inflammation and neovascularization [142, 143]. Formation of EpFA, particularly if induced, is primarily accomplished by CYP2C and CYP2J subfamily; however, other CYP enzymes can generate EETs [144]. The EpFA act as homeostatic regulators to other metabolites in this pathway by stabilizing mitochondria, reducing ROS, decreasing ER-stress [20] and inflammation [15], regulating the vascular endothelium [88] and increasing bronchodilation [145]

however, with the proper precursors, PUFAs can be altered and interconverted. Depletion of either omega-3 or −6 PUFA result in serious side effects such as neuronal and vision impairment, skin anomalies, thrombocytopenia and intellectual disability [3]. PUFA in the body are either circulating as free fatty acids or incorporated into glycerides and cellular membranes. Upon injury or stress, PUFA are released from cell membranes by phospholipase acetyltransferases (PLA) and other enzymes. PLA2 liberates PUFA from the triglycerides to release free fatty acid (FFA), but also PLC and PLD further act to increase FFA in circulation (Fig. 5.2). Once freed from the plasma membrane, FFA are rapidly metabolized by β-oxidation or other enzymatic metabolism [4] (Fig. 5.1). As mentioned above, the three main enzymes responsible for non-catabolic metabolism of PUFA are COX, LO and CYP450. The COX and LO enzymes form primarily inflammatory mediators, and therapeutic interventions focus on blocking the formation of

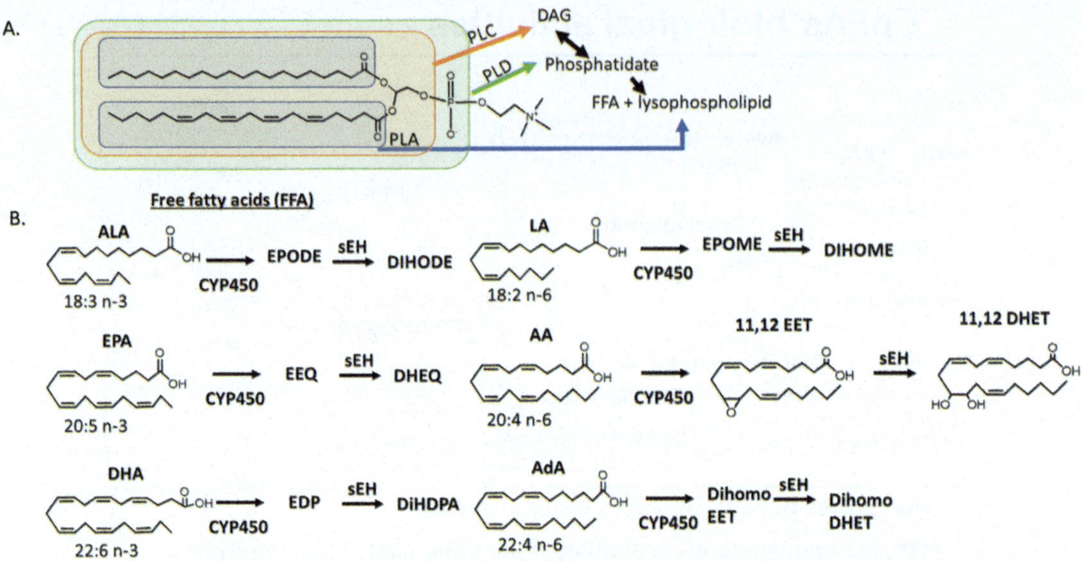

Fig. 5.2 Formation of EpFA
Polyunsaturated fatty acids (PUFAs) differ in both structure and function based on number of carbons and location of double bonds. They are incorporated as glycerides in fat cells, cellular membranes or circulating micelles and are liberated to free fatty acids by different phospholipases (PL) that act upon different areas of the glyceride or phospholipid (A). EpFA are formed through the oxygenation of FFA by CYP450. The metabolism of Arachidonic Acid (AA, 20:4 n-6) by cytochrome P450 yields EpFAs, epoxyeicosatrienoic acids (EET), which are further degraded by the soluble epoxide hydrolase into dihydroxyeicosatrienoic acids (DHET). The epoxide and diol on the 11,12 position is shown, but similar regioisomers are possible on all the double bonds in the PUFA. The n-6 fatty acid linoleic acid, LA, has been attributed to largely inflammatory epoxides, EPOMES; however, recent studies show that they are only toxic in the presence of sEH, suggesting that the diols of LA, DIHOMES, are responsible for this inflammatory action [137]. The 18:3 omega-3 fatty acid, linolenic acid (ALA), does not seem to have this same inflammatory action. The omega-6 fatty Adrenic Acid, AdA, is named for its abundance in the adrenal gland. Less is known about this PUFA and its metabolites, although the AdA EpFA, dihomoEETs are thought to regulate blood flow to the adrenal gland [138]. n-3 fatty acids alpha eicosapentaenoic acid (EPA) and docosahexaenoic acid (DHA) form EpFA epoxyeicosatetraenoic acids (EEQs) and epoxydocosapentaenoic acids (EDPs) respectively. Omega-3 EpFA are largely beneficial with anti-inflammatory and pain resolving properties in vitro and in vivo. Little biological activity has been associated with the diols of Ada (dihomoDHET) or the diols of EPA and DHA, dihydroxyeicosatetraenoic acids (DHEQs) and dihydroxydocosapentaenoic acids (DiHDPAs) respectively. (see [86] for review)

these compounds. In contrast, the CYP450 metabolism results in the formation of both hypertensive and inflammatory hydroxylated compounds, as well as and anti-hypertensive and anti-inflammatory epoxide compounds. Therapeutic interventions discussed here focus on increasing the concentration of the anti-inflammatory epoxide metabolites.

5.2 Epoxy Fatty Acids Are Therapeutic Targets for Disease

5.2.1 Epoxy Fatty Acids (EpFA) Are Important Signaling Molecules That Are Regulated by Their Metabolism

EpFA are formed from the activity of CYP450, a large class of metabolizing enzymes that oxidize fatty acids as well as xenobiotics using heme as a co-factor. The addition of molecular oxygen results in the formation of compounds from the epoxidation of double bounds, end terminal hydroxylation or allylic oxidation [5] each with unique biologies and roles in disease. EpFA have positive beneficial effects on maintaining endothelium function, inflammation and cellular oxidative stress. As early as 1986 and continuing through present day, CYP450 metabolites were identified as influencing blood pressure and renal function by regulating vascular smooth muscle proliferation through MAP kinase signaling pathways [6], anti-aggregation of platelets by decreasing leukocyte adhesion to endothelial cells [7], and regulating vascular tone. Vascular tone, in part, is regulated by the balance of hydroxylated and epoxygenated CYP450 PUFA metabolites. The 20-hydroxy-fatty acids (20-HETE) act as vasoconstrictors by hyperpolarizing vascular smooth muscle cells through activation of the protein kinase C pathway, while the epoxy fatty acids hyperpolarize vascular endothelium through activation of voltage-gated potassium (BK) channels. In this way, the two metabolites act as homeostatic regulators to prevent pathological changes in vascular tone [8, 9]. Although increased 20-HETE results in endothelium dysfunction and increased hypertension, increased EpFA decrease blood pressure, but has not been shown to cause hypotension [10, 11]. This provides strong evidence that EpFA are homeostatic regulators of endothelium function [12, 13]. EpFA also show other beneficial biologies, including potent anti-inflammatory effects and decreasing the endoplasmic-reticulum (ER) stress response pathway [14]. EpFA act as anti-inflammatory agents by reducing the nuclear translocation of NF-kappaβ, thereby preventing the transcription of a number of inflammatory cytokines and synthesis of inflammatory eicosanoids [15, 16]. Furthermore, the diol metabolite of EpFA metabolism drives monocyte chemotaxis in response to monocyte chemoattractant protein, MCP-1. Thus preventing the metabolism of EpFA by the sEH further regulates inflammation by decreasing monocyte infiltration [17]. EpFA have demonstrated a wide range of beneficial effects in animal models of disease [18] which is not fully explained by anti-inflammatory or endothelial homeostasis, but possibly explained by the regulation of ER-stress by EpFA. ER-stress is a protective mechanism deployed by the cell to overcome cellular stress and prevent deleterious mutations. The ER-stress pathway is a homeostatic mechanism regulating cellular responses to the presence of increasing misfolded protein resulting from oxidative, physiological or pathological stressors. Although a fundamental mechanism in maintaining homeostasis, this response is often upregulated in disease, and if not controlled, will result in increased inflammation or activation of cell death pathways. Preventing or reducing the inflammatory and apoptotic branches of ER-stress is associated with many beneficial disease treatments [19], and EpFA have demonstrated the ability to reduce ER-stress [20]. These ubiquitous mechanisms (inflammation, endothelial function, and ER-stress) underlie many diseases and are regulated by EpFA, which may explain why preclinical models investigating the role of EpFA indicate beneficial results in many different diseases when these lipid mediators are maintained.

In normal cell systems, and increasingly in disease, EpFA are rapidly removed from circulation either because they are re-incorporated back in cellular membranes, metabolized by beta-oxidation, or further degraded by epoxide hydrolases [21]. Metabolism enzymes generally increase polarity of compounds to aid in their elimination from the body, and the metabolism of PUFA are no different. After metabolism by the CYP450, the EpFA are further degraded by epoxide hydrolases into the corresponding polar vicinal diols which diffuse from the cells or are rapidly conjugated and removed from the body. In many diseases, the sEH activity is upregulated compared to healthy controls, thus inhibiting its activity in disease settings indicate a potential target for increasing EpFA as a potential therapy.

The epoxide hydrolases are catalytically active dimers that convert xenobiotic or PUFA epoxides to corresponding diols through an exothermic, 2-step hydrolysis reaction (Fig. 5.3) [22]. There are four structurally related isozymes in the epoxide hydrolase family of enzymes

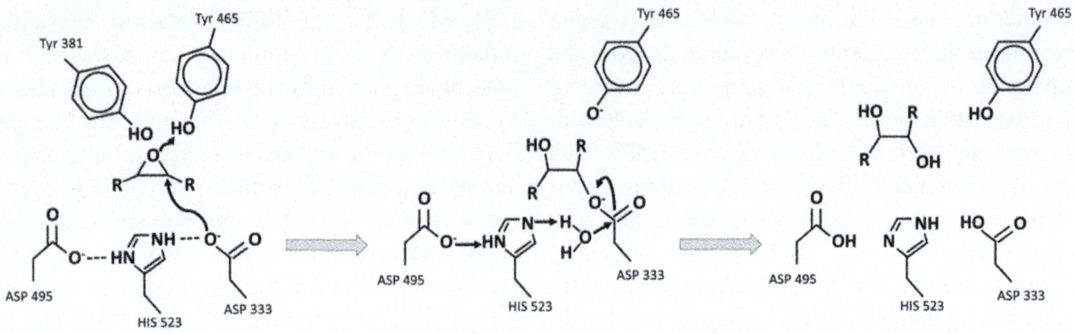

Fig. 5.3 Mechanism of epoxide to diol metabolism by the soluble epoxide hydrolase (sEH)
Two acidic tyrosines are oriented by pi-stacking to bind to and polarize the epoxide moiety. sEH inhibitors functionally mimic transient intermediates and the transition state of the enzyme (middle panel). The NH groups of the urea, amide, carbamate or other electronegative groups possibly encourage a salt bridge formation, for example between a polarized urea and the catalytic aspartic acid, likely stabilized by hydrogen bonds with the tyrosine. [146]

(EPHX1–4) in mammals, in addition to other enzymes that have hydrolase activity such as the leukotriene A_4 hydrolase (LTA_4H), cholesterol hydrolase, and Peg1/MEST. LTA_4H metabolizes leukotriene A_4 into the leukocyte recruiter, leukotriene B4, to help in the resolution of inflammation. While the functions of PEG/MEST and EPHX4 are not known, isozymes EPHX 1–3 are capable of metabolizing EpFA and differ in cellular location and substrate preference. EPHX1 (mEH), is named for its cellular location: it is found in microsomal membrane fractions in tissue and is active on polyaromatic hydrocarbons found in a variety of xenobiotic epoxides and is also capable of metabolizing EpFA. Conversely, EPHX2 is similarly named as the soluble epoxide hydrolase (sEH) after its predominant location in the cytosolic and peroxisomal fractions of the cell. The EH activity is located at the C-terminus portion of the protein, and the N-terminal portion of sEH has phosphatase activity while the N-terminus of mEH is anchored to the membrane. While mEH is capable of metabolizing both aromatic and aliphatic epoxides, catalytic turnover by the sEH is greater for mono and disubstituted epoxides, making EpFA a good substrate for this enzyme. However, trans, di, tri and even tetra-substituted epoxides can be turned over sometimes with a low Km [23]. Cholesterol epoxides and the squalene epoxide precursor to lanosterol are also substrates of mEH, although the cholesterol epoxide hydrolase and lanosterol synthase are the primary routes of metabolism for these compounds [24]. EPHX3 is also capable of hydrolyzing EpFA with similar efficiency (Kcat/Km) as that of sEH, but a > 10x higher Km indicates that EpFA of arachidonic acid (AA) and presumably other PUFA are a weak substrate for EPHX3 compared to their affinity for sEH or mEH [25].

5.2.2 Human Polymorphisms Altering Epoxide Hydrolase Activity Affect EpFA Concentrations and Correlate Epoxides with Disease States

Most EH related mutations in humans that are associated with disease states are associated with EPHX2 (sEH). There are four main mutations that affect enzyme activity, two that do not alter function, and as many as 20 less characterized non-coding mutations that can occur in up to 20% in human populations [26, 27]. The four most common coding mutations result in two gain of function mutations (K55R and C154Y) and two reduced function mutations (R287Q and R103C mutations results in decreased dimerization). These mutations affect the velocity of epoxide hydrolysis while leaving the selectivity for substrate binding and phosphatase activity apparently unchanged [28, 29] Considering that the concentration of the sEH enzyme (ranging from 3 nM in lung to 400 nM in liver) is often higher than that of the EpFA substrate (low nM in tissue [30–32]), and the concentration of EH enzymes, such as EPHX1 and 3, can convert epoxides to diols, it should be taken into consideration that mutations that only slightly affect enzyme efficiency may not significantly affect EpFA concentration. However, mutations in the sEH and prevalence of SNPs associated with different diseases suggest that this enzyme plays an important role in disease, as described below.

Mutations in sEH Are Biomarkers of Anorexia Nervosa In a genome wide association study of patients with anorexia nervosa (AN), investigators identified rare mutations in non-coding regions of the EPHX2 that correlate with disease susceptibility. The activity of the mutations was not determined, but investigations in the levels of epoxide:diol ratios found they were decreased in ill AN patients compared to recovered AN or healthy populations, suggesting an upregulation in the conversion of epoxides to diols. Furthermore, the loss of function mutation, R287Q, occurred less frequently in AN patients compared to healthy populations, further correlating increased sEH activity to higher risk of severe AN [33–36]. Although the exact mechanism for the involvement of EpFA in AN is unknown, considering the impacts of diet and aversion to food in this disease, understanding the role of altered lipid signaling in this disease is of increasing interest.

Reduced sEH Activity Protects Against Familial Hypercholesterolemia Familial hypercholesterolemia (FH) is a hereditary disease that causes increased plasma cholesterol concentrations resulting from a defective hepatic low-density lipoprotein receptor (LDLR). In a targeted analysis approach that studied the prevalence of the reduced function sEH mutation, R287Q, in 8 generations of families with FH, Sato et al. discovered that family members with the R287Q mutation had normal cholesterol levels compared to members with the normal 287R allele. Cholesterol was unchanged in healthy humans with the R287Q mutation [37]. Preclinical investigations further support the role that sEH has in reverse cholesterol transport, and possibly explain the protective effects on inhibiting its activity in LDLR related disease. For example, in $Ldlr^{-/-}$ mice that mimic the FH disease, treatment with the sEH inhibitor (sEHI) *t*-AUCB decreased atherosclerosis plaques through increasing HDL synthesis and efflux of cholesterol from adipose lesions compared to vehicle treated controls. Further investigation demonstrated that sEHI treated mice had increased ATP binding cassette transporter A1, which is responsible for HDL synthesis through efflux of cellular cholesterol to extracellular apoA1 [38].

sEH Mutations Are Associated with Outcome Measures in Vascular Disease Impaired vascular function contributes to heart disease, age-related vascular decline, blood pressure and stroke. EpFA act as vasodilators by opening calcium-activated potassium channels that relax the vascular smooth muscle and are thought to mitigate vascular disease [39]; therefore, it is hypothesized that decreased EpFA would contribute to heart disease, and in fact gain-of-function sEH polymorphisms are associated with increased cardiovascular diseases. For example, pregnant women with preeclampsia had higher frequency of the K55R gain of function mutation and lower methylation of the EPHX2 promotor region, causing higher expression, than healthy pregnant women [40]. In another study that analyzed sEH polymorphisms with frequency of stroke in African American and Caucasian populations participating in the Atherosclerosis Risk in Communities study, investigators identified a rare EPHX2 mutation in African Americans that was not found in Caucasian populations. The mutation resulted in increased sEH activity and a twofold to threefold increased risk of stroke. Although the infrequency of the mutation complicated statistical analysis, and larger sample sizes were needed to determine significance. Additional haplotypes in the sEH gene correlated with both increased and decreased risk of stroke in both races [26]. Although the effect of these gene sequences on enzyme activity is not known, further studies analyzed site-directed mutagenesis on survival rates of ischemic cells *in vivo* and found that decreased sEH activity with the R287Q mutant conferred to increased survival after inducing a simulated stroke environment of depleted glucose and oxygen [41]. This mutation was also associated with lower risk of stroke in Europeans [42]. Although a large Danish study failed to identify correlations in sEH SNPs and stroke, myocardial infarction or ischemic heart disease [43]. A large study in Swedish men found correlations between the gain-of-function mutation, K55R, with increased risk of stroke [44]. Additionally, Hawaiian Asians with dementia but without prior ischemic injury compared to healthy age-matched controls had increased 14,15 DHET in cortical brain tissue from patients with dementia; however, heterozygous carriers of the reduced function R287Q SNP had increased markers for plaques compared to healthy patients [45]. These data are at odds for understanding if sEH activity protects or contributes to the progress of AD especially considering that dementia correlates with a decreased EET:DHET ratio, suggesting a negative role of sEH in this disease. However, because the R287Q mutation affects dimerization, it's possible that these heterozygous carriers still have functional sEH protein.

In addition to the sEH, the mEH also hydrolyses EpFA into corresponding diols, although with much less efficiency. As discussed later, the association of the mEH in the lipophilic endoplasmic reticulum membrane and with the cytochrome P450 that oxidize epoxy fatty acids may result in greater mEH contribution to diol formation than anticipated from the concentration and kinetic constants of the mEH. Mutations in mEH are also associated with several diseases, including cancer, preeclampsia, seizure, neurological disease, drug-dependence and COPD. However, because this enzyme is primarily responsible for metabolizing aromatic xenobiotic epoxides, it is uncertain if these associations are a result of decreased epoxides in the body, or from accumulation of toxic xenobiotics (reviewed in [46]). Interestingly, sEH expression in the brain is mostly localized in glial cells of the brain, while mEH is found more throughout the brain, suggesting either a specific role of sEH or more significant contribution of mEH to EpFA metabolism in the brain [47]. Recent studies demonstrate that sEH activity outside the brain can influence depression. In this study, overexpression of sEH in the liver resulted in increased depression in mice, and genetic deletion of sEH in the liver had a positive effect in treating stress-induced depression. This study suggests that peripheral-acting EpFA can influence CNS diseases [48].

5.2.3 Laboratory Knockout Models Identify Beneficial Effects of Increasing EpFA in Treating Preclinical Models of Disease

A homozygous mouse model deleting EPHX2 stabilize EpFA and further demonstrate biological activity of increasing EpFA in disease. The first incidence using sEH−/− mice observed decreased blood pressure on a high-salt diet compared to wild-type mice [49], and $sEH^{-/-}$ mice are used to further demonstrate that increasing EpFA and decreasing diol formation resolves multiple disease states. For example, sEH null mice demonstrate beneficial effects in decreasing inflammation, maintaining the vascular endothelium, and resolving neuroinflammatory diseases. Specifically, genetic sEH −/− mice demonstrated accelerated wound healing, reduced inflammation in inflammatory bowel disease and LPS-induced inflammatory models, improved insulin signaling in a Type II model of diabetes, reduced cisplatin-induced kidney damage, reduced niacin flushing, reduced arteriosclerosis, smaller infarct size in cerebral artery occlusion/reperfusion injury, decreased hepatic and arterial fibrotic disease, decreased autism, depression, and Parkinson's disease (see [50, 51] for review). Recently, tissue specific sEH knockout models demonstrate the local effect that EpFA have on tissues. Mice with podocyte specific sEH knockout were protected from hyperglycemia-induced renal injury resulting from high fat diet or STZ-induced diabetic hyperglycemia [52]. In further support of the importance of this pathway in disease, the analgesic effects of morphine were attenuated in CYP-null mice that lack the ability to generate EpFA, and in mice administered compounds that inhibit fatty acid oxidizing CYPs [53–55]. Thus genetic models that increase EpFA concentrations through deletion of their metabolism support the beneficial effects of EpFA in treating disease, and models that decrease the ability of animals to form EpFA reduce these beneficial effects, suggesting an essential role of EpFA in disease.

Table 5.1 Altered ratios of EpFA: diol correlate to disease outcomes

Alzheimers disease (AD) [130]
AD patients with and without type 2 diabetes (T2D) had increased DHET compared to healthy controls. However, there were no effects following adjustments for multiple comparisons.
Arthritis [131]
In synovial fluid of arthritic vs. normal joints, 11,12-DHET and 14,15-DHET were higher in affected joints of people with unilateral osteoarthritis. In addition, these and 8,9-DHET were associated with worse progression over 3.3 years.
Anorexia Nervosa [34, 35]
Ill anorexia nervosa patients have higher DHA diol metabolites 19,20 DiHDPE:EpDPE compared to either recovered AN patients or healthy human subjects, while both ill and recovered AN patients have higher ALA diol metabolites 15,16 DiHODE:EpODE ratios compared to healthy subjects.
Peripheral arterial disease [132]
Increased 8,9 DHET correlated with increased risk of coronary and cerebrovascular events in patients with peripheral arterial disease.
Coronary artery disease [133]
Decreased EETs in patients with obstructive coronary artery disease compared to healthy controls
Depression [134]
In patients with major seasonal depression syndrome, sEH-derived oxylipins (12,13 DiHOME, 7,8- and 19,20 DiHDPE), in addition to other eicosanoids, increased in winter compared to summer-fall, while 14,15 EET and corresponding diol both decreased in the winter.
Preeclampsia [135]
In preeclamptic women 14,15-DHET was higher in urine samples compared to healthy pregnant women.
Vascular dementia [136]
In patients with cognitive impairment, an increase in 9,10- and 12,13 DiHOME: EpOME was associated with poor performance in function but not memory.

5.3 Clinical Approaches for Increasing EpFA Concentrations

The EpFA to diol ratio are reduced in several disease sates suggesting that increasing the EpFA concentrations in the body could provide beneficial effects in preventing or treating disease (Table 5.1). There are many different approaches for increasing concentrations of fatty acid epoxides for the treatment of disease. Strategies include direct administration of PUFA or EpFA, inhibition of EpFA metabolism through chemical inhibitors of sEH (sEHI), induction of EpFA formation through CYP modulators, or directly mimicking the fatty acid epoxide.

5.3.1 PUFA Supplementation

Omega-3 supplementation is a widely investigated approach for improving health or treating disease, and the resulting fatty acid epoxides are thought to account for some of the beneficial effects observed with their supplementation [31]. However, randomized clinical trials often fail to support the efficacy observed in meta-analysis of diet behaviors and disease risk [56]. Increased sEH activity in disease would increase EpFA metabolism, thus limiting efficacy and may account for the inconsistent therapeutic benefits reported with omega-3 supplementation. This is possibly due to lipid peroxidation products in some omega-3 lipids. Omega-3 lipids are inherently unstable and subject to oxidation, often as much as 200% increase in PUFA peroxides after only 22 days of storage [57], and further oxidation results in aldehydes that are not tested for in commercial settings [58]. The oxidized products are associated with cellular stress and increased cellular toxicities and are hard to detect or control outside of sophisticated analytical labs and thus not a well-controlled approach for improving health or treating disease [59]. Similarly, supplementation with EpFA is also subject to oxidation and provides further challenges because EpFA are also rapidly degraded by acid hydrolysis in the stomach or are eliminated through first-pass metabolism in the liver. An alternative administration route, for example direct application to affected tissue, would avoid first-pass metabolism; however, other metabolizing enzymes would still contribute to rapid elimination. For example, enteric coating to bypass the acid in the gut may provide options for treatment of intestinal diseases; however, metabolizing enzymes in the gut or microbiota containing epoxide hydrolases would further limit their effectiveness. Ocular therapy provides another application to avoid first-pass metabolism; however, recent studies have implicated increased sEH expression as contributing to disease progression in a mouse model of diabetic retinopathy as well as in samples from humans with diabetic retinopathy [60]. Thus, ocular application to treat retinopathy would be susceptible to sEH metabolism possibly demonstrating that disease treatment by direct EpFA supplementation will be difficult due to sample stability and metabolic instability.

5.3.2 Increase Formation of EpFA

Clinical and preclinical strategies for investigating biological activity of EpFA have focused on preventing their metabolism either through inhibition of the sEH enzyme or through synthesizing stable EpFA mimics. A third approach involves increasing their formation through CYP450 activity. Preclinical data indirectly suggests that increased CYP activity could reduce certain pathologies. In a mouse model of inflammatory pain, co-administration of a CYP modulator, omeprazole, and the sEHI, TPPU, resulted in increased efficacy compared to either compound administered alone. Omeprazole induces CYA1A1, 1A2, 2B1 and 3A1 and inhibits 2D2 and 2C. CYP2C is a more efficient at forming epoxides of AA (12/8 ratio of epoxidation/hydroxylation) compared to the other CYP isoforms which form hydroxylated metabolites in greater amounts than EpFA; however, because the other isoforms also form EpFA, increasing their activity would also increase EpFA as well as hydroxylated PUFA metabolites [61]. This is consistent with the results published in this study showing that the EpFA and hydroxylated metabolites were increased in the plasma of mice treated with both compounds [62]. Hydroxylated metabolites, especially hydroxylated metabolites of 18:2 linoleic acid (LA), are often associated with increased inflammation and oxidative stress [63–66], which brings up another complication with targeting CYP induction to increase EpFA in that it could also increase the production of inflammatory PUFA metabolites that could negate the beneficial effects of EpFA. Preclinical studies demonstrate that administration of ketoconazole, an inhibitor of CYP2A6, 2C19 and 3A4, reduces pain in an inflammatory mouse model by reducing the formation of hydroxy octadecadienoic acid (HODE), the monohydroxy metabolite of LA that causes hyperalgesia through activation of TRPV1 channels [61, 67]. These studies demonstrate that increasing CYP activity is a complicated approach given the promiscuity of the CYP enzymes for forming both anti-inflammatory EpFA and pre-inflammatory hydroxylated PUFA metabolites, as well as influencing the metabolism of other xenobiotics. Thus, increasing CYP activity does not present a viable strategy for clinical utility, and clinical approaches designed to increase EpFA should take into consideration drug-drug interaction potentially affecting EpFA formation if patients are currently taking CYP modulators.

5.3.3 Stabilizing EpFA Through Small Molecule Inhibitors of SEH

To overcome metabolic instability of EpFA, sEH inhibitors (sEHI) were identified that mimic the transition state of the epoxide ring opening. The pharmacophore of sEHI consist of a urea, carbamate, or amine heterocycle that mimic the interaction of the substrate and enzyme by forming one or more hydrogen bonds between the NH and aspartate (Fig. 5.3.) The inhibitors contain hydrophobic regions that allow interaction with the enzyme and maintain high specificity over the mEH, while specifically placed polar side groups increase druggable properties such as water solubility and metabolic stability and increased oral bioavailability. The first generation of sEH inhibitor, 12-(3-adamantan-1-yl-ureido)-dodecanoic acid (AUDA) [68], was a small molecule designed to mimic to 14,15 EET with a dodecanoic acid to mimic the aliphatic chain and carboxylic acid present at the α end of the fatty acid, an adamantine to mimic the *ω* hydrophobic end of the fatty acid, and urea to act as a mimic to the epoxide and also inhibitor to the sEH enzyme [69]. AUDA is a potent sEHI, with an IC_{50} of 3–60 nM, but is short lived in the body ($T_{1/2} = 2.9$ h in canine PK) due to oxidation of the adamantine and beta oxidation of the alkyl chain, thus limiting its use *in vivo* to conditions where long half-life is not required. Since AUDA also mimics epoxy fatty acids, it shows biological activity in tissues lacking sEH. Such dual action may improve efficacy, but it complicates interpretation when AUDA is used to investigate physiology [70]. Second generation sEHI replaced the adamantine with a more stable benzyl-trifluoromethoxy and alkyl chain with a benzenesulfonamide to maintain potency with the sEH enzyme while improving solubility and PK stability. ([71] for review). The most potent inhibitors of the sEH are reversible and tight binding with a slow off-rate from the enzyme and sub-nanomolar potency. They also show improvements in PK for increased chances of testing clinical relevance for treating disease. Kinetic studies demonstrate optimization of new inhibitors by targeting drug occupancy time on the enzyme and demonstrate that inhibitors remain bound to the enzyme long after they are detectable by classical PK techniques. These data correlate with in vitro half-life measurements and provide novel techniques for selecting potent lead compounds for clinical development [72].

5.3.4 Mimics of EpFA

As an alternative to blocking the metabolism of EpFA in order to increase their concentrations, another strategy focuses on creating more stable mimics of EpFA while retaining the active moiety of the original compound. Epoxy fatty acid mimics investigated in clinical trials often replace the epoxide with functional groups that retain beneficial effects *in vivo* while also blocking β-oxidation by adding functional groups to the α-hydroxy portion of the fatty acid. Mimics of epoxy fatty acids remove the complexity of enzyme potency and off-rate kinetic parameters of small molecule sEHI that can complicate translation from the lab bench to clinical efficacy, but due to the unknown target for EpFA activity and complex structure diversity of EpFA, selecting one isomer for development could be overly simplistic. Many of the mimics were based on earlier studies in agricultural chemistry where various groups resistant to epoxide hydration were used to replace the epoxide of natural juvenile hormone while presumably retaining efficacy at the putative receptor [73].

5.4 Clinical Development of Compounds That Alter EpFA Concentrations

One of the first identified biological targets of EpFA action was modulation of the microvasculature and inflammation which underlies many pathological disease states. Clinical targets testing efficacy of increasing EpFA include hypertension, cardiovascular pulmonary disease, stroke, diabetes and pain. Preclinical studies provide evidence that altering concentrations of regulatory lipid mediators formed from metabolism of PUFA could provide beneficial effects in many other diseases, including neurological disease and bone degeneration. Despite the unknown target, multiple clinical trials investigating the therapeutic potential of both small molecule inhibitors of the sEH as well as EpFA mimics have been initiated. These trials are highlighted in Table 5.2 and the mechanisms are explained in detail below along with the benefits and liabilities of these approaches.

Table 5.2 Clinical development candidates of small molecule sEHI and EpFA mimics

Small molecule sEH inhibitors		Clinical Trials
AUDA	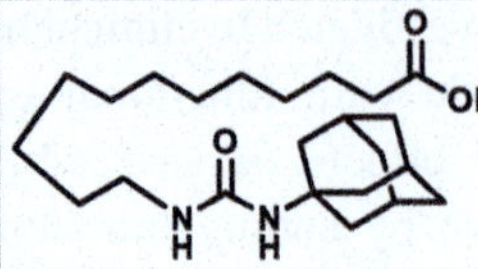	*Clinical trial NCT00654966: Microvessel Tone in patients with heart failure.* Status: Complete. Healthy humans and patients with heart failure challenged with topical urotensin II, a potent vasoconstrictor, and treated with sEHI [74].
TPPU	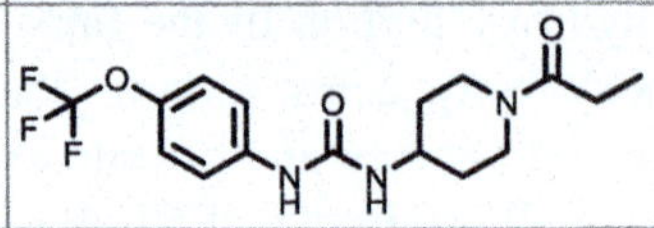	This compound has not been investigated in clinical trials but is the most frequently used compound in preclinical research.
AR9281	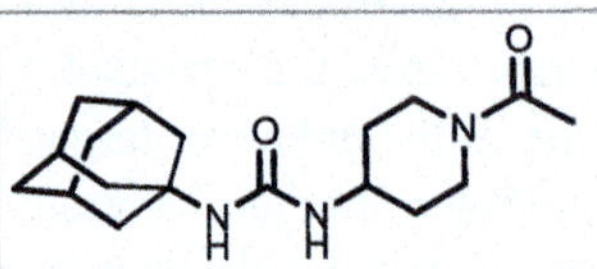	*Clinical trial NCT00847899: Evaluation of sEHI in Patients with Hypertension and Impaired Glucose Tolerance.* Status: Complete. Single and multiple day testing up to 8 days at doses up to 1.2 g/day (400 mg every 8 h) were well tolerated [129]. Data from Phase II clinical trials not published.
GSK2256294	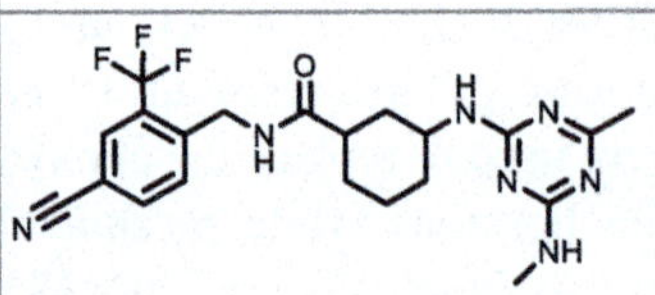	*Clinical trial NCT01762774: A Study to Assess the Safety, Tolerability, PK and PD of Single and Repeat Doses of GSK2256294 in Healthy Volunteers and Adulate Male Moderately Obese Smokers.* Status: Complete. Doses were well tolerated and attenuated smoking related endothelial dysfunction.
EC1728 (*t*-TUCB)	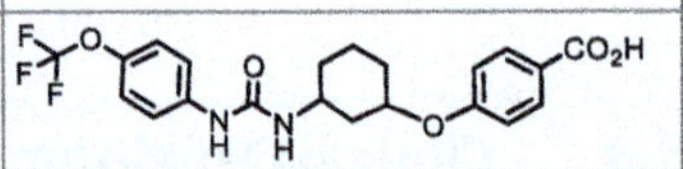	Clinical trials are in development for companion animals.
EC5026	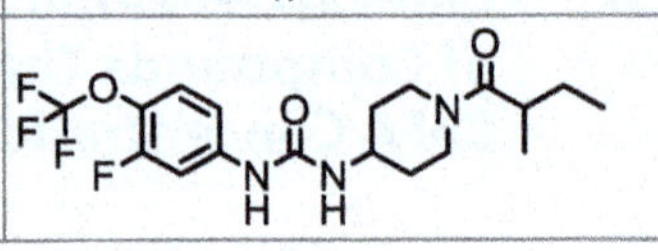	*Clinical trial NCT04228302: Safety, Tolerability, and Pharmacokinetics of Oral EC5026 in Healthy Subjects.* Status: Enrolling
EpFA Mimics		**Summary of results**
CMX020	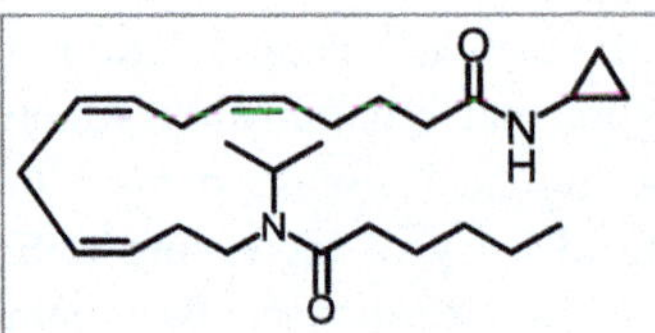	*Clinical trial ACTRN12615000885594: A Study to Evaluate the Safety and Analgesic Efficacy of Oral CMX-020 in Subjects with Symptoms of Sciatica Resulting from Lumbosacral Radiculopathy.* *ACTRN12616001435471: A Phase 2 Study to Assess the Efficacy and Safety of CMX-020 in Treating Osteoarthritis.* Status: Phase 1 studies complete. Enrolling for Phase 2.
OMT-28	Structure not disclosed	*Clinical trial NCT03906799: Study on OMT-28 in Maintenance of Sinus Rhythm in Patients with Persistent Atrial Fibrillation.* Status: Enrolling
Icosabutate	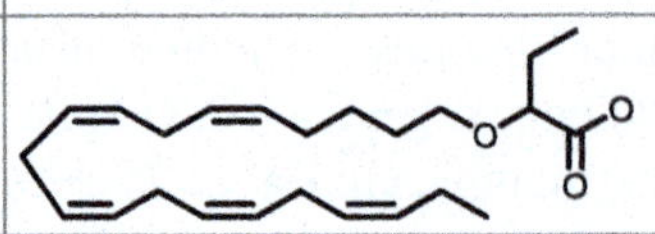	*Clinical trial NCT04052516: A Phase 2b Study of Icosabutate in Fatty Liver Disease.* Status: Phase 1 studies complete. Enrolling for Phase 2b.
Vascepa (ethyl-EPA)	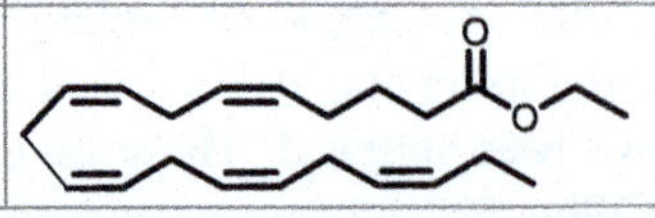	Status: approved for the treatment of hypertriglyceridemia.

5.4.1 Small Molecule sEHI in Clinical Development

The clinical trials registry lists six clinical trials completed with small molecule soluble epoxide hydrolase inhibitors tested in diseases mostly affected by endothelial dysfunction. The following sections will highlight the small molecule sEHI investigated in these clinical strategies to increase EpFA as a mechanism for treating disease.

AUDA Based on the observation that EpFA open potassium channels and hyperpolarize vascular endothelium resulting in vasodilation [39], investigators in Australia conducted clinical trial NCT00654966 as an exploratory study in humans to determine if AUDA increases EpFA to protect against heart failure by increasing blood flow in the microvasculature. After topical challenge with a potent vasoactive peptide, urotensin II (UII), blood flow was measured in healthy humans or patients with heart failure (HF) treated with and without topical AUDA. UII causes vasodilation and increased blood flow in healthy humans but vasoconstriction and reduced blood flow in patients with HF. AUDA alone caused increase blood flow when administered at the intermediate dose of 0.1 μM in both healthy and HF patient populations. When administered with UII, AUDA was able to reverse the reduced blood flow observed in HF patients, although not back to levels observed in healthy subjects, and significantly increased blood flow in healthy subjects more than when UII was administered alone [74]. The increased vasodilation observed from this study indicate many potential benefits in addition to heart failure that increasing EpFA would have in patients; for example, diabetics and hypertensive patients would benefit from increased vasodilation; however, poor PK probably prevented this molecule from being a viable drug candidate.

AR9281 (UC1153) AR9281 offered some advantages as a clinical candidate including low IC_{50} on the rodent sEH, easy synthesis, a surprisingly high water solubility, high selectivity for the sEH, and low mammalian toxicity; however, its poor target occupancy and the speed with which it was metabolized to synthetically complex metabolites were clear liabilities. In preclinical models, AR9281 reduced hypertension and renal injury in an angiotensin induced model of hypertension in rats [75]. Multiple clinical trials were launched by Arete Therapeutics to test the v and efficacy of AR9281 for treating hypertension. Published results from the Phase 1 study found that AR9281 was well tolerated at doses up to 1000 g/day but was rapidly cleared from the body, with a half-life of only 3–5 h in humans. However, high plasma concentrations well above the IC_{50}, and *ex vivo* assays monitoring sEH activity showing 90% inhibition of the enzyme up to 24 h at the top dose, justified advancing the compound to Phase 2 human efficacy studies. The results of the Phase 2 efficacy studies in hypertensive patients with glucose intolerance were not published, and Arete Therapeutics closed shortly after completing the study, so one assumes that the compound failed to show efficacy. One possible speculation for lack of efficacy was the high metabolic clearance of the compound. For example, the *ex vivo* determination of the IC_{90} was based on a 60 min assay; however, AR9281 is a reversible inhibitor that has a kinetic $T_{1/2}$ on the enzyme of 6 min [76]. Thus, the ex vivo data may not be representative of the in vivo environment where the inhibitor is rapidly metabolized and cleared from the body. Additional measurements of diols in the urine of subjects were combined from all treatments and compared to pretreatment levels and placebo treated subjects as a marker of target engagement. The diols of human subjects treated with AR9281 modestly decreased compared to pretreated controls, while placebo treated subjects remained unchanged. Epoxide concentrations were not detectable in the urine; thus, the ratio of epoxide:diol change was not monitored and thus impossible to determine if the decrease in diol levels was due to treatment effect or other independent variables. Overall, the data from the Phase 1 trial did not demonstrate convincing data that AR9281 effectively inhibited sEH activity, possibly explaining the lack of efficacy in phase 2 trials. There are situations where rapid clearance such as that demonstrated with adamantane containing compounds like AR9281 and AUDA offer clinical advantages; however, these have not been explored.

GSK225629A4 GSK225629A4 is a potent sEHI that attenuated leukocyte infiltration in mice after exposure to cigarette smoke [77]. Based on this information, clinical trials were initiated to test safety and endothelial dysfunction in both healthy humans and obese smokers at doses of 2–20 mg administered as a single or repeat oral dose for 14-days. Endpoints included safety and endothelial dysfunction measured by forearm venous occlusion plethysmography after intra-articular challenge with vasodilator, bradykinin. Overall, GSK225629A4 was well tolerated with favorable PK and $T_{1/2}$ equal to 19–30 h. in healthy humans and 41–49 h in overweight smokers. One significant adverse event was reported, but ultimately considered non-drug related as the subject had a history of this event prior to the clinical trial. Otherwise, only mild-moderate adverse events of headache and contact dermatitis were reported. Headache occurred with similar occurrence between the placebo and treated groups, and contact dermatitis around the site of ECG electrode placement was reported in nine healthy patients receiving active compound and none in the placebo or obese smokers receiving active or placebo compound. Ex-vivo sEH activity was measured in the plasma of treated patients as a function of EET hydrolysis after 30-min incubations with 14,15 EET, and investigators found that >80% inhibition of EET to diol hydrolysis was observed after repeat dosing in all dose cohorts. In obese smokers, vasodilation after bradykinin injection was reduced compared to healthy subjects, as expected. After administration of GSK225629A4 vasodilation improved in a dose and time dependent manner in obese smokers after 1 and 14-d administration compared to placebo control. Although some limitations need to be considered, such as small sample size and non-smoking obese control subjects. Furthermore, as an added marker of safety, the sponsor measured VEGF and plasma fibrinogen after single and repeat dosing in response to potential safety concerns from increased angiogenesis [78, 79] and tumor metastasis [80] in preclinical settings. Both plasma VEGF and fibrinogen were similar across all treatment groups potentially indicating that preclinical angiogenesis observations may not be clinically relevant. Overall, these data suggest that GSK225629A4 is a safe compound with potential efficacy in patients with COPD, a disease exacerbated by cigarette smoke, or other disease affected by dysfunction of the microvasculature.

Given the successful safety profile of GSK225629A4, independent investigators have secured rights to initiate clinical trials with the GSK compound to investigate insulin resistance and stroke as described below.

Insulin Resistance Preclinical data suggests that anti-inflammatory properties of EETs reduce inflammation in adipose tissue resulting in decreased insulin resistance in several, but not all, rodent models [81–83]. Based on this information, a group at Vanderbilt University in Nashville, Tennessee initiated a Phase 2 clinical trial to investigate the therapeutic potential of GSK225629A4 ability to improve insulin sensitivity in response to glucose infusion. The study is currently recruiting patients, and there are no reported data from the study.

Subarachnoid Hemorrhage (SAH) Numerous preclinical studies have investigated the role of soluble epoxide hydrolase in exacerbating stroke symptoms and the utility of inhibiting the sEH as a potential treatment option. (Reviewed in [84]) As a result of these data, investigators at the Oregon Health and Science University initiated a Phase 1 and 2 clinical trial with GSK225629A4 in patients with aneurysmal SAH to evaluate effects on length of hospital stay, incidence of new stroke, disposition upon discharge and outcome measures. The study recently finished recruiting patients, but no results have been published.

Neuropathic Pain As new mechanisms for explaining EpFA beneficial effects has been discovered, clinical activities are targeting diseases with complicated etiologies. For example, the newest clinical development programs that are scheduled to start at the end of 2019 focus on treating neuropathic pain, a disease attributed to pathological ER-stress. EC5026 is the newest sEHI being developed for clinical utility. Recent press releases announced FDA approval of an IND application from EicOsis to test their small molecule inhibitor, EC5026, for the treatment of neuropathic pain in humans, and EC1728 for treating inflammatory and neuropathic pain in companion animals. Many review papers have described the therapeutic potential of increasing epoxy fatty acids for the treatment of both neuropathic and inflammatory pain (reviewed in [18, 85, 86]). Considering the devastating impacts of the opioid epidemic and lack of effective and safe pain medications, there is considerable interest in the outcomes of these clinical trials.

5.4.2 EpFA Mimics in Clinical Development

CMX-020 is a mimic similar to AUDA that is also designed around 14,15 EET. This compound maintains most of the original structure except the olefin on the 14,15 carbon is replaced with a dimethylcarbamoyl and the α hydroxy group is replaced with a cyclopropyl amine to prevent beta-oxidation. In preclinical studies, CMX-020 undergoes rapid elimination, with most of the drug eliminated 1-h after administration; however, the compound exhibits potent pain-relieving properties preclinically in the acetic assay writhing assay and tail flick assay, comparable to that of morphine [87]. It is unknown if CMX-020 metabolism results in epoxy-fatty acid mimics as a result of CYP450 epoxidation, or if CMX-020 acts to inhibit the sEH, similar to carbamates such as GSK225629A4. Based on the potent analgesic effects of CMX-020 and despite the rapid elimination profile, multiple clinical studies were initiated to test the safety and efficacy of CMX-020 in treating painful conditions such as osteoarthritis and sciatic nerve pain in Australia. Results have not been published.

Focused mimics based on the structure of one regioisomer of one PUFA describe the challenges of developing a mimic to a fatty acid epoxide. While CMX-20 chose to mimic the omega-6 AA epoxides, other companies, described below, selected omega-3 fatty acids as the background for their mimics. The epoxides of omega-3 fatty acids are thought to account for some of the beneficial effects of omega-3 supplementation in the diet, and animal models show that they have more potent activity in *in vitro* and animal models than the omega-6 epoxides [31]. In addition to choosing which omega position of the fatty acids to model the mimic, further complications arise when identifying which regioisomer to target. For example, 11,12 and 14,15 EET are commonly associated with having vasodilator activity compared to the other regioisomers of arachidonic acid [88].

Omeicos is focused on developing another epoxy mimic by creating a transition state mimics of the 17,18 EPA omega-3 epoxide. Their lead compound, OMT-28, is actively recruiting patents for a Phase 2 clinical trial to treat atrial fibrillation. Although the exact structure is not disclosed, detailed structure activity relationship studies identified important characteristics needed to exert antiarrhythmic effects. Atrial fibrillation is a type of arrhythmia initiating from the top chambers of the heart and can lead to increased risk of blood clot, stroke, and heart failure. *In vitro* models using neonatal rat cardiomyocytes have been used to investigate the anti- or arrhythmic effects of certain drugs and demonstrate that omega-3 fatty acids and the R,S isomer of the 17,18 epoxide of EPA decreased the contraction rate in these cells [89], indicating a potential ability to alleviate atrial fibrillation. These studies further demonstrate the complexity of EpFA biology that could potentially complicate the identification of an active mimic.

Other approaches that mimic the PUFA of omega-3 or −6 fatty acids have been developed. For example, Icosabutate, developed by Northsea Therapeutics, and Vascepa (icosapent ethyl, or ethyl-EPA), an approved product sold by Amarin, are structurally engineered fatty acids both being developed to lower triglycerides in the body. Icosabutate mimics EPA with the addition of 2-bromo butyric acid on the α end, while Vascepa adds a methyl group to the α end to prevent β-oxidation; otherwise, both structures are unaltered compared to EPA. Northsea data presented at the International Liver Congress in 2018 show the compound remains in the non-esterified form longer than EPA as expected, but was metabolized primarily by CYP2C enzymes, suggesting that epoxides are likely formed in the metabolism of this compound [90]. Multiple clinical trials were completed with Icosabutate testing the safety and potential drug-drug interactions with CYP inducers and inhibitors, as well as efficacy in hypertriglyceridemia (NCT01893515) and NASH (NCT04052516). Patients with hypertriglyceridemia receiving 600 mg once daily for 12-weeks had significantly lower triglyceride, very low-density lipoprotein cholesterol, and Apo C-III levels [91]. The NASH study recently started dosing and results are not yet available. There is more published information in icosapent ethyl considering that it is an approved drug for reducing the risk of cardiovascular disease in patients with hypertriglyceridemia. In a placebo controlled clinical trial enrolling 8179 patients, administration of 2 g twice daily of icosapent ethyl significantly reduced ischemic events, including reduced incidence of death [92, 93].

Similar to risks in choosing a mimic to omega-3 or 6 fatty acid, identifying the regioisomer and stereoisomer to mimic further complicates the process. Omeicos is in a unique position by having structure-activity relationship information for atrial fibrillation, but without knowing the target of EH activity, translating this information to other disease areas should not be assumed.

5.4.3 New Therapeutic Approaches Targeting EpFA Through Polypharmaceutical Approaches

Pharmaceutical targets to the COX and LO enzymes focus on preventing the formation of inflammatory prostaglandins and leukotrienes, respectively. NSAID drugs inhibit COX-1 and-2 enzymes and are potent anti-inflammatory compounds by preventing the formation of inflammatory prostaglandins; however, these inhibitors are often accompanied by toxic side effects associated with enzyme distribution. COX-1 enzymes are in most cells and protect the GI mucosa by regulating acid secretion through the EP3 receptor. Inhibition of COX-1 increases gastric release, thus causing GI-toxicity and increased ulcer formation. COX-2 enzymes are found in lymphocytes, red blood cells and synovial cells and are thought to regulate pain and inflammation. Selectively inhibiting COX-2 was thought of as a desirable approach to reduce pain and inflammation while avoiding GI complications. However, while selective inhibition of COX-2 avoided gastro-intestinal ulcer formation, chronic use increased the risk of stroke and cardiovascular disease. Further investigation into the mechanism of this toxicity found that COX-1 inhibition decreases the vasoconstrictor thromboxane A2 synthesis; while COX-2 inhibition decreases the vasodilator, prostacyclin. Non-selective inhibitors would decrease both vasoregulators and maintain homeostasis, while selective COX-2 inhibitors would decrease only prostacyclin, resulting in platelet aggregation and vasoconstriction [94, 95]. Vasodilatory effects of EpFA suggest further protection against endothelial dysfunction associated with COX-2 inhibition, but recent studies suggest an additional mechanism for decreasing COX toxicities results from the reduction of ER-stress [96]. NSAIDs increase ER-stress which has been attributed to the toxicities associated with NSAID use, including cardiovascular toxicity [97–99]. ER-stress also increases COX transcription, which creates a feedback inflammatory mechanism perpetuated by COX-2 activation of IRE1a, a key protein involved in activating the ER-stress pathway. Blocking ER-stress, as has been demonstrated with EpFA, prevent the upregulation of COX, and dual inhibition of sEH and COX are an attractive therapeutic strategy for increasing the benefit of NSAIDs while reducing their toxicity. Many preclinical studies demonstrate the advantages of dual COX-sEH inhibition. For example, the sEH inhibitor, t-AUCB, or NSAID inhibitor, Celecoxib, administered alone do not affect tumor volume or metastasis in a Lewis lung carcinoma mouse model of cancer while treatment with both significantly reduced both tumor volume and metastasis. Previous studies demonstrated that sEH null mice have increased metastasis and tumor volume [100], presumably through increased angiogenesis and VEGF expression; however, clinical trials with GSK's sEHI failed to show translation of VEGF increases in mice to human patients [101]. Furthermore, preclinical studies identify that EETs are angiogenic and induce endothelial cell proliferation [102–104] through metabolism by COX that produces a potent angiogenic metabolite [105]. Thus, dual inhibition would prevent the formation of this metabolite and could explain the added benefit in tumor models. Additional benefits have been observed in improving survival after tumor-induced cytokine surge in mice. Inflammation is a protective mechanism to eliminate foreign toxicants from the body; however, an intense and rapid inflammatory response, as occurs in sepsis, can be deadly. CAR-T therapy is a promising treatment of cancer that activates the body's immune system to recognize tumor cells as foreign material. The body mounts an aggressive immune attack to eliminate these tumor cells; however, if too effective, the outcomes also result in sepsis and death. Preclinical studies demonstrate that the dual COX/sEHI, p-TUPB, improves survival rate in a cancer cell debris model of cytokine-surge in mice [106].

Dual inhibition with COX and sEH is an attractive target for treating disease by minimizing the toxicities associated with NSAIDs while also increasing the efficacy of both compounds [107]. Many NSAIDs and coxibs lead to mitochondrial dysfunction, which in turn increases the production of reactive oxygen species. They have also been shown to increase ER-stress [99], which can induce the transcription of COX2 and prostaglandin synthase [108] leading to an increase in inflammatory eicosanoids and cytokines. Thus, the most widely used drugs in the world are themselves inflammatory. Anti-inflammatory sEHI block this inflammation axis both at the level of the mitochondria and the endoplasmic reticulum. Thus, when used with NSAIDS and coxibs should make these drugs not only safer but also improve their anti-inflammatory activity. Multiple rodent studies have shown that sEHI synergize with NSAIDs and coxibs to reduce inflammatory pain and extend the utility of NSAIDs and coxibs to include neuropathic pain. The sEHI also reduce gastrointestinal and cardiovascular side effects of cyclooxygenase inhibitors. Since prescription and over the counter COX inhibitors are so commonly used in pain management, this interaction is likely to be seen in the clinic if patients in trials are continued on standard of care treatment.

Exploiting this interaction commercially will be more complex. Regulatory pathways require that novel therapeutics demonstrate stand-alone efficacy before being combined with other targets, thus the pathway toward human treatment with NSAID-sEHI combinations is complicated. A polypharmaceutical approach: where one compound inhibits both COX and sEH, such as PTUBP, would lessen this challenge; however, the FDA closely monitors NSAID cardiovascular toxicity and requires extensive cardiovascular safety studies and lengthy clinical trials prior to testing in humans, thus developing this dual target under the current regulatory environment would likely be more expensive and risky than other targets. For example, this regulatory scrutiny resulted in the FDA not approving a potent NSAID, etoricoxib, after the FDA advisory committee recommended further cardiovascular risk assessment, despite etoricoxib approval in other countries [109].

Other targets in the fatty acid metabolism pathway provide attractive opportunities for treatments. For example, inhibition of LO is currently approved for the treatment of asthma. Leukotrienes, the main target of LO metabolism, are potent inflammatory agents that potentiate bronchoconstriction and exacerbate an asthma attack. Inhibiting LO activity through direct inhibition (e.g. Zileuton) or through the inhibition of 5-lipoxygnase activating protein (FLAP) is an approved treatment for asthma [110]. sEHI has also shown efficacy for treating asthma [111]. Interestingly, Zileuton and the FLAP inhibitors, Zafirlukast and Montelukast, also have weak activity for inhibiting the sEH (0.9–1.95 μM), but at concentrations likely achieved at therapeutic doses [112]. Identifying improved dual LO-sEHI with greater potency for the sEH is an attractive target for improving asthma treatments.

The fatty acid amide hydrolase enzyme (FAAH) is a complicated target for drug therapy. The endocannabinoids, a bioactive class of lipids formed through the addition of an amide to the carboxylic acid of PUFA. For example, N-arachidonoyl phosphatidylethanolamine (NAPE), and related amides are associated with beneficial effects in a variety of disease models including inflammation, pain, asthma, epilepsy, neurological disease, among others (see [113] for review). Because FAAH metabolizes endocannabinoids, therapeutic approaches of inhibiting this enzyme were considered for treatments of some diseases. Over ten clinical trials have been initiated to test efficacy of FAAH inhibitors, but after significant toxicities and one death were reported in a Phase 1 study of a claimed FAAH inhibitor in France, all clinical trials were put on hold. Although the exact cause of the toxicity is unknown, recent reports suggest that off-target effects on lipases caused the toxicity [114]. Renewed interests in FAAH as a target for pain came after a case study identified a woman with a knock-out mutation in FAAH were thought to result in her inability to feel pain [115]. Administration of both sEH and FAAH inhibitors show synergistic effects in animal models of pain [116] and continued efforts to develop a dual inhibitor suitable for clinical development are ongoing [117].

sEH inhibitors are more effective in reducing pain in the presence of cAMP. Dual phosphodiesterase 4 (PDE4) and sEH inhibition also show synergistic effects in decreasing pain in animal models [118]. PDE inhibitors (PDEi) have been developed for the treatment of inflammation by preventing the metabolism of cyclic adenosine monophosphate (cAMP); however, increasing cAMP has also been associated with increased pain states. Interestingly, PDE4 and 5 inhibitors increase EpFA through lipase activity, releasing arachidonate and possibly EETs from phospholipids, and could possibly explain the mechanism for analgesic effects of PDEi despite increases in cAMP [118]. In laboratory settings, investigators demonstrated that decreasing EpFA concentrations through CYP450 inhibition while treating with PDEi resulted in increased pain states. Conversely, increasing EpFA concentrations through sEHI resulted in synergistic increases in pain relief. Efforts are ongoing to identify dual inhibitors of PDE4 and sEH, and recent advances identified MPPA, a compound with potent inhibition on both enzymes that is also efficacious in an LPS model of inflammatory pain [119].

While some compounds are synthesized specifically to inhibit an enzyme, others are identified after approval as having off-target effects against desirable or undesirable enzymes. For example, sorafenib and regorafenib are receptor tyrosine kinases that have anti-angiogenic properties by inhibiting VEGFR-2 and other kinases. They are approved for the treatment of hepatocellular carcinoma and renal cell carcinoma, and both are potent inhibitors of the sEH (12 and 0.5 nM IC_{50}, respectively). The other approved small molecule inhibitor of VEGFR, sunitinib, is not an sEH inhibitor. Although the biological relevance of this dual activity is difficult to compare with other kinases that do not affect sEH because kinases are in general promiscuous inhibitors of many kinases. Both sorafenib and regorafenib are known for being difficult to formulate and for often serious side effects of their use. Clinical doses of these Raf-1 and pan-kinase inhibitors are likely to inhibit most sEH activity and increase EpFA *in vivo*. It is likely that without the sEH inhibition, these kinase inhibitors would have even worse side effects. It follows that the side effects of these two kinase inhibitors could be dramatically reduced by increasing dietary ω−3 and limiting dietary ω−6 lipids. The kinase inhibition is not ubiquitous to the urea pharmacophore that inhibits sEH; however, Sorafenib and regorafenib are unique compared to other published sEH inhibitors used in preclinical or clinical studies in their ability to inhibit kinases, as can be predicted from structure activity relationships [120]. Although recent publications indicate that TPPU is a weak inhibitor of p38 kinase (IC_{50} = 0.98 μM). The p38 kinase is a mitogen-activated protein kinase and activates inflammatory cytokines, and inhibition of this kinase would explain many of the beneficial effects seen in the preclinical studies using this compound; however, p38 is also regulated by oxidative stress, and the study failed to show if the activity was an indirect result of reducing oxidative stress vs. direct inhibition of the kinase given that the *in vitro* assays were conducted below the IC_{50} of 100 nM [121].

5.5 Summary and Future Directions for Targeting EpFA as Treatments for Disease

The newest innovations in inhibiting the sEH focus on identifying natural compounds with inhibitory activity, or polypharmaceutical approaches targeting multiple enzyme pathways. The inhibitors being tested for sEH inhibition focus on a central pharmacophore of either a urea, amide, or carbamate, and few new patents have been filed with different pharmacophores. Instead, new technologies focus on identifying new targets of disease, poly-pharmaceutical approaches developing a single compound to inhibit multiple targets [122], or inhibitors isolated from natural products. Inhibiting the sEH affects a regulatory pathway, ER stress, and is a relatively safe target that could offer benefits when combining with other targets. In some cases, efforts for developing dual targets to provide added benefits or safety have been intentional, while others have been discovered as off-target effects after developments. Both approaches are described below.

Current clinical approaches are focused on developing small molecule inhibitors or EpFA mimics; however, future approaches are investigating polypharmaceutical approaches, thus capitalizing on the low toxicity of alerting this pathway. Yet despite widespread interest in targeting this pathway, few new pharmacophores are identified that lack a carbamate or urea as the active moiety. Recently, natural products have been identified as a new source of sEH inhibitors that hold promise for future nutraceutical development; however, these products have relatively poor potency, with IC_{50}s in the μM range, which would require g/day amounts of dosing. Techniques are being considered to improve production and isolation of natural sEHI or identify more potent natural sEHI [123, 124].

Many early sEHI, such as AUDA, were designed as mimics of specific EpFA. This dual action was clearly demonstrated in earlier publications [125]. Subsequent synthesis led to compounds that inhibited the EH without mimicking EpFA or compounds that mimicked specific EpFA without inhibiting sEH. Shen and Hammock discussed the advantages and limitations of each of these approaches [126]. The resulting molecules made biological data much easier to interpret when they were used as physiological probes, and likely simplified patent positions as well. Possibly it is worth reconsidering dual acting compounds. For example, one could have an sEHI pharmacophore which would increase EpFA systemically, while also targeting mimicry of 17,18 EEQ or 19,20 EDP to control atrial fibrillation [127]. Such compounds would offer the advantages of stabilizing all EpFA while allowing direct action of the EpFA mimic in tissue with a reduced ability to epoxidize PUFA.

In summary, lipid epoxides and their mimics or stabilizers are of increasing interest in treating disease due to their many beneficial and homeostatic functions as well as limited observed toxicities. However, their multimodal mechanism of action, including effects on angiogenesis and cancer, raises caution for monitoring safety in clinical trials. Furthermore, few studies report toxicities associated either with direct supplementation of EpFA, or inhibition of sEH; however, one study reported that sEH null mice displayed reduced survival in a cardiac arrest model compared to wild type mice [128] demonstrating one of the few examples of increased toxicities in preclinical models modulating the sEH. While the most recent advances are focused on investigating the polypharmaceutical advantages of increasing EpFA, other approaches that could increase EpFA include altering the release and reincorporation into the cell membrane and could provide attractive benefits to the current clinical approaches being investigated (Fig. 5.4.). Currently small molecule inhibitors of sEH are being tested in clinical settings to investigate the effects of increasing EpFA concentrations in the body as a way of treating pain, diabetes, and reducing the symptoms of subarachnoid hemor-

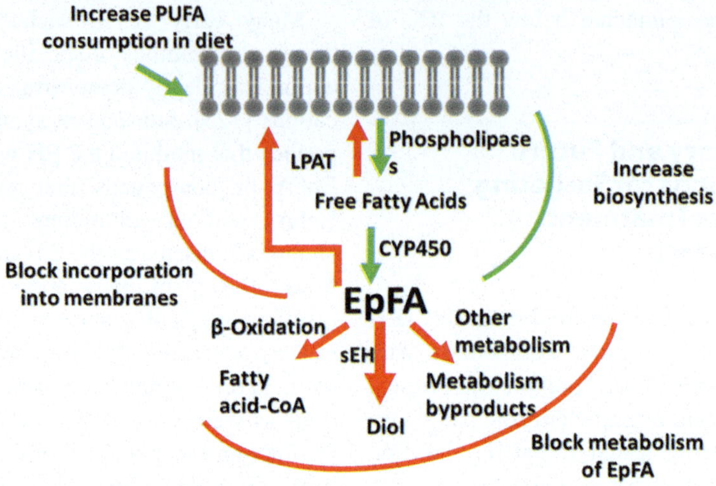

Fig. 5.4 Possible ways of increasing natural epoxy fatty acid chemical mediators to prevent and treat disease
Increasing EpFA concentrations in the body have been associated with treating many diseases. The figure above demonstrates potential pathways of increasing their concentrations through increasing their biosynthesis or through preventing their metabolism. EpFA can be increased by increased PUFA consumption, increasing release from lipid membranes, or increased formation through CYP450 activity. Inhibiting the metabolic enzymes that convert EpFA to more polar compounds that are rapidly eliminated from the body, or supplementing the diet with EpFA mimics that prevent degradation by β oxidation are other ways in increasing the concentrations of the beneficial fatty acids

rhage. Other strategies focus on mimicking the beneficial EpFA and are targeted in clinical investigations for treating pain, fatty liver disease, hypertriglyceridemia, and atrial fibrillation. Overall, previous clinical trials targeting strategies to increase EpFA lack on-target toxicities, and most studies using sEH knock-out animals lack toxic side effects, which increases confidence in the safety of increasing EpFA concentrations in the body. The combined safety profile and wide-ranging efficacy in treating disease encourage continued investigation in the beneficial effects of this pathway as a treatment option.

Acknowledgements This work was supported by the National Institute of Environmental Health Sciences (NIEHS) Grant R35ES030443-01, NIEHS Superfund Research Program P42 ES004699, and T32GM113770 (to CBM). Partial support for clinical development of sEH inhibitors for human medicine comes from the NIEHS SBIR Program R44ES025598, the NIH NINDS Blueprint Neurotherapeutics Network UH2NS094258, and the National Institute On Drug Abuse Award Number UG3DA048767. The content is solely the responsibility of the authors and does not necessarily represent the official views of the National Institutes of Health. The authors would also like to thank Jogen Atone for input on fig. 5.3.

Conflict of interest statement The University of California holds patents on the sEH inhibitor used in this study as well as their use to treat inflammation, inflammatory pain, and neuropathic pain. KM Wagner, CB McReynolds, and BD Hammock are employees of EicOsis L.L.C., a startup company advancing sEH inhibitors into the clinic.

References

1. Vane JR (1971) Inhibition of prostaglandin synthesis as a mechanism of action for aspirin-like drugs. Nat New Biol 231(25):232–235. Epub 1971/06/23. https://doi.org/10.1038/newbio231232a0. PubMed PMID: 5284360
2. Liu MC, Dube LM, Lancaster J (1996) Acute and chronic effects of a 5-lipoxygenase inhibitor in asthma: a 6-month randomized multicenter trial. Zileuton Study Group. J Allergy Clin Immunol 98(5 Pt 1):859–871. Epub 1996/11/01. https://doi.org/10.1016/s0091-6749(96)80002-9. PubMed PMID: 8939149

3. J H. Essential fatty acids. Linus Pauling Institute, 2003 [cited 2019]. Available from: https://lpi.oregonstate.edu/mic/other-nutrients/essential-fatty-acids
4. Zorn K, Oroz-Guinea I, Brundiek H, Bornscheuer UT (2016) Engineering and application of enzymes for lipid modification, an update. Prog Lipid Res 63:153–164. Epub 2016/06/16. https://doi.org/10.1016/j.plipres.2016.06.001. PubMed PMID: 27301784
5. Konkel A, Schunck WH (2011) Role of cytochrome P450 enzymes in the bioactivation of polyunsaturated fatty acids. Biochim Biophys Acta 1814(1):210–222. Epub 2010/09/28. S1570-9639(10)00258-X [pii], https://doi.org/10.1016/j.bbapap.2010.09.009. PubMed PMID: 20869469
6. Potente M, Michaelis UR, Fisslthaler B, Busse R, Fleming I (2002) Cytochrome P450 2C9-induced endothelial cell proliferation involves induction of mitogen-activated protein (MAP) kinase phosphatase-1, inhibition of the c-Jun N-terminal kinase, and up-regulation of cyclin D1. J Biol Chem 277(18):15671–15676. Epub 2002/02/28. https://doi.org/10.1074/jbc.M110806200. PubMed PMID: 11867622
7. Fitzpatrick FA, Ennis MD, Baze ME, Wynalda MA, McGee JE, Liggett WF (1986) Inhibition of cyclooxygenase activity and platelet aggregation by epoxyeicosatrienoic acids. Influence of stereochemistry. J Biol Chem 261(32):15334–15338. Epub 1986/11/15. PubMed PMID: 3095326
8. Fan F, Muroya Y, Roman RJ (2015) Cytochrome P450 eicosanoids in hypertension and renal disease. Curr Opin Nephrol Hypertens 24(1):37–46. Epub 2014/11/27. https://doi.org/10.1097/mnh.0000000000000088. PubMed PMID: 25427230; PMCID: PMC4260681
9. Fleming I, Busse R (2006) Endothelium-derived epoxyeicosatrienoic acids and vascular function. Hypertension 47(4):629–633. Epub 2006/02/24. https://doi.org/10.1161/01.hyp.0000208597.87957.89. PubMed PMID: 16490839
10. Roman RJ, Fan F (2018) 20-HETE: hypertension and beyond. Hypertension 72(1):12–18. Epub 2018/05/16. https://doi.org/10.1161/hypertensionaha.118.10269. PubMed PMID: 29760152; PMCID: PMC6002933
11. Ulu A, Inceoglu B, Yang J, Singh V, Vito S, Wulff H, Hammock BD (2016) Inhibition of soluble epoxide hydrolase as a novel approach to high dose diazepam induced hypotension. J Clin Toxicol 6(3). Epub 2017/03/04. https://doi.org/10.4172/2161-0495.1000300. PubMed PMID: 28255523; PMCID: PMC5328659
12. McGiff JC, Carroll MA (1991) Cytochrome P450-dependent arachidonate metabolites, renal function and blood pressure regulation. Adv Prostaglandin Thromboxane Leukot Res 21b:675–682. Epub 1991/01/01
13. Imig JD, Pham BT, LeBlanc EA, Reddy KM, Falck JR, Inscho EW (2000) Cytochrome P450 and cyclooxygenase metabolites contribute to the endothelin-1 afferent arteriolar vasoconstrictor and calcium responses. Hypertension 35(1 Pt 2):307–312. Epub 2000/01/21. https://doi.org/10.1161/01.hyp.35.1.307
14. Liu L, Chen C, Gong W, Li Y, Edin ML, Zeldin DC, Wang DW (2011) Epoxyeicosatrienoic acids attenuate reactive oxygen species level, mitochondrial dysfunction, caspase activation, and apoptosis in carcinoma cells treated with arsenic trioxide. J Pharmacol Exp Ther 339(2):451–463. Epub 2011/08/19. https://doi.org/10.1124/jpet.111.180505. PubMed PMID: 21846841; PMCID: PMC3199997
15. Node K, Huo Y, Ruan X, Yang B, Spiecker M, Ley K, Zeldin DC, Liao JK (1999) Anti-inflammatory properties of cytochrome P450 epoxygenase-derived eicosanoids. Science 285(5431):1276–1279. Epub 1999/08/24. https://doi.org/10.1126/science.285.5431.1276. PubMed PMID: 10455056; PMCID: PMC2720027
16. Xu D, Li N, He Y, Timofeyev V, Lu L, Tsai HJ, Kim IH, Tuteja D, Mateo RK, Singapuri A, Davis BB, Low R, Hammock BD, Chiamvimonvat N (2006) Prevention and reversal of cardiac hypertrophy by soluble epoxide hydrolase inhibitors. Proc Natl Acad Sci U S A 103(49):18733–18738. Epub 2006/11/30. https://doi.org/10.1073/pnas.0609158103. PubMed PMID: 17130447; PMCID: PMC1693731
17. Kundu S, Roome T, Bhattacharjee A, Carnevale KA, Yakubenko VP, Zhang R, Hwang SH, Hammock BD, Cathcart MK (2013) Metabolic products of soluble epoxide hydrolase are essential for monocyte chemotaxis to MCP-1 in vitro and in vivo. J Lipid Res 54(2):436–447. https://doi.org/10.1194/jlr.M031914. PubMed PMID: 23160182; PMCID: 3588870
18. Wagner KM, McReynolds CB, Schmidt WK, Hammock BD (2017) Soluble epoxide hydrolase as a therapeutic target for pain, inflammatory and neurodegenerative diseases. Pharmacol Ther 180:62–76. Epub 2017/06/24. https://doi.org/10.1016/j.pharmthera.2017.06.006. PubMed PMID: 28642117; PMCID: PMC5677555
19. Lin JH, Walter P, Yen TS (2008) Endoplasmic reticulum stress in disease pathogenesis. Ann Rev Pathol 3:399–425. Epub 2007/11/28. https://doi.org/10.1146/annurev.pathmechdis.3.121806.151434. PubMed PMID: 18039139; PMCID: PMC3653419
20. Inceoglu B, Bettaieb A, Trindade da Silva CA, Lee KS, Haj FG, Hammock BD (2015) Endoplasmic reticulum stress in the peripheral nervous system is a significant driver of neuropathic pain. Proc Natl Acad Sci U S A 112(29):9082–9087. Epub 2015/07/08. https://doi.org/10.1073/pnas.1510137112. PubMed PMID: 26150506; PMCID: Pmc4517273

21. Fang X, Kaduce TL, Weintraub NL, Harmon S, Teesch LM, Morisseau C, Thompson DA, Hammock BD, Spector AA (2001) Pathways of epoxyeicosatrienoic acid metabolism in endothelial cells. Implications for the vascular effects of soluble epoxide hydrolase inhibition. J Biol Chem 276(18):14867–14874. Epub 2001/03/30. https://doi.org/10.1074/jbc.M011761200
22. Nelson JW, Subrahmanyan RM, Summers SA, Xiao X, Alkayed NJ (2013) Soluble epoxide hydrolase dimerization is required for hydrolase activity. J Biol Chem 288(11):7697–7703. Epub 2013/01/31. https://doi.org/10.1074/jbc.M112.429258. PubMed PMID: 23362272; PMCID: PMC3597810
23. Morisseau C (2013) Role of epoxide hydrolases in lipid metabolism. Biochimie 95(1):91–95. Epub 2012/06/23. https://doi.org/10.1016/j.biochi.2012.06.011. PubMed PMID: 22722082; PMCID: PMC3495083
24. Botham KM, Mayes PA (2016) Cholesterol synthesis, transport, & excretion. In: Rodwell VW, Bender DA, Botham KM, Kennelly PJ, Weil PA (eds) Harper's illustrated biochemistry, 30th edn. McGraw-Hill Education, New York
25. Decker M, Adamska M, Cronin A, Di Giallonardo F, Burgener J, Marowsky A, Falck JR, Morisseau C, Hammock BD, Gruzdev A, Zeldin DC, Arand M (2012) EH3 (ABHD9): the first member of a new epoxide hydrolase family with high activity for fatty acid epoxides. J Lipid Res 53(10):2038–2045. Epub 2012/07/17. https://doi.org/10.1194/jlr.M024448. PubMed PMID: 22798687; PMCID: PMC3435537
26. Fornage M, Lee CR, Doris PA, Bray MS, Heiss G, Zeldin DC, Boerwinkle E (2005) The soluble epoxide hydrolase gene harbors sequence variation associated with susceptibility to and protection from incident ischemic stroke. Hum Mol Genet 14(19):2829–2837. Epub 2005/08/24. https://doi.org/10.1093/hmg/ddi315. PubMed PMID: 16115816; PMCID: PMC1343524
27. Martini RP, Ward J, Siler DA, Eastman JM, Nelson JW, Borkar RN, Alkayed NJ, Dogan A, Cetas JS (2014) Genetic variation in soluble epoxide hydrolase: association with outcome after aneurysmal subarachnoid hemorrhage. J Neurosurg 121(6):1359–1366. Epub 2014/09/13. https://doi.org/10.3171/2014.7.jns131990. PubMed PMID: 25216066; PMCID: PMC4370510
28. Morisseau C, Wecksler AT, Deng C, Dong H, Yang J, Lee KS, Kodani SD, Hammock BD (2014) Effect of soluble epoxide hydrolase polymorphism on substrate and inhibitor selectivity and dimer formation. J Lipid Res. 55(6):1131–1138. Epub 2014/04/29. https://doi.org/10.1194/jlr.M049718. PubMed PMID: 24771868; PMCID: PMC4031944
29. Przybyla-Zawislak BD, Srivastava PK, Vazquez-Matias J, Mohrenweiser HW, Maxwell JE, Hammock BD, Bradbury JA, Enayetallah AE, Zeldin DC, Grant DF (2003) Polymorphisms in human soluble epoxide hydrolase. Mol Pharmacol 64(2):482–490. Epub 2003/07/19. https://doi.org/10.1124/mol.64.2.482
30. Morisseau C, Hammock BD (2013) Impact of soluble Epoxide Hydrolase and Epoxyeicosanoids on human health. Annu Rev Pharmacol 53:37–58. https://doi.org/10.1146/annurev-pharmtox-011112-140244. PubMed PMID: WOS:000323040100003
31. Morisseau C, Inceoglu B, Schmelzer K, Tsai HJ, Jinks SL, Hegedus CM, Hammock BD (2010) Naturally occurring monoepoxides of eicosapentaenoic acid and docosahexaenoic acid are bioactive antihyperalgesic lipids. J Lipid Res 51(12):3481–3490. Epub 2010/07/29. https://doi.org/10.1194/jlr.M006007. PubMed PMID: 20664072; PMCID: PMC2975720
32. Ulu A, Harris TR, Morisseau C, Miyabe C, Inoue H, Schuster G, Dong H, Iosif AM, Liu JY, Weiss RH, Chiamvimonvat N, Imig JD, Hammock BD (2013) Anti-inflammatory effects of omega-3 polyunsaturated fatty acids and soluble epoxide hydrolase inhibitors in angiotensin-II-dependent hypertension. J Cardiovasc Pharmacol 62(3):285–297. https://doi.org/10.1097/FJC.0b013e318298e460. PubMed PMID: 23676336; PMCID: 3773051
33. Scott-Van Zeeland AA, Bloss CS, Tewhey R, Bansal V, Torkamani A, Libiger O, Duvvuri V, Wineinger N, Galvez L, Darst BF, Smith EN, Carson A, Pham P, Phillips T, Villarasa N, Tisch R, Zhang G, Levy S, Murray S, Chen W, Srinivasan S, Berenson G, Brandt H, Crawford S, Crow S, Fichter MM, Halmi KA, Johnson C, Kaplan AS, La Via M, Mitchell JE, Strober M, Rotondo A, Treasure J, Woodside DB, Bulik CM, Keel P, Klump KL, Lilenfeld L, Plotnicov K, Topol EJ, Shih PB, Magistretti P, Bergen AW, Berrettini W, Kaye W, Schork NJ (2014) Evidence for the role of EPHX2 gene variants in anorexia nervosa. Mol Psychiatry 19(6):724–732. Epub 2013/09/04. https://doi.org/10.1038/mp.2013.91. PubMed PMID: 23999524; PMCID: PMC3852189
34. Shih PB (2017) Integrating multi-omics biomarkers and postprandial metabolism to develop personalized treatment for anorexia nervosa. Prostaglandins Other Lipid Mediat 132:69–76. Epub 2017/02/25. https://doi.org/10.1016/j.prostaglandins.2017.02.002. PubMed PMID: 28232135; PMCID: PMC5565718
35. Shih PB, Yang J, Morisseau C, German JB, Zeeland AA, Armando AM, Quehenberger O, Bergen AW, Magistretti P, Berrettini W, Halmi KA, Schork N, Hammock BD, Kaye W (2016) Dysregulation of soluble epoxide hydrolase and lipidomic profiles in anorexia nervosa. Mol Psychiatry 21(4):537–546. Epub 2015/04/01. https://doi.org/10.1038/mp.2015.26. PubMed PMID: 25824304; PMCID: PMC4591075
36. Yang J, Shih PB (2018) Fasting and postprandial soluble epoxide hydrolase-associated eicosanoids of remitted patients with eating disorder. Data Brief 17:334–338. Epub 2018/06/08. https://doi.org/10.1016/j.dib.2018.01.028. PubMed PMID: 29876402; PMCID: PMC5988286

37. Sato K, Emi M, Ezura Y, Fujita Y, Takada D, Ishigami T, Umemura S, Xin Y, Wu LL, Larrinaga-Shum S, Stephenson SH, Hunt SC, Hopkins PN (2004) Soluble epoxide hydrolase variant (Glu287Arg) modifies plasma total cholesterol and triglyceride phenotype in familial hypercholesterolemia: intrafamilial association study in an eight-generation hyperlipidemic kindred. J Hum Genet 49(1):29–34. Epub 2003/12/16. https://doi.org/10.1007/s10038-003-0103-6
38. Shen L, Peng H, Peng R, Fan Q, Zhao S, Xu D, Morisseau C, Chiamvimonvat N, Hammock BD (2015) Inhibition of soluble epoxide hydrolase in mice promotes reverse cholesterol transport and regression of atherosclerosis. Atherosclerosis 239(2):557–565. Epub 2015/03/04. https://doi.org/10.1016/j.atherosclerosis.2015.02.014. PubMed PMID: 25733327; PMCID: PMC4527317
39. Campbell WB, Falck JR, Gauthier K (2001) Role of epoxyeicosatrienoic acids as endothelium-derived hyperpolarizing factor in bovine coronary arteries. Med Sci Monit Int Med J Exp Clin Res 7(4):578–584. Epub 2001/07/04
40. Sari I, Pinarbasi H, Pinarbasi E, Yildiz C (2017) Association between the soluble epoxide hydrolase gene and preeclampsia. Hypertens Pregnancy 36(4):315–325. Epub 2017/10/24. https://doi.org/10.1080/10641955.2017.1388390
41. Koerner IP, Jacks R, DeBarber AE, Koop D, Mao P, Grant DF, Alkayed NJ (2007) Polymorphisms in the human soluble epoxide hydrolase gene EPHX2 linked to neuronal survival after ischemic injury. J Neurosci 27(17):4642–4649. https://doi.org/10.1523/JNEUROSCI.0056-07.2007. PubMed PMID: 17460077; PMCID: PMC6672984
42. Gschwendtner A, Ripke S, Freilinger T, Lichtner P, Muller-Myhsok B, Wichmann HE, Meitinger T, Dichgans M (2008). Genetic variation in soluble epoxide hydrolase (EPHX2) is associated with an increased risk of ischemic stroke in white Europeans. Stroke 39(5):1593–1596. Epub 2008/03/08. https://doi.org/10.1161/strokeaha.107.502179
43. Lee J, Dahl M, Grande P, Tybjaerg-Hansen A, Nordestgaard BG (2010) Genetically reduced soluble epoxide hydrolase activity and risk of stroke and other cardiovascular disease. Stroke 41(1):27–33. Epub 2009/11/27. https://doi.org/10.1161/strokeaha.109.567768
44. Fava C, Montagnana M, Danese E, Almgren P, Hedblad B, Engstrom G, Berglund G, Minuz P, Melander O (2010) Homozygosity for the EPHX2 K55R polymorphism increases the long-term risk of ischemic stroke in men: a study in swedes. Pharmacogenet Genomics 20(2):94–103. Epub 2010/01/13. https://doi.org/10.1097/FPC.0b013e3283349ec9
45. Nelson JW, Young JM, Borkar RN, Woltjer RL, Quinn JF, Silbert LC, Grafe MR, Alkayed NJ (2014) Role of soluble epoxide hydrolase in age-related vascular cognitive decline. Prostaglandins Other Lipid Mediat 113–115:30–37. Epub 2014/10/04. https://doi.org/10.1016/j.prostaglandins.2014.09.003. PubMed PMID: 25277097; PMCID: PMC4254026
46. Václavíková R, Hughes DJ, Souček P (2015) Microsomal epoxide hydrolase 1 (EPHX1): gene, structure, function, and role in human disease. Gene 571(1):1–8. Epub 07/26. https://doi.org/10.1016/j.gene.2015.07.071
47. Marowsky A, Burgener J, Falck JR, Fritschy JM, Arand M (2009) Distribution of soluble and microsomal epoxide hydrolase in the mouse brain and its contribution to cerebral epoxyeicosatrienoic acid metabolism. Neuroscience 163(2):646–661. Epub 06/18. https://doi.org/10.1016/j.neuroscience.2009.06.033
48. Qin XH, Wu Z, Dong JH, Zeng YN, Xiong WC, Liu C, Wang MY, Zhu MZ, Chen WJ, Zhang Y, Huang QY, Zhu XH (2019). Liver soluble Epoxide Hydrolase regulates behavioral and cellular effects of chronic stress. Cell Rep. 2019;29(10):3223–3234.e6. Epub 2019/12/05. https://doi.org/10.1016/j.celrep.2019.11.006
49. Sinal CJ, Miyata M, Tohkin M, Nagata K, Bend JR, Gonzalez FJ (2000) Targeted disruption of soluble epoxide hydrolase reveals a role in blood pressure regulation. J Biol Chem 275(51):40504–40510. Epub 2000/09/26. https://doi.org/10.1074/jbc.M008106200
50. Harris TR, Hammock BD (2013) Soluble epoxide hydrolase: gene structure, expression and deletion. Gene 526(2):61–74. https://doi.org/10.1016/j.gene.2013.05.008. PubMed PMID: 23701967; PMCID: 3733540
51. Atone J, Wagner K, Hashimoto K, Hammock B (2019) Prostaglandins and other lipid mediators cytochrome P450 derived epoxidized fatty acids as a therapeutic tool against neuroinflammatory diseases. Prostaglandins Other Lipid Mediat 106385. Epub 2019/11/08. https://doi.org/10.1016/j.prostaglandins.2019.106385
52. Bettaieb A, Koike S, Hsu MF, Ito Y, Chahed S, Bachaalany S, Gruzdev A, Calvo-Rubio M, Lee KSS, Inceoglu B, Imig JD, Villalba JM, Zeldin DC, Hammock BD, Haj FG (2017) Soluble epoxide hydrolase in podocytes is a significant contributor to renal function under hyperglycemia. Biochimica Biophysica Acta Gen Subj 1861(11 Pt A):2758–2765. Epub 2017/08/02. https://doi.org/10.1016/j.bbagen.2017.07.021. PubMed PMID: 28757338; PMCID: PMC5873293
53. Heinricher MM, Maire JJ, Lee D, Nalwalk JW, Hough LB (2010) Physiological basis for inhibition of morphine and improgan antinociception by CC12, a P450 epoxygenase inhibitor. J Neurophysiol 104(6):3222–3230. Epub 2010/10/12. https://doi.org/10.1152/jn.00681.2010. PubMed PMID: 20926616; PMCID: PMC3007650
54. Conroy JL, Fang C, Gu J, Zeitlin SO, Yang W, Yang J, VanAlstine MA, Nalwalk JW, Albrecht PJ, Mazurkiewicz JE, Snyder-Keller A, Shan Z, Zhang

SZ, Wentland MP, Behr M, Knapp BI, Bidlack JM, Zuiderveld OP, Leurs R, Ding X, Hough LB (2010) Opioids activate brain analgesic circuits through cytochrome P450/epoxygenase signaling. Nat Neurosci 13(3):284–286. https://doi.org/10.1038/nn.2497. PubMed PMID: 20139973; PMCID: PMC2828325

55. Hough LB, Nalwalk JW, Yang J, Conroy JL, VanAlstine MA, Yang W, Gargano J, Shan Z, Zhang SZ, Wentland MP, Phillips JG, Knapp BI, Bidlack JM, Zuiderveld OP, Leurs R, Ding X (2011) Brain P450 epoxygenase activity is required for the antinociceptive effects of improgan, a nonopioid analgesic. Pain 152(4):878–887. Epub 2011/02/15. https://doi.org/10.1016/j.pain.2011.01.001. PubMed PMID: 21316152; PMCID: PMC3065546
56. Grey A, Bolland M (2014) Clinical trial evidence and use of fish oil supplements. JAMA Intern Med 174(3):460–462. https://doi.org/10.1001/jamainternmed.2013.12765. PubMed PMID: 24352849
57. Rymer C, Gibbs RA, Givens DI (2010) Comparison of algal and fish sources on the oxidative stability of poultry meat and its enrichment with omega-3 polyunsaturated fatty acids. Poult Sci 89(1):150–159. https://doi.org/10.3382/ps.2009-00232. PubMed PMID: 20008813
58. Jackowski SA, Alvi AZ, Mirajkar A, Imani Z, Gamalevych Y, Shaikh NA, Jackowski G (2015) Oxidation levels of North American over-the-counter n-3 (omega-3) supplements and the influence of supplement formulation and delivery form on evaluating oxidative safety. J Nutr Sci 4:e30. https://doi.org/10.1017/jns.2015.21. PubMed PMID: 26688721
59. Albert BB, Cameron-Smith D, Hofman PL, Cutfield WS (2013) Oxidation of marine omega-3 supplements and human health. BioMed Res Int 2013:464921. Epub 04/30. https://doi.org/10.1155/2013/464921. PubMed PMID: 23738326
60. Hu J, Dziumbla S, Lin J, Bibli SI, Zukunft S, de Mos J, Awwad K, Fromel T, Jungmann A, Devraj K, Cheng Z, Wang L, Fauser S, Eberhart CG, Sodhi A, Hammock BD, Liebner S, Muller OJ, Glaubitz C, Hammes HP, Popp R, Fleming I (2017) Inhibition of soluble epoxide hydrolase prevents diabetic retinopathy. Nature 552(7684):248–252. Epub 2017/12/07. https://doi.org/10.1038/nature25013. PubMed PMID: 29211719; PMCID: PMC5828869
61. Ruparel S, Green D, Chen P, Hargreaves KM (2012) The cytochrome P450 inhibitor, ketoconazole, inhibits oxidized linoleic acid metabolite-mediated peripheral inflammatory pain. Mol Pain 8:73. Epub 2012/09/26. https://doi.org/10.1186/1744-8069-8-73. PubMed PMID: 23006841; PMCID: PMC3488501
62. Goswami SK, Inceoglu B, Yang J, Wan D, Kodani SD, da Silva CA, Morisseau C, Hammock BD (2015). Omeprazole increases the efficacy of a soluble epoxide hydrolase inhibitor in a PGE(2) induced pain model. Toxicol Appl Pharmacol. 289(3):419–427. Epub 2015/11/03. https://doi.org/10.1016/j.taap.2015.10.018. PubMed PMID: 26522832; PMCID: PMC4666679
63. Hennebelle M, Zhang Z, Metherel AH, Kitson AP, Otoki Y, Richardson CE, Yang J, Lee KSS, Hammock BD, Zhang L, Bazinet RP, Taha AY (2017) Linoleic acid participates in the response to ischemic brain injury through oxidized metabolites that regulate neurotransmission. Sci Rep 7(1):4342. Epub 2017/07/01. https://doi.org/10.1038/s41598-017-02914-7. PubMed PMID: 28659576; PMCID: PMC5489485
64. Ramsden CE, Hennebelle M, Schuster S, Keyes GS, Johnson CD, Kirpich IA, Dahlen JE, Horowitz MS, Zamora D, Feldstein AE, McClain CJ, Muhlhausler BS, Makrides M, Gibson RA, Taha AY (2018) Effects of diets enriched in linoleic acid and its peroxidation products on brain fatty acids, oxylipins, and aldehydes in mice. Biochimica Biophysica Acta Mol Cell Biol Lipid 1863(10):1206–1213. Epub 2018/07/28. https://doi.org/10.1016/j.bbalip.2018.07.007. PubMed PMID: 30053599; PMCID: PMC6180905
65. Ramsden CE, Ringel A, Feldstein AE, Taha AY, MacIntosh BA, Hibbeln JR, Majchrzak-Hong SF, Faurot KR, Rapoport SI, Cheon Y, Chung YM, Berk M, Mann JD (2012) Lowering dietary linoleic acid reduces bioactive oxidized linoleic acid metabolites in humans. Prostaglandins Leukot Essent Fatty Acids 87(4–5):135–141. Epub 2012/09/11. https://doi.org/10.1016/j.plefa.2012.08.004. PubMed PMID: 22959954; PMCID: PMC3467319
66. Vangaveti VN, Jansen H, Kennedy RL, Malabu UH (2016) Hydroxyoctadecadienoic acids: oxidised derivatives of linoleic acid and their role in inflammation associated with metabolic syndrome and cancer. Eur J Pharmacol 785:70–76. Epub 2015/05/20. https://doi.org/10.1016/j.ejphar.2015.03.096. PubMed PMID: 25987423.
67. Patwardhan AM, Scotland PE, Akopian AN, Hargreaves KM (2009) Activation of TRPV1 in the spinal cord by oxidized linoleic acid metabolites contributes to inflammatory hyperalgesia. Proc Natl Acad Sci U S A 106(44):18820–18824. Epub 2009/10/22. https://doi.org/10.1073/pnas.0905415106. PubMed PMID: 19843694; PMCID: PMC2764734
68. Henry Krum MUAH. Evaluation of the effects of urotensin-II and soluble epoxide hydrolase inhibitors on skin microvessel tone in patients with heart failure, and in healthy volunteers (NCT00654966) 2011 [cited 2019 October 1]. Available from: https://clinicaltrials.gov/ct2/show/NCT00654966?term=soluble+epoxide+hydrolase&rank=1
69. Morisseau C, Goodrow MH, Newman JW, Wheelock CE, Dowdy DL, Hammock BD (2002) Structural refinement of inhibitors of urea-based soluble epoxide hydrolases. Biochem Pharmacol 63(9):1599–1608. Epub 2002/05/15. https://doi.org/10.1016/s0006-2952(02)00952-8. PubMed PMID: 12007563.

70. Tsai HJ, Hwang SH, Morisseau C, Yang J, Jones PD, Kasagami T, Kim IH, Hammock BD (2010) Pharmacokinetic screening of soluble epoxide hydrolase inhibitors in dogs. Eur J Pharm Sci. 40(3):222–238. Epub 2010/04/03. https://doi.org/10.1016/j.ejps.2010.03.018. PubMed PMID: 20359531; PMCID: PMC3285443
71. Kodani SD, Hammock BD (2015) The 2014 Bernard B. Brodie Award Lecture Epoxide Hydrolases: drug metabolism to therapeutics for chronic pain. Drug Metab Dispos. https://doi.org/10.1124/dmd.115.063339. PubMed PMID: 25762541
72. Lee KS, Liu JY, Wagner KM, Pakhomova S, Dong H, Morisseau C, Fu SH, Yang J, Wang P, Ulu A, Mate CA, Nguyen LV, Hwang SH, Edin ML, Mara AA, Wulff H, Newcomer ME, Zeldin DC, Hammock BD (2014) Optimized inhibitors of soluble epoxide hydrolase improve in vitro target residence time and in vivo efficacy. J Med Chem 57(16):7016–7030. https://doi.org/10.1021/jm500694p. PubMed PMID: 25079952; PMCID: 4148150
73. Henrick CA, Staal GB, Siddall JB (1973) Alkyl 3,7,11-trimethyl-2,4-dodecadienoates, a new class of potent insect growth regulators with juvenile hormone activity. J Agric Food Chem 21(3):354–359. Epub 1973/05/01. https://doi.org/10.1021/jf60187a043. PubMed PMID: 4708794
74. Tran L, Kompa AR, Wang BH, Krum H (2012) Evaluation of the effects of urotensin II and soluble epoxide hydrolase inhibitor on skin microvessel tone in healthy controls and heart failure patients. Cardiovasc Ther 30(5):295–300. Epub 2011/09/03. https://doi.org/10.1111/j.1755-5922.2011.00282.x. PubMed PMID: 21884016
75. Imig JD, Carpenter MA, Shaw S (2009) The soluble Epoxide Hydrolase inhibitor AR9281 decreases blood pressure, ameliorates renal injury and improves vascular function in Hypertension. Pharmaceuticals (Basel, Switzerland) 2(3):217–227. Epub 2009/12/18. https://doi.org/10.3390/ph2030217. PubMed PMID: 27713235; PMCID: PMC3978544
76. Lee KS, Morisseau C, Yang J, Wang P, Hwang SH, Hammock BD (2013) Forster resonance energy transfer competitive displacement assay for human soluble epoxide hydrolase. Anal Biochem 434(2):259–268. Epub 2012/12/12. https://doi.org/10.1016/j.ab.2012.11.015. PubMed PMID: 23219719; PMCID: PMC3632402
77. Podolin PL, Bolognese BJ, Foley JF, Long E, 3rd, Peck B, Umbrecht S, Zhang X, Zhu P, Schwartz B, Xie W, Quinn C, Qi H, Sweitzer S, Chen S, Galop M, Ding Y, Belyanskaya SL, Israel DI, Morgan BA, Behm DJ, Marino JP, Jr., Kurali E, Barnette MS, Mayer RJ, Booth-Genthe CL, Callahan JF (2013) In vitro and in vivo characterization of a novel soluble epoxide hydrolase inhibitor. Prostaglandins Other Lipid Mediat 104–105:25–31. https://doi.org/10.1016/j.prostaglandins.2013.02.001. PubMed PMID: 23434473
78. Fleming I, Rueben A, Popp R, Fisslthaler B, Schrodt S, Sander A, Haendeler J, Falck JR, Morisseau C, Hammock BD, Busse R (2007) Epoxyeicosatrienoic acids regulate Trp channel dependent Ca2+ signaling and hyperpolarization in endothelial cells. Arterioscler Thromb Vasc Biol 27(12):2612–2618. https://doi.org/10.1161/ATVBAHA.107.152074. PubMed PMID: 17872452
79. Pozzi A, Macias-Perez I, Abair T, Wei S, Su Y, Zent R, Falck JR, Capdevila JH (2005) Characterization of 5,6- and 8,9-epoxyeicosatrienoic acids (5,6- and 8,9-EET) as potent in vivo angiogenic lipids. J Biol Chem 280(29):27138–27146. Epub 2005/05/27. doi: https://doi.org/10.1074/jbc.M501730200. PubMed PMID: 15917237
80. Panigrahy D, Edin ML, Lee CR, Huang S, Bielenberg DR, Butterfield CE, Barnes CM, Mammoto A, Mammoto T, Luria A, Benny O, Chaponis DM, Dudley AC, Greene ER, Vergilio JA, Pietramaggiori G, Scherer-Pietramaggiori SS, Short SM, Seth M, Lih FB, Tomer KB, Yang J, Schwendener RA, Hammock BD, Falck JR, Manthati VL, Ingber DE, Kaipainen A, D'Amore PA, Kieran MW, Zeldin DC. Epoxyeicosanoids stimulate multiorgan metastasis and tumor dormancy escape in mice. J Clin Investig. 2012;122(1):178–191. https://doi.org/10.1172/JCI58128. PubMed PMID: 22182838; PMCID: 3248288
81. Dai M, Wu L, Wang P, Wen Z, Xu X, Wang DW (2017) CYP2J2 and its metabolites EETs attenuate insulin resistance via regulating macrophage polarization in adipose tissue. Sci Rep 7:46743. Epub 2017/04/26. https://doi.org/10.1038/srep46743. PubMed PMID: 28440284; PMCID: PMC5404269
82. Li R, Xu X, Chen C, Wang Y, Gruzdev A, Zeldin DC, Wang DW (2015) CYP2J2 attenuates metabolic dysfunction in diabetic mice by reducing hepatic inflammation via the PPARgamma. Am J Physiol Endocrinol Metab 308(4):E270–E282. Epub 2014/11/13. https://doi.org/10.1152/ajpendo.00118.2014. PubMed PMID: 25389363; PMCID: PMC4329496
83. Yang Y, Dong R, Chen Z, Hu D, Fu M, Tang Y, Wang DW, Xu X, Tu L (2018) Endothelium-specific CYP2J2 overexpression attenuates age-related insulin resistance. Aging Cell 17(2). Epub 2018/01/11. https://doi.org/10.1111/acel.12718. PubMed PMID: 29318723; PMCID: PMC5847864
84. Zarriello S, Tuazon JP, Corey S, Schimmel S, Rajani M, Gorsky A, Incontri D, Hammock BD, Borlongan CV (2019) Humble beginnings with big goals: small molecule soluble epoxide hydrolase inhibitors for treating CNS disorders. Prog Neurobiol 172:23–39. Epub 2018/11/18. https://doi.org/10.1016/j.pneurobio.2018.11.001. PubMed PMID: 30447256
85. Wagner K, Inceoglu B, Hammock BD (2011) Soluble epoxide hydrolase inhibition, epoxygenated fatty acids and nociception. Prostaglandins Other Lipid Mediat 96(1–4):76–83. Epub 2011/08/23. https://

doi.org/10.1016/j.prostaglandins.2011.08.001. PubMed PMID: 21854866; PMCID: PMC3215909
86. Wagner K, Vito S, Inceoglu B, Hammock BD (2014) The role of long chain fatty acids and their epoxide metabolites in nociceptive signaling. Prostaglandins Other Lipid Mediat 113–115:2–12. Epub 2014/09/23. https://doi.org/10.1016/j.prostaglandins.2014.09.001. PubMed PMID: 25240260; PMCID: PMC4254344
87. Brostrom L, Falck JR, inventor Arachidonic Acid Analogs and Methods for Analgesi Treatment Using Same patent 8,658,632 B2. 2012
88. Campbell WB, Gebremedhin D, Pratt PF, Harder DR (1996) Identification of epoxyeicosatrienoic acids as endothelium-derived hyperpolarizing factors. Circ Res 78(3):415–423. Epub 1996/03/01. https://doi.org/10.1161/01.res.78.3.415. PubMed PMID: 8593700
89. Falck JR, Wallukat G, Puli N, Goli M, Arnold C, Konkel A, Rothe M, Fischer R, Muller DN, Schunck WH (2011) 17(R),18(S)-epoxyeicosatetraenoic acid, a potent eicosapentaenoic acid (EPA) derived regulator of cardiomyocyte contraction: structure-activity relationships and stable analogues. J Med Chem 54(12):4109–4118. Epub 2011/05/20. https://doi.org/10.1021/jm200132q. PubMed PMID: 21591683; PMCID: PMC3122156
90. Fraser DA, Wang XY, Skjaeret T, Kastelein JP, Schuppan D (2018) A structurally engineered fatty acid, icosabutate, displays optimised absorption, distribution and metabolism properties for targeting hepatic inflammation and normalises elevated liver enzymes in dyslipidemic patients. Available from: https://www.northseatherapeutics.com/wp-content/uploads/2018/04/EASL-ILC2018-NST.pdf
91. Bays HE, Hallen J, Vige R, Fraser D, Zhou R, Hustvedt SO, Orloff DG, Kastelein JJ (2016) Icosabutate for the treatment of very high triglycerides: A placebo-controlled, randomized, double-blind, 12-week clinical trial. J Clin Lipidol 10(1):181–91.e1–2. Epub 2016/02/20. https://doi.org/10.1016/j.jacl.2015.10.012. PubMed PMID: 26892135
92. Bhatt DL, Steg PG, Miller M, Brinton EA, Jacobson TA, Ketchum SB, Doyle RT, Jr., Juliano RA, Jiao L, Granowitz C, Tardif JC, Ballantyne CM (2019) Cardiovascular risk reduction with Icosapent ethyl for hypertriglyceridemia. N Engl J Med 380(1):11–22. Epub 2018/11/13. https://doi.org/10.1056/NEJMoa1812792. PubMed PMID: 30415628.
93. Bhatt DL, Steg PG, Miller M, Brinton EA, Jacobson TA, Ketchum SB, Doyle RT, Jr., Juliano RA, Jiao L, Granowitz C, Tardif JC, Gregson J, Pocock SJ, Ballantyne CM (2019) Effects of Icosapent ethyl on Total ischemic events: from REDUCE-IT. J Am Coll Cardiol 73(22):2791–2802. Epub 2019/03/23. https://doi.org/10.1016/j.jacc.2019.02.032. PubMed PMID: 30898607
94. Wallace JL (2000) How do NSAIDs cause ulcer disease? Bailliere Best Pract Res Clin Gastroenterol 14(1):147–159. Epub 2000/04/05. PubMed PMID: 10749095
95. Becker RC (2005) COX-2 inhibitors. Tex Heart Inst J 32(3):380–383. Epub 2006/01/06. PubMed PMID: 16392224; PMCID: PMC1336714
96. Groenendyk J, Paskevicius T, Urra H, Viricel C, Wang K, Barakat K, Hetz C, Kurgan L, Agellon LB, Michalak M (2018) Cyclosporine A binding to COX-2 reveals a novel signaling pathway that activates the IRE1alpha unfolded protein response sensor. Sci Rep 8(1):16678. Epub 2018/11/14. https://doi.org/10.1038/s41598-018-34891-w. PubMed PMID: 30420769; PMCID: PMC6232179
97. Ghosh R, Alajbegovic A, Gomes AV (2015) NSAIDs and cardiovascular diseases: role of reactive oxygen species. Oxidative Med Cell Longev 2015:536962. Epub 2015/10/13. https://doi.org/10.1155/2015/536962. PubMed PMID: 26457127; PMCID: PMC4592725
98. Ohyama K, Shiokawa A, Ito K, Masuyama R, Ichibangase T, Kishikawa N, Imai K, Kuroda N (2012) Toxicoproteomic analysis of a mouse model of nonsteroidal anti-inflammatory drug-induced gastric ulcers. Biochem Biophys Res Commun 420(1):210–215. Epub 2012/03/20. https://doi.org/10.1016/j.bbrc.2012.03.009. PubMed PMID: 22426477
99. Tsutsumi S, Gotoh T, Tomisato W, Mima S, Hoshino T, Hwang HJ, Takenaka H, Tsuchiya T, Mori M, Mizushima T (2004) Endoplasmic reticulum stress response is involved in nonsteroidal anti-inflammatory drug-induced apoptosis. Cell Death Differ 11(9):1009–1016. Epub 2004/05/08. https://doi.org/10.1038/sj.cdd.4401436. PubMed PMID: 15131590
100. Panigrahy D, Greene ER, Pozzi A, Wang DW, Zeldin DC (2011) EET signaling in cancer. Cancer Metastasis Rev 30(3–4):525–540. Epub 2011/10/20. https://doi.org/10.1007/s10555-011-9315-y. PubMed PMID: 22009066; PMCID: PMC3804913
101. Lazaar AL, Yang L, Boardley RL, Goyal NS, Robertson J, Baldwin SJ, Newby DE, Wilkinson IB, Tal-Singer R, Mayer RJ, Cheriyan J (2016) Pharmacokinetics, pharmacodynamics and adverse event profile of GSK2256294, a novel soluble epoxide hydrolase inhibitor. Br J Clin Pharm 81(5):971–979. Epub 2015/12/02. https://doi.org/10.1111/bcp.12855. PubMed PMID: 26620151; PMCID: PMC4834590
102. Askari A, Thomson SJ, Edin ML, Zeldin DC, Bishop-Bailey D (2013) Roles of the epoxygenase CYP2J2 in the endothelium. Prostaglandins Other Lipid Mediat. 107:56–63. Epub 2013/03/12. https://doi.org/10.1016/j.prostaglandins.2013.02.003. PubMed PMID: 23474289; PMCID: PMC3711961
103. Fleming I (2011) The cytochrome P450 pathway in angiogenesis and endothelial cell biology. Cancer Metastasis Rev 30(3–4):541–555. Epub 2011/10/20. doi: https://doi.org/10.1007/s10555-011-9302-3. PubMed PMID: 22009065

104. Imig JD (2012) Epoxides and soluble epoxide hydrolase in cardiovascular physiology. Physiol Rev 92(1):101–130. Epub 2012/02/03. https://doi.org/10.1152/physrev.00021.2011. PubMed PMID: 22298653; PMCID: PMC3613253
105. Rand AA, Rajamani A, Kodani SD, Harris TR, Schlatt L, Barnych B, Passerini AG, Hammock BD (2019) Epoxyeicosatrienoic acid (EET)-stimulated angiogenesis is mediated by epoxy hydroxyeicosatrienoic acids (EHETs) formed from COX-2. J Lipid Res. Epub 2019/10/24. https://doi.org/10.1194/jlr.M094219. PubMed PMID: 31641036
106. Gartung A, Yang J, Sukhatme VP, Bielenberg DR, Fernandes D, Chang J, Schmidt BA, Hwang SH, Zurakowski D, Huang S, Kieran MW, Hammock BD, Panigrahy D (2019) Suppression of chemotherapy-induced cytokine/lipid mediator surge and ovarian cancer by a dual COX-2/sEH inhibitor. Proc Natl Acad Sci U S A 116(5):1698–1703. Epub 2019/01/17. https://doi.org/10.1073/pnas.1803999116. PubMed PMID: 30647111; PMCID: PMC6358686
107. Schmelzer KR, Inceoglu B, Kubala L, Kim IH, Jinks SL, Eiserich JP, Hammock BD (2006) Enhancement of antinociception by coadministration of nonsteroidal anti-inflammatory drugs and soluble epoxide hydrolase inhibitors. Proc Natl Acad Sci U S A 103(37):13646–13651. Epub 2006/09/05. https://doi.org/10.1073/pnas.0605908103. PubMed PMID: 16950874; PMCID: PMC1564210
108. Chopra S, Giovanelli P, Alvarado-Vazquez PA, Alonso S, Song M, Sandoval TA, Chae CS, Tan C, Fonseca MM, Gutierrez S, Jimenez L, Subbaramaiah K, Iwawaki T, Kingsley PJ, Marnett LJ, Kossenkov AV, Crespo MS, Dannenberg AJ, Glimcher LH, Romero-Sandoval EA, Cubillos-Ruiz JR (2019) IRE1alpha-XBP1 signaling in leukocytes controls prostaglandin biosynthesis and pain. Science 365(6450). Epub 2019/07/20. https://doi.org/10.1126/science.aau6499. PubMed PMID: 31320508.
109. Sasich LD, Barasain MA, Al Kudsi MA (2008) The cardiovascular risks of etoricoxib (Arcoxia). Ann Saudi Med 28(2):141–142. Epub 2008/04/10. https://doi.org/10.5144/0256-4947.2008.141. PubMed PMID: 18398284; PMCID: PMC6074531
110. Bruno F, Spaziano G, Liparulo A, Roviezzo F, Nabavi SM, Sureda A, Filosa R, D'Agostino B (2018) Recent advances in the search for novel 5-lipoxygenase inhibitors for the treatment of asthma. Eur J Med Chem 153:65–72. Epub 2017/11/15. https://doi.org/10.1016/j.ejmech.2017.10.020. PubMed PMID: 29133059
111. Yang J, Bratt J, Franzi L, Liu JY, Zhang G, Zeki AA, Vogel CF, Williams K, Dong H, Lin Y, Hwang SH, Kenyon NJ, Hammock BD (2015) Soluble epoxide hydrolase inhibitor attenuates inflammation and airway hyperresponsiveness in mice. Am J Respir Cell Mol Biol 52(1):46–55. https://doi.org/10.1165/rcmb.2013-0440OC. PubMed PMID: 24922186
112. Gobel T, Diehl O, Heering J, Merk D, Angioni C, Wittmann SK, Buscato E, Kottke R, Weizel L, Schader T, Maier TJ, Geisslinger G, Schubert-Zsilavecz M, Steinhilber D, Proschak E, Kahnt AS (2019) Zafirlukast is a dual modulator of human soluble Epoxide Hydrolase and Peroxisome Proliferator-activated receptor gamma. Front Pharmacol 10:263. Epub 2019/04/06. https://doi.org/10.3389/fphar.2019.00263. PubMed PMID: 30949053; PMCID: PMC6435570
113. Watkins BA (2019) Diet, endocannabinoids, and health. Nutr Res (New York, NY) 70:32–39. Epub 2019/07/10. https://doi.org/10.1016/j.nutres.2019.06.003. PubMed PMID: 31280882
114. van Esbroeck ACM, Janssen APA, Cognetta AB, Ogasawara D, Shpak G, van der Kroeg M, Kantae V, Baggelaar MP, de Vrij FMS, Deng H, Allara M, Fezza F, Lin Z, van der Wel T, Soethoudt M, Mock ED, den Dulk H, Baak IL, Florea BI, Hendriks G, DePetrocellis L, Overkleeft HS, Hankemeier T, De Zeeuw CI, Di Marzo V, Maccarrone M, Cravatt BF, Kushner SA, van der Stelt M (2017) Activity-based protein profiling reveals off-target proteins of the FAAH inhibitor BIA 10-2474. Science 356(6342):1084–1087. Epub 2017/06/10. https://doi.org/10.1126/science.aaf7497. PubMed PMID: 28596366; PMCID: PMC5641481
115. Habib AM, Okorokov AL, Hill MN, Bras JT, Lee MC, Li S, Gossage SJ, van Drimmelen M, Morena M, Houlden H, Ramirez JD, Bennett DLH, Srivastava D, Cox JJ (2019) Microdeletion in a FAAH pseudogene identified in a patient with high anandamide concentrations and pain insensitivity. Br J Anaesth 123(2):e249-e253. Epub 2019/04/02. https://doi.org/10.1016/j.bja.2019.02.019. PubMed PMID: 30929760; PMCID: PMC6676009
116. Sasso O, Wagner K, Morisseau C, Inceoglu B, Hammock BD, Piomelli D (2015) Peripheral FAAH and soluble epoxide hydrolase inhibitors are synergistically antinociceptive. Pharmacol Res 97:7–15. Epub 2015/04/18. https://doi.org/10.1016/j.phrs.2015.04.001. PubMed PMID: 25882247; PMCID: PMC4464910
117. Kodani SD, Bhakta S, Hwang SH, Pakhomova S, Newcomer ME, Morisseau C, Hammock BD (2018) Identification and optimization of soluble epoxide hydrolase inhibitors with dual potency towards fatty acid amide hydrolase. Bioorg Med Chem Lett. Epub 2018/01/26. https://doi.org/10.1016/j.bmcl.2018.01.003. PubMed PMID: 29366648
118. Inceoglu B, Wagner K, Schebb NH, Morisseau C, Jinks SL, Ulu A, Hegedus C, Rose T, Brosnan R, Hammock BD (2011) Analgesia mediated by soluble epoxide hydrolase inhibitors is dependent on cAMP. Proc Natl Acad Sci U S A 108(12):5093–5097. https://doi.org/10.1073/pnas.1101073108. PubMed PMID: 21383170; PMCID: 3064364
119. Blocher R, Wagner KM, Gopireddy RR, Harris TR, Wu H, Barnych B, Hwang SH, Xiang YK, Proschak E, Morisseau C, Hammock BD (2018) Orally avail-

able soluble Epoxide Hydrolase/Phosphodiesterase 4 dual inhibitor treats inflammatory pain. J Med Chem 61(8):3541–3450. Epub 2018/04/04. https://doi.org/10.1021/acs.jmedchem.7b01804. PubMed PMID: 29614224; PMCID: PMC5933862

120. Hwang SH, Wecksler AT, Zhang G, Morisseau C, Nguyen LV, Fu SH, Hammock BD (2013) Synthesis and biological evaluation of sorafenib- and regorafenib-like sEH inhibitors. Bioorg Med Chem Lett 23(13):3732–3737. Epub 05/15. https://doi.org/10.1016/j.bmcl.2013.05.011. PubMed PMID: 23726028

121. Liang Z, Zhang B, Xu M, Morisseau C, Hwang SH, Hammock BD, Li QX (2019) 1-Trifluoromethoxyphenyl-3-(1-propionylpiperidin-4-yl) urea, a selective and potent dual inhibitor of soluble epoxide hydrolase and p38 kinase intervenes in Alzheimer's Signaling in human nerve cells. ACS Chem Neurosci 10(9):4018–4030. Epub 2019/08/06. https://doi.org/10.1021/acschemneuro.9b00271. PubMed PMID: 31378059

122. Reddy AS, Zhang S (2013) Polypharmacology: drug discovery for the future. Expert Rev Clin Pharmacol 6(1):41–47. Epub 2013/01/01. https://doi.org/10.1586/ecp.12.74. PubMed PMID: 23272792; PMCID: PMC3809828

123. Kim JH, Morgan AM, Tai BH, Van DT, Cuong NM, Kim YH (2016) Inhibition of soluble epoxide hydrolase activity by compounds isolated from the aerial parts of Glycosmis stenocarpa. J Enzyme Inhib Med Chem 31(4):640–644. Epub 2015/10/08. https://doi.org/10.3109/14756366.2015.1057719. PubMed PMID: 26444316

124. Sun YN, Kim JH, Li W, Jo AR, Yan XT, Yang SY, Kim YH (2015) Soluble epoxide hydrolase inhibitory activity of anthraquinone components from aloe. Bioorg Med Chem 23(20):6659–6665. Epub 2015/09/16. https://doi.org/10.1016/j.bmc.2015.09.003. PubMed PMID: 26372074

125. Dimitropoulou C, West L, Field MB, White RE, Reddy LM, Falck JR, Imig JD (2007) Protein phosphatase 2A and Ca2+–activated K+ channels contribute to 11,12-epoxyeicosatrienoic acid analog mediated mesenteric arterial relaxation. Prostaglandins Other Lipid Mediat 83(1–2):50–61. Epub 2007/01/30. https://doi.org/10.1016/j.prostaglandins.2006.09.008. PubMed PMID: 17259072

126. Shen HC, Hammock BD (2012) Discovery of inhibitors of soluble epoxide hydrolase: a target with multiple potential therapeutic indications. J Med Chem 55(5):1789–1808. https://doi.org/10.1021/jm201468j. PubMed PMID: 22168898; PMCID: 3420824

127. Arnold C, Markovic M, Blossey K, Wallukat G, Fischer R, Dechend R, Konkel A, von Schacky C, Luft FC, Muller DN, Rothe M, Schunck WH (2010) Arachidonic acid-metabolizing cytochrome P450 enzymes are targets of {omega}-3 fatty acids. J Biol Chem 285(43):32720–32733. Epub 2010/08/25. https://doi.org/10.1074/jbc.M110.118406. PubMed PMID: 20732876; PMCID: PMC2963419

128. Hutchens MP, Nakano T, Dunlap J, Traystman RJ, Hurn PD, Alkayed NJ (2008) Soluble epoxide hydrolase gene deletion reduces survival after cardiac arrest and cardiopulmonary resuscitation. Resuscitation 76(1):89–94. Epub 2007/08/31. https://doi.org/10.1016/j.resuscitation.2007.06.031. PubMed PMID: 17728042; PMCID: PMC2585367

129. Chen D, Whitcomb R, MacIntyre E, Tran V, Do ZN, Sabry J, Patel DV, Anandan SK, Gless R, Webb HK (2012) Pharmacokinetics and pharmacodynamics of AR9281, an inhibitor of soluble epoxide hydrolase, in single- and multiple-dose studies in healthy human subjects. J Clin Pharmacol 52(3):319–328. Epub 2011/03/23. https://doi.org/10.1177/0091270010397049. PubMed PMID: 21422238

130. Morris JK, Piccolo BD, John CS, Green ZD, Thyfault JP, Adams SH (2019) Oxylipin profiling of alzheimer's disease in nondiabetic and type 2 diabetic elderly. Metabolites 9(9). Epub 2019/09/08. https://doi.org/10.3390/metabo9090177. PubMed PMID: 31491971; PMCID: PMC6780570

131. Valdes AM, Ravipati S, Pousinis P, Menni C, Mangino M, Abhishek A, Chapman V, Barrett DA, Doherty M(2018) Omega-6 oxylipins generated by soluble epoxide hydrolase are associated with knee osteoarthritis. J Lipid Res 59(9):1763–1770. Epub 2018/07/11. https://doi.org/10.1194/jlr.P085118. PubMed PMID: 29986999; PMCID: PMC6121933

132. Caligiuri SPB, Aukema HM, Ravandi A, Lavallee R, Guzman R, Pierce GN (2017) Specific plasma oxylipins increase the odds of cardiovascular and cerebrovascular events in patients with peripheral artery disease. Can J Physiol Pharmacol 95(8):961–968. Epub 2017/07/18. https://doi.org/10.1139/cjpp-2016-0615. PubMed PMID: 28714336

133. Theken KN, Schuck RN, Edin ML, Tran B, Ellis K, Bass A, Lih FB, Tomer KB, Poloyac SM, Wu MC, Hinderliter AL, Zeldin DC, Stouffer GA, Lee CR (2012) Evaluation of cytochrome P450-derived eicosanoids in humans with stable atherosclerotic cardiovascular disease. Atherosclerosis 222(2):530–536. Epub 2012/04/17. https://doi.org/10.1016/j.atherosclerosis.2012.03.022. PubMed PMID: 22503544; PMCID: PMC3361525

134. Hennebelle M, Otoki Y, Yang J, Hammock BD, Levitt AJ, Taha AY, Swardfager W (2017) Altered soluble epoxide hydrolase-derived oxylipins in patients with seasonal major depression: an exploratory study. Psychiatry Res 252:94–101. Epub 2017/03/05. https://doi.org/10.1016/j.psychres.2017.02.056. PubMed PMID: 28259037; PMCID: PMC5611448

135. Santos JM, Park JA, Joiakim A, Putt DA, Taylor RN, Kim H (2017) The role of soluble epoxide hydrolase in preeclampsia. Med Hypotheses 108:81–85. Epub 2017/10/23. https://doi.org/10.1016/j.mehy.2017.07.033. PubMed PMID: 29055406

136. Yu D, Hennebelle M, Sahlas DJ, Ramirez J, Gao F, Masellis M, Cogo-Moreira H, Swartz RH, Herrmann N, Chan PC, Pettersen JA, Stuss DT, Black SE, Taha AY, Swardfager W (2019) Soluble epoxide hydrolase-derived linoleic acid Oxylipins in serum are associated with periventricular White matter Hyperintensities and vascular cognitive impairment. Transl Stroke Res 0(5):522–533. Epub 2018/11/18. https://doi.org/10.1007/s12975-018-0672-5. PubMed PMID: 30443886
137. Moghaddam MF, Grant DF, Cheek JM, Greene JF, Williamson KC, Hammock BD (1997) Bioactivation of leukotoxins to their toxic diols by epoxide hydrolase. Nat Med 3(5):562–5666. Epub 1997/05/01. PubMed PMID: 9142128
138. Kopf PG, Zhang DX, Gauthier KM, Nithipatikom K, Yi XY, Falck JR, Campbell WB (2010) Adrenic acid metabolites as endogenous endothelium-derived and zona glomerulosa-derived hyperpolarizing factors. Hypertension 55(2):547–554. Epub 2009/12/30. https://doi.org/10.1161/hypertensionaha.109.144147. PubMed PMID: 20038752; PMCID: PMC2819927
139. Serhan CN (2014) Pro-resolving lipid mediators are leads for resolution physiology. Nature 510(7503):92–101. Epub 2014/06/06. https://doi.org/10.1038/nature13479. PubMed PMID: 24899309; PMCID: PMC4263681
140. McGiff JC, Quilley J (1999) 20-HETE and the kidney: resolution of old problems and new beginnings. Am J Phys 277(3 Pt 2):R607–R623. Epub 1999/09/14. PubMed PMID: 10484476
141. Roman RJ (2002) P-450 metabolites of arachidonic acid in the control of cardiovascular function. Physiol Rev 82(1):131–185. Epub 2002/01/05. https://doi.org/10.1152/physrev.00021.2001. PubMed PMID: 11773611
142. Mezentsev A, Mastyugin V, Seta F, Ashkar S, Kemp R, Reddy DS, Falck JR, Dunn MW, Laniado-Schwartzman M (2005) Transfection of cytochrome P4504B1 into the cornea increases angiogenic activity of the limbal vessels. J Pharmacol Exp Ther 315(1):42–50. Epub 2005/07/13. https://doi.org/10.1124/jpet.105.088211. PubMed PMID: 16009741
143. Seta F, Patil K, Bellner L, Mezentsev A, Kemp R, Dunn MW, Schwartzman ML (2007) Inhibition of VEGF expression and corneal neovascularization by siRNA targeting cytochrome P450 4B1. Prostaglandins Other Lipid Mediat 84(3–4):116–127. Epub 2007/11/10. https://doi.org/10.1016/j.prostaglandins.2007.05.001. PubMed PMID: 17991614; PMCID: PMC2128778
144. Fromel T, Kohlstedt K, Popp R, Yin X, Awwad K, Barbosa-Sicard E, Thomas AC, Lieberz R, Mayr M, Fleming I (2013) Cytochrome P4502S1: a novel monocyte/macrophage fatty acid epoxygenase in human atherosclerotic plaques. Basic Res Cardiol 108(1):319. Epub 2012/12/12. https://doi.org/10.1007/s00395-012-0319-8. PubMed PMID: 23224081
145. Weintraub NL, Fang X, Kaduce TL, VanRollins M, Chatterjee P, Spector AA (1999) Epoxide hydrolases regulate epoxyeicosatrienoic acid incorporation into coronary endothelial phospholipids. Am J Phys; 277(5):H2098–H2108. Epub 1999/11/24. https://doi.org/10.1152/ajpheart.1999.277.5.H2098. PubMed PMID: 10564166
146. Morisseau C, Hammock BD (2005) Epoxide hydrolases: mechanisms, inhibitor designs, and biological roles. Annu Rev Pharmacol Toxicol; 45:311–33. https://doi.org/10.1146/annurev.pharmtox.45.120403.095920. PubMed PMID: 15822179

Druggable Sphingolipid Pathways: Experimental Models and Clinical Opportunities

6

Victoria A. Blaho

Abstract

Intensive research in the field of sphingolipids has revealed diverse roles in cell biological responses and human health and disease. This immense molecular family is primarily represented by the bioactive molecules ceramide, sphingosine, and sphingosine 1-phosphate (S1P). The flux of sphingolipid metabolism at both the subcellular and extracellular levels provides multiple opportunities for pharmacological intervention. The caveat is that perturbation of any single node of this highly regulated flux may have effects that propagate throughout the metabolic network in a dramatic and sometimes unexpected manner. Beginning with S1P, the receptors for which have thus far been the most clinically tractable pharmacological targets, this review will describe recent advances in therapeutic modulators targeting sphingolipids, their chaperones, transporters, and metabolic enzymes.

Keywords

Sphingolipid · S1P · Fingolimod · Siponimod · Ozanimod

V. A. Blaho (✉)
Immunity and Pathogenesis Program, Sanford Burnham Prebys Medical Discovery Institute, La Jolla, CA, USA
e-mail: vblaho@sbpdiscovery.org

6.1 Introduction

The family of bioactive sphingolipid molecules is immense, but all are characterized by the same core component, the sphingoid backbone, which is simply an amino alcohol with a long carbon chain [1]. The canonical sphingolipid, sphingosine, can be enzymatically modified to add fatty acids, phosphorous-containing head groups (e.g., phosphocholine), sugar moieties, and changes in acyl chain saturation [1–3]. Flux through the sphingolipid pathway has far-ranging effects, from cellular architecture to multi-organ system coordination [4, 5]. The ubiquity of sphingolipids presents both challenges to and opportunities for their manipulation. Unlike other bioactive lipids with proposed shunting to different enzymatic pathways, the flow of sphingolipids rarely has an alternative for degradation or synthesis other than reversal [6–11].

Subsequently, while inhibition of a particular enzyme stops generation of a specific product, the biological outcome could be the result of increased concentrations of upstream precursors and not necessarily the most immediate parent molecule. For instance, a great deal of effort has focused on the inhibition of two enzymes, the sphingosine kinases (Sphk1/2), for the treatment of cancer. Although the desired decrease in product may be achieved, an increase in the parent molecule of the Sphk substrate is commonly credited with affecting the biological outcome.

Y. Kihara (ed.), *Druggable Lipid Signaling Pathways*, Advances in Experimental Medicine and Biology 1274, https://doi.org/10.1007/978-3-030-50621-6_6

Like a malfunctioning traffic light, a blockage at one sphingolipid node can often have repercussions throughout the metabolic pathway (Fig. 6.1). This review will present known mechanisms for therapies that target sphingolipid signaling pathways, including our current understanding of sphingosine 1-phosphate (S1P) receptor modulators, S1P chaperones and transporters, and sphingolipid metabolic enzymes [12–14].

6.1.1 Sphingosine 1-Phosphate (S1P) and Its Receptors, $S1P_{1-5}$

Unlike the on-demand production of other bioactive lipids, the signaling molecule sphingosine 1-phosphate (S1P) is omnipresent in blood and lymph circulation, with both human and murine concentrations of 18:1 S1P in the mid-nanomolar to low micromolar range [15–18]. Blood plasma concentrations of other powerful bioactive lipids, such as prostaglandin E_2 (PGE_2) and leukotriene B_4 (LTB_4), are approximately 10 times less than S1P [19]. Despite high plasma concentrations, tissue concentrations of S1P are far lower than circulating levels, setting up an "S1P gradient" in which small alterations in local S1P concentrations are sensed by finely-tuned mechanisms within specific cell types, potentially triggering dramatic effects [20–22].

The five S1P receptors (S1PR), $S1P_{1-5}$, are members of the rhodopsin class of G protein-coupled receptors (GPCR). This family of proteins constitutes 40% of drug targets [23–27]. The first S1PR, $S1P_1$, was cloned from endothelial cells (EC) owing to its critical role in endothelial cell and vascular biology [28, 29]. Lack of *S1pr1* results in embryonic lethality due to hemorrhage from immature vasculature [30]. Induced *S1pr1*$^{-/-}$ in endothelial cells (EC) causes increased vascular leakage and inflammatory molecule expression, and *S1pr1*$^{-/-}$ tumor vessels have disordered architecture and poor perfusion [31–33]. Conversely, overexpression of EC $S1P_1$ reinforces the vascular barrier, increasing perfusion of tumor vessels [31, 33]. As one of the genes regulated by the transcription factor KLF2, a hemodynamic responsive protein, EC *S1pr1* also signals in response to fluid shear stress [31, 32, 34].

In the vasculature, as well as in other cell and tissues types, $S1P_1$ and $S1P_2$ signaling counteract each other, with $S1P_3$ signaling acting as a modifier [35–37]. Effects on the vasculature are of great importance in understanding the mechanisms and potential side effects of S1PR modulating drugs. The literature regarding S1PR and the vascular system is vast, and the reader is referred to several recent excellent reviews on the topic for more detailed descriptions [33, 38, 39].

Although the first S1PR was cloned from EC, S1PR modulating drugs work primarily through immune cell modification [23, 28]. There are currently three FDA-approved S1PR modulating drugs: the first-in-class FTY720 (fingolimod/Gilenya), BAF312 (siponimod/Mayzent), and RPC1063 (ozanimod/Zeposia) are approved for treatment of the relapsing-remitting form of multiple sclerosis (RRMS) (Fig. 6.2) [12, 40–42]. A New Drug Application (NDA) has been submitted for a fourth compound, ACT128800 (ponesimod), also for the treatment of RRMS [43, 44]. Siponimod is also approved for secondary progressive MS (SPMS), the stage of MS disease progression after RRMS. While numerous S1PR modulators are utilized as pharmacological tools in the laboratory, most were not tested in the clinical setting due to poor solubility, *in vivo* stability, half-life, or specificity. The side effects of fingolimod and siponimod illustrate why increasing target specificity is a driving factor behind the continued development of S1PR modulators.

FTY720 is a sphingosine analogue derivative of a fungal metabolite and must be phosphorylated (FTY720P) for recognition by S1PRs [45, 46]. Although FTY720P activates all S1PRs except $S1P_2$, $S1P_1$ binding FTY720P causes its polyubiquitination and degradation, resulting in "functional antagonism" [47]. In humans, FTY720 has a long *in vivo* half-life of greater than 100 hours, combined with low oral clearance and a high volume of distribution that is likely due to absorbance into lipid-rich tissues and cell membranes [48–51]. Lymphopenia is the most striking effect of FTY720 treatment and the

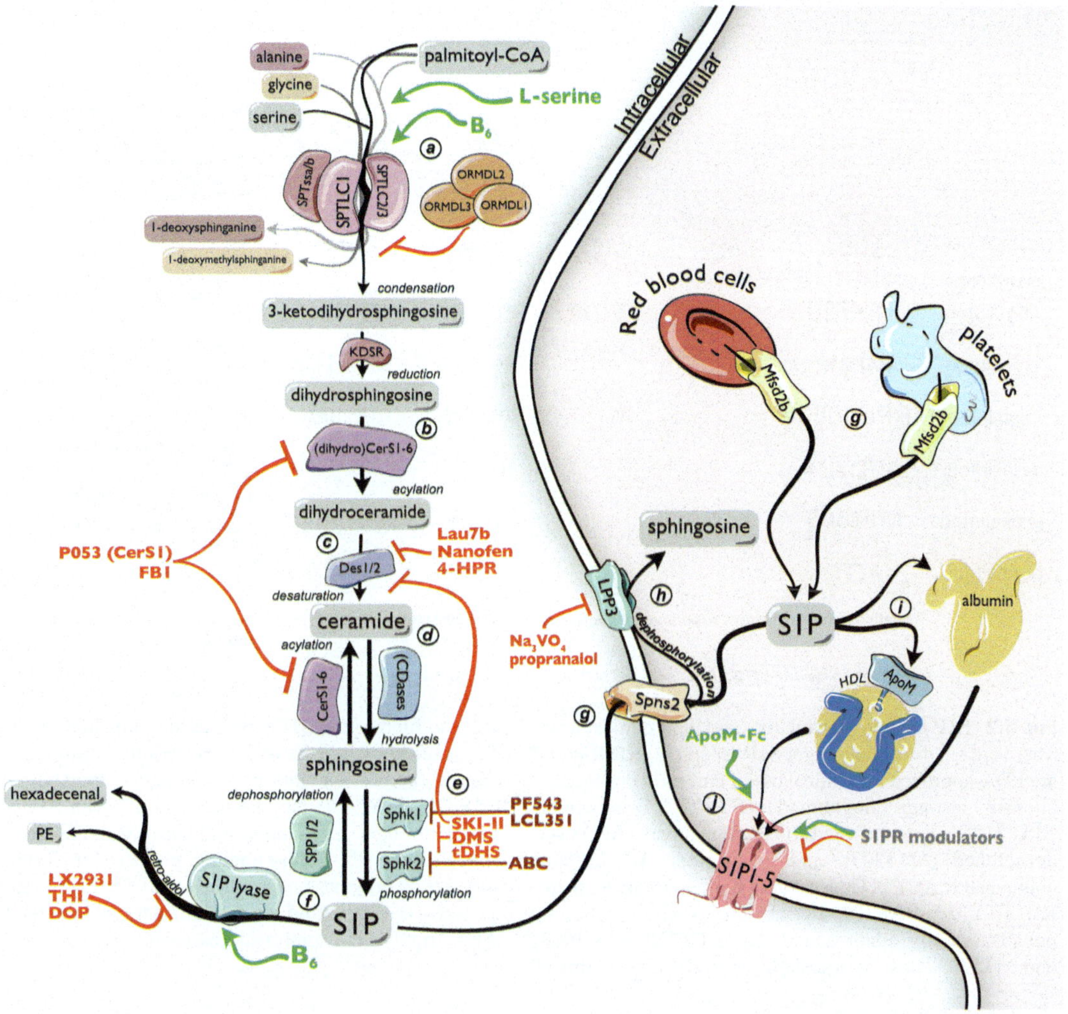

Fig. 6.1 Sphingolipid metabolism, signaling pathways, and pharmacological interventions
(**a**) *De novo* sphingolipid synthesis begins in the endoplasmic reticulum (ER) membrane with the condensation of the fatty acyl-CoA palmitoyl-CoA and the amino acid serine to form 3-ketodihydrosphingosine. The reaction is catalyzed by the pyridoxal 5′-phosphate (PLP)-dependent heterodimeric enzyme serine palmitoyl-CoA transferase (SPT), which consists of the SPTLC1 subunit and either SPTLC2 or SPTLC3. SPT activity is enhanced by SPTssa or SPTssb and additional vitamin B_6 (B_6) and is homeostatically inhibited by the ORMDL1-3 proteins. If alanine or glycine are utilized instead of serine, toxic 1-deoxysphinganine or 1-deoxymethylsphinganine, respectively, are produced. Generation of these toxic deoxysphingolipids can be reduced by supplementation with L-serine. 3-ketodigydrosphingosine is then reduced by 3-ketodihydrosphingosine reductase (KDSR) to dihydrosphingosine. (**b**) Acylation of dihydrosphingosine by the (dihydro) ceramide synthases (CerS) 1-6 generates dihydroceramide. All CerS can be inhibited by fumonisin B1 (FB1) and P053 specifically inhibits CerS1. (**c**) Dihydroceramide is then desaturated by dihydroceramide desaturase (Des1/2) to ceramide. This reaction can be inhibited by Lau7b, Nanofen, or fenretinide (4-HPR). (**d**) Ceramide is hydrolyzed to sphingosine by ceramidases (CDases) according to subcellular location. Sphingosine can be converted back to ceramide by CerS1-6 or (**e**) phosphorylated by sphingosine kinases (Sphk1/2) to sphingsosine 1-phosphate (S1P). Sphk1 and 2 are inhibited by SKI-II, dimethylsphingosine (DMS) or *D,L*-threo-dihydrosphingosine (tDHS). Sphk1 is specifically inhibited by PF543 or LCL351 and Sphk2 is specifically inhibited by ABC294640 (ABC). (**f**) Intracellular S1P can be dephosphorylated by S1P phosphatases (SPP1/2) or irreversibly degraded by retro-aldol cleavage of the C2-C3 bond by S1P lyase to form (2E)-hexadecenal and phosphoethanolamine (PE). S1P lyase is a PLP-dependent enzyme whose activity can be increased by B_6 supplementation or inhibited by LC2951, 2-acetyl-4-tetrahydroxybutylimidazole (THI), or 4-deoxypyridoxine (DOP). (**g**) S1P can be actively transported out of cells by Spns2 or from red blood cells and platelets by Mfsd2b. (**h**) Extracellular S1P can be dephosphorylated by lipid phosphate phosphatase 3 (LPP), which is non-specifically inhibited by sodium orthovanadate (Na_3VO_4) or propranolol. (**i**) S1P in circulation is carried by albumin or the HDL-bound S1P-specific chaperone apolipoprotein M (ApoM). (**j**) Albumin-S1P, ApoM-S1P, or synthetic ApoM-Fc-S1P can activate cell surface S1P receptors, $S1P_{1–5}$. The multiple S1PR agonists and antagonists in clinical use or testing are described in more detail in Fig. 6.2

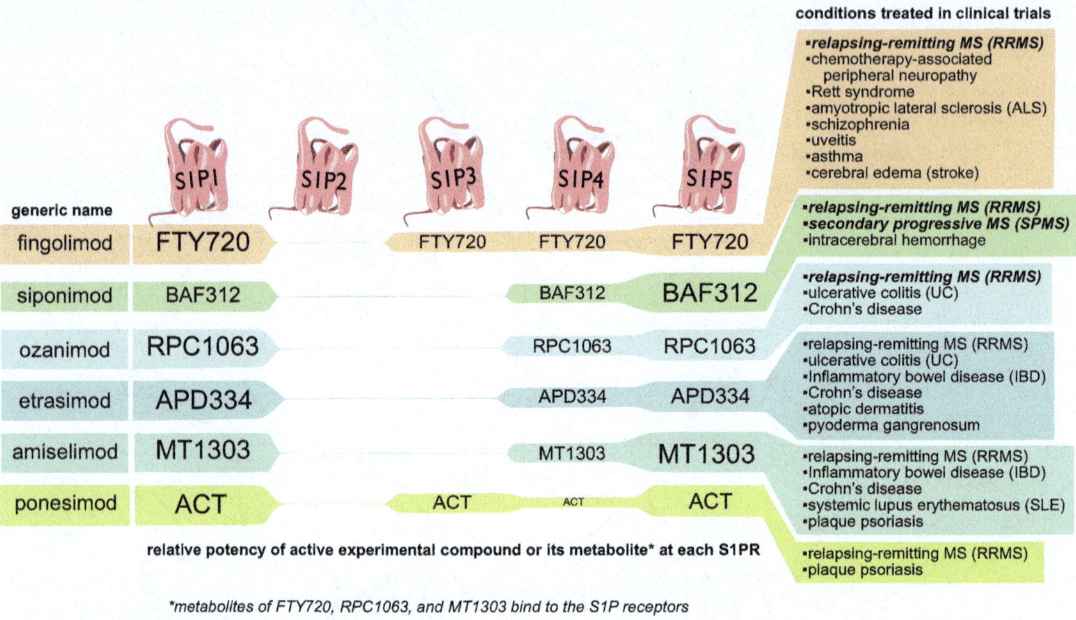

Fig. 6.2 S1P receptor modulators and their clinical uses There are six drugs that target S1P receptors (S1PRs) and are FDA-approved for clinical use or currently undergoing clinical testing: fingolimod (FTY720), siponimod (BAF312), ozanimod (RPC1063), etrasimod (APD334), amiselimod (MT1303), and ponesimod (ACT128800; abbreviated "ACT"). Generic drug names are given on the left. FTY720, BAF312, etc. are the experimental compounds actually administered, but FTY720, RPC1063, and MT1303 must be metabolized to an active form in order to bind to the S1PR. For each compound, the relative potency of it or its active metabolite is shown below each receptor. For instance, the active metabolite of FTY720, FTY720P, has the greatest effect on $S1P_1$, the second strongest effect on $S1P_5$, relatively similar potency at $S1P_3$ and $S1P_4$, and no binding to $S1P_2$. Conditions for which each drug has undergone clinical trials are listed on the right. Listed in **bold** are conditions for which that drug is a currently FDA-approved treatment

greatest contributor to its mechanism of action: it restricts lymphocytes from exiting the lymphoid organs, preventing them from trafficking to attack other organs [15, 52–55]. Lymphopenia is largely the result of lymphocyte-expressed $S1P_1$ functional antagonism, since mature B and T cells use $S1P_1$ to sense and migrate toward the high circulating S1P concentrations [56]. In the animal model of MS, experimental autoimmune encephalomyelitis (EAE), and in MS patients treated with FTY720, autoreactive lymphocytes are prevented from entering circulation and reaching the central nervous system (CNS) [45, 57].

A phase III trial comparing FTY720 to cyclosporine in the prevention of renal transplantation rejection found that although higher doses are generally well-tolerated compared to high doses of other classical immunosuppressants, kidney function was consistently lower in FTY720-treated cohorts and small increases in pulmonary resistance were also documented [58]. The most frequently observed adverse effect in clinical studies of FTY720/fingolimod was macular edema, which, along with the other adverse events, directly correlated with dose, whereas efficacy did not [50, 59, 60]. This indicates that the escalating dose results in increased engagement of the S1PR in order of their affinity for FTY720P: $S1P_1 > S1P_3 > S1P_4 \geq S1P_5$ [61, 62]. With increasing S1PR engagement comes greater signaling complexity, since cells are likely to express multiple S1PR [23].

S1PR modulators may also find use in the treatment of neurological disorders other than MS. In the mSOD1^{G93A} model of amyotrophic lateral sclerosis (ALS), treatment of mice with 0.1 mg/kg FTY720 significantly delayed progression of neurological deterioration and extended survival [63]. In brains of Alzheimer's disease patients (AD), *S1PR1* mRNA was significantly decreased, but *S1PR3* was increased [64]. An increase in brain $S1P_3$ may partially explain why FTY720 treatment could effectively reduce brain Aβ plaques and neuroinflammation and improve neurological function in multiple mouse models of AD [65–67]. In a mouse model of AD, FTY720 treatment reduced the numbers of activated microglia and astrocytes [67]. Astrocyte-intrinsic $S1P_1$ may be key to FTY720's efficacy in neurological disorders, since a subset of astrocytes are also activated during EAE and reduced with FTY720 treatment [68].

The possibility of using FTY720 as an anti-cancer therapy has been enthusiastically pursued since early reports of *in vitro* and *in vivo* pro-apoptotic activity [69–71]. The immunosuppressive effect of FTY720 was originally attributed to induction of T cell apoptosis [69, 70, 72, 73]. Subsequently, FTY720 has been tested for efficacy in cancer cells derived from, and cancer models affecting, almost every organ system including leukemia, prostate, glioma, breast, mesothelioma, non-small cell lung carcinoma, pancreatic, and colorectal [70, 71, 73–81]. Rather than affecting S1PR signaling, the proposed anti-tumoral/pro-apoptotic effect of FTY720 is inhibition of sphingosine kinases and activation of protein phosphatase 2A (PP2A) [81–84]. A recent study used NMR spectroscopy to characterize the direct interaction of FTY720 with SET, a PP2A inhibitor [85].

The recommended dosage for Gilenya (FTY720, fingolimod) is 0.5 mg once daily, and multiple clinical trials examining the long-term effects of FTY720 administration in patients have been completed or are ongoing (NCT01201356, NCT02720107, NCT00662649, NCT01281657, NCT02307877, NCT03216915, NCT02232061, NCT01442194). Clinical trials have confirmed that adverse events, both common (acute bradycardia, macular edema, renal dysfunction) and uncommon (fatal infection, seizure, lymphoma) are dose-dependent while efficacy and effectiveness are not [59, 60, 86–88]. Adverse events are not commonly recapitulated in mice, and mouse studies investigating the use of FTY720 in cancer models have frequently (although not always) used extended dosing of 10 mg/kg. However, the approximate conversion of a 10 mg/kg dose for a mouse reveals that this is equivalent to a single oral dose of 56 mg in a human – greater than 100 times the approved dose and 10 times the highest dose administered in clinical trials [89, 90]. Currently, two clinical trials are investigating the use of FTY720 treatment during cancer, but with the goal of reducing chemotherapy-induced neuropathy (NCT03943498, NCT03941743). No other S1PR modulators are undergoing clinical trials for cancer as the targeted condition.

There is potential for modulating S1PR signaling for the treatment of viral infections, particularly HIV. T cells from the lymph nodes (LN) of HIV patients showed impaired migratory activity, including toward S1P, possibly explaining HIV lymphadenopathy [91]. CD69 expression usually inversely correlates with that of $S1P_1$ because their physical interaction down-regulates surface $S1P_1$ and $S1P_1$ expression suppresses that of CD69, yet both CD69 and $S1P_1$ were decreased in cells from viremic HIV patients compared to control cells [91, 92]. KLF2, the primary transcription factor responsible for T cell *S1PR1* gene activation, was also down regulated in LN, but surprisingly, mRNAs for *S1PR1* and *KLF2* from purified T cells were *not* decreased and anti-retroviral therapy (ART) improved responsiveness to S1P [91, 93]. *In vitro*, cells expressing $S1P_1$ allowed greater HIV replication and treatment with the $S1P_1$ agonist SEW2871 resulted in reversal of latency and reactivation of viral replication in peripheral blood cells and LN, whereas FTY720P decreased *in vitro* virus production by monocyte-derived dendritic cells [94]. An early study found that FTY720 administration to simian immunodeficiency virus (SIV)-infected macaques did not affect viremia and proviral DNA [95]. However, a more recent study reported that FTY720 treatment of SIV-infected rhesus macaques initiated after at least 4 months of combination ART (cART) both increased LN numbers of cytotoxic T cells and decreased the infection of LN T follicular helper (Tfh) cells, as determined by level of proviral DNA [96]. The authors suggested that FTY720 treatment could be started concurrently with cART or cytotoxic T cell activating IL-2 or IL-15 therapies, and by decreasing the number of circulating cells with viral RNA, the viral reservoir allowed into circulation would be decreased. FTY720 may also protect from HIV-associated dementia in the context of inflammation. Human neural progenitor cells (hNP1) exposed to HIV and treated with FTY720P had decreased expression of immune and inflammatory response-related genes [97]. Thus, FTY720 treatment at doses relevant to human patients, particularly in the context of cART, has potential for increasing immune responses that would be beneficial for controlling HIV replication and suppressing inflammatory responses that would be neurotoxic.

$S1P_2$ expression is also necessary for regulating migration of many cell types and coordinates S1P responsiveness with $S1P_1$ for proper positioning of various B and T cell subsets within the lymphoid organs, including germinal center B cell (GCB) confinement, and suppression of their over-proliferation and apoptosis [98]. Tonsil-derived T central memory (TCM) and T resident memory (TRM) cells are chemorepulsed *in vitro* by S1P via $S1P_2$ signaling, which counters pro-migratory CXCL12 responses [99]. Both TCM and TRM are defined by markers known to counter-regulate $S1P_1$ responses: TCM are $CCR7^+$ and TRM are $CD69^+$ [100, 101]. Immunohistochemically, expression of $S1P_1$ and $S1P_2$ are mutually exclusive and vary between different diffuse large B cell lymphoma (DLBCL) types [102]. *S1PR2* mutations were present in more than 25% of diffuse large B cell lymphoma (DLBCL) patients, and the *S1PR2* mutations present in GCB-DLBCL resulted in altered protein expression or an inability to bind the G protein $G\alpha_{13}$, leading to loss of GCB confinement [103, 104]. Expression of a negative regulator of *S1PR2*, the transcription factor FOXP1, correlates with poor survival in activated B cell (ABC)-DLBCL patients [105]. A subset of Tfh with high PD-1 expression coordinates $S1P_2$ and CXCR5 signaling to localize to the GC, indicating that $S1P_2$ mutations could impair antibody production and lymphomagenesis through altered T-B cell interactions [106].

Follicular B cells (FBC) utilize $S1P_1$ to migrate to the marginal zone (MZ) from the follicle, whereas MZ B cells (MZB) use $S1P_1$ signals to shuttle between the MZ and follicle [107, 108]. MZB $S1P_1$ phosphorylation by G protein-coupled receptor kinase 2 (GRK2) results in their desensitization to S1P, permitting them to respond to other chemotactic signals [108]. GRK2 also inhibits MALT1, a protease and scaffold protein positive regulator of NF-κB, impairing the survival of DLBCL cells [109, 110]. Since ABC-DLBCL expression of GRK2 positively correlated with patient survival, altered S1P-directed migratory patterns may be related to anti-apoptotic signaling [110].

In vitro studies repeatedly indicated that $S1P_3$ would be the primary regulator of bone marrow (BM) B cell egress; however, *in vivo* studies determined that agonism and subsequent down-regulation of $S1P_1$ allowed $S1P_3$ signals to be misinterpreted as the dominant BM egress signal [111, 112]. In a mouse model of autoantibody production, immature B cells in the BM use $S1P_3$ to migrate from the parenchyma to the sinusoids, but only if they are not autoreactive [112]. Subsequently, $S1P_1$ signaling draws them from BM into circulation [111]. Mature B cells also use $S1P_3$ to position themselves within the MZ, in addition to using $S1P_1$ as their cue to migrate to the MZ [113].

In human leukemia cells (CLL (chronic lymphocytic leukemia), pre-B-ALL (pre-B cell acute lymphoblastic leukemia), and CLL (chronic lymphocytic leukemia)), $S1P_1$ was down regulated by 10-fold and had impaired *in vitro* chemotaxis toward S1P as compared to control cells [114]. *S1PR4* mRNA was co-expressed with that of *S1PR1* and its over-expression appeared to modulate $S1P_1$-directed chemotaxis [114]. Highly expressed by lymphocytes, $S1P_4$ appears to have only modulatory effects on $S1P_1$-induced migration in most lymphocyte types, but may be critical for the regulation of lymphocyte proliferation and activation. T cell $S1P_4$ can suppress proliferation and IL-2 and IL-4 production initiated by anti-CD3/CD28 activation and signaling, providing a partial explanation for the skewing of *S1pr4*$^{-/-}$ immune response to Th2 and away from Th17 [115, 116]. $S1P_4$ is not a critical regulator of B cell migration or activation in the spleen, but some peritoneal B cell subsets rely on $S1P_4$ signaling, rather than $S1P_1$. Peritoneal B1a, B1b, and B2 B cells all expressed *S1pr4* mRNA, although only B1a and B1b cells appeared to use either $S1P_1$ or $S1P_4$ for migratory cues [117]. Although *S1pr4*$^{-/-}$ animals had normal numbers of B cells in circulation, the numbers of B1a and B1b peritoneal cells were significantly reduced, as was the level of secretory IgA in the small intestine [116, 117].

$S1P_4$ signaling may play a greater role in the innate immune system by regulating inflammatory responses. A missense variant of the human *S1PR4* was discovered that correlated with decreased numbers of neutrophils in circulation [118]. Dendritic cells (DC) utilize $S1P_4$ for their migration to lymph nodes in cooperation with CCL21 [116]. High concentrations of S1P *in vitro* (>3 μM) inhibited neutrophil and macrophage 5-lipoxygenase (5-LO), a leukotriene biosynthetic enzyme critical for innate immune function and activated in multiple inflammatory and autoimmune disorders, including arthritis and asthma [119]. Whole blood cell mRNA profiling of patients after aneurysmal subarachnoid hemorrhage found higher *S1PR4* transcript levels correlated with a greater risk of vasospasm, a major cause of severe cognitive defects and mortality, the pathogenesis of which may be linked to neutrophil recruitment and activation [120, 121]. Neutrophils are also of interest in psoriasis because of their induction of Th17 responses as well as direct pro-inflammatory activities [122]. *S1pr4*$^{-/-}$ mice had less inflammation in an imiquimod model of psoriasis as a result of decreased production of the macrophage and neutrophil chemokines CCL2 (MCP-1) and CXCL1 (KC), possibly because of decreased NF-κB activation [123].

S1pr4$^{-/-}$ mice had decreased pathology in the dextran sulfate sodium (DSS) colitis model, which correlated with decreased IL-6 production and a skewing from Th17 to increased Th2 differentiation, although their CD4$^+$ and CD8$^+$ cells had increased *in vitro* migration toward an intermediate S1P concentration (0.1 μM), and their DC had increased migration toward draining lymph nodes [116]. The S1PR agonist etrasimod activates $S1P_1 > S1P_5 > S1P_4$ with no activity at $S1P_2$ or $S1P_3$ and decreased inflammation in a T cell transfer model of colitis [124]. Etrasimod has also shown efficacy at a dose of 2 mg in clinical trials for treatment of moderate to severe active ulcerative colitis (UC), including histologic remission [125].

6.2 S1P Chaperone: Apolipoprotein M (APOM)

An ideal pharmacological target is one that has minimal impact on other components of the pathway. With this qualifier, the S1PR present the best option, since agonism or antagonism of a single receptor could be anticipated to impact only the signaling of the remaining S1PR, rather than changing flux within the entre sphingolipid metabolic pathway. The target with the second lowest possibility of large pathway perturbation is the primary S1P chaperone, apolipoprotein M (ApoM), the other chaperone being the nonspecific lipid transporter albumin [126, 127].

High concentrations of S1P can be found in plasmas of both blood and lymph, known to be bound to protein and lipoprotein fractions [128]. Although ApoM is present in the lymph, *Apom*$^{-/-}$ mice do not have significantly altered lymph S1P concentrations compared to the 60–70% drop in blood S1P, indicating that either all lymph S1P is bound to albumin or there is another lymph-specific S1P chaperone [16]. In the blood of mice constitutively lacking both albumin and ApoM, S1P is bound to ApoA4 [129]. A recent study of human males described a third pool of ApoM, neither lipoprotein nor protein associated, in blood, although it is unclear what the purpose is of this ApoM population [130].

In mice, the 60–70% ApoM-S1P bound to lipoprotein is usually associated with HDL; however, human lipoprotein studies have found that while S1P is usually highly correlated to ApoM, the lipoprotein class that ApoM is found on may vary based on several factors, including sex, race, age, and disease. The correlation of HDL anti-inflammatory activity with S1P content has driven the search for diseases in which ApoM could be outcome predictive, including coronary artery disease (CAD), type 1 (TI) and type 2 diabetes (T2D), metabolic syndrome (MetS), lupus, IgA nephropathy, and insulin resistance [131–139].

There are diseases where altered ApoM or ApoM-HDL concentrations appear to correlate with increased disease severity or mortality. ApoM and HDL-S1P were both decreased in T2D and mortality of African Americans with T2D inversely correlated with ApoM or S1P levels [135, 140]. MetS patients without diabetes have both higher TG and lower S1P [130, 133]. Although the ratio of ApoM to ApoA1, an HDL-specific protein, was unchanged in MetS patients, the molar ratio of S1P:ApoM was 30% decreased compared to controls [132]. CAD patients also have normal HDL but decreased plasma S1P [141, 142]. In post-menopausal women, although plasma S1P concentrations are the same as in pre-menopausal women, they have increased ApoM, resulting in a greater than 25% decrease in the S1P:ApoM ratio, a characteristic accompanied by endothelial dysfunction and metabolic syndrome [143]. Interestingly, women with very low TG and low LDL had an increased risk of hemorrhagic stroke, but there was no significant correlation between total cholesterol or HDL and hemorrhagic stroke risk [144].

Alternatively, although the absolute concentration of ApoM may be the same, a shift in the lipoprotein particle that it associates with can be indicative of a loss of anti-inflammatory or atheroprotective quality. T1D patients can have normal or above normal HDL concentrations, but are still at increased risk of cardiovascular disease (CVD) [145, 146]. Female T1D patients had ApoM on less dense HDL particles which are believed to be less atheroprotective and in some cases, pro-inflammatory [147, 148]. While still capable of activating the $S1P_1$-ERK pathway and inducing $S1P_1$ internalization in endothelial cells, T1D HDL was ineffective at activating AKT, a major inducer of endothelial nitric oxide synthase (eNOS), the enzyme responsible for production of homeostatic nitric oxide production induced by $S1P_1$ signaling [149]. MetS patient ApoM was more frequently found on LDL, having been transferred from the HDL particles [150]. A study investigating the impact of a hypercholesterolemic diet on HDL content in a porcine model of ischemia reperfusion found that less ApoM was present in HDL particles from hypercholesterolemic pigs versus controls [151].

MetS patient HDL, which has lower S1P in addition to other lipidome alterations, was also less effective at activating eNOS [150, 152, 153]. Once S1P was loaded exogenously onto MetS HDL, the ability to activate eNOS was restored [132]. Loading S1P onto S1P-poor HDL of CAD patients also restored S1PR signaling to levels achieved with HDL from control patients [154]. Another group found that recombinant HDL without sphingomyelin (SM), a metabolic S1P precursor present in high concentrations in the HDL particle, is not sufficient to activate eNOS, but the phosphorylation of eNOS is not directly proportional to HDL SM content, since too much SM will decrease eNOS phosphorylation [155]. These effects may be due to providing the S1P precursor as well as the biophysical effect SM has on the HDL particle itself. High SM content in a lipid layer reduces fluidity, which would alter the flexibility of the HDL particle and subsequently cell membranes to which it transferred lipid cargo to [156, 157]. In control patients, S1P and ApoM are usually enriched on the smaller HDL3 particles, in which the ratio of S1P to SM is over 30 times greater than that of HDL2 [158, 159].

The link between ApoM and metabolism has recently become even more complex than attempting to correlate plasma ApoM concentrations with S1P and lipoproteins. The *Ldlr*$^{-/-}$ (LDL receptor) mouse is a key mouse model of atherosclerosis, and the impact of ApoM-S1P on vascular integrity and endothelial cell health would imply that *Apom*$^{-/-}$ mice would be more prone to atherosclerotic disease, but

Apom$^{-/-}$*Ldlr*$^{-/-}$ double knockout mice are protected from atherosclerosis [160]. Our current understanding of the roles ApoM may play in lipoprotein metabolism indicate that defective LDL regulation, as in the setting of LDLR deficiency, lead to a compensatory increase in plasma ApoM, which in turn, resulted in increased circulating LDL, promoting a pro-atherogenic phenotype that was mitigated by concomitant ApoM deficiency [161]. The impact of ApoM expression in other models of atherosclerosis, such as *Apoe* or LDLR-related protein 1 (*Lrp1*) knockouts, is dependent on the LDL metabolism in each model [160–162].

Soon after its discovery, *APOM* was identified as a leptin-modulated gene [163], the expression of which positively correlated with leptin measurements in human plasma and was suppressed in the obesity model, leptin receptor-deficient (*ob/ob*) mice [164]. More recently, it was reported that *Apom*$^{-/-}$ mice had increased brown adipose tissue (BAT) and were protected from diet-induced obesity, a phenotype reversed by $S1P_1$ agonist administration [165]. A study of tissue from almost 500 human patients found adipose ApoM was produced by adipocytes and secreted to plasma, with ApoM levels inversely correlating with obesity, metabolic syndrome, and T2D [166]; however, it is unclear if S1P is involved in this adipocyte-derived ApoM signaling. In a mouse model of diabetes, insulin administration reversed the decrease in ApoM levels [167]. It is likely that ApoM and insulin cross regulate each other, since *APOM*^{Tg} mice have increased circulating insulin, which can be reduced by treatment with the $S1P_{1/3}$ antagonist VPC23019 [168].

In systemic lupus erythmatosus (SLE), low plasma ApoM correlated with the presence of disease activity markers, including nephritis, leukopenia, and anti-double stranded DNA antibodies (anti-dsDNA) [138, 169]. In an *in vivo* model of immune complex deposition, the reverse Arthus reaction (RAR), mice lacking endothelial cell (EC) $S1P_1$ developed a stronger response [170]. Surprisingly, *Apom* knockout alone did not impact the magnitude of the RAR generated, but when treated with a low dose of $S1P_1$ antagonist the response was significantly greater. Conversely, patients with IgA vasculitis had increased serum ApoM but those with nephritis as a complication had lower ApoM levels than those without [171]. The authors suggest that renal tubular epithelial cell destruction triggered by renal inflammation may have led to decreased ApoM production.

The vascular role of ApoM impacts the development of inflammation and responses to infection in addition to direct signaling on immune cells. In sepsis patients, ApoM produced in the liver drastically drops within 12 h and S1P and ApoM in plasma drop at 6–12 h in both human patients and a baboon model of lethal sepsis [172, 173]. The most severe cases of infection tend to have the lowest ApoM levels, and this drop in ApoM-S1P may contribute to the defects in vascular barrier function that occur in sepsis. ApoM deficiency does not result in gross vascular permeability in the same way that loss of EC $S1P_1$ does [16, 31, 174]. While larger molecules cannot diffuse freely across the BBB of *Apom*$^{-/-}$, paracellular transport of much smaller molecules (<0.07 kDa) was increased in specific vessel types, as was transcytosis; however, not all of the vessels were responsive to $S1P_1$ agonist rescue [175]. The differences in regulation of vascular bed permeability are dramatically illustrated by the pulmonary vascular leakage seen in $S1P_1$ EC-specific knockout (ECKO) animals as compared to the small effect on the BBB of the same animals [16, 174, 176]. In sepsis, effects on barrier integrity within organs such as those in $S1P_1$ ECKO versus ApoM KO mice highlight the necessity for more detailed characterization of the effects of both the S1PR and its chaperones [177].

APOM mRNA has been identified in EC but protein, if produced, is below the limit of detection and the inflammatory stimulus TNFα does not change this mRNA production [178]. The authors put forth the hypothesis that this distinctive expression may imply a purpose for endothelial versus hepatic ApoM, particularly in an inflammatory context such as sepsis, where the plasma levels of ApoM and S1P decrease, as does HDL [173, 179]. The ability of EC to make and retain their own ApoM while producing and secreting their own S1P would allow for some tissue-intrinsic control in response to the drastic systemic decreases in ApoM-S1P seen in sepsis.

6.2.1 ApoM-Fc

Disease modulation through targeting of ApoM signaling pathways has recently been demonstrated by administration of a recombinant ApoM fused to a modified immunoglobulin Fc domain (ApoM-Fc) [180]. ApoM-Fc has improved *in vivo* stability versus a traditional recombinant ApoM protein and has similar *in vitro* properties as ApoM-HDL. The effects of ApoM-Fc appear to be endothelium-centric: it reduced infarct size in the middle cerebral artery occlusion (MCAO) model of stroke, preserved cardiac function in a model of myocardial ischemia/reperfusion, and reduced pulmonary inflammation in the RAR model of immune complex injury [170, 180]. Since immune cell numbers were not affected, particularly lymphocytes, it is possible that ApoM-Fc cannot access the hematopoietic compartment and therefore may provide a tool for differential delivery of ApoM-S1P to EC, sparing the immune system. Subsequent iterations of ApoM-Fc may aim to target endothelial subtypes or the specific S1P receptors they express. However, since ApoM prevents excessive bone marrow lymphopoiesis and HDL is known to affect survival of mature T cells, T regulatory cell (Treg) differentiation, and antigen presentation to T cells, modification of ApoM-Fc or development of a novel ApoM mimetic that targets lymphocytes and/or their progenitors, could be beneficial for direct immunosuppression [16, 181–183].

6.2.2 ApoM Receptor Megalin

To date, the lipocalin receptor megalin is the only known ApoM receptor but it does not appear to be involved in recognition of ApoM outside of the kidney [184]. In this context, megalin is believed to rescue only locally synthesized ApoM from secretion in the urine, but it is also involved in the resorption of albumin from urine, a role that can be modified by a high glucose diet that reduces megalin expression [184–187]. Reported competitive inhibitors of megalin include cilastatin and receptor-associated protein (RAP) [188–190]. Although modulation of megalin expression or activity may provide an indirect mechanism for altering the S1P signaling axis, since this could impact both albumin and ApoM metabolism, without identifying ways of creating specificity for these two proteins the impact on the resorption of other megalin binding partners makes this a less attractive therapeutic target.

6.2.3 Megalin and Vitamin D_3

Vitamin D_3 is either ingested or synthesized in the skin from 7-dehydrocholesterol, then undergoes sequential metabolism to 25(OH)D_3 (calcidiol) in the liver and is then converted to the active metabolite, 1,25-dihydroxyvitamin D (1,25$(OH)_2D_3$; calcitriol) in the kidney [191, 192]. Megalin, also known as LDL receptor-related protein 2 (LRP2), is involved in renal uptake of 25(OH)D_3 for conversion to 1,25$(OH)_2D_3$ through binding vitamin D binding protein (DBP) and may be expressed in other tissues, allowing their vitamin D metabolism [188, 193, 194]. Low D_3 has been correlated to glucose intolerance and increased risk of diabetes and CVD, although the effect of 1,25$(OH)_2D_3$ or D_3 on circulating lipid profiles is unclear [195–197]. A recent meta-analysis of 41 randomized controlled trials found that D_3 supplementation lowered LDL and TG, but in many studies D_3 had no effect on circulating HDL, in others it raised HDL; however, the trend was actually toward decreased HDL in response to D_3 supplementation [198]. One group reported that increased deoxysphingolipids, particularly deoxysphinganine (deoxydihydrosphingosine) are predictive of T2D development in non-obese individuals and correlate with increased TG and glucose, whereas another found that total dihydroceramides (dhCer), particularly C18:0, were elevated at least five years before T2D onset [199, 200]. In another T2D patient cohort, after 6 months of D_3

supplementation plasma 25(OH)D_3 was increased, as were C18 dhCer (d18:0/18:0) and C18 Cer (d18:1/18:0), but there was no effect on plasma S1P or dihydrosphingosine 1-phosphate (dhS1P, dihydrosphinganene) [201]. Yet monocytes from T2D patients stimulated *ex vivo* in the presence of 1,25$(OH)_2D_3$ secreted less S1P, had decreased mRNA *S1PR1* and S1PR2 mRNA and increased *S1PR3* and *S1PR4* mRNA [202]. A study of overweight and obese Asian-Australians found that individuals with low D_3 may benefit from D_3 supplementation by developing increased glucose tolerance; however, supplementation does not offer added protection in patients that are at risk of developing diabetes but already have levels of circulating D_3 that are within the normal range [203].

As the chaperone for S1P, ApoM has emerged as not only a critical component of the signals triggered by S1P binding to its receptors, but also as a potential target for pharmaceutical manipulation. GPCRs are the most popular pharmacologic targets but the inability to selectively bind a single S1PR or a group of S1PR expressed by a specific tissue or cell type has led to a search for other points of modification within the S1P-S1PR signaling pathway. Manipulating the S1P chaperone is an unconventional approach that may provide the opportunity to target subpopulations of cells that are more likely to be exposed to blood or lymph plasma, such as endothelial or immune cells [129, 170, 180].

6.3 S1P Transporters

6.3.1 Spinster 2 (Spns2)

Although the ABC transporters may be involved in the subcellular localization of S1P or its secretion by a limited subset of cells, multiple studies have determined that they are not involved in efflux from the major cell sources of S1P: EC, red blood cells (RBC), and platelets [176, 204–206]. There are now two confirmed S1P transporters, spinster 2 (Spns2) and Mfsd2b. Spns2 is the EC S1P transporter and the secreted S1P modulates specific biological effects of blood and lymph [21, 207]. *Spns2*$^{-/-}$ are lymphopenic, although not to the same extent as mice lacking the S1P biosynthetic enzymes, Sphk1/2. S1P produced by Sphk and secreted by lymphatic EC Spns2 binds naïve T cell $S1P_1$ and promotes their migration and survival through maintenance of mitochondrial numbers [22, 208]. High endothelial venules (HEV) are blood vessels specialized in regulating lymphocyte trafficking into secondary lymphoid organs at homeostasis and tertiary lymphoid organs under inflammatory and disease conditions, including cancer [209]. DC recruited to HEV by CCL21 produce lymphotoxin-β receptor (LTβR) ligands to activate HEV LTβR, which in turn promotes HEV function and EC survival [209]. HEV EC require Spns2 to secrete S1P, which then acts in an autocrine fashion, activating $S1P_1$ signaling and increasing CCL21 production to recruit DC [18].

6.3.2 Major Facilitator Superfamily Domain Containing 2b (Mfsd2b)

Two groups recently reported the characterization of major facilitator superfamily domain containing 2b (Mfsd2b) as the S1P transporter in RBC and platelets [210, 211]. Unlike *Spns2*$^{-/-}$, *Mfsd2b*$^{-/-}$ mice were not lymphopenic, despite a 50% drop in blood plasma S1P, approximately the same decrease as seen in *Spns2*$^{-/-}$. Intriguingly, *Apom*$^{-/-}$ mice had a 65% decrease in blood plasma S1P and albumin/ApoM double knockout mice had a 75% decease in blood plasma S1P and both have significantly more lymphocytes in circulation, further emphasizing the need to better understand the microenvironmental regulation of S1P concentrations by key cell types through production, secretion, and degradation [16, 129]. Mfsd2b also transports docasahexanoic acid (DHA) in the form of lysophosphatidic acid (LPA) precursor molecule lysophosphatidylcholine (LPC), the binding of which is dependent upon the phosphocholine head group and transports LPC from the plasma into brain parenchyma [212]. Mice lacking Mfsd2b also have increased transcytosis by EC involved in the BBB, resulting in leaky CNS vasculature without a breakdown in tight junction (TJ), a phenotype also reported in *Apom*$^{-/-}$ mice [175, 213]. Mfsd2b also suppressed endocytic vesicle formation without affecting TJ by altering the lipid composition of the cell membranes themselves, making assembly of caveolae domains less favorable for vesicle formation [214].

6.4 S1P Metabolism

6.4.1 S1P Lyase (SPL)

Terminal metabolism of intracellular S1P occurs through cleavage of the C2–3 bond by the ER membrane-bound S1P lyase (SPL), yielding phosphoethanolamide (PE) and (2E)-hexadecenal [215]. Almost all mammalian cells express some level of SPL. For example, the brain, kidneys, and splenic and thymic stromal cells have high expression, whereas splenic and thymic immune cells have low expression [216, 217]. Platelets lack SPL and RBC appear to have little to no SPL activity, allowing these two cell types to carry S1P cargo without the danger of degradation [218, 219].

Unlike most of the S1PR, mutations in the SPL gene *SGPL1* are known to cause human disease, such as a form of Charcot-Marie-Tooth disease, the most common hereditary peripheral neuropathy, with the earliest manifestations including weakness in the feet and lower legs [220–222]. More commonly, *SGPL1* mutations manifest as nephrotic syndromes, such as congenital nephrotic syndrome with adrenal calcification and steroid-resistant nephrotic syndrome (SRNS) and adrenal insufficiency, often present with comorbidities of icthyosis, immunodeficiency, gastrointestinal disorders, and neurological deterioration, recently named nephrotic syndrome type 14 (NPHS14) or SPL Insufficiency Syndrome (SPLIS) [223–227].

SPL inhibitors targeted for the clinic have been developed based on the chemical structure of 2-acetyl-4-(tetrahydroxybutyl)imidazole (THI), an SPL inhibitor commonly used in experimental settings and a component of caramel food coloring [228, 229]. Most known SPL inhibitors, including parental compounds THI and 4-deoxypyridoxine (DOP), act by blocking the binding site of cofactor pyridoxal 5′ phosphate (PLP), the active form of vitamin B_6 [228, 230]. Both DOP and THI must be metabolized to a form compatible with the PLP site of SPL. Ohtoyo et al. hypothesized that THI is metabolized by the gut microbiota to an intermediate form before phosphorylation and SPL binding, providing an explanation for why THI will not block SPL activity *in vitro* or under B_6-rich conditions [231, 232]. There are over 100 PLP-dependent enzymes in eukaryotes, one of which is serine palmitoyltransferase (SPT), a complication for data interpretation and the design of SPL inhibitors because SPT is the initiating enzyme in the *de novo* biosynthesis of sphingolipids [232, 233]. Most recently, creation of the compound RBM10-8 was reported as an SPL inhibitor structurally based on S1P that acts as an enzyme substrate, irreversibly binding in the active site by forming a covalent bond [234]. Although RBM10-8 would not require metabolism for activity, its *in vivo* utility and specificity have not yet been studied.

SPL activity is critical for normal and pathological development of the nervous system. Neuron-specific knockdown of the *Drosophila* SPL, *sply,* caused progressive axonal degradation similar to that seen with *SMN* deletion, the gene responsible for spinal muscular atrophy (SMA) in humans [220, 235]. Mice with *Sgpl1* deleted in neuronal progenitors, ependymal cells, and oligodendrocytes with *Nestin-Cre* (*Sgpl*$^{Nes\text{-}Cre}$) had accumulation of brain S1P, decreased PE, and behavioral abnormalities concomitant with alterations in the hippocampus, increased microglial activation, and decreased neuronal autophagosome formation [236–239]. Microglia from these mice also had decreased expression of beclin-1, ATG7, and LC3-II, rendering them defective in autophagy induction [239, 240]. Similarly, autophagic flux in neurons from *Sgpl1*$^{Nes\text{-}Cre}$ mice could be restored *ex vivo* by incubation with exogenous PE [237].

The effect of SPL on neuronal autophagy is notable in the context of several neurodegenerative diseases, since inducers of autophagy are being investigated as possible therapeutics to degrade toxic protein plaques or aggregates [241]. Mice transgenic for mutant human FUS protein (*FUS (1-359)*), an RNA/DNA-binding protein responsible for altered splicing and cytoplasmic aggregation of target mRNAs in amyotrophic lateral sclerosis (ALS), had significantly increased *Sgpl1* mRNA and decreased *Sphk2* [242, 243]. Sph was significantly increased in brains and spinal cords of *FUS (1-359)* mice, and although S1P concentrations were not significantly different, this may have been due to a dramatic increase in SPL activity [243]. Loss-of-function mutations in a FUS target gene, *MECP2*, are the genetic cause of the X-linked neuroregressive disorder Rett syndrome [242, 244]. Plasma from Rett syndrome patients had significantly higher S1P and dhS1P, as well as Sph and dhSph [245]. Rett syndrome patient fibroblasts had defective autophagosome formation and mice lacking *Mecp2* (*Mecp2*$^{-/y}$) developed cerebellar intracellular aggregates as they aged, concurrent with clinical phenotype development [246]. When incubated *in vitro* with THI, mouse primary neurons expressing exon 1 of mutant *huntingtin* (mHTT), the same abnormal splice product seen in brains of Huntington's disease patients, had increased autophagy and consequently, longer survival compared to control-treated cells [247]. In brain tissue from AD patients, Sphk1 was decreased and SPL1 was increased, potentially resulting in a net decrease in S1P concentrations [248]. SPL expression correlated histologically with focal amyloid β deposits in the entorhinal cortex and changes in *SGPL1* mRNA were already significantly increased in brains of patients with the lowest clinical dementia ratings [248, 249].

Model organisms have helped to clarify how the loss of SPL activity can manifest in various organ systems besides the nervous system. *Drosophila* lacking functional Sply are flightless and have a decreased number of dorsal longitudinal flight muscles (DLM) [250]. In a mouse model of post-menopausal osteoporosis, administration of LX2931 (LX3305) restored bone volume by increasing osteoblast activity, subsequently increasing cortical bone thickness and mechanical bone strength [251]. In a collagen-induced arthritis mouse model, LX2931 prevented development of clinical disease and had a small ameliorative effect on joint swelling and inflammation without affecting anti-collagen antibody titers [252]. Phase I and II clinical trials were conducted where rheumatoid arthritis patients received oral LX3305, but results have not published (NCT00847886, NCT00903383, NCT01417052).

SPL also has roles in immune homeostasis because of its control over S1P concentrations. Global *Sgpl1*$^{-/-}$ mice have increased circulating and tissue S1P as well as pro-inflammatory cytokine production [253]. *Drosophila* lacking methyltransferase 2 (Mt2) activity had reduced Sply activity as they aged, resulting in increased total S1P and Cer concentrations, altered hematopoiesis and immune cell morphology, and defective antibacterial immune responses [254]. Although present in very low numbers, thymic parenchymal CD11c^{+} DC with SPL activity were found to be major regulators of the S1P gradient required for T cell egress [255]. While mature T cell-intrinsic SPL also affected thymic S1P concentrations and had a moderate effeon egress, thymic stromal epithelial cells expressing large amounts of SPL protein were not involved in maintaining the S1P gradient required for T cell egress. Treatment of mice with SPL inhibitor for only three days showed a trend toward decreased numbers of double positive thymocytes (CD4^{+}CD8^{+}), and prolonged SPL inhibitor administration resulted in significant depletion of CD4^{+}CD8^{+} cells in the thymus [20, 256].

Subcutaneous administration of an SPL inhibitor in an imiquimod-induced mouse model of psoriasis was as effective as cyclosporine at decreasing redness and epidermal thickness [257]. In the DSS/azoxymethane (DSS/AOM) mouse model of colitis-associated cancer, gut epithelium-specific knockout of *Sgpl1* resulted in increased disease severity and tumor formation accompanied by increased numbers of macrophages and Th17 T cells [258]. Loss of SPL in bone marrow immune cells also resulted in increased colonic inflammatory lesions composed of infiltrating myeloid cells and T cells and loss of crypt architecture; however, *Sgpl1*$^{-/-}$ mice that received wild-type bone marrow cells developed more severe colitis, more tumors, and had increased mortality [259]. Conversely, administration of DOP or THI decreased inflammation, T cell recruitment, and Crohn's disease-like pathology in mice that overexpress TNFα in intestinal epithelium (*TNF*$^{i\Delta ARE}$) [256, 260]. Results from studies of SPL in colitis models corroborate the hypothesis of a key role for S1P in the numerous clinical studies assessing the efficacy of S1PR modulating drugs etrasimod (APD334), ozanimod (RPC-1063), and amiselimod (MT-1303) in ulcerative colitis (UC) and Crohn's disease [125, 261–263].

The vascular nature of the lung and its role as an interface between the host and microorganisms indicates that regulation of S1P would be crucial in this organ. For instance, severity of cystic fibrosis (CF) is potentially influenced by altered regulation of S1P. CF varies in the number of organ systems it affects, but the greatest morbidity and mortality are due to progressive lung inflammation and dysfunction arising from defective cystic fibrosis transmembrane conductance regulator (CFTR) protein [264]. 352 of the greater than two thousand mutations in the *CFTR* gene are causative for CF, although severity of symptoms is typically based on the mutation, the most common of which is ΔF508 (https://cftr2.org/) [265]. Although bacterial infections are usually associated with CF, viral infections can also capitalize on defective mucus clearance and host immunity [266]. Mice expressing ΔF508 showed increased lethality when infected with an enterovirus (coxsackievirus B3 Nancy) despite viremia levels similar to controls [267]. Yet ΔF508 mice had increased viral titers in lymphoid tissues, decreased IFNα production, and decreased virus-specific IgM and IgG titers.

Defects in antiviral activation could be linked to SPL activity. IKKε-mediated inhibition of influenza A (IVA) virus replication was regulated by SPL, which enhanced type I interferon (IFN) responses *in vitro* in response to viral RNA [268]. However, the authors stated that enzymatic activity was not necessary for the antiviral effect. Another group found ΔF508 mice had lungs with decreased S1P but high basal innate and adaptive immune cell infiltration [269]. Administration of LX2931 reduced immune cell numbers in the lungs of ΔF508 mice, including inducible NOS (iNOS)$^{+}$ granulocytes, reducing the cellular infiltrate to wild-type levels. Work from a different group contradicted the implication that SPL inhibition would have beneficial anti-inflammatory effects in CF, reporting that S1P was a negative regulator of CFTR cell surface expression and activity, making SPL inhibition particularly counterproductive in CF patients [270]. A study of patients with community-acquired pneumonia (CAP) found elevated circulating S1P compared to patients without pneumonia [271]. This was not a general response to lung dysfunction, because chronic obstructive pulmonary disease (COPD) patients only had increased S1P concentrations if they developed pneumonia. A clinical trial (NCT03473119) is currently underway to determine if plasma S1P concentrations could serve as a reliable biomarker in CAP cases.

Infection with the malaria parasite *Plasmodium spp.* may also engage the sphingolipid pathway. In a malaria model using infection with *Plasmodium berghei*, humanized mice with decreased SPL activity (h*SGPL1*$^{-/-}$) had approximately the same level of parasitemia but almost no mortality. Curiously, administration of LX2931 did not have a significant effect on survival in experimental cerebral malaria, but FTY720 treatment of h*SGPL*$^{-/-}$ mice had a small survival benefit, particularly when co-administered with the anti-malarial drug artesamate [272]. Compared to children with uncomplicated malaria, children with cerebral malaria have decreased plasma S1P, possibly due to a drop in platelet numbers [272].

The potential for therapeutic manipulation of SPL-derived disorders is dependent upon how the lyase deficiency manifests. For instance, administration of the SPT inhibitor L-cycloserine decreases the amount of substrate entering the S1P biosynthetic pathway, reducing the amount of SPL substrate available, but reportedly inhibits SPL in addition to SPT [227]. However, patients with non-functional SPL could potentially benefit from such a compound. Conversely, as in the autophagy studies cited earlier, if a lack of SPL products is causing disease, it may be possible to administer those molecules to patients. A recent retrospective study of SPLIS patients supplemented with vitamin B_6 found 30% of patients responded positively with increased SPL activity and decreased sphingolipids, emphasizing that SPL variant should determine treatment regimen [273].

6.4.2 Sphingosine 1-Phosphate Phosphatases (SPP1/2)

There are other, nondegradative enzymes that metabolize S1P by dephosphorylation: the S1P phosphohydrolases, SPP1 and SPP2, are specific for long chain sphingoid bases, and LPP3 (discussed below), which is nonspecific and better characterized with regard to LPA metabolism [274]. Both SPP1 and SPP2 localize to the ER but are expressed in different cell types [275–277]. For instance, *SGPP1* mRNA is highly expressed in human placenta and has moderate expression in liver and skeletal muscle, whereas *SGPP2* mRNA is highest in heart [277]. Both *SGPP1* and *SGPP2* mRNAs are highly expressed in human kidney. In mice, northern blotting showed *Sgpp1* to be highly expressed in liver with little expression in skeletal muscle [275]. Differential expression was evident in the endometrium of women with endometriosis, where *SGPP1* was increased and *SGPP2*, along with *SGPL1*, were decreased in endometriosis [278]. SPP1 over-expression *in vitro* led to Cer accumulation and apoptosis, whereas knockdown of *SGPP1* in MCF7 cells caused S1P and dhS1P accumulation and induced an ER stress response, leading to autophagosome formation [276].

Mice lacking *Sgpp1* on a pure C57Bl/6 genetic background are viable, but have severely stunted growth and detachment of an abnormally thin stratum corneum from thickened subcorneal layers [279]. Loss of SPP1 activity caused almost no significant changes in the epidermal sphingolipidome profile, with only C26 Cer significantly decreased and trends toward decreased C24 Cer and increased S1P and dhS1P. Keratinocytes were hyperproliferative and differentiated abnormally in response to increased epidermal Ca2+ concentrations. *Sgpp1*$^{-/-}$ mice maintained on a mixed C57/129sv genetic background reportedly did not have the same epidermal defects [280]. In studies utilizing these *Sgpp1*$^{-/-}$ and *Sgpp2*$^{-/-}$ mice on a mixed background, compared to wild-type controls, *Sgpp1*$^{-/-}$ animals developed increased DSS-induced colitis and their colons had increased proinflammatory cytokines TNF, IL-6, and IL-1b. Conversely, cytokines in colons of *Sgpp2*$^{-/-}$ animals did not increase with DSS administration, resulting in decreased colitis severity. Both *Sgpp1* and *Sgpl1* were downregulated in bone marrow-derived DC stimulated with lipopolysaccharide (LPS) [281]. While naïve DC showed nuclear SPP1 staining, LPS stimulation caused its translocation from the nucleus to the cytosolic compartment. There are currently no drugs that can specifically target either SPP1 or SPP2.

6.4.3 Lipid Phosphate Phosphatase 3 (LPP3)

Lipid phosphate phosphatase (LPP3) also dephosphorylates S1P, but shows substrate promiscuity, metabolizing various extracellular phospholipids [282]. LPP3 has a cytoplasmic motif targeting it to the basolateral membrane [283]. Global loss of *Plpp3*, the LPP3 gene, is embryonic lethal because of abnormal Wnt/β--catenin signaling generating vascular defects [284]. Constitutive endothelial/hematopoietic-specific deletion using the *Tie2-Cre* yields a similar phenotype [285]. Transcriptional regulation of *PLPP3* by NF-κB can be induced by inflammation in the monocyte-like cell line THP-1 [274]. However, a SNP in a regulatory element for endothelial *PLPP3* was identified as protective against coronary artery disease (CAD) and ischemia/stroke [286]. The protective allele sequence created a binding site for KLF2, the same transcription factor regulating *S1pr1*. Modulation of LPP3 expression in primary human aortic EC *in vitro* altered extracellular and intracellular S1P concentrations [287]. In the thymus, LPP3 produced by both endothelial and epithelial cells is necessary for maintaining the S1P gradient utilized by T cells for egress into circulation [285]. Despite difficulties in characterizing the physiological roles of LPP3-mediated S1P metabolism, it is believed to be the primary dephosphorylating enzyme for FTY720P, although SPP1 is also able to do so [288]. There are no published LPP3-specific inhibitors and compounds that have been used, such as sodium orthovanadate and propranolol, are too nonspecific for clinical use [289].

6.5 Biosynthetic Enzymes

6.5.1 Sphingosine Kinases 1 and 2 (Sphk1/2)

The sphingosine kinases, Sphk1 and Sphk2, are differentially expressed by cell and tissue type and although both generate S1P from sphingosine, they show different subcellular localization patterns [5, 290, 291]. The topic of Sphk is quite broad, so this review will address biology directly relevant to common inhibitors and the development of new compounds [14, 292–295]. The Sphks have long been targets for inhibition but isoform specificity, which is particularly important, has been lacking. Double null *Sphk1/2*$^{-/-}$ is embryonic lethal because of massive hemorrhage [296]. When *Sphk1* alone is deleted, circulating S1P concentrations go down, although compensatory activity by Sphk2 prevents tissue S1P concentrations from changing substantially [297]. *Sphk1*$^{-/-}$ animals also develop lymphopenia when administered FTY720, demonstrating that Sphk2 is the kinase responsible for the majority of FTY720 phosphorylation *in vivo* [297]. Contrary to expectation, deletion of *Sphk2* resulted in increased, rather than decreased, circulating S1P concentrations [298]. The proposed mechanism for this increase is a regulatory pathway where S1P is dephosphorylated by LPP3 and some of the Sph is subsequently taken up by cells and rephosphorylated by Sphk2 [298, 299]. The Sphks can also display a differential substrate preference: Sphk1 will preferentially metabolize dhSph over Sph [300]. When cells overexpressing Sphk1 were incubated with FTY720, intracellular dhS1P and S1P increased, but dhSph and Sph concentrations remained the same [300]. This occurred because Sphk2 is 30 times more efficient at phosphorylating FTY720 and FTY720 itself can act as a Sphk inhibitor at micromolar concentrations [46]. Of note is that Spns2 expression is high in the same tissues where Sphk2 is expressed and where FTY720P is secreted, illustrating the interconnectedness of the entire sphingolipid metabolic system [300].

While the search for S1PR inhibitors tends to focus on vascular and autoimmune diseases, the development of Sphk inhibitors is driven by cancer research [301]. The primary anti-tumoral mechanism of action of Sphk inhibition is believed to be an increase in intracellular Cer, triggering cancer cell apoptosis. *D,L*-threo-dihydrosphingosine (tDHS) or *N,N*, dimethylsphingosine (DMS) act as competitive nonspecific inhibitors [290, 302]. The most frequently used inhibitor is SKi (also referred to as SKI-II), which inhibits both Sphk1 and 2 [303]. Like many Sphk inhibitors subsequently developed, it must be present in micromolar concentrations for full potency and since it inhibits both Sphks there are unwanted and unexpected effects [304]. The block on catalytic activity by Sphk1 inhibitors correlates with its degradation, a unique mechanism where inhibitor binding induces polyubiquitnylation, targeting Sphk1 to the proteasomal degradation pathway [305–307]. The binding of inhibitors induces a conformational change in Sphk1, allowing ubiquitnylation of Lys183 and subsequent binding of the Kelch-like protein 5 (KLHL5)-cullin 3 ubiquitin ligase complex [308]. While *Klhl5* knockdown reduced Sphk1 degradation, KLHL5 expression was correlated with decreased chemotherapy sensitivity [309]. SKi also inhibits another enzyme in the sphingolipid metabolic pathway, Des1, in a noncompetitive manner at submicromolar concentrations, which may account for the cellular accumulation of dhCer with the use of this inhibitor [310].

Several Sphk1-specific inhibitors with improved specificity and/or potency have been described. LCL351 is a sphingosine analogue that has 10 times greater inhibitory activity for Sphk1 versus Sphk2, but still requires low micromolar concentrations for inhibition [311]. Although not clinically viable, LCL351 is a useful Sphk1-specific tool. Patients with UC have increased Sphk1 expression and *Sphk1*$^{-/-}$ mice develop less severe DSS-induced colitis [312]. LCL351 administration prevented the development of DSS-induced colitis, reducing colonic neutrophil recruitment in combination with a decrease in colon S1P concentrations, although circulating S1P concentrations were slightly elevated [311].

PF543 is a Sph competitive Sphk1 inhibitor at low nanomolar concentrations and was used in the crystallization of Sphk1 [313, 314]. Incubation of 1483 (squamous cell carcinoma) cells with nanomolar concentrations of PF543 did not increase Cer concentrations but did increase Sph in direct correlation with the S1P decrease [314]. Since PF543 did not induce apoptosis in cancer cell lines A549 (lung adenocarcinoma), Jurkat (T cell acute lymphoblastic leukemia), LN229 (glioblastoma), MCF7 (invasive ductal carcinoma), or U937 (acute monocytic leukemia) but did inhibit S1P generation, this illustrated that the pro-apoptotic effect of Sphk inhibition was not a result of lost S1P, but the increase in Cer [314]. PF543 also decreased the severity of DSS-induced colitis, concurrent with the decrease in S1P concentrations [315, 316]. EC in isolated rat aortic and coronary arteries transiently exposed to hypoxic conditions upregulated Sphk1, increasing S1P and vasodilation that was blocked by PF543 [317]. *In vivo*, PF543 blocked S1P production and increased cardiac Sph concentrations in the angiotensin II (AngII)-dependent model of arterial hypertension and cardiac remodeling [318]. At a dose of 1 mg/kg, PF543 blocked development of cardiac hypertrophy and decreased $S1P_1$ protein in the heart with a subsequent decrease in activated STAT3 and ERK1/2 [318]. Despite the intense interest in Sphk1 inhibitors, none have made it to clinical trials.

However, ABC294640 (ABC), an Sphk2 inhibitor, has been in multiple clinical trials since it was first reported in 2010 [319]. The first clinical trial, in patients with solid tumors (cholangiosarcoma, colon, pancreatic), found a dose of 500 mg twice a day was well tolerated [320]. An important finding was that after trial initiation, a protocol amendment was needed requiring fasting blood glucose below 160 mg/dL because of dose-limiting hyperglycemia [320]. Experimentally, ABC suppressed the development of cancer in the DSS/AOM colon cancer model, decreased chemoresistance in breast and ovarian cancer models, and suppressed inflammation in models of arthritis and lupus [257, 321–327]. Some ABC efficacy may be due to its accumulation in tumor tissues, since the half-life in human plasma is only 5.5 h [319]. Currently, ABC (Opaganib/Yeliva) is being investigated in two clinical trials: ABC plus androgen antagonist in metastatic castration-resistant prostate cancer (NCT04207255) and alone or in combination with hydroxychloroquine sulfate in advanced cholangiosarcoma (NCT03377179). So far, no changes have been posted for the second trial with regard to possible hydroxychloroquine shortages during the COVID-19 pandemic [328].

Sphks could also be targets for inhibition during some viral infections, although efficacy is likely to be pathogen- and manifestation-specific. Influenza A virus (IAV) infection increased Sphk2 expression and activation *in vitro* and treatment with ABC during IAV infection increased survival and decreased lung viral titers [329]. Inhibition with non-specific DMS had the same effect, and although SKi treatment did not result in the same magnitude of survival increase, it did significantly decrease viral titers better than ABC, indicating that Sphk1 and Sphk2 may be responsible for different aspects of IAV viral reproduction and host response [329, 330]. Dengue virus type 2 (DENV2), a positive-sense single-stranded RNA virus, actively down-regulated *Sphk1* transcription, decreasing the activation of IFN-responsive genes [331]. *In vitro*, DENV2 replicates less efficiently in *Sphk2*$^{-/-}$ mouse embryonic fibroblasts, which did not produce IFNβ in response to viral infection, but lack of Sphk2 did not impact viral replication *in vivo* or survival [332]. IAV and DENV2 have different modes of transmission, different cycles, and are not related, but similar responses involving Sphks *in vitro* implicate a more general role in the anti-viral immune response [333].

6.5.2 Ceramide Synthases (CerS)

Production of Cer can occur through two pathways, one of which is the salvage pathway: reacylation of Sph by Ceramide synthases (CerS) [334, 335]. Alternatively, Cer is produced *de novo* from dhCer, which will be covered in the following section. There are six CerS, which are ER membrane-bound enzymes that catalyze the *N*-acylation of sphingoid bases and require phosphorylation of C-terminal residues for catalytic activity [336–338]. Each of the CerS exhibit different cellular expression patterns and acyl chain preferences, and inhibition or deletion of one CerS typically results in up-regulation of another and production of different Cer species [336, 339]. CerS2 is the most ubiquitously expressed and produces C20–C26 Cer [334, 340]. Knockout or knockdown in MCF7 cells resulted in accumulation of dhSph and Sph, increases in *CERS4*, *5*, and *6*, and decreased very long chain (VLC) Cer [336, 341]. CerS2 overexpression increases VLC Cer production, causing insulin resistance and oxidative stress in cardiomyocytes [342]. Conversely, CerS1 has the most restricted expression and is highest in the CNS, skeletal muscle (SkM), and testis and generates only C18 Cer, which decreases with CerS1 knockdown [334, 340, 343]. Mutations in *CERS1* have been linked to progressive myoclonus epilepsy and CerS1 interactions with mutant heat shock protein (Hsp27) result in decreased mitochondrial Cer, leading to neurodegeneration in Charcot-Marie-Tooth variant 2F disease [344–346].

There are two inhibitors of CerS, the most specific of which is P053, an FTY720 derivative and selective noncompetitive inhibitor of CerS1 [347]. P053 selectively decreased C18 Cer in SkM while liver and adipose Cer concentrations were not affected. P053 also decreased triacylglycerol (TAG) by 50% in SkM of HFD-fed mice but did not affect TAG in SkM of normal chow-fed mice [347]. P053 may not effectively cross the BBB since it is found in much lower concentrations in brain tissue versus SkM and has less of an effect on brain C18 Cer production. CerS1-specific inhibition also increased mitochondrial capacity and enhanced fatty acid oxidation in SkM while decreasing whole body fat mass, despite HFD consumption and no effect on insulin resistance [347]. However, genetic deletion of *Cers1* in SkM did show increased insulin and glucose tolerance with HFD feeding, in addition to reduced adiposity [348].

The other CerS inhibitor is fumonisin B1 (FB1), a fungal toxin with a deoxysphingoid base structure and known carcinogenic activity [349–351]. FB1 inhibits all six CerS, leading to increased S1P, dhS1P, Sph, and dhSph, and decreased Cer and dhCer [11, 352]. It also causes accumulation of 1-deoxysphinganine, possibly compounding neurological sequelae of CerS inhibition, similar to production of deoxysphingolipids associated with hereditary sensory and autonomic neuropathy type 1 (HSAN1) [351, 353].

6.5.3 Dihydroceramide Desaturase (Des1 and 2)

De novo Cer synthesis occurs by insertion of a 4,5 *trans* double bond into the sphingoid backbone of dihydroceramide (dhCer) by the dihydroceramide desaturases, Des1 and Des2 [354, 355]. CerS are responsible for the production of the dhCer substrate, so their inhibition affects both salvage and *de novo* pathways [356]. Compared to Des1, far less is known about Des2, which in addition to desaturase activity can also exhibit C4-hydroxylase activity and synthesize phytoceramides (phytoCer) [357]. Membrane-bound cytochrome b_5 affinity and complex formation may determine which of these enzymatic activities Des2 engages in [358]. Des2 is highly expressed in the digestive tract, kidneys, and skin, where phytoCer are critical [355, 358]. *DEGS2*, the Des2 gene, is also expressed in the adult brain and was significantly upregulated in brains of schizophrenia patients and downregulated in major depressive disorder patients [359]. A *DEGS2* missense mutation also correlated with cognitive deficits in schizophrenia patients [360].

Des1 is ubiquitously expressed and its activity is most associated with insulin resistance and cancer [361]. Palmitate upregulates *Degs1* mRNA in SkM myoblasts, increasing Cer and subsequently inducing insulin resistance, which was reversible by oleate incubation [362]. Cells and mice lacking *Degs1* have increased dihydroxysphingolipids and uncoupled nutrient and apoptosis signaling [363]. *Degs1*$^{-/-}$ mice crossed with the obesity model *ob/ob* mice had significantly increased dhCer and decreased Cer in liver, white adipose tissue, and serum [364]. *ob/ob Degs1*$^{-/-}$ animals subsequently had lower fat mass, blood glucose, and improved liver function. Accumulation of dhCer in plasma had previously been proposed as a biomarker for diabetes progression [199].

Des1 is the target of the synthetic retinoid chemotherapeutic, fenretinide (4-hydroxyphenyl retinamide (4-HPR)) [365–367]. 4-HPR also increases activity of SPT, leading to the accumulation of cytotoxic dhCer [14, 368, 369]. In HEK293 cells, 4-HPR induced polyubiqutinylation of Des1, increasing enzymatic activity but targeting it for degradation, making Des1 activity dependent upon the rate of degradation induced by polyUb [310, 370]. Metabolites of 4-HPR differentially affect the 4-HPR target enzymes [371]. The 3-keto-HPR metabolite inhibits all targets, stearoyl CoA desaturase (SCD1), β-carotene oxyegnase (BCO1), and Des1, but the metabolite N-[4-methoxyphenyl]retinamide (MPR) specifically affects BCO1. SCD1 converts saturated fatty acids (FA) to monounsaturated FA, particularly palmitic acid or stearic acid to palmitoleic or oleic acid, respectively, but was also reported to decrease Cer in cardiomyocytes [372–374]. However, SCD1 deficiency decreased Cer and mRNA for SPT components in SkM, so its inhibition by 4-HPR may contribute to effects attributed to Des1 inhibition *in vivo* [373, 375].

This target combination may also explain results of 4-HPR treatment in CF. 4-HPR decreased inflammation and corrected the FA imbalance and Cer deficiencies seen in *CFTR*$^{-/-}$ mouse models and in CF patients [376–378]. A new oral formulation of 4-HPR by Laurent Pharmaceuticals, Lau-7b, is currently in phase II trials for CF (APPLAUD, NCT03265288) and reportedly normalized blood and lung polyunsaturated FA (PUFA) and Cer concentrations in mouse models of asthma [379]. Similar to the *Degs1*$^{-/-}$ mice, 4-HPR has been shown to prevent or partially reverse obesity, insulin resistance, and hepatic steatosis by blocking Cer synthesis [380–382]. The effects of 4-HPR in obese patients have been investigated in a clinical trial, with results under review by the FDA in January 2020 (NCT00546455).

The most extensive clinical testing of 4-HPR has been in clinical trials for a wide range of cancers: neuroblastoma, glioblastoma, lymphomas, leukemias, recurrent ovarian and prostate, lung, bladder, head and neck, and breast (Clinicaltrials.gov search "fenretinide"). A new formulation of 4-HPR complexed with 2 hydroxypropyl-β-cyclodextrin (Nanofen) was recently described [383]. Nanofen had improved bioavailability and efficacy in lung tumor xenograft models, having led to C18 dhCer accumulation and inducing tumor apoptosis.

6.5.4 Ceramidases (CDases)

The enzymes that convert dhCer to dhSph or Cer to Sph are ceramidases (CDases), which hydrolyze the N acyl linkage between the FA and the sphingoid base [14, 384]. The CDases are categorized based on their optimal catalytic pH: neutral ceramidase (NCDase), alkaline ceramidases (Acer), and acid ceramidase (ACDase). Like the CerS, the CDases have substrate specificity and subcellular localization. Sph can be converted to Cer by ACDase, NCDase, and Acer1-3, whereas dhSph and phytoSph are converted from their dhCer and phytoCer precursors by Acer2 & 3 [384, 385].

Defects in ACDase lead to Faber disease and SMA with progressive myoclonic epilepsy [386]. mRNA for two CDases, *Asah1* (ACDase) and *Asah2* (NCDase), are altered in the brains of ALS model *FUS (1-359)* mice [243]. Knockout of ACDase (*Asah1*$^{-/-}$) is embryonic lethal, and conditional *Asah1*$^{-/-}$ mice have elevated ovarian Cer levels, leading to decreased fertility [387, 388].

NCDases are present in humans, mice, and pathogens, including *Pseudomonas aeruginosa* and *Mycobacterium tuberculosis* [389, 390]. Mice lacking NCDase (*Asah2*$^{-/-}$) were relatively normal and demonstrated the critical role of NCDase in the intestines [391]. *Asah2*$^{-/-}$ have high circulating endotoxin and inflammation of the gut epithelium in the DSS colitis model [327].

Loss of each of the alkaline CDases (*Acer1–3*) manifests in different tissues. Acer1 is critical for epidermal Cer regulation, and *Acer1*$^{-/-}$ mice have progressive alopecia due to altered hair follicle cycling and have increased energy expenditure with decreased body fat [392, 393]. *Acer2*$^{-/-}$ mice have significantly decreased circulating Sph, dhSPh, S1P, and dhS1P [394]. *In vitro*, Acer2 displays broad substrate specificity and is upregulated by serum deprivation [395]. Acer3 preferentially hydrolyzes C18:1 Cer and is highly expressed in brain, increasing with age [396]. *Acer3*$^{-/-}$ mice appear mostly normal until they age beyond 8 months, at which point C18:1 Cer is significantly decreased in addition to Sph and S1P. As their brain sphingolipid composition changes, *Acer3*$^{-/-}$ mice develop impaired balance, motor coordination, and grip strength due to Purkinje cell degeneration [396]. In humans, *ACER3* deficiency manifests in childhood as progressive leukodytrophy [397]. Although CDase inhibitors exist for laboratory use, none have been advanced to the clinic [398–400].

6.5.5 Serine Palmitoyl-CoA Transferase (SPT)

The initial reaction in the *de novo* sphingolipid biosynthetic pathway is the condensation of the amino acid serine and palmitoyl-CoA by serine palmitoyltransferase (SPT) to 3-ketodihydrosphingosine [401, 402]. The essential components of SPT are the protein subunits SPTLC1 and SPTLC2 or 3, with SPTLC1 being ubiquitously expressed and SPTLC2 and 3 showing some tissue specificity [403, 404]. SPTLC2 and 3 contain the PLP consensus motif for cofactor binding and whichever is included in the heterodimer (SPTLC1 plus SPTLC2 versus 3) determines whether longer fatty acyl-CoAs (palmitoyl and larger) are incorporated (SPTLC2) or shorter myristoyl or lauroyl (SPTLC3) are selected [405, 406]. Other components of the enzyme complex are the proteins SPT small subunit a and b (SPTssa/b), which bind to the SPTLC complex conferring optimal catalytic activity [407]. A mutation in *SPTssb* results in increased SPT activity and over-production of C20 long chain bases, resulting in abnormal membranes and vacuoles in the brain, leading to ataxia and early death [408]. SPT activity is negatively regulated by the ORMDL proteins, which are always complexed with the SPT holoenzyme and act through conformational changes in response to sphingolipid concentrations, particularly D-erythro Cer [409–411].

ORMDL deficiency in mice is not lethal and null animals appear normal at weaning [412]. *Ormdl3*$^{-/-}$ mice have significantly increased brain sphingolipids, particularly Cer and Sph, but dhCer is increased only in *Ormdl1/3*$^{-/-}$ double-null mice. These mice have smaller body weights, exhibit neurological defects, and their sciatic nerves contain significantly greater concentrations of dhSph, dhCer, Cer, and Sph. Altered sphingolipid concentrations manifest as abnormal sciatic nerve morphology with excessive (redundant) myelination, a phenotype recapitulated in mice with an inducible constitutive SPT [412]. SNPs in *ORMDL3* and a *cis* gene, *GSDML*, are linked to non-allergic childhood asthma [413]. DNA methylation sites in *ORMDL3* were also independently correlated with childhood asthma and DNA methylation regions in the 5′ UTR of *ORMDL3* were significantly less methylated in CD8^{+} T cells and children with asthma [414].

Mutations in *SPTLC1* or *2* are responsible for hereditary sensory and autonomic neuropathy (HSAN1) types 1A and 1C, respectively, characterized by damage to peripheral neurons leading to progressive neuropathy, ulcerations, and weakness [415–417]. The most prevalent mutations change SPT substrate amino acid preference, but others change affinity for acyl-CoAs of different chain lengths and increase basal activity [405, 418–420]. Mutations causing substrate preference to change from serine to alanine or glycine result in accumulation of neurotoxic 1-deoxysphingolipids, such as 1-deoxysphinganine and 1-deoxymethylsphinganine [418–421]. HSAN1C patients also have decreased $CD8^+$ T cell sphingolipid synthesis upon activation and impaired proliferation and survival [422].

Dietary supplementation of 10% L-serine decreased plasma deoxysphingolipids and improvement in motor and coordination testing in mice with the C133W HSAN1 mutation, whereas L-alanine supplementation led to accumulation of deoxysphingolipids and motor function deterioration [423]. In a randomized controlled trial, 400 mg/kg/day of L-serine resulted in improvement of disease scores concomitant with dramatic, significant decreases in plasma deoxysphingolipids [423, 424]. Surprisingly, despite improvement in neuropathy, L-serine-treated patients had no decrease in ulcers and a higher frequency of skin infections and osteomyelitis, which the authors suggested could be due to permanent nerve damage occurring before supplementation [424]. L-serine supplementation has been shown to increase D-serine in plasma and CSF in a mouse model of *GRIN2B* encephalopathy, a Rett-like syndrome [425]. Considering the frequent overlap of substrate specificity for various sphingolipid enzyme inhibitors, a relatively simple diet modification that is efficacious in severe diseases related to SPT activity would be a preferred treatment.

6.6 Glycosphingolipids

The glycosphingolipids (GSL) are a large subfamily of sphingolipid molecules created by attachment of glycans to a ceramide moiety that anchors them in the lipid bilayer, primarily in the plasma membrane [1, 426]. Galactosylceramides (GalCer) and glucosylceramides (GlcCer) are synthesized in the Golgi by **β**-linkage of the respective galactose or glucose sugar moiety to the primary hydroxyl of a ceramide [427, 428]. GlcCer can then be metabolized to lactosylceramide (LacCer), which serves as the base molecule for the more complex GSL: globo/isoglobo-, ganglio/isoganglio-, and lacto/neolacto-series [428, 429]. Clinically, the GSL are most widely recognized for their roles in lipid storage diseases: Fabry, Tay-Sachs, Sandhoff, Gaucher, Krabbe, Niemann-Pick C, and GM1 and GM2 gangliosidosis [430, 431]. The complexities of the glycosphingolipid metabolic pathways are such that interested readers are directed to the detailed reviews referenced in this section.

6.7 Summary

The known contributions of sphingolipids to all aspects of biological homeostasis and pathogenesis have continuously expanded since their first description almost a century and a half earlier [432]. The generation of animal models and the increasing depth of genetic sequencing have allowed researchers and clinicians to discover unexpected phenotypes and rare mutations in the sphingolipid metabolic and signaling pathways responsible for diseases in every biological system. Although the greatest successes in sphingolipid pharmaceutical targeting have been S1PR modulating drugs, the development of new compounds and modification of old ones have generated promising results. Complex rules governing sphingolipid flux combine with cellular and subcellular specialization to create a network that is, unfortunately, at times irreducible. However, the restricted utilization of sphingolipid biosynthetic and signaling pathways also provides opportunities for targeted therapeutic exploitation.

Acknowledgements The author thanks Danielle Jones, Nemekh Tsogtbaatar, and Daniel Charette for critical comments and editorial assistance. This work was funded by the NHLBI (R01 HL141880) and the American Heart Association (16SDG27020014).

References

1. Merrill AH (2011) Sphingolipid and glycosphingolipid metabolic pathways in the era of sphingolipidomics. Chem Rev 111:6387–6422
2. Carreira AC et al (2019) Mammalian sphingoid bases: biophysical, physiological and pathological properties. Prog Lipid Res 75:100988
3. Hannun YA, Obeid LM (2011) Many ceramides. J Biol Chem 286:1–9
4. Castro BM, Prieto M, Silva LC (2014) Ceramide: a simple sphingolipid with unique biophysical properties. Prog Lipid Res 54:53–67
5. Pyne S, Adams DR, Pyne NJ (2016) Sphingosine 1-phosphate and sphingosine kinases in health and disease: recent advances. Prog Lipid Res 62:1–63
6. Cianchi F et al (2006) Inhibition of 5-lipoxygenase by MK886 augments the antitumor activity of celecoxib in human colon cancer cells. Mol Cancer Ther 5:2716–2726
7. Maxis K et al (2006) The shunt from the cyclooxygenase to lipoxygenase pathway in human osteoarthritic subchondral osteoblasts is linked with a variable expression of the 5-lipoxygenase-activating protein. Arthritis Res Ther 8:R181
8. Marnett LJ (2009) Mechanisms of cyclooxygenase-2 inhibition and cardiovascular side effects: the plot thickens. Cancer Prev Res (Phila) 2:288–290
9. Ganesh R, Marks DJB, Sales K, Winslet MC, Seifalian AM (2012) Cyclooxygenase/lipoxygenase shunting lowers the anti-cancer effect of cyclooxygenase-2 inhibition in colorectal cancer cells. World J Surg Oncol 10:200
10. Hagen-Euteneuer N, Lütjohann D, Park H, Merrill AH, van Echten-Deckert G (2012) Sphingosine 1-phosphate (S1P) lyase deficiency increases sphingolipid formation via recycling at the expense of de novo biosynthesis in neurons. J Biol Chem 287:9128–9136
11. Riley RT, Merrill AH (2019) Ceramide synthase inhibition by fumonisins: a perfect storm of perturbed sphingolipid metabolism, signaling, and disease. J Lipid Res 60:1183–1189
12. Chun J, Kihara Y, Jonnalagadda D, Blaho VA (2019) Fingolimod: lessons learned and new opportunities for treating multiple sclerosis and other disorders. Annu Rev Pharmacol Toxicol 59:149–170
13. Proia RL, Hla T (2015) Emerging biology of sphingosine-1-phosphate: its role in pathogenesis and therapy. J Clin Invest 125:1379–1387
14. Hannun YA, Obeid LM (2018) Sphingolipids and their metabolism in physiology and disease. Nat Rev Mol Cell Biol 19:175–191
15. Allende ML, Dreier JL, Mandala S, Proia RL (2004) Expression of the sphingosine 1-phosphate receptor, S1P1, on T-cells controls thymic emigration. J Biol Chem 279:15396–15401
16. Blaho VA et al (2015) HDL-bound sphingosine-1-phosphate restrains lymphopoiesis and neuroinflammation. Nature 523:1–18
17. Nagahashi M et al (2016) Sphingosine-1-phosphate in the lymphatic fluid determined by novel methods. Heliyon 2:e00219
18. Simmons S et al (2019) High-endothelial cell-derived S1P regulates dendritic cell localization and vascular integrity in the lymph node. elife 8:464
19. Quehenberger O et al (2010) Lipidomics reveals a remarkable diversity of lipids in human plasma. J Lipid Res 51:3299–3305
20. Schwab SR et al (2005) Lymphocyte sequestration through S1P lyase inhibition and disruption of S1P gradients. Science 309:1735–1739
21. Mendoza A et al (2012) The transporter Spns2 is required for secretion of lymph but not plasma sphingosine-1-phosphate. Cell Rep 2:1104–1110
22. Mendoza A et al (2017) Lymphatic endothelial S1P promotes mitochondrial function and survival in naive T cells. Nature 28:1–20
23. Blaho VA, Hla T (2014) An update on the biology of sphingosine 1-phosphate receptors. J Lipid Res 55:1596–1608
24. Blaho VA, Chun J (2018) 'Crystal' clear? Lysophospholipid receptor structure insights and controversies. Trends Pharmacol Sci 39:953–966
25. Kihara Y (2019) Systematic understanding of bioactive lipids in neuro-immune interactions: lessons from an animal model of multiple sclerosis. Adv Exp Med Biol 1161:133–148
26. Bethany AR, Huiqun WAYZ (2019) Recent advances in the drug discovery and development of dualsteric/bitopic activators of G protein-coupled receptors. Curr Top Med Chem 19:2378–2392
27. Insel PA et al (2019) GPCRomics: an approach to discover GPCR drug targets. Trends Pharmacol Sci 40:378–387
28. Hla T, Maciag T (1990) An abundant transcript induced in differentiating human endothelial cells encodes a polypeptide with structural similarities to G-protein-coupled receptors. J Biol Chem 265:9308–9313
29. Lee MJ et al (1998) Sphingosine-1-phosphate as a ligand for the G protein-coupled receptor EDG-1. Science 279:1552–1555
30. Liu Y et al (2000) Edg-1, the G protein-coupled receptor for sphingosine-1-phosphate, is essential for vascular maturation. J Clin Invest 106:951–961
31. Jung B et al (2012) Flow-regulated endothelial S1P receptor-1 signaling sustains vascular development. Dev Cell 23:600–610
32. Galvani S et al (2015) HDL-bound sphingosine 1-phosphate acts as a biased agonist for the endothelial cell receptor S1P1 to limit vascular inflammation. Sci Signal 8:ra79
33. Cartier A, Leigh T, Liu CH, Hla T (2020) Endothelial sphingosine 1-phosphate receptors promote vascular normalization and antitumor therapy. Proc Natl Acad Sci U S A 117(6):3157–3166

34. Lee JS et al (2006) Klf2 is an essential regulator of vascular hemodynamic forces in vivo. Dev Cell 11:845–857
35. Kono M et al (2004) The sphingosine-1-phosphate receptors S1P1, S1P2, and S1P3 function coordinately during embryonic angiogenesis. J Biol Chem 279:29367–29373
36. Sanchez T et al (2007) Induction of vascular permeability by the sphingosine-1-phosphate receptor-2 (S1P2R) and its downstream effectors ROCK and PTEN. Arterioscler Thromb Vasc Biol 27:1312–1318
37. Yanagida K et al (2020) Sphingosine 1-phosphate receptor signaling establishes AP-1 gradients to allow for retinal endothelial cell specialization. Dev Cell 52:779–793.e7
38. Yanagida K, Hla T (2017) Vascular and immunobiology of the circulatory sphingosine 1-phosphate gradient. Annu Rev Physiol 79:67–91
39. Hisano Y, Hla T (2019) Bioactive lysolipids in cancer and angiogenesis. Pharmacol Ther 193:91–98
40. Adachi K, Chiba K (2007) FTY720 story. Its discovery and the following accelerated development of sphingosine 1-phosphate receptor agonists as immunomodulators based on reverse pharmacology. Perspect Medicin Chem 1:11–23
41. Faissner S, Gold R (2019) Oral therapies for multiple sclerosis. CSH Perspect Med 9:a032011
42. Derfuss T et al (2020) Advances in oral immunomodulating therapies in relapsing multiple sclerosis. Lancet Neurol 19:336–347
43. Bolli MH et al (2010) 2-imino-thiazolidin-4-one derivatives as potent, orally active S1P1 receptor agonists. J Med Chem 53:4198–4211
44. D'Ambrosio D, Freedman MS, Prinz J (2016) Ponesimod, a selective S1P1 receptor modulator: a potential treatment for multiple sclerosis and other immune-mediated diseases. Ther Adv Chronic Dis 7:18–33
45. Brinkmann V et al (2002) The immune modulator FTY720 targets sphingosine 1-phosphate receptors. J Biol Chem 277:21453–21457
46. Billich A et al (2003) Phosphorylation of the immunomodulatory drug FTY720 by sphingosine kinases. J Biol Chem 278:47408–47415
47. Oo ML et al (2011) Engagement of S1P1-degradative mechanisms leads to vascular leak in mice. J Clin Invest 121:2290–2300
48. Budde K et al (2002) First human trial of FTY720, a novel immunomodulator, in stable renal transplant patients. J Am Soc Nephrol 13:1073–1083
49. Kovarik JM et al (2004) Multiple-dose FTY720: tolerability, pharmacokinetics, and lymphocyte responses in healthy subjects. J Clin Pharmacol 44:532–537
50. Skerjanec A et al (2005) FTY720, a novel immunomodulator in de novo kidney transplant patients: pharmacokinetics and exposure-response relationship. J Clin Pharmacol 45:1268–1278
51. Foster CA et al (2007) Brain penetration of the oral immunomodulatory drug FTY720 and its phosphorylation in the central nervous system during experimental autoimmune encephalomyelitis: consequences for mode of action in multiple sclerosis. J Pharmacol Exp Ther 323:469–475
52. Yagi H et al (2000) Immunosuppressant FTY720 inhibits thymocyte emigration. Eur J Immunol 30:1435–1444
53. Graeler M, Goetzl EJ (2002) Activation-regulated expression and chemotactic function of sphingosine 1-phosphate receptors in mouse splenic T cells. FASEB J 16:1874–1878
54. Mandala S et al (2002) Alteration of lymphocyte trafficking by sphingosine-1-phosphate receptor agonists. Science 296:346–349
55. Matloubian M et al (2004) Lymphocyte egress from thymus and peripheral lymphoid organs is dependent on S1P receptor 1. Nature 427:355–360
56. Baeyens A, Fang V, Chen C, Schwab SR (2015) Exit strategies: S1P signaling and T cell migration. Trends Immunol 36:1–10
57. Brinkmann V (2009) FTY720 (fingolimod) in multiple sclerosis: therapeutic effects in the immune and the central nervous system. Br J Pharmacol 158:1173–1182
58. Salvadori M et al (2006) FTY720 versus MMF with cyclosporine in de novo renal transplantation: a 1-year, randomized controlled trial in Europe and Australasia. Am J Transplant 6:2912–2921
59. Kappos L et al (2010) A placebo-controlled trial of oral fingolimod in relapsing multiple sclerosis. N Engl J Med 362:387–401
60. Jain N, Bhatti MT (2012) Fingolimod-associated macular edema: incidence, detection, and management. Neurology 78:672–680
61. Chiba K (2005) FTY720, a new class of immunomodulator, inhibits lymphocyte egress from secondary lymphoid tissues and thymus by agonistic activity at sphingosine 1-phosphate receptors. Pharmacol Ther 108:308–319
62. Valentine WJ et al (2008) Subtype-specific residues involved in ligand activation of the endothelial differentiation gene family lysophosphatidic acid receptors. J Biol Chem 283:12175–12187
63. Potenza RL et al (2016) Fingolimod: a disease-modifier drug in a mouse model of amyotrophic lateral sclerosis. Neurotherapeutics 13:918–927
64. Jęśko H, Wencel PL, Lukiw WJ, Strosznajder RP (2019) Modulatory effects of fingolimod (FTY720) on the expression of sphingolipid metabolism-related genes in an animal model of Alzheimer's disease. Mol Neurobiol 56:174–185
65. Takasugi N et al (2013) FTY720/fingolimod, a sphingosine analogue, reduces amyloid-β production in neurons. PLoS One 8:e64050
66. Aytan N et al (2016) Fingolimod modulates multiple neuroinflammatory markers in a mouse model of Alzheimer's disease. Sci Rep-UK 6:24939
67. Carreras I et al (2019) Dual dose-dependent effects of fingolimod in a mouse model of Alzheimer's disease. Sci Rep-UK 9:10972

68. Groves A et al (2018) A functionally defined in vivo astrocyte population identified by c-Fos activation in a mouse model of multiple sclerosis modulated by S1P Signaling: immediate-early astrocytes (ieAstrocytes). eNeuro 5:ENEURO.0239-18.2018
69. Enosawa S, Suzuki S, Kakefuda T, Li XK, Amemiya H (1996) Induction of selective cell death targeting on mature T-lymphocytes in rats by a novel immunosuppressant, FTY720. Immunopharmacology 34:171–179
70. Suzuki S, Li XK, Enosawa S, Shinomiya T (1996) A new immunosuppressant, FTY720, induces bcl-2-associated apoptotic cell death in human lymphocytes. Immunology 89:518–523
71. Wang JD et al (1999) Early induction of apoptosis in androgen-independent prostate cancer cell line by FTY720 requires caspase-3 activation. Prostate 40:50–55
72. Suzuki S et al (1997) The in vivo induction of lymphocyte apoptosis in MRL-lpr/lpr mice treated with FTY720. Clin Exp Immunol 107:103–111
73. Shinomiya T, Li XK, Amemiya H, Suzuki S (1997) An immunosuppressive agent, FTY720, increases intracellular concentration of calcium ion and induces apoptosis in HL-60. Immunology 91:594–600
74. Neviani P et al (2007) FTY720, a new alternative for treating blast crisis chronic myelogenous leukemia and Philadelphia chromosome-positive acute lymphocytic leukemia. J Clin Invest 117:2408–2421
75. Wallington-Beddoe CT, Hewson J, Bradstock KF, Bendall LJ (2011) FTY720 produces caspase-independent cell death of acute lymphoblastic leukemia cells. Autophagy 7:1–10
76. Estrada-Bernal A, Palanichamy K, Ray Chaudhury A, Van Brocklyn JR (2012) Induction of brain tumor stem cell apoptosis by FTY720: a potential therapeutic agent for glioblastoma. Neuro-Oncology 14:405–415
77. Azuma H et al (2002) Marked prevention of tumor growth and metastasis by a novel immunosuppressive agent, FTY720, in mouse breast cancer models. Cancer Res 62:1410–1419
78. Szymiczek A et al (2017) FTY720 inhibits mesothelioma growth in vitro and in a syngeneic mouse model. J Transl Med 15:58
79. Booth L, Roberts JL, Spiegel S, Poklepovic A, Dent P (2019) Fingolimod augments pemetrexed killing of non-small cell lung cancer and overcomes resistance to ERBB inhibition. Cancer Biol Ther 20:597–607
80. Lankadasari MB et al (2018) Targeting S1PR1/STAT3 loop abrogates desmoplasia and chemosensitizes pancreatic cancer to gemcitabine. Theranostics 8:3824–3840
81. Shrestha J et al (2018) Synthesis of novel FTY720 analogs with anticancer activity through PP2A activation. Molecules 23(11):2750
82. McCracken AN et al (2017) Phosphorylation of a constrained azacyclic FTY720 analog enhances anti-leukemic activity without inducing S1P receptor activation. Leukemia 31:669–677
83. Pippa R et al (2014) Effect of FTY720 on the SET-PP2A complex in acute myeloid leukemia; SET binding drugs have antagonistic activity. Leukemia 28:1915–1918
84. Oaks JJ et al (2013) Antagonistic activities of the immunomodulator and PP2A-activating drug FTY720 (fingolimod, Gilenya) in Jak2-driven hematologic malignancies. Blood 122:1923–1934
85. De Palma RM et al (2019) The NMR-based characterization of the FTY720-SET complex reveals an alternative mechanism for the attenuation of the inhibitory SET-PP2A interaction. FASEB J 33:7647–7666
86. Cohen JA et al (2010) Oral fingolimod or intramuscular interferon for relapsing multiple sclerosis. N Engl J Med 362:402–415
87. Longbrake EE et al (2018) Effectiveness of alternative dose fingolimod for multiple sclerosis. Neurol Clin Pract 8:102–107
88. Fonseca J (2015) Fingolimod real world experience: efficacy and safety in clinical practice. Neurosci J 2015:1–7
89. Copland DA et al (2012) Therapeutic dosing of fingolimod (FTY720) prevents cell infiltration, rapidly suppresses ocular inflammation, and maintains the blood-ocular barrier. Am J Pathol 180:672–681
90. Raveney BJE, Copland DA, Nicholson LB, Dick AD (2008) Fingolimod (FTY720) as an acute rescue therapy for intraocular inflammatory disease. Arch Ophthalmol (Chicago, Ill) 126:1390–1395
91. Mudd JC et al (2013) Impaired T cell responses to sphingosine-1-phosphate in HIV-1 infected lymph nodes. Blood 121(15):2914–2922
92. Bankovich AJ, Shiow LR, Cyster JG (2010) CD69 suppresses sphingosine 1-phosophate receptor-1 (S1P1) function through interaction with membrane helix 4. J Biol Chem 285:22328–22337
93. Takada K et al (2011) Kruppel-like factor 2 is required for trafficking but not quiescence in postactivated T cells. J Immunol 186:775–783
94. Duquenne C et al (2017) Reversing HIV latency via sphingosine-1-phosphate receptor 1 signaling. AIDS 31:2443–2454
95. Kersh EN et al (2009) Evaluation of the lymphocyte trafficking drug FTY720 in SHIVSF162P3-infected rhesus macaques. J Antimicrob Chemother 63:758–762
96. Pino M et al (2019) Fingolimod retains cytolytic T cells and limits T follicular helper cell infection in lymphoid sites of SIV persistence. PLoS Pathog 15:e1008081–e1008024
97. Geffin R, Martinez R, de Las Pozas A, Issac B, McCarthy M (2017) Fingolimod induces neuronal-specific gene expression with potential neuroprotective outcomes in maturing neuronal progenitor cells exposed to HIV. J Neurovirol 23:808–824
98. Green JA et al (2011) The sphingosine 1-phosphate receptor S1P2 maintains the homeostasis of germi-

nal center B cells and promotes niche confinement. Nat Immunol 12:672–680
99. Drouillard A et al (2018) Human naive and memory T cells display opposite migratory responses to sphingosine-1 phosphate. J Immunol 200:551–557
100. Shannon LA et al (2012) CCR7/CCL19 controls expression of EDG-1 in T cells. J Biol Chem 287:11656–11664
101. Pham THM, Okada T, Matloubian M, Lo CG, Cyster JG (2008) S1P1 receptor signaling overrides retention mediated by G alpha i-coupled receptors to promote T cell egress. Immunity 28:122–133
102. Al-Kawaaz M, Sanchez T, Kluk MJ (2019) Evaluation of S1PR1, pSTAT3, S1PR2, FOXP1 expression in aggressive, mature B cell lymphomas. J Hematop 12:57–65
103. Cattoretti G et al (2009) Targeted disruption of the S1P2 sphingosine 1-phosphate receptor gene leads to diffuse large B-cell lymphoma formation. Cancer Res 69:8686–8692
104. Muppidi JR et al (2014) Loss of signalling via Gα13 in germinal centre B-cell-derived lymphoma. Nature 516:254–258
105. Flori M et al (2016) The hematopoietic oncoprotein FOXP1 promotes tumor cell survival in diffuse large B-cell lymphoma by repressing S1PR2 signaling. Blood 127:1438–1448
106. Moriyama S et al (2014) Sphingosine-1-phosphate receptor 2 is critical for follicular helper T cell retention in germinal centers. J Exp Med 211:1297–1305
107. Arnon TI, Horton RM, Grigorova IL, Cyster JG (2012) Visualization of splenic marginal zone B-cell shuttling and follicular B-cell egress. Nature 493:1–7
108. Arnon TI et al (2011) GRK2-dependent S1PR1 desensitization is required for lymphocytes to overcome their attraction to blood. Science 333:1898–1903
109. Gehring T et al (2019) MALT1 phosphorylation controls activation of T lymphocytes and survival of ABC-DLBCL tumor cells. Cell Rep 29:873–888.e10
110. Cheng J et al (2020) GRK2 suppresses lymphomagenesis by inhibiting the MALT1 proto-oncoprotein. J Clin Invest 130:1036–1051
111. Pereira JP, Xu Y, Cyster JG (2010) A role for S1P and S1P1 in immature-B cell egress from mouse bone marrow. PLoS One 5:e9277
112. Donovan EE, Pelanda R, Torres RM (2010) S1P3 confers differential S1P-induced migration by autoreactive and non-autoreactive immature B cells and is required for normal B-cell development. Eur J Immunol 40:688–698
113. Cinamon G, Zachariah MA, Lam OM, Foss FW, Cyster JG (2008) Follicular shuttling of marginal zone B cells facilitates antigen transport. Nat Immunol 9:54–62
114. Sic H et al (2014) Sphingosine-1-phosphate receptors control B-cell migration through signaling components associated with primary immunodeficiencies, chronic lymphocytic leukemia, and multiple sclerosis. J Allergy Clin Immunol 134:420–428
115. Wang W, Graeler MH, Goetzl EJ (2005) Type 4 sphingosine 1-phosphate G protein-coupled receptor (S1P4) transduces S1P effects on T cell proliferation and cytokine secretion without signaling migration. FASEB J 19:1731–1733
116. Schulze T et al (2011) Sphingosine-1-phospate receptor 4 ($S1P_4$) deficiency profoundly affects dendritic cell function and TH17-cell differentiation in a murine model. FASEB J 25:4024–4036
117. Kleinwort A, Lührs F, Heidecke CD, Lipp M, Schulze T (2018) S1P signalling differentially affects migration of peritoneal B cell populations in vitro and influences the production of intestinal IgA in vivo. Int J Mol Sci 19(2):391
118. Schick UM et al (2016) Meta-analysis of rare and common exome chip variants identifies S1PR4 and other loci influencing blood cell traits. Nat Genet 48:1–12
119. Fettel J et al (2018) Sphingosine-1-phosphate (S1P) induces potent anti-inflammatory effects in vitroand in vivoby S1P receptor 4-mediated suppression of 5-lipoxygenase activity. FASEB J. https://doi.org/10.1096/fj.201800221R
120. Pulcrano-Nicolas A-S et al (2019) Whole blood levels of S1PR4 mRNA associated with cerebral vasospasm after aneurysmal subarachnoid hemorrhage. J Neurosurg:1–5. https://doi.org/10.3171/2019.9.JNS191305
121. Provencio JJ et al (2010) CSF neutrophils are implicated in the development of vasospasm in subarachnoid hemorrhage. Neurocrit Care 12:244–251
122. Schön MP, Broekaert SMC, Erpenbeck L (2017) Sexy again: the renaissance of neutrophils in psoriasis. Exp Dermatol 26:305–311
123. Schuster C et al (2020) S1PR4-dependent CCL2 production promotes macrophage recruitment in a muine psoriasis model. Eur J Immunol
124. Al-Shamma H et al (2019) The selective sphingosine 1-phosphate receptor modulator etrasimod regulates lymphocyte trafficking and alleviates experimental colitis. J Pharmacol Exp Ther 369:311–317
125. Sandborn WJ et al (2019) Efficacy and safety of etrasimod in a phase 2 randomized trial of patients with ulcerative colitis. Gastroenterology 158:1–68
126. Xu N, Dahlbäck B (1999) A novel human apolipoprotein (apoM). J Biol Chem 274:31286–31290
127. Christoffersen C et al (2011) Endothelium-protective sphingosine-1-phosphate provided by HDL-associated apolipoprotein M. Proc Natl Acad Sci U S A 108:9613–9618
128. Murata N et al (2000) Interaction of sphingosine 1-phosphate with plasma components, including lipoproteins, regulates the lipid receptor-mediated actions. Biochem J 352(Pt 3):809–815
129. Obinata H et al (2019) Identification of ApoA4 as a sphingosine 1-phosphate chaperone in ApoM- and albumin-deficient mice. J Lipid Res 60. https://doi.org/10.1194/jlr.RA119000277

130. Croyal M et al (2018) Stable isotope kinetic study of ApoM (apolipoprotein M). Arterioscl Thromb Vasc Biol 38:255–261
131. Kurano M et al (2020) Protection against insulin resistance by apolipoprotein M/sphingosine 1-phosphate. Diabetes 69(5):867–881
132. Denimal D et al (2017) Impairment of the ability of HDL from patients with metabolic syndrome but without diabetes mellitus to activate eNOS: correction by S1P enrichment. Arterioscler Thromb Vasc Biol 37:804–811
133. Denimal D et al (2015) Significant abnormalities of the HDL phosphosphingolipidome in type 1 diabetes despite normal HDL cholesterol concentration. Atherosclerosis 241:752–760
134. Sattler K et al (2015) Defects of high-density lipoproteins in coronary artery disease caused by low sphingosine-1-phosphate content. J Am Coll Cardiol 66:1470–1485
135. Brinck JW et al (2016) Diabetes mellitus is associated with reduced high-density lipoprotein sphingosine-1-phosphate content and impaired high-density lipoprotein cardiac cell protection. Arterioscler Thromb Vasc Biol 36:817–824
136. Ruiz M et al (2017) High-density lipoprotein-associated apolipoprotein M limits endothelial inflammation by delivering sphingosine-1-phosphate to the sphingosine-1-phosphate receptor 1. Arterioscler Thromb Vasc Biol 37:118–129
137. Kurano M, Tsuneyama K, Morimoto Y, Nishikawa M, Yatomi Y (2019) Apolipoprotein M suppresses the phenotypes of IgA nephropathy in hyper-IgA mice. FASEB J 33:5181–5195
138. Du W et al (2017) Low apolipoprotein M serum levels correlate with systemic lupus erythematosus disease activity and apolipoprotein M gene polymorphisms with lupus. Lipids Health Dis 16:88
139. Kim SY et al (2020) High-density lipoprotein in lupus: disease biomarkers and potential therapeutic strategy. Arthritis Rheumatol 72:20–30
140. Yang L, Li T, Zhao S, Zhang S (2019) Niacin regulates apolipoprotein M expression via liver X receptor-α. Mol Med Rep 20:1–7
141. Sattler KJE et al (2010) Sphingosine 1-phosphate levels in plasma and HDL are altered in coronary artery disease. Basic Res Cardiol 105:821–832
142. Sattler K et al (2014) HDL-bound sphingosine 1-phosphate (S1P) predicts the severity of coronary artery atherosclerosis. Cell Physiol Biochem 34:172–184
143. Yafasova A et al (2019) Effect of menopause and exercise training on plasma apolipoprotein M and sphingosine-1-phosphate. J Appl Physiol (1985) 126:214–220
144. Rist PM et al (2019) Lipid levels and the risk of hemorrhagic stroke among women. Neurology 92:e2286–e2294
145. He Y, Kothari V, Bornfeldt KE (2018) High-density lipoprotein function in cardiovascular disease and diabetes mellitus. Arterioscler Thromb Vasc Biol 38:e10–e16
146. Gourgari E et al (2019) Proteomic alterations of HDL in youth with type 1 diabetes and their associations with glycemic control: a case-control study. Cardiovasc Diabetol 18:43
147. Kontush A (2015) HDL particle number and size as predictors of cardiovascular disease. Front Pharmacol 6:218
148. Kim DS et al (2016) Concentration of smaller high-density lipoprotein particle (HDL-P) is inversely correlated with carotid intima media thickening after confounder adjustment: the multi ethnic study of atherosclerosis (MESA). J Am Heart Assoc 5:e002977
149. Igarashi J, Miyoshi M, Hashimoto T, Kubota Y, Kosaka H (2007) Hydrogen peroxide induces S1P1 receptors and sensitizes vascular endothelial cells to sphingosine 1-phosphate, a platelet-derived lipid mediator. Am J Physiol Cell Physiol 292:C740–C748
150. Khan AA et al (2018) Weight loss and exercise alter the high-density lipoprotein lipidome and improve high-density lipoprotein functionality in metabolic syndrome. Arterioscl Thromb Vasc Biol 38:438–447
151. Padró T et al (2017) Detrimental effect of hypercholesterolemia on high-density lipoprotein particle remodeling in pigs. J Am Coll Cardiol 70:165–178
152. Qian J, Fulton D (2013) Post-translational regulation of endothelial nitric oxide synthase in vascular endothelium. Front Physiol 4:347
153. Heiss C, Rodriguez-Mateos A, Kelm M (2015) Central role of eNOS in the maintenance of endothelial homeostasis. Antioxid Redox Signal 22:1230–1242
154. Keul P et al (2018) Potent anti-inflammatory properties of HDL in vascular smooth muscle cells mediated by HDL-S1P and their impairment in coronary artery disease due to lower HDL-S1P: a new aspect of HDL dysfunction and its therapy. FASEB J 33:1482–1495
155. Muñoz-Vega M et al (2018) HDL-mediated lipid influx to endothelial cells contributes to regulating intercellular adhesion molecule (ICAM)-1 expression and eNOS phosphorylation. Int J Mol Sci 19(11):3394
156. Plochberger B et al (2018) Direct observation of cargo transfer from HDL particles to the plasma membrane. Atherosclerosis 277:53–59
157. Harayama T, Riezman H (2018) Understanding the diversity of membrane lipid composition. Nat Rev Mol Cell Biol 19:281–296
158. Hammad SM, Al Gadban MM, Semler AJ, Klein RL (2012) Sphingosine 1-phosphate distribution in human plasma: associations with lipid profiles. J Lipids 2012:180705–180708
159. Kontush A et al (2007) Preferential sphingosine-1-phosphate enrichment and sphingomyelin depletion are key features of small dense HDL3 particles. Arterioscl Thromb Vasc Biol 27:1843–1849
160. Christoffersen C et al (2008) Effect of apolipoprotein M on high density lipoprotein metabolism and atherosclerosis in low density lipoprotein receptor knock-out mice. J Biol Chem 283:1839–1847

161. Christoffersen C et al (2010) Opposing effects of apolipoprotein m on catabolism of apolipoprotein B-containing lipoproteins and atherosclerosis. Circ Res 106:1624–1634
162. Wolfrum C, Poy MN, Stoffel M (2005) Apolipoprotein M is required for prebeta-HDL formation and cholesterol efflux to HDL and protects against atherosclerosis. Nat Med 11:418–422
163. Liang CP, Tall AR (2001) Transcriptional profiling reveals global defects in energy metabolism, lipoprotein, and bile acid synthesis and transport with reversal by leptin treatment in ob/ob mouse liver. J Biol Chem 276:49066–49076
164. Xu N, Nilsson-Ehle P, Hurtig M, Ahrén B (2004) Both leptin and leptin-receptor are essential for apolipoprotein M expression in vivo. Biochem Biophys Res Commun 321:916–921
165. Christoffersen C et al (2018) The apolipoprotein M/S1P axis controls triglyceride metabolism and brown fat activity. Cell Rep 22:175–188
166. Sramkova V et al (2019) Apolipoprotein M: a novel adipokine decreasing with obesity and upregulated by calorie restriction. Am J Clin Nutr 109:1499–1510
167. Xu N, Nilsson-Ehle P, Ahrén B (2006) Suppression of apolipoprotein M expression and secretion in alloxan-diabetic mouse: partial reversal by insulin. Biochem Biophys Res Commun 342:1174–1177
168. Kurano M et al (2014) Induction of insulin secretion by apolipoprotein M, a carrier for sphingosine 1-phosphate. BBA Mol Cell Biol Lipids 1841:1–44
169. Tydén H et al (2019) Low plasma concentrations of apolipoprotein M are associated with disease activity and endothelial dysfunction in systemic lupus erythematosus. Arthritis Res Ther 21:1–9
170. Burg N, Swendeman S, Worgall S, Hla T, Salmon JE (2018) Sphingosine 1-phosphate receptor 1 signaling maintains endothelial cell barrier function and protects against immune complex-induced vascular injury. Arthritis Rheumatol 70:1879–1889
171. Wu J, He L, Bai L, Tan L, Hu M, Apolipoprotein M (2019) Serum levels correlate with IgA vasculitis and IgA vasculitis nephritis. Dis Markers 2019:1825849
172. Palmiere C, Bonsignore A, Augsburger M (2015) Measurement of apolipoprotein M in sepsis-related deaths. Clin Chem Lab Med 53:e93–e96
173. Kumaraswamy SB, Linder A, Åkesson P, Dahlbäck B (2012) Decreased plasma concentrations of apolipoprotein M in sepsis and systemic inflammatory response syndromes. Crit Care (London) 16:R60
174. Yanagida K et al (2017) Size-selective opening of the blood-brain barrier by targeting endothelial sphingosine 1-phosphate receptor 1. Proc Natl Acad Sci U S A 114:4531–4536
175. Mathiesen Janiurek M, Soylu-Kucharz R, Christoffersen C, Kucharz K, Lauritzen M (2019) Apolipoprotein M-bound sphingosine-1-phosphate regulates blood-brain barrier paracellular permeability and transcytosis. elife 8:e49405
176. Christensen PM, Bosteen MH, Hajny S, Nielsen LB, Christoffersen C (2017) Apolipoprotein M mediates sphingosine-1-phosphate efflux from erythrocytes. Sci Rep 7:14983
177. Poojary S, Shah M (2017) Sjögren-Larsson syndrome: definitive diagnosis on magnetic resonance spectroscopy. Cutis 100:452–455
178. Ruiz M, Okada H, Dahlbäck B (2017) HDL-associated ApoM is anti-apoptotic by delivering sphingosine 1-phosphate to S1P1 & S1P3 receptors on vascular endothelium. Lipids Health Dis 16:36
179. Frej C et al (2016) Sphingosine 1-phosphate and its carrier apolipoprotein M in human sepsis and in Escherichia coli sepsis in baboons. J Cell Mol Med 20:1–12
180. Swendeman SL et al (2017) An engineered S1P chaperone attenuates hypertension and ischemic injury. Sci Signal 10:eaal2722-30
181. Rueda CM et al (2017) High density lipoproteins selectively promote the survival of human regulatory T cells. J Lipid Res 58:1514–1523
182. Wang S-H, Yuan S-G, Peng D-Q, Zhao S-P (2012) HDL and ApoA-I inhibit antigen presentation-mediated T cell activation by disrupting lipid rafts in antigen presenting cells. Atherosclerosis 225(1):105–114
183. Cheng H-Y et al (2016) Loss of ABCG1 influences regulatory T cell differentiation and atherosclerosis. J Clin Invest 126:3236–3246
184. Faber K, Hvidberg V, Moestrup SK, Dahlbäck B, Nielsen LB (2006) Megalin is a receptor for apolipoprotein M, and kidney-specific megalin-deficiency confers urinary excretion of apolipoprotein M. Mol Endocrinol 20:212–218
185. Zhai XY et al (2000) Cubilin- and megalin-mediated uptake of albumin in cultured proximal tubule cells of opossum kidney. Kidney Int 58:1523–1533
186. Christoffersen C, Dahlbäck B, Nielsen LB (2006) Apolipoprotein M: progress in understanding its regulation and metabolic functions. Scand J Clin Lab Invest 66:631–637
187. Peruchetti DDB, Silva-Aguiar RP, Siqueira GM, Dias WB, Caruso-Neves C (2018) High glucose reduces megalin-mediated albumin endocytosis in renal proximal tubule cells through protein kinase BO-GlcNAcylation. J Biol Chem 293:11388–11400
188. Rowling MJ, Kemmis CM, Taffany DA, Welsh J (2006) Megalin-mediated endocytosis of vitamin D binding protein correlates with 25-hydroxycholecalciferol actions in human mammary cells. J Nutr 136:2754–2759
189. Hori Y et al (2017) Megalin blockade with cilastatin suppresses drug-induced nephrotoxicity. J Am Soc Nephrol 28:1783–1791
190. Sengul S, Erturk S, Khan AM, Batuman V (2013) Receptor-associated protein blocks internalization and cytotoxicity of myeloma light chain in cultured human proximal tubular cells. PLoS One 8:e70276
191. Demay MB (2018) The good and the bad of vitamin D inactivation. J Clin Invest 128:3736–3738
192. Šimoliūnas E, Rinkūnaitė I, Bukelskienė Ž, Bukelskienė V (2019) Bioavailability of different

vitamin D Oral supplements in laboratory animal model. Medicina (Kaunas) 55(6):265
193. Nykjaer A et al (1999) An endocytic pathway essential for renal uptake and activation of the steroid 25-(OH) vitamin D3. Cell 96:507–515
194. Chapron BD et al (2018) Reevaluating the role of megalin in renal vitamin D homeostasis using a human cell-derived microphysiological system. ALTEX 35:504–515
195. Jaimungal S, Wehmeier K, Mooradian AD, Haas MJ (2011) The emerging evidence for vitamin D-mediated regulation of apolipoprotein A-I synthesis. Nutr Res 31:805–812
196. Abbasi F et al (2015) Low circulating 25-hydroxyvitamin D concentrations are associated with defects in insulin action and insulin secretion in persons with prediabetes. J Nutr 145:714–719
197. Scott D et al (2019) Vitamin D supplementation improves waist-to-hip ratio and fasting blood glucose in vitamin D deficient, overweight or obese Asians: a pilot secondary analysis of a randomised controlled trial. J Steroid Biochem Mol Biol 186:136–141
198. Dibaba DT (2019) Effect of vitamin D supplementation on serum lipid profiles: a systematic review and meta-analysis. Nutr Rev 77:890–902
199. Wigger L et al (2017) Plasma dihydroceramides are diabetes susceptibility biomarker candidates in mice and humans. Cell Rep 18:2269–2279
200. Mwinyi J et al (2017) Plasma 1-deoxysphingolipids are early predictors of incident type 2 diabetes mellitus. PLoS One 12:e0175776
201. Koch A et al (2017) Vitamin D supplementation enhances C18(dihydro)ceramide levels in type 2 diabetes patients. Int J Mol Sci 18(7):1532
202. Nejatian N et al (2019) Vitamin D effects on sphingosine 1-phosphate signaling and metabolism in monocytes from type 2 diabetes patients and controls. J Steroid Biochem Mol Biol 186:130–135
203. Pittas AG et al (2019) Vitamin D supplementation and prevention of type 2 diabetes. N Engl J Med 381:520–530
204. Lee Y, Venkataraman K, HWANG S, HAN D, Hla T (2007) A novel method to quantify sphingosine 1-phosphate by immobilized metal affinity chromatography (IMAC). Prostaglandins Other Lipid Mediat 84:154–162
205. Decouture B, Becker PH, Therond P, Gaussem P, Bachelot-Loza C (2018) Evidence that MRP4 is only partly involved in S1P secretion during platelet activation. Thromb Haemost 118:1116–1118
206. Vogt K et al (2018) Release of platelet-derived sphingosine-1-phosphate involves multidrug resistance protein 4 (MRP4/ABCC4) and is inhibited by statins. Thromb Haemost 118:132–142
207. Hisano Y, Kobayashi N, Yamaguchi A, Nishi T (2012) Mouse SPNS2 functions as a sphingosine-1-phosphate transporter in vascular endothelial cells. PLoS One 7:e38941
208. Pham THM et al (2010) Lymphatic endothelial cell sphingosine kinase activity is required for lymphocyte egress and lymphatic patterning. J Exp Med 207:17–27
209. Girard J-P, Moussion C, Förster R (2012) HEVs, lymphatics and homeostatic immune cell trafficking in lymph nodes. Nat Rev Immunol 12:762–773
210. Kobayashi N et al (2018) MFSD2B is a sphingosine 1-phosphate transporter in erythroid cells. Sci Rep 8:4969
211. Vu TM et al (2017) Mfsd2b is essential for the sphingosine-1-phosphate export in erythrocytes and platelets. Nature 550:524–528
212. Nguyen LN et al (2014) Mfsd2a is a transporter for the essential omega-3 fatty acid docosahexaenoic acid. Nature 509:503–506
213. Ben-Zvi A et al (2014) Mfsd2a is critical for the formation and function of the blood-brain barrier. Nature 509:507–511
214. Andreone BJ et al (2017) Blood-brain barrier permeability is regulated by lipid transport-dependent suppression of caveolae-mediated Transcytosis. Neuron 94:581–594.e5
215. Zhou J, Saba JD (1998) Identification of the first mammalian sphingosine phosphate lyase gene and its functional expression in yeast. Biochem Biophys Res Commun 242:502–507
216. Hagen N et al (2009) Subcellular origin of sphingosine 1-phosphate is essential for its toxic effect in lyase-deficient neurons. J Biol Chem 284:11346–11353
217. Borowsky AD et al (2012) Sphingosine-1-phosphate lyase expression in embryonic and adult murine tissues. J Lipid Res 53:1920–1931
218. Yatomi Y et al (2004) Sphingosine 1-phosphate breakdown in platelets. J Biochem 136:495–502
219. Selim S et al (2011) Plasma levels of sphingosine 1-phosphate are strongly correlated with haematocrit, but variably restored by red blood cell transfusions. Clin Sci (Lond) 121:565–572
220. Atkinson D et al (2017) Sphingosine 1-phosphate lyase deficiency causes Charcot-Marie-Tooth neuropathy. Neurology 88:533–542
221. Pareyson D, Saveri P, Pisciotta C (2017) New developments in Charcot-Marie-Tooth neuropathy and related diseases. Curr Opin Neurol 30:471–480
222. Barreto LCLS et al (2016) Epidemiologic study of Charcot-Marie-Tooth disease: a systematic review. Neuroepidemiology 46:157–165
223. Prasad R et al (2017) Sphingosine-1-phosphate lyase mutations cause primary adrenal insufficiency and steroid-resistant nephrotic syndrome. J Clin Invest 127:942–953
224. Lovric S et al (2017) Mutations in sphingosine-1-phosphate lyase cause nephrosis with ichthyosis and adrenal insufficiency. J Clin Invest 127:912–928
225. Janecke AR et al (2017) Deficiency of the sphingosine-1-phosphate lyase SGPL1 is associated with congenital nephrotic syndrome and congenital adrenal calcifications. Hum Mutat 38:365–372
226. Linhares ND, Arantes RR, Araujo SA, Pena SDJ (2018) Nephrotic syndrome and adrenal insuffi-

ciency caused by a variant in SGPL1. Clin Kidney J 11:462–467
227. Choi YJ, Saba JD (2019) Sphingosine phosphate lyase insufficiency syndrome (SPLIS): a novel inborn error of sphingolipid metabolism. Adv Biol Regul 71:128–140
228. Ohtoyo M, Tamura M, Machinaga N, Muro F, Hashimoto R (2014) Sphingosine 1-phosphate lyase inhibition by 2-acetyl-4-(tetrahydroxybutyl)imidazole (THI) under conditions of vitamin B6 deficiency. Mol Cell Biochem 400:1–9
229. Sengar G, Sharma HK (2014) Food caramels: a review. J Food Sci Technol 51:1686–1696
230. Gailani SD, Holland JF, Nussbaum A, Olson KB (1968) Clinical and biochemical studies of pyridoxine deficiency in patients with neoplastic diseases. Cancer 21:975–988
231. Ohtoyo M et al (2016) Component of caramel food coloring, THI, causes lymphopenia indirectly via a key metabolic intermediate. Cell Chem Biol 23:555–560
232. Richts B, Rosenberg J, Commichau FM (2019) A survey of pyridoxal 5′-phosphate-dependent proteins in the gram-positive model bacterium Bacillus subtilis. Front Mol Biosci 6:32
233. Percudani R, Peracchi A (2009) The B6 database: a tool for the description and classification of vitamin B6-dependent enzymatic activities and of the corresponding protein families. BMC Bioinformatics 10:273
234. Pons G et al (2020) A mechanism-based sphingosine-1-phosphate lyase inhibitor. J Org Chem 85:419–429
235. Coady TH, Lorson CL (2011) SMN in spinal muscular atrophy and snRNP biogenesis. WIREs RNA 2:546–564
236. Mitroi DN et al (2016) Sphingosine 1-phosphate lyase ablation disrupts presynaptic architecture and function via an ubiquitin- proteasome mediated mechanism. Sci Rep 6:37064
237. Mitroi DN et al (2017) SGPL1 (sphingosine phosphate lyase 1) modulates neuronal autophagy via phosphatidylethanolamine production. Autophagy 13:885–899
238. Bernal A, Arranz L (2018) Nestin-expressing progenitor cells: function, identity and therapeutic implications. Cell Mol Life Sci 75:2177–2195
239. Karunakaran I et al (2019) Neural sphingosine 1-phosphate accumulation activates microglia and links impaired autophagy and inflammation. Glia 288:27667–27614
240. Karaca I et al (2014) Deficiency of sphingosine-1-phosphate lyase impairs lysosomal metabolism of the amyloid precursor protein. J Biol Chem 289:16761–16772
241. Djajadikerta A et al (2019) Autophagy induction as a therapeutic strategy for neurodegenerative diseases. J Mol Biol 432(8):2799–2821
242. Coady TH, Manley JL (2015) ALS mutations in TLS/FUS disrupt target gene expression. Genes Dev 29:1696–1706
243. Gutner UA et al (2019) Changes in the metabolism of sphingoid bases in the brain and spinal cord of transgenic FUS(1-359) mice, a model of amyotrophic lateral sclerosis. Biochemistry (Mosc) 84:1166–1176
244. Lombardi LM, Baker SA, Zoghbi HY (2015) MECP2 disorders: from the clinic to mice and back. J Clin Invest 125:2914–2923
245. Cappuccio G et al (2019) Sphingolipid metabolism perturbations in Rett syndrome. Meta 9(10):221
246. Sbardella D et al (2017) Retention of mitochondria in mature human red blood cells as the result of autophagy impairment in Rett syndrome. Sci Rep 7:12297
247. Moruno Manchon JF et al (2015) Cytoplasmic sphingosine-1-phosphate pathway modulates neuronal autophagy. Sci Rep 5:15213
248. Ceccom J et al (2014) Reduced sphingosine kinase-1 and enhanced sphingosine 1-phosphate lyase expression demonstrate deregulated sphingosine 1-phosphate signaling in Alzheimer's disease. Acta Neuropathol Commun 2:12
249. Katsel P, Li C, Haroutunian V (2007) Gene expression alterations in the sphingolipid metabolism pathways during progression of dementia and Alzheimer's disease: a shift toward ceramide accumulation at the earliest recognizable stages of Alzheimer's disease. Neurochem Res 32:845–856
250. Herr DR et al (2003) Sply regulation of sphingolipid signaling molecules is essential for Drosophila development. Development 130:2443–2453
251. Weske S et al (2019) Agonist-induced activation of the S1P receptor 2 constitutes a novel osteoanabolic therapy for the treatment of osteoporosis in mice. Bone 125:1–7
252. Bagdanoff JT et al (2010) Inhibition of sphingosine 1-phosphate lyase for the treatment of rheumatoid arthritis: discovery of (E)-1-(4-((1R,2S,3R)-1,2,3,4-tetrahydroxybutyl)-1H-imidazol-2-yl)ethanone oxime (LX2931) and (1R,2S,3R)-1-(2-(isoxazol-3-yl)-1H-imidazol-4-yl)butane-1,2,3,4-tetraol (LX2932). J Med Chem 53:8650–8662
253. Allende ML et al (2011) Sphingosine-1-phosphate lyase deficiency produces a pro-inflammatory response while impairing neutrophil trafficking. J Biol Chem 286:7348–7358
254. Abhyankar V, Kaduskar B, Kamat SS, Deobagkar D, Ratnaparkhi GS (2018) *Drosophila* DNA/RNA methyltransferase contributes to robust host defense in aging animals by regulating sphingolipid metabolism. J Exp Biol 221:jeb187989
255. Zamora-Pineda J, Kumar A, Suh JH, Zhang M, Saba JD (2016) Dendritic cell sphingosine-1-phosphate lyase regulates thymic egress. J Exp Med 213:2773–2791
256. Karuppuchamy T et al (2020) Sphingosine-1-phosphate lyase inhibition alters the S1P gradient and ameliorates Crohn's-like ileitis by suppress-

ing thymocyte maturation. Inflamm Bowel Dis 26:216–228
257. Jeon S et al (2019) Inhibition of sphingosine 1phosphate lyase activates human keratinocyte differentiation and attenuates psoriasis in mice. J Lipid Res 61:20–32
258. Degagné E et al (2014) Sphingosine-1-phosphate lyase downregulation promotes colon carcinogenesis through STAT3-activated microRNAs. J Clin Invest 124:5368–5384
259. Schwiebs A et al (2019) Cancer-induced inflammation and inflammation-induced cancer in colon: a role for S1P lyase. Oncogene 38:4788–4803
260. Bamias G et al (2013) Intestinal-specific TNFα overexpression induces Crohn's-like ileitis in mice. PLoS One 8:e72594
261. Peyrin-Biroulet L, Christopher R, Behan D, Lassen C (2017) Modulation of sphingosine-1-phosphate in inflammatory bowel disease. Autoimmun Rev 16:495–503
262. Shimano K et al (2019) Amiselimod (MT-1303), a novel sphingosine 1-phosphate receptor-1 functional antagonist, inhibits progress of chronic colitis induced by transfer of CD4+CD45RBhigh T cells. PLoS One 14:e0226154
263. Sandborn WJ et al (2016) Ozanimod induction and maintenance treatment for ulcerative colitis. N Engl J Med 374:1754–1762
264. Pranke I, Golec A, Hinzpeter A, Edelman A, Sermet-Gaudelus I (2019) Emerging therapeutic approaches for cystic fibrosis. From gene editing to personalized medicine. Front Pharmacol 10:121
265. Lopes-Pacheco M (2019) CFTR modulators: the changing face of cystic fibrosis in the era of precision medicine. Front Pharmacol 10:1662
266. Rossi GA, Morelli P, Galietta LJ, Colin AA (2019) Airway microenvironment alterations and pathogen growth in cystic fibrosis. Pediatr Pulmonol 54:497–506
267. Svedin E et al (2017) A link between a common mutation in CFTR and impaired innate and adaptive viral defense. J Infect Dis 216:1308–1317
268. Vijayan M et al (2017) Sphingosine 1-phosphate lyase enhances the activation of IKKε to promote type I IFN-mediated innate immune responses to influenza a virus infection. J Immunol 199:677–687
269. Veltman M et al (2016) Correction of lung inflammation in a F508del CFTR murine cystic fibrosis model by the sphingosine-1-phosphate lyase inhibitor LX2931. Am J Physiol Lung Cell Mol Physiol 311:L1000–L1014
270. Malik FA et al (2015) Sphingosine-1-phosphate is a novel regulator of cystic fibrosis transmembrane conductance regulator (CFTR) activity. PLoS One 10:e0130313
271. Hsu SC et al (2019) Circulating sphingosine-1-phosphate as a prognostic biomarker for community-acquired pneumonia. PLoS One 14:e0216963
272. Finney CA et al (2011) S1P is associated with protection in human and experimental cerebral malaria. Mol Med 17:717–725
273. Zhao P et al (2020) Responsiveness of sphingosine phosphate lyase insufficiency syndrome (SPLIS) to vitamin B6 cofactor supplementation. J Inherit Metab Dis. https://doi.org/10.1002/jimd.12238
274. Mao G, Smyth SS, Morris AJ (2019) Regulation of PLPP3 gene expression by NF-κB family transcription factors. J Biol Chem 294:14009–14019
275. Mandala SM et al (2000) Molecular cloning and characterization of a lipid phosphohydrolase that degrades sphingosine-1- phosphate and induces cell death. Proc Natl Acad Sci U S A 97:7859–7864
276. Le Stunff H, Galve-Roperh I, Peterson C, Milstien S, Spiegel S (2002) Sphingosine-1-phosphate phosphohydrolase in regulation of sphingolipid metabolism and apoptosis. J Cell Biol 158:1039–1049
277. Ogawa C, Kihara A, Gokoh M, Igarashi Y (2003) Identification and characterization of a novel human sphingosine-1-phosphate phosphohydrolase, hSPP2. J Biol Chem 278:1268–1272
278. Santulli P et al (2012) Sphingosine pathway deregulation in endometriotic tissues. Fertil Steril 97:904–911
279. Allende ML et al (2013) Sphingosine-1-phosphate phosphatase 1 regulates keratinocyte differentiation and epidermal homeostasis. J Biol Chem 288:18381–18391
280. Huang WC et al (2016) Sphingosine-1-phosphate phosphatase 2 promotes disruption of mucosal integrity, and contributes to ulcerative colitis in mice and humans. FASEB J 30:2945–2958
281. Schwiebs A, Thomas D, Kleuser B, Pfeilschifter JM, Radeke HH (2017) Nuclear translocation of SGPP-1 and decrease of SGPL-1 activity contribute to sphingolipid rheostat regulation of inflammatory dendritic cells. Mediat Inflamm 2017:5187368
282. Sigal YJ, McDermott MI, Morris AJ (2005) Integral membrane lipid phosphatases/phosphotransferases: common structure and diverse functions. Biochem J 387:281–293
283. Jia YJ, Kai M, Wada I, Sakane F, Kanoh H (2003) Differential localization of lipid phosphate phosphatases 1 and 3 to cell surface subdomains in polarized MDCK cells. FEBS Lett 552:240–246
284. Escalante-Alcalde D et al (2003) The lipid phosphatase LPP3 regulates extra-embryonic vasculogenesis and axis patterning. Development 130:4623–4637
285. Bréart B et al (2011) Lipid phosphate phosphatase 3 enables efficient thymic egress. J Exp Med 208(6):1267–1278
286. Krause MD et al (2018) Genetic variant at coronary artery disease and ischemic stroke locus 1p32.2 regulates endothelial responses to hemodynamics. Proc Natl Acad Sci U S A 115:E11349–E11358
287. Touat-Hamici Z et al (2016) Role of lipid phosphate phosphatase 3 in human aortic endothelial cell function. Cardiovasc Res 112:702–713

288. Mechtcheriakova D et al (2007) FTY720-phosphate is dephosphorylated by lipid phosphate phosphatase 3. FEBS Lett 581:3063–3068
289. Aaltonen N, Lehtonen M, Varonen K, Goterris GA, Laitinen JT (2012) Lipid phosphate phosphatase inhibitors locally amplify lysophosphatidic acid LPA1 receptor signalling in rat brain cryosections without affecting global LPA degradation. BMC Pharmacol 12:7
290. Kohama T et al (1998) Molecular cloning and functional characterization of murine sphingosine kinase. J Biol Chem 273:23722–23728
291. Liu H et al (2000) Molecular cloning and functional characterization of a novel mammalian sphingosine kinase type 2 isoform. J Biol Chem 275:19513–19520
292. Trayssac M, Hannun YA, Obeid LM (2018) Role of sphingolipids in senescence: implication in aging and age-related diseases. J Clin Invest 128:2702–2712
293. Wolf JJ, Studstill CJ, Hahm B (2019) Emerging connections of S1P-metabolizing enzymes with host defense and immunity during virus infections. Viruses 11:1097–1014
294. Pulkoski-Gross MJ, Obeid LM (2018) Molecular mechanisms of regulation of sphingosine kinase 1. Biochim Biophys Acta Mol Cell Biol Lipids 1863:1413–1422
295. Lewis CS, Voelkel-Johnson C, Smith CD (2018) Targeting sphingosine kinases for the treatment of cancer. Adv Cancer Res 140:295–325
296. Mizugishi K et al (2005) Essential role for sphingosine kinases in neural and vascular development. Mol Cell Biol 25:11113–11121
297. Allende ML et al (2004) Mice deficient in sphingosine kinase 1 are rendered lymphopenic by FTY720. J Biol Chem 279:52487–52492
298. Sensken S-C et al (2010) Redistribution of sphingosine 1-phosphate by sphingosine kinase 2 contributes to lymphopenia. J Immunol 184:4133–4142
299. Kharel Y et al (2020) Mechanism of sphingosine 1-phosphate clearance from blood. Biochem J 77(5):925–935
300. Hisano Y, Kobayashi N, Kawahara A, Yamaguchi A, Nishi T (2011) The sphingosine 1-phosphate transporter, SPNS2, functions as a transporter of the phosphorylated form of the immunomodulating agent FTY720. J Biol Chem 286:1758–1766
301. Pyne S, Bittman R, Pyne NJ (2011) Sphingosine kinase inhibitors and cancer: seeking the golden sword of Hercules. Cancer Res 71:6576–6582
302. Gude DR et al (2008) Apoptosis induces expression of sphingosine kinase 1 to release sphingosine-1-phosphate as a 'come-and-get-me' signal. FASEB J 22:2629–2638
303. French KJ et al (2003) Discovery and evaluation of inhibitors of human sphingosine kinase. Cancer Res 63:5962–5969
304. French KJ et al (2006) Antitumor activity of sphingosine kinase inhibitors. J Pharmacol Exp Ther 318:596–603
305. Loveridge C et al (2010) The sphingosine kinase 1 inhibitor 2-(p-hydroxyanilino)-4-(p-chlorophenyl) thiazole induces proteasomal degradation of sphingosine kinase 1 in mammalian cells. J Biol Chem 285:38841–38852
306. Tonelli F et al (2010) FTY720 and (S)-FTY720 vinylphosphonate inhibit sphingosine kinase 1 and promote its proteasomal degradation in human pulmonary artery smooth muscle, breast cancer and androgen-independent prostate cancer cells. Cell Signal 22:1536–1542
307. McNaughton M, Pitman M, Pitson SM, Pyne NJ, Pyne S (2016) Proteasomal degradation of sphingosine kinase 1 and inhibition of dihydroceramide desaturase by the sphingosine kinase inhibitors, SKi or ABC294640, induces growth arrest in androgen-independent LNCaP-AI prostate cancer cells. Oncotarget 7:16663–16675
308. Powell JA et al (2019) Kelch-like protein 5-mediated ubiquitination of lysine 183 promotes proteasomal degradation of sphingosine kinase 1. Biochem J 476:3211–3226
309. Schleifer RJ et al (2018) KLHL5 knockdown increases cellular sensitivity to anticancer drugs. Oncotarget 9:37429–37438
310. Cingolani F et al (2014) Inhibition of dihydroceramide desaturase activity by the sphingosine kinase inhibitor SKI II. J Lipid Res 55:1711–1720
311. Pulkoski-Gross MJ et al (2017) Novel sphingosine kinase-1 inhibitor, LCL351, reduces immune responses in murine DSS-induced colitis. Prostaglandins Other Lipid Mediat 130:47–56
312. Snider AJ et al (2009) A role for sphingosine kinase 1 in dextran sulfate sodium-induced colitis. FASEB J 23:143–152
313. Wang C et al (2014) Depletion of Sf3b1 impairs proliferative capacity of hematopoietic stem cells but is not sufficient to induce myelodysplasia. Blood 123:3336–3343
314. Schnute ME et al (2012) Modulation of cellular S1P levels with a novel, potent and specific inhibitor of sphingosine kinase-1. Biochem J 444:79–88
315. Sun M, Zhou Y, Shi Y, Liu B (2019) Effect of the sphingosine kinase 1 selective inhibitor, PF543 on dextran sodium sulfate-induced colitis in mice. DNA Cell Biol 38:1338–1345
316. Liu J, Jiang B (2020) Sphk1 promotes ulcerative colitis via activating JAK2/STAT3 signaling pathway. Hum Cell 33:57–66
317. Alganga H et al (2019) Short periods of hypoxia upregulate sphingosine kinase 1 and increase vasodilation of arteries to sphingosine 1-phosphate (S1P) via S1P(3). J Pharmacol Exp Ther 371:63–74
318. Józefczuk E et al (2020) Cardiovascular effects of pharmacological targeting of sphingosine kinase 1. Hypertension 75:383–392
319. French KJ et al (2010) Pharmacology and antitumor activity of ABC294640, a selective inhibitor of sphingosine kinase-2. J Pharmacol Exp Ther 333:129–139

320. Britten CD et al (2017) A phase I study of ABC294640, a first-in-class sphingosine kinase-2 inhibitor, in patients with advanced solid tumors. Clin Cancer Res 23:4642–4650
321. Chumanevich AA et al (2010) Suppression of colitis-driven colon cancer in mice by a novel small molecule inhibitor of sphingosine kinase. Carcinogenesis 31:1787–1793
322. Antoon JW et al (2010) Antiestrogenic effects of the novel sphingosine kinase-2 inhibitor ABC294640. Endocrinology 151:5124–5135
323. Antoon JW, White MD, Driver JL, Burow ME, Beckman BS (2012) Sphingosine kinase isoforms as a therapeutic target in endocrine therapy resistant luminal and basal-A breast cancer. Exp Biol Med 237:832–844
324. Gomez-Muñoz A et al (2013) New insights on the role of ceramide 1-phosphate in inflammation. Biochim Biophys Acta 1831(6):1060
325. Fitzpatrick LR et al (2011) Attenuation of arthritis in rodents by a novel orally-available inhibitor of sphingosine kinase. Inflammopharmacology 19:75–87
326. Baker DA, Eudaly J, Smith CD, Obeid LM, Gilkeson GS (2013) Impact of sphingosine kinase 2 deficiency on the development of TNF-alpha-induced inflammatory arthritis. Rheumatol Int 33:2677–2681
327. Snider AJ et al (2012) Loss of neutral ceramidase increases inflammation in a mouse model of inflammatory bowel disease. Prostaglandins Other Lipid Mediat 99:124–130
328. Kim AHJ et al (2020) A rush to judgment? Rapid reporting and dissemination of results and its consequences regarding the use of hydroxychloroquine for COVID-19. Ann Intern Med. https://doi.org/10.7326/M20-1223
329. Xia C et al (2018) Transient inhibition of sphingosine kinases confers protection to influenza A virus infected mice. Antivir Res 158:171–177
330. Seo YJ, Blake C, Alexander S, Hahm B (2010) Sphingosine 1-phosphate-metabolizing enzymes control influenza virus propagation and viral cytopathogenicity. J Virol 84:8124–8131
331. Morris EJ et al (2013) Discovery of a novel ERK inhibitor with activity in models of acquired resistance to BRAF and MEK inhibitors. Cancer Discov 3:1–10
332. Al-Shujairi WH et al (2019) In vitro and in vivo roles of sphingosine kinase 2 during dengue virus infection. J Gen Virol 100:629–641
333. Heaton NS et al (2016) Targeting viral proteostasis limits influenza virus, HIV, and dengue virus infection. Immunity 44:46–58
334. Zelnik ID, Rozman B, Rosenfeld-Gur E, Ben-Dor S, Futerman AH (2019) A stroll down the CerS lane. Adv Exp Med Biol 1159:49–63
335. Fang Z, Pyne S, Pyne NJ (2019) Ceramide and sphingosine 1-phosphate in adipose dysfunction. Prog Lipid Res 74:145–159
336. Mullen TD et al (2011) Selective knockdown of ceramide synthases reveals complex interregulation of sphingolipid metabolism. J Lipid Res 52:68–77
337. Russo SB, Tidhar R, Futerman AH, Cowart LA (2013) Myristate-derived d16:0 sphingolipids constitute a cardiac sphingolipid pool with distinct synthetic routes and functional properties. J Biol Chem 288:13397–13409
338. Sassa T, Hirayama T, Kihara A (2016) Enzyme activities of the ceramide synthases CERS2-6 are regulated by phosphorylation in the C-terminal region. J Biol Chem 291:7477–7487
339. Wegner M-S, Schiffmann S, Parnham MJ, Geisslinger G, Grösch S (2016) The enigma of ceramide synthase regulation in mammalian cells. Prog Lipid Res 63:93–119
340. Levy M, Futerman AH (2010) Mammalian ceramide synthases. IUBMB Life 62:347–356
341. Laviad EL et al (2008) Characterization of ceramide synthase 2: tissue distribution, substrate specificity, and inhibition by sphingosine 1-phosphate. J Biol Chem 283:5677–5684
342. Law BA et al (2018) Lipotoxic very-long-chain ceramides cause mitochondrial dysfunction, oxidative stress, and cell death in cardiomyocytes. FASEB J 32:1403–1416
343. Ginkel C et al (2012) Ablation of neuronal ceramide synthase 1 in mice decreases ganglioside levels and expression of myelin-associated glycoprotein in oligodendrocytes. J Biol Chem 287:41888–41902
344. Vanni N et al (2014) Impairment of ceramide synthesis causes a novel progressive myoclonus epilepsy. Ann Neurol 76:206–212
345. Godeiro Junior CO et al (2018) Progressive myoclonic epilepsy type 8 due to CERS1 deficiency: a novel mutation with prominent ataxia. Mov Disord Clin Pract 5:330–332
346. Schwartz NU et al (2018) Decreased ceramide underlies mitochondrial dysfunction in Charcot-Marie-Tooth 2F. FASEB J 32:1716–1728
347. Turner N et al (2018) A selective inhibitor of ceramide synthase 1 reveals a novel role in fat metabolism. Nat Commun 9:3165
348. Turpin-Nolan SM et al (2019) CerS1-derived $C_{18:0}$ ceramide in skeletal muscle promotes obesity-induced insulin resistance. Cell Rep 26:1–10.e7
349. Gelderblom WC et al (1988) Fumonisins--novel mycotoxins with cancer-promoting activity produced by *Fusarium moniliforme*. Appl Environ Microbiol 54:1806–1811
350. Marasas WF (2001) Discovery and occurrence of the fumonisins: a historical perspective. Environ Health Perspect 109(Suppl 2):239–243
351. Zitomer NC et al (2009) Ceramide synthase inhibition by fumonisin B1 causes accumulation of 1-deoxysphinganine: a novel category of bioactive 1-deoxysphingoid bases and 1-deoxydihydroceramides biosynthesized by mammalian cell lines and animals. J Biol Chem 284:4786–4795

352. Wang E, Norred WP, Bacon CW, Riley RT, Merrill AH (1991) Inhibition of sphingolipid biosynthesis by fumonisins. Implications for diseases associated with Fusarium moniliforme. J Biol Chem 266:14486–14490
353. Steiner R et al (2016) Elucidating the chemical structure of native 1-deoxysphingosine. J Lipid Res 57:1194–1203
354. Siddique MM, Li Y, Chaurasia B, Kaddai VA, Summers SA (2015) Dihydroceramides: from bit players to lead actors. J Biol Chem 290:15371–15379
355. Fabrias G et al (2012) Dihydroceramide desaturase and dihydrosphingolipids: debutant players in the sphingolipid arena. Prog Lipid Res 51:82–94
356. Petrache I et al (2013) Ceramide synthases expression and role of ceramide synthase-2 in the lung: insight from human lung cells and mouse models. PLoS One 8:e62968
357. Mao C et al (2001) Cloning and characterization of a novel human alkaline ceramidase. A mammalian enzyme that hydrolyzes phytoceramide. J Biol Chem 276:26577–26588
358. Enomoto A et al (2006) Dihydroceramide:sphinganine C-4-hydroxylation requires Des2 hydroxylase and the membrane form of cytochrome b5. Biochem J 397:289–295
359. Ohi K et al (2015) DEGS2 polymorphism associated with cognition in schizophrenia is associated with gene expression in brain. Transl Psychiatry 5:e550–e550
360. Hashimoto R et al (2013) Genome-wide association study of cognitive decline in schizophrenia. Am J Psychiatry 170:683–684
361. Casasampere M, Ordoñez YF, Pou A, Casas J (2016) Inhibitors of dihydroceramide desaturase 1: therapeutic agents and pharmacological tools to decipher the role of dihydroceramides in cell biology. Chem Phys Lipids 197:33–44
362. Hu W, Ross J, Geng T, Brice SE, Cowart LA (2011) Differential regulation of dihydroceramide desaturase by palmitate versus monounsaturated fatty acids. J Biol Chem 286:16596–16605
363. Siddique MM et al (2013) Ablation of dihydroceramide desaturase 1, a therapeutic target for the treatment of metabolic diseases, simultaneously stimulates anabolic and catabolic signaling. Mol Cell Biol 33:2353–2369
364. Chaurasia B et al (2019) Targeting a ceramide double bond improves insulin resistance and hepatic steatosis. Science 365:386–392
365. Rahmaniyan M, Curley RW, Obeid LM, Hannun YA, Kraveka JM (2011) Identification of dihydroceramide desaturase as a direct in vitro target for fenretinide. J Biol Chem 286:24754–24764
366. Cooper JP, Reynolds CP, Cho H, Kang MH (2017) Clinical development of fenretinide as an antineoplastic drug: pharmacology perspectives. Exp Biol Med (Maywood) 242:1178–1184
367. Voelkel-Johnson C, Norris JS, White-Gilbertson S (2018) Interdiction of sphingolipid metabolism revisited: focus on prostate cancer. Adv Cancer Res 140:265–293
368. Wang H, Maurer BJ, Reynolds CP, Cabot MC (2001) N-(4-hydroxyphenyl)retinamide elevates ceramide in neuroblastoma cell lines by coordinate activation of serine palmitoyltransferase and ceramide synthase. Cancer Res 61:5102–5105
369. Holliday MW, Cox SB, Kang MH, Maurer BJ (2013) C22:0- and C24:0-dihydroceramides confer mixed cytotoxicity in T-cell acute lymphoblastic leukemia cell lines. PLoS One 8:e74768
370. Alsanafi M et al (2018) Native and polyubiquitinated forms of dihydroceramide desaturase are differentially linked to human embryonic kidney cell survival. Mol Cell Biol 38:1–45
371. Poliakov E et al (2017) Inhibitory effects of fenretinide metabolites N-[4-methoxyphenyl]retinamide (MPR) and 4-oxo-N-(4-hydroxyphenyl) retinamide (3-keto-HPR) on fenretinide molecular targets β-carotene oxygenase 1, stearoyl-CoA desaturase 1 and dihydroceramide Δ4-desaturase 1. PLoS One 12:e0176487
372. Chen L et al (2016) Stearoyl-CoA desaturase-1 mediated cell apoptosis in colorectal cancer by promoting ceramide synthesis. Sci Rep 6:19665
373. Igal R, Stearoyl A (2016) CoA desaturase-1: new insights into a central regulator of cancer metabolism. BBA Mol Cell Biol Lipids 1861:1865–1880
374. Matsui H et al (2012) Stearoyl-CoA desaturase-1 (SCD1) augments saturated fatty acid-induced lipid accumulation and inhibits apoptosis in cardiac myocytes. PLoS One 7:e33283
375. Dobrzyn A et al (2005) Stearoyl-CoA desaturase-1 deficiency reduces ceramide synthesis by downregulating serine palmitoyltransferase and increasing beta-oxidation in skeletal muscle. Am J Physiol Endocrinol Metab 288:E599–E607
376. Vilela RM, Lands LC, Meehan B, Kubow S (2006) Inhibition of IL-8 release from CFTR-deficient lung epithelial cells following pre-treatment with fenretinide. Int Immunopharmacol 6:1651–1664
377. Guilbault C et al (2009) Cystic fibrosis fatty acid imbalance is linked to ceramide deficiency and corrected by fenretinide. Am J Respir Cell Mol Biol 41:100–106
378. Lachance C et al (2013) Fenretinide corrects the imbalance between omega-6 to omega-3 polyunsaturated fatty acids and inhibits macrophage inflammatory mediators via the ERK pathway. PLoS One 8:e74875
379. Youssef M et al (2020) Efficacy of optimized treatment protocol using LAU-7b formulation against ovalbumin (OVA) and house dust mite (HDM)-induced allergic asthma in atopic hyperresponsive A/J mice. Pharm Res 37:31
380. Bikman BT et al (2012) Fenretinide prevents lipid-induced insulin resistance by blocking ceramide biosynthesis. J Biol Chem 287:17426–17437
381. Preitner F, Mody N, Graham TE, Peroni OD, Kahn BB (2009) Long-term fenretinide treatment prevents

high-fat diet-induced obesity, insulin resistance, and hepatic steatosis. Am J Physiol Endocrinol Metab 297:E1420–E1429
382. Mcilroy GD et al (2013) Fenretinide treatment prevents diet-induced obesity in association with major alterations in retinoid homeostatic gene expression in adipose, liver, and hypothalamus. Diabetes 62:825–836
383. Orienti I et al (2019) A new bioavailable fenretinide formulation with antiproliferative, antimetabolic, and cytotoxic effects on solid tumors. Cell Death Dis 10:529
384. Coant N, Sakamoto W, Mao C, Hannun YA (2017) Ceramidases, roles in sphingolipid metabolism and in health and disease. Adv Biol Regul 63:122–131
385. Mao C, Obeid LM (2008) Ceramidases: regulators of cellular responses mediated by ceramide, sphingosine, and sphingosine-1-phosphate. Biochim Biophys Acta 1781:424–434
386. Zhou J et al (2012) Spinal muscular atrophy associated with progressive myoclonic epilepsy is caused by mutations in ASAH1. Am J Hum Genet 91:5–14
387. Park JH, Schuchman EH (2006) Acid ceramidase and human disease. Biochim Biophys Acta 1758:2133–2138
388. Eliyahu E, Shtraizent N, Shalgi R, Schuchman EH (2012) Construction of conditional acid ceramidase knockout mice and in vivo effects on oocyte development and fertility. Cell Physiol Biochem 30:735–748
389. Okino N, Tani M, Imayama S, Ito M (1998) Purification and characterization of a novel ceramidase from Pseudomonas aeruginosa. J Biol Chem 273:14368–14373
390. Okino N, Ikeda R, Ito M (2010) Expression, purification, and characterization of a recombinant neutral ceramidase from Mycobacterium tuberculosis. Biosci Biotechnol Biochem 74:316–321
391. Kono M et al (2006) Neutral ceramidase encoded by the Asah2 gene is essential for the intestinal degradation of sphingolipids. J Biol Chem 281:7324–7331
392. Lin CL et al (2017) Alkaline ceramidase 1 protects mice from premature hair loss by maintaining the homeostasis of hair follicle stem cells. Stem Cell Rep 9:1488–1500
393. Liakath-Ali K et al (2016) Alkaline ceramidase 1 is essential for mammalian skin homeostasis and regulating whole-body energy expenditure. J Pathol 239:374–383
394. Li F et al (2018) Alkaline ceramidase 2 is essential for the homeostasis of plasma sphingoid bases and their phosphates. FASEB J 32:3058–3069
395. Sun W et al (2010) Substrate specificity, membrane topology, and activity regulation of human alkaline ceramidase 2 (ACER2). J Biol Chem 285:8995–9007
396. Wang K et al (2015) Alkaline ceramidase 3 deficiency results in Purkinje cell degeneration and cerebellar ataxia due to dyshomeostasis of sphingolipids in the brain. PLoS Genet 11:e1005591
397. Edvardson S et al (2016) Deficiency of the alkaline ceramidase ACER3 manifests in early childhood by progressive leukodystrophy. J Med Genet 53:389–396
398. Saied EM, Arenz C (2016) Inhibitors of ceramidases. Chem Phys Lipids 197:60–68
399. Realini N et al (2013) Discovery of highly potent acid ceramidase inhibitors with in vitro tumor chemosensitizing activity. Sci Rep 3:1035
400. Morad SAF, Davis TS, Kester M, Loughran TP Jr, Cabot MC (2015) Dynamics of ceramide generation and metabolism in response to fenretinide – diversity within and among leukemia. Leuk Res 39:1071–1078
401. Hanada K, Hara T, Nishijima M (2000) Purification of the serine palmitoyltransferase complex responsible for sphingoid base synthesis by using affinity peptide chromatography techniques. J Biol Chem 275:8409–8415
402. Hanada K (2003) Serine palmitoyltransferase, a key enzyme of sphingolipid metabolism. BBA-Mol Cell Biol Lipids 1632:16–30
403. Hornemann T, Richard S, Rütti MF, Wei Y, von Eckardstein A (2006) Cloning and initial characterization of a new subunit for mammalian serine-palmitoyltransferase. J Biol Chem 281:37275–37281
404. Dunn TM, Tifft CJ, Proia RL (2019) A perilous path: the inborn errors of sphingolipid metabolism. J Lipid Res 60:475–483
405. Hornemann T et al (2009) The SPTLC3 subunit of serine palmitoyltransferase generates short chain sphingoid bases. J Biol Chem 284:26322–26330
406. Beattie AE et al (2013) The pyridoxal 5′-phosphate (PLP)-dependent enzyme serine palmitoyltransferase (SPT): effects of the small subunits and insights from bacterial mimics of human hLCB2a HSAN1 mutations. BioMed Res Int 2013:1–13
407. Han G et al (2009) Identification of small subunits of mammalian serine palmitoyltransferase that confer distinct acyl-CoA substrate specificities. Proc Natl Acad Sci U S A 106:8186–8191
408. Zhao L et al (2015) Elevation of 20-carbon long chain bases due to a mutation in serine palmitoyltransferase small subunit b results in neurodegeneration. Proc Natl Acad Sci U S A 112:12962–12967
409. Breslow DK et al (2010) Orm family proteins mediate sphingolipid homeostasis. Nature 463:1048–1053
410. Davis DL, Gable K, Suemitsu J, Dunn TM, Wattenberg BW (2019) The ORMDL/Orm–serine palmitoyltransferase (SPT) complex is directly regulated by ceramide: reconstitution of SPT regulation in isolated membranes. J Biol Chem 294:5146–5156
411. Gupta SD et al (2015) Expression of the ORMDLS, modulators of serine palmitoyltransferase, is regulated by sphingolipids in mammalian cells. J Biol Chem 290:90–98
412. Clarke BA et al (2019) The Ormdl genes regulate the sphingolipid synthesis pathway to ensure proper

myelination and neurologic function in mice. elife 8:e51067
413. Moffatt MF et al (2007) Genetic variants regulating ORMDL3 expression contribute to the risk of childhood asthma. Nature 448:470–473
414. Acevedo N et al (2015) Risk of childhood asthma is associated with CpG-site polymorphisms, regional DNA methylation and mRNA levels at the GSDMB/ORMDL3 locus. Hum Mol Genet 24:875–890
415. Scherer SS (2011) The debut of a rational treatment for an inherited neuropathy? J Clin Invest 121:4624–4627
416. Bode H et al (2016) HSAN1 mutations in serine palmitoyltransferase reveal a close structure–function–phenotype relationship. Hum Mol Genet 25:853–865
417. Dawkins JL, Hulme DJ, Brahmbhatt SB, Auer-Grumbach M, Nicholson GA (2001) Mutations in SPTLC1, encoding serine palmitoyltransferase, long chain base subunit-1, cause hereditary sensory neuropathy type I. Nat Genet 27:309–312
418. Suriyanarayanan S et al (2019) A novel variant (Asn177Asp) in SPTLC2 causing hereditary sensory autonomic neuropathy type 1C. NeuroMolecular Med 21:182–191
419. Ernst D et al (2015) Novel HSAN1 mutation in serine palmitoyltransferase resides at a putative phosphorylation site that is involved in regulating substrate specificity. NeuroMolecular Med 17:47–57
420. Gable K et al (2010) A disease-causing mutation in the active site of serine palmitoyltransferase causes catalytic promiscuity. J Biol Chem 285:22846–22852
421. Penno A et al (2010) Hereditary sensory neuropathy type 1 is caused by the accumulation of two neurotoxic sphingolipids. J Biol Chem 285:11178–11187
422. Wu J et al (2019) Loss of neurological disease HSAN-I-associated gene SPTLC2 impairs CD8$^+$ T cell responses to infection by inhibiting T cell metabolic fitness. Immunity 50:1218–1231.e5
423. Garofalo K et al (2011) Oral L-serine supplementation reduces production of neurotoxic deoxysphingolipids in mice and humans with hereditary sensory autonomic neuropathy type 1. J Clin Invest 121:4735–4745
424. Fridman V et al (2019) Randomized trial of l-serine in patients with hereditary sensory and autonomic neuropathy type 1. Neurology 92:e359–e370
425. Soto D et al (2019) L-serine dietary supplementation is associated with clinical improvement of loss-of-function GRIN2B-related pediatric encephalopathy. Sci Signal 12(586):eaaw0936
426. Aureli M et al (2014) Gangliosides and cell surface ganglioside glycohydrolases in the nervous system. In: Yu RK, Schengrund C-L (ed) Glycobiology of the nervous system. Springer, New York, pp 223–244
427. Shayman JA (2016) Targeting glycosphingolipid metabolism to treat kidney disease. Nephron 134:37–42
428. D'Angelo G et al (2013) Vesicular and non-vesicular transport feed distinct glycosylation pathways in the Golgi. Nature 501:116–120
429. Allende ML, Proia RL (2014) Simplifying complexity: genetically resculpting glycosphingolipid synthesis pathways in mice to reveal function. Glycoconj J 31:1–10
430. Jeyakumar M, Butters TD, Dwek RA, Platt FM (2002) Glycosphingolipid lysosomal storage diseases: therapy and pathogenesis. Neuropathol Appl Neurobiol 28:343–357
431. Breiden B, Sandhoff K (2019) Lysosomal glycosphingolipid storage diseases. Annu Rev Biochem 88:461–485
432. Carter HE, Haines WJ (1947) Biochemistry of the sphingolipides; preparation of sphingolipides from beef brain and spinal cord. J Biol Chem 169:77–82

Druggable Lysophospholipid Signaling Pathways

7

Keisuke Yanagida and William J. Valentine

Abstract

Lysophosphatidic acid (LPA) has major roles as a bioactive signaling molecule, with multiple physiological and pathological roles being described in almost every major organ system. In this review we discuss LPA signaling pathways as emerging drug targets for multiple conditions relevant to human health and disease. LPA signals through the six G protein-coupled receptors LPA_{1-6}, and several of these receptors along with the LPA-producing enzyme including autotaxin (ATX) are now established as therapeutic targets with potential to treat various human diseases as exemplified by several LPA signaling targeting compounds now in clinical trials for idiopathic pulmonary fibrosis and systemic sclerosis. Several crystal structures of LPA receptors and ATX have been solved, which will accelerate development of highly selective and effective LPA signaling targeting compounds. We also review additional bioactive lysophospholipid (LPL) signaling molecules including lysophosphatidylserine and lysophosphatidylinositol, which represent the next wave of LPL druggable targets. An emerging theme in bioactive LPL signaling is that where the ligand is produced and how it is delivered to the cognate receptor are critical determinants of the biological responses. We will also discuss how connecting the production and function of bioactive LPLs will identify new therapeutic strategies to effectively target LPL signaling pathways.

Keywords

GPCR · LysoPS · LPI · Lysophosphatidylglucoside · Lysophosphatidylcholine · GPR55 · GPR34 · P2Y10 · GPR174

K. Yanagida (✉)
Department of Lipid Signaling, National Center for Global Health and Medicine, Tokyo, Japan
e-mail: kyanagida@ri.ncgm.go.jp

W. J. Valentine
Department of Lipid Signaling, National Center for Global Health and Medicine, Tokyo, Japan

Department of Molecular Therapy, National Center of Neurology and Psychiatry, Tokyo, Japan
e-mail: wvalentine@ncnp.go.jp

7.1 Introduction

Lysophospholipids (LPLs), glycerophospholipids with just one fatty chain, include lysophosphatidic acid (LPA), lysophosphatidylcholine (LPC), lysophosphatidylethanolamine (LPE), lysophosphatidylserine (lysoPS) and lysophosphatidylinositol (LPI) (Fig. 7.1). LPLs were recognized in the early 1900s as cytolytic substances derived from

Y. Kihara (ed.), *Druggable Lipid Signaling Pathways*, Advances in Experimental Medicine and Biology 1274, https://doi.org/10.1007/978-3-030-50621-6_7

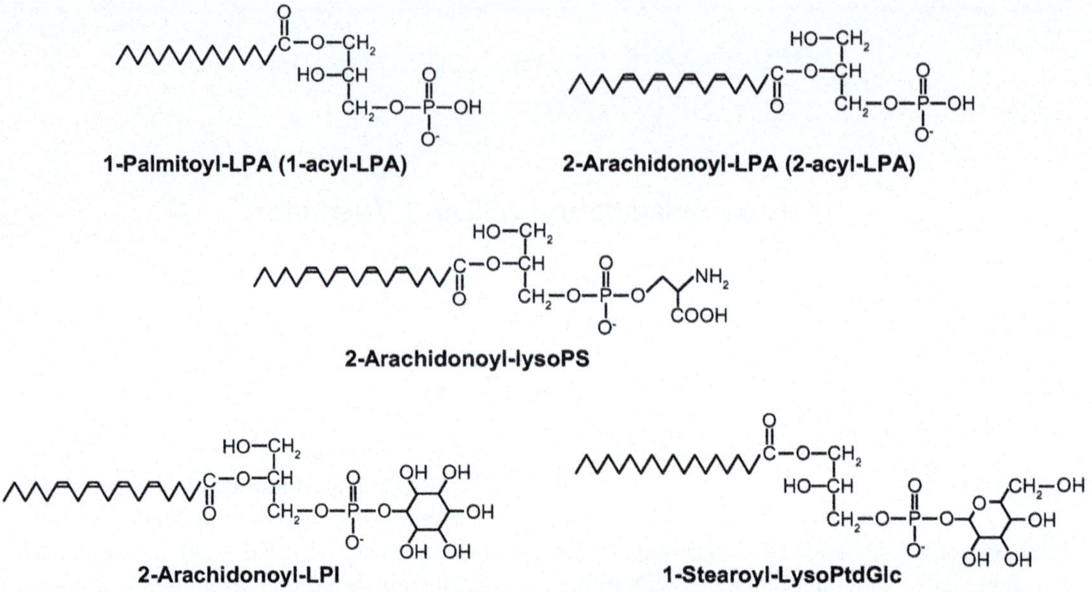

Fig. 7.1 Chemical structures of LPL ligands
Representative structures of LPLs including LPA, lysoPS, LPI and LysoPtdGlc are shown. Note that the acyl chain can be linked to either the *sn*-1 or *sn*-2 position of the glycerol backbone

phospholipids which were incubated with snake venom; their production was later shown to be through an enzymatic phospholipase A (PLA) activity in the venom. Indeed, the prefix "lyso-" comes from this cytolytic activity of LPLs, and their actions on cells were initially associated primarily with their detergent-like properties. However, by the middle of the twentieth century, researchers started to become aware of potential roles for LPLs as bioactive molecules which, when applied to cells or tissues, evoked physiological responses that were independent of their lytic and other detergent-like properties. In 1960, Vogt reported the contraction of isolated rabbit duodenum by LPA [1], and subsequently Tokumura et al. reported vasopressor activities of several molecular species of LPA and their structure-activity relationships [2]. In 1989, Moolenaar and colleagues found that the mitogenic action of LPA involves G proteins [3], and the existence of a specific G protein-coupled receptor (GPCR) for LPA was proposed. From the late 1990s, pioneered by studies by Chun and colleagues [4], various receptor targets have been identified for bioactive LPLs. Since then, numerous physiological and pathological roles of LPL signaling have been elucidated, and these extensive "bench" works are now about to be translated into the "bedside"; as exemplified by ongoing clinical drug trials targeting LPA signaling in order to treat fibrotic diseases. Moreover, recent successes in crystallization of LPA receptors and LPA-producing enzymes will further accelerate rational design of novel potent modulators of LPA and other LPL signaling pathways.

In this chapter, we will discuss the druggable-potential of LPL signaling pathways, especially by reviewing the physiological and pathological roles of LPA signaling that have been revealed by studies utilizing genetic and pharmacological tools *in vivo*. We will also discuss findings from crystal structures of LPA receptors and an LPA-producing enzyme. Lastly, we will discuss other bioactive LPLs, such as lysoPS and LPI, which may be the "next-generation" of LPL signaling targets for drug discovery.

7.2 Bioactive LPA Synthesis Pathways *In Vivo*

LPA is not a single molecular entity; many LPA species exist which differ in the chemistry of their hydrocarbon chains and linkages to the glycerol backbone. The biological activities of these LPA species also differ, largely because multiple LPA receptors (LPA_{1-6}) exist which differ in their selectivities for the different LPA species [5]. For example, LPA_3 [6] and LPA_6 [7] receptors are more potently activated by LPA with their acyl chain at the *sn-2* position (2-acyl-LPA) rather than at the *sn-1* position (1-acyl-LPA). In addition, LPA_5 receptor shows preference for alkyl-ether-linked LPA (alkyl-LPA) rather than acyl-LPA [5]. Therefore, not only the LPA levels but also where and how the LPA was produced will determine LPA receptor activation and the biological actions of LPA in various settings.

LPA is an intermediate in phospholipid biosynthesis and can be synthesized via multiple pathways, and at least two pathways are known to have important roles in the production of LPA which functions as a bioactive lipid mediator [8]. In one pathway, LPA is extracellularly produced from other LPLs such as LPC by autotaxin (ATX), a secreted enzyme with lysophospholipase D activity [9, 10]. In the other pathway, LPA is produced from intramembrane phosphatidic acid (PA) by the actions of membrane-bound, PA-selective $PLA_{1\alpha}$ (mPA-$PLA_{1\alpha}$) [11]. These two pathways produce compartmentally distinct pools of LPA (extracellular production by ATX *versus* intramembrane production by mPA-$PLA_{1\alpha}$), and the produced LPA species also differ between these pathways. ATX produces both 1- and 2-acyl-LPA, while only 2-acyl-LPA is produced by mPA-$PLA_{1\alpha}$, indicating that the LPA species produced by these two pathways may differ in their activities on various LPA receptors and therefore have distinct biological roles. In addition to ATX- and mPA-$PLA_{1\alpha}$-dependent LPA production, other pathways have also been suggested to contribute to production of LPA as a bioactive lipid mediator, including conversion of PA to LPA via secretory-type or cytosolic PLA2; however, *in vivo* evidence is lacking and awaits future validation [12].

7.3 LPA Receptor Repertoire

In 1996, Chun and colleagues provided the first report that a member of the endothelial differentiation gene (EDG) family of GPCRs represents a functional receptor for LPA [4], which is now referred to as LPA_1. Subsequently, two additional EDG family GPCRs were also identified as LPA receptors, LPA_2 [13] and LPA_3 [14]. LPA_{1-3} share 50–58% amino acid identity with each other and comprise the EDG family of LPA receptors. In 2003, Shimizu and colleagues discovered a fourth LPA receptor, LPA_4, from an orphan GPCR, GPR23/p2y9, that has low homology to EDG family LPA receptors but is in the P2Y purinergic GPCR cluster [15]. This has led to the identification of two additional LPA receptors from the P2Y cluster, LPA_5 [16] and LPA_6 [7]. LPA_{4-6} share 35–55% amino acid identity with each other and constitute the non-EDG family of LPA receptors. Depending on the functional coupling of a given LPA receptor to different G proteins, namely $G\alpha_s$, $G\alpha_i$, $G\alpha_q$, and $G\alpha_{13}$ proteins, LPA binding activates diverse signaling cascades involving adenylyl cyclase, phosphoinositide 3-kinase, phospholipase C, and Rho-family GTPase. The G protein-coupling selectivity of each receptor is summarized in Table 7.1.

7.4 Crystal Structures of ATX and LPA Receptors

Rational drug design frequently relies on computer modeling techniques, which are greatly aided when a high-resolution structure of the target protein is known. Recent studies have successfully resolved crystal structures of ATX, LPA_1, and LPA_6. Their high-resolution structures will increase not only our understanding of substrate/ligand recognition but also provide precious information for the rational design of novel potent modulators of these targets.

Table 7.1 LPL receptor repertoire

Ligand	Receptor	Previous names	Gene name (human/mouse)	Coupling Gα protein family
LPA	LPA_1	EDG2, Vzg-1	*LPAR1/Lpar1*	$G\alpha_i$, $G\alpha_q$, $G\alpha_{13}$
	LPA_2	EDG4	*LPAR2/Lpar2*	$G\alpha_i$, $G\alpha_q$, $G\alpha_{13}$
	LPA_3	EDG7	*LPAR3/Lpar3*	$G\alpha_i$, $G\alpha_q$
	LPA_4	P2Y9, GPR23	*LPAR4/Lpar4*	$G\alpha_s$, $G\alpha_i$, $G\alpha_q$, $G\alpha_{13}$
	LPA_5	GPR92, GPR93	*LPAR5/Lpar5*	$G\alpha_q$, $G\alpha_{13}$
	LPA_6	P2Y5	*LPAR6/Lpar6*	$G\alpha_{13}$, $G\alpha_s$[a], $G\alpha_i$[a]
lysoPS	$lysoPS_1$	GPR34	*LYSOPSR1/Lysopsr1*	$G\alpha_i$
	$lysoPS_2$	P2Y10	*LYSOPSR2/Lysopsr2*	$G\alpha_{13}$
	$lysoPS_{2L}$	A630033H20	pseudo gene/not determined	$G\alpha_{13}$
	$lysoPS_3$	GPR174, FKSG79	*LYSOPSR3/Lysopsr3*	$G\alpha_s$, $G\alpha_{13}$
LPI	LPI_1	GPR55	*LPIR1/Lpir1*	$G\alpha_{13}$

[a]Has not been validated in multiple studies

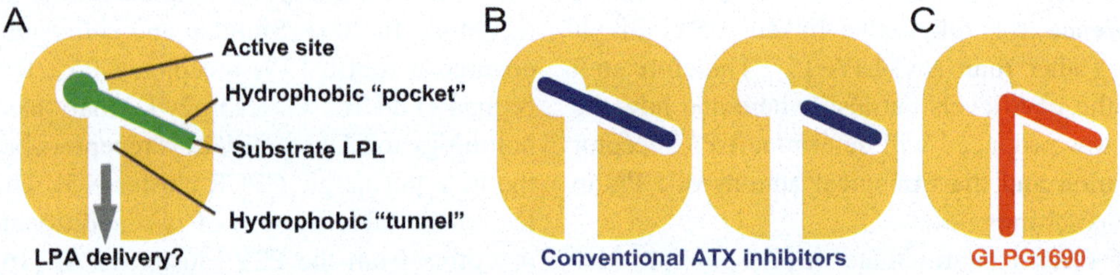

Fig. 7.2 Schematic views of ATX structure revealed by crystallographic studies
(**a**) Schematic presentation of the active site, the hydrophobic "pocket" and "tunnel" structure, with bound LPL substrate. The hydrophobic "tunnel" structure is proposed to be essential for effective delivery of LPA to receptor targets
(**b**) Binding modes of conventional ATX inhibitors to ATX. Many conventional ATX inhibitors occupy either the orthosteric site (model on left) or the hydrophobic "pocket" (model on right) and thereby inhibit the enzymatic activity
(**c**) Binding mode of GLPG1690 to ATX. GLPG1690 is the most successful ATX inhibitor in clinical trials at present and targets both the hydrophobic "pocket" and "tunnel" structures

7.4.1 Structure of ATX

Crystal structures of ATX were first reported in 2011 by two independent research groups [17, 18]. ATX was found to have a unique structural component, a hydrophobic "tunnel" structure, in addition to the hydrophobic "pocket" that accepts LPL substrates (Fig. 7.2a). The structures revealed that ATX probably contains LPA molecules in the hydrophobic "tunnel", which would serve as a second binding site of LPA. Significantly impaired cell motility-stimulating activity was observed in an ATX mutant with the "tunnel" precluded, even though it showed similar activity to wild-type (WT) ATX in enzymatic assays [18]. Addition of ATX to cultured cells had been shown to induce cell motility even without production of detectable levels of LPA in the culture medias [10], which supports the possibility that ATX-derived LPA might directly access LPA receptors via the hydrophobic "tunnel" structures even without diffusion into the media. Thus, ATX may function not only as an LPA-producing enzyme, but also as an LPA carrier or chaperone protein to facilitate effective stimulation of LPA receptors.

The resolved structures of ATX enabled a design of a wave of selective inhibitors based on the three-dimensional architecture of the enzyme; and additional crystallographic studies followed, resolving structures of ATX with various inhibitors [19–32]. Currently we can find about 40 different crystal structures of ATX deposited in the Protein Data Bank (http://www.rcsb.org), providing further insights in their various binding sites and unique inhibitory mechanisms. Most of the classical ATX inhibitors target the hydrophobic substrate binding sites or solely inhibit the catalytic activity [33] (Fig. 7.2b). Given that ATX has at least two important functional domains, the hydrophobic "pocket" that is the site of LPL substrate acceptance and the "tunnel" structure which may be involved in LPA delivery, development of inhibitors that target both structures may be expected to yield more efficacious drugs. GLPG1690 is one such molecule [30] (Fig. 7.2c) and the most successful ATX inhibitor in clinical trials at present, and it is suggested that the clinical success of GLPG1690 is not independent from its unique inhibitory mode; blocking both "pocket" and "tunnel" [33].

7.4.2 Structures of LPA_1 and LPA_6

Following the pioneering work on crystal structure of the antagonist-bound sphingosine 1-phosphate (S1P) receptor $S1P_1$ [34], crystal structures for various GPCRs with lipid ligands have now been reported, including LPA_1 and LPA_6, which are EDG- and non-EDG-LPA receptors, respectively [35]. These two receptors share low amino acid sequence homology, and their crystal structures have revealed contrasting features including the access route of LPA into the binding site (Fig. 7.3). Here we briefly discuss the structural characteristics of LPA_1 and LPA_6, including comparisons with other lipid-binding GPCRs.

In 2015, the crystal structures of LPA_1 bound to three different antagonist compounds were reported [36]. Recent structural analyses on lipid GPCRs have revealed that many of them have a "gap" structure that is embedded in the membrane lipid bilayer, which is proposed to provide a route for the lipid ligand to access the binding pocket of the receptor by lateral diffusion within the membrane [35]. However, structural analyses with molecular dynamics simulations indicate absence of a "gap" structure in LPA_1 and support an open access route for ligand entry from the extracellular space to the binding pocket [36]. The unique open access route from the extracellular space might also facilitate effective accommodation of LPA extracellularly delivered by lipid chaperones including ATX. Absence of an intramembrane "gap" structure was also observed in antagonist-bound BLT_1 [37], agonist-bound EP3 [38], and agonist-bound CB_1 receptors [39]. However, a "gap" structure was present in CB_1 when it was bound with an antagonist [40] or inverse agonist [41], suggesting that the "gap" structure formed by transmembrane helices is not fixed and its presence may depend on the recep-

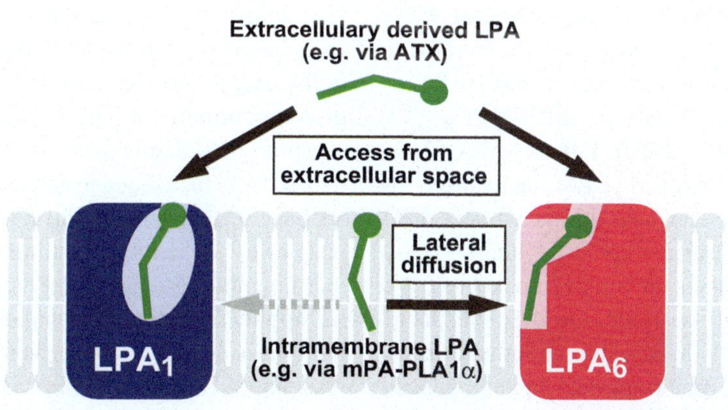

Fig. 7.3 Models for LPA access to LPA_1 and LPA_6 based on their crystallographic structures
LPA_1 lacks an intramembrane "gap" structure, supporting a model of open access route from the extracellular space. On the other hand, LPA_6 potentially accommodates LPA delivered either by open access from the extracellular space or lateral diffusion from the lipid bilayer

tor's ligand-dependent conformational state [39]. A similar scenario was also suggested for agonist-bound EP_3, which showed a totally enclosed ligand binding pocket in its crystal structure [38]. Therefore, further crystallization of LPA_1 in different ligand-bound states may be necessary to determine whether LPA_1 actually lacks the lateral diffusion access route. The structure of LPA_1 is also characterized by its "baggy" large globular ligand binding pocket, which may facilitate broader ligand recognition not only among various LPA species but also phosphorylated endocannabinoids [36]. Many lipid GPCRs, including LPA_1, possess an extracellular "lid" structure over the ligand pocket formed by the N-terminus and/or second extracellular loop of the receptors, which is proposed to regulate or limit extracellular access of lipid ligands [35]. However, the extracellular "lid" structure of LPA_1 is proposed to be relatively flexible and less restrictive, which might facilitate access of LPA delivered by carrier or chaperone proteins including ATX [36].

The crystal structure of LPA_6 was resolved in 2017, and, in contrast to LPA_1, revealed an intramembrane "gap" structure, suggesting a lateral diffusion mode of LPA access to the ligand binding site [42]. LPA_6 notably lacks the above-mentioned "lid" structure, suggesting that LPA could have open access from the extracellular space, in addition to the lateral diffusion access [42]. Currently, the biological significance of the potential "dual" access routes of LPA into the ligand binding pocket of LPA_6 is unclear. LPA_6 signaling in hair follicles is induced by 2-acyl-LPA derived from membrane-embedded PA by the enzymatic activity of mPA-$PLA_{1\alpha}$ [43] (described in Sect. 7.5.10). Indeed, mPA-$PLA_{1\alpha}$ over-expression induced LPA_6 activation by increasing membrane LPA levels without LPA secretion [42], suggesting that mPA-$PLA_{1\alpha}$-derived LPA accesses the binding pocket of LPA_6 via the lateral diffuse route. On the other hand, ATX-derived LPA delivered to endothelial LPA_6 is known to have important roles in angiogenesis [44] and lymphocyte homing into secondary lymphoid organs [45]. Delivery of LPA by ATX to LPA receptors is proposed to occur from the extracellular space, suggesting that the open access route rather than lateral diffusion may be utilized to deliver ATX-derived LPA to LPA_6 on the plasma membrane of endothelium.

Since LPA_1 and LPA_6 belongs to distinct families of LPA receptors, it follows that their contrasting structural features may reflect the common characteristics either of EDG or non-EDG family LPA receptors. However, there are several concerns about simply applying their properties as family-specific (EDG *versus* non-EDG) characteristics of LPA receptors. For example, $S1P_1$, an EDG-family S1P receptor with high sequence homology to LPA_1, has a lateral diffusion ligand route and no ligand access from the extracellular space [34]. Furthermore, in contrast to the broad ligand recognition of LPA_1 with its "baggy" ligand binding pocket, LPA_3, also an EDG-family receptor, has a rather strict ligand selectivity and clearly prefers 2-acyl LPA [6], possibly suggesting a more restrictive agonist binding pocket. Determining the crystal structures of additional LPA receptors will elucidate their mechanisms of ligand binding and facilitate rational design of drugs to effectively and selectively target LPA receptors.

7.5 Pathophysiology of LPA Signaling

After the cloning of the first LPA receptor, numerous studies have assessed the roles of LPA signaling both *in vitro* and *in vivo*. Especially, *in vivo* genetic tools including knock-out (KO) animals have revealed roles of LPA signaling in virtually every organ system. In addition, a rapidly growing number of ATX inhibitors and LPA receptor modulators have been developed, and their effects have been tested *in vivo*. These studies have further revealed diverse pathophysiological roles of LPA signaling and have been instrumental in targeting LPA pathways in clinical applications. Indeed, studies with *Lpar1*-KO mice have led to clinical trials of LPA_1 antagonists and ATX inhibitors for fibrosis patients. In this section, we discuss roles of LPA signaling in pathophysiology across organ systems, especially focusing on the knowledge obtained from studies that utilized genetic and pharmacological approaches.

7.5.1 Fibrosis

Fibrosis is a major feature and cause of morbidity for a number of diseases including idiopathic pulmonary fibrosis (IPF), systemic sclerosis, progressive kidney diseases, and diabetes; thus effective treatment of fibrotic disorders represents a highly unmet and urgent medical need. LPA signaling pathway is a potential therapeutic target for fibrosis, and indeed multiple small molecule drugs are currently being evaluated in clinical trials. Profibrotic roles of LPA have been extensively studied since the first reports on the roles of LPA_1 in lung [46] and renal [47] fibrosis in 2007–2008. In the bleomycin-induced lung fibrosis model, *Lpar1*-KO mice were protected from disease progression and mortality. The LPA-LPA_1 axis seems to be involved in multiple pathological aspects of fibrosis including fibroblast recruitment/proliferation, extracellular matrix accumulation, epithelial apoptosis, myofibroblast differentiation, and vascular leak [46, 48]. Of note, it seems that the profibrotic action of LPA involves induction of connective tissue growth factor, another important player and potential drug target for fibrosis, through the LPA_1-$G\alpha_{13}$-ROCK-serum response factor axis. Indeed, genetic deletion or pharmacological inhibition of the players in this axis successfully diminished fibrosis development in murine fibrosis models [47, 49–51].

The profibrotic role of LPA_1 is not limited to the bleomycin-induced lung fibrosis model, and *Lpar1*-KO mice also displayed significant protection from lung fibrosis induced by radiation [52] or lung allograft [52] as well as various models of renal [47, 49], skin [53] and peritoneum [50] fibrosis, indicating that LPA-LPA_1 signaling has a fundamental role in the process of fibrosis. Moreover, beneficial effects of LPA_1 antagonists have also been observed in various rodent models of fibrosis including fibrosis in lung [54–59], kidney [55, 60], skin [53, 61], and peritoneum [50]. Although LPA_1 seems to be the primary player mediating the profibrotic LPA action, other LPA receptors may also have roles in fibrosis. For example, *Lpar2*-KO mice showed protection from bleomycin-induced lung fibrosis [62] but not from dermal fibrosis [53].

It is likely that profibrotic LPA is mainly derived from the ATX-dependent pathway, and elevated ATX levels were observed in lungs of bleomycin-challenged mice and also IPF patients [63]. Indeed, conditional genetic deletion of *Enpp2*, which encodes ATX, specifically in bronchiolar epithelial cells or myeloid cells resulted in protection from bleomycin-induced lung fibrosis [63]. Of note, systemic ATX levels do not seem to contribute to pulmonary fibrosis, because liver-specific *Enpp2* over-expression increased systemic ATX levels without affecting fibrosis development [63]. On the other hand, hepatocyte-specific *Enpp2*-KO mice displayed diminished disease progression in liver fibrosis models [64, 65]. Therefore, profibrotic LPA may be locally supplied at the site of fibrosis rather than systemically supplied. In addition to LPA_1 antagonists, many studies have shown the beneficial effects of ATX inhibitors on different types of fibrosis including lung [59, 63, 66], kidney [60, 67], skin [68, 69], liver [27], and synovial joint [70] fibrosis.

As mentioned above, the LPA-LPA_1 profibrotic pathway is a promising druggable target to treat fibrotic diseases. Treatment with BMS-986020, an LPA_1 antagonist from Bristol-Myers Squibb, significantly slowed the rate of respiratory function decline in Phase 2 clinical trials for IPF patients [71]. However, the appearance of cholecystitis in three patients receiving the drug lead to early termination of the study. The cholecystitis was determined to be a drug-specific side effect not related to LPA_1 antagonism, and the study suggested it was possible to slow decline of lung function in IPF patients by targeting LPA_1. Later, Bristol-Myers Squibb developed a novel LPA_1 antagonist BMS-986278 that does not have hepatobiliary toxicity [72], and it entered into Phase 1 clinical trials with healthy volunteers that were ongoing or completed in 2019 (NCT03429933, NCT03712540, and NCT03981094). Another LPA_1 antagonist, SAR100842 from Sanofi, was well tolerated with an acceptable safety profile in a Phase 2 trial for systemic sclerosis patients, and was associated with numerically improved clinical scores suggestive of potential clinical benefits [73].

Several ATX inhibitors are also being tested in clinical trials to slow or halt progression of fibrotic diseases. One such compound, GLPG1690 (Ziritaxestat) from Galapagos NV is currently under clinical trials for patients with fibrotic diseases. In a Phase 2 study for a small cohort of IPF patients (FLORA study), GLPG1690 showed a halt of disease progression and favorable safety profile and tolerability [74]. Two larger Phase 3 clinical trials started from December 2018 to test the efficacy of GLPG1690 in IPF patients, which combined have a total of ~1,500 patients and more than 200 clinical sites worldwide (ISABELA1 and ISABELA2 studies) [75]. In 2019, Galapagos NV also started a Phase 2 trial with GLPG1690 in systemic sclerosis patients (NOVESA study, NCT03798366). BBT-877 from Bridge Biotherapeutics is another ATX inhibitor in clinical trial, and was found to be well-tolerated and safe in a Phase 1 trial with healthy volunteers [76]. In July 2019, Boehringer Ingelheim obtained a license from Bridge Biotherapeutics to develop this drug beyond Phase 1. Lastly, BLD-0409 (PAT-409) from Blade Therapeutics (previously ATXCo, a spin-off of PharmAkea Therapeutics) is a Phase 1-ready ATX inhibitor for patients with fibrotic diseases, and the trial will start from January 2020 (NCT04146805).

7.5.2 Central Nervous System

The gene encoding LPA_1 was originally called ventricular zone gene-1 due to its enriched expression in the ventricular zone of the developing brain [4]. Indeed, LPA_1 is highly expressed in neural progenitor cells (NPCs) in developing brain, and *Lpar1*-KO mice have defects in cortical development including a reduction in proliferation, premature NPC maturation, and increased apoptosis [77]. Importantly, *ex vivo* or *in vivo* LPA application studies have revealed an important role for LPA signaling in pre-natal and newborn brain damage, a serious health issue that affects millions of babies each year and may be caused by hypoxia, bleeding, or infection. In *ex vivo* mouse brain slice culture experiments, hypoxia-induced cerebral cortical disruption including displacement of mitotic NPCs was blocked by either genetic depletion of *Lpar1* or application of LPA_1 antagonists [78]. Furthermore, *in vivo* injection of LPA to either fetal or neonatal brains produced hydrocephalus phenotype, which was ameliorated in *Lpar1*- or *Lpar3*-KO mice and mice treated with LPA_1 antagonists [79, 80]. Since intraventricular bleeding often precedes hydrocephalus formation, and LPA is abundant in blood, LPA may be one of the key causative mediators of the post-hemorrhagic hydrocephalus. Prenatal intracerebral hemorrhage is an identified risk factor for schizophrenia, and prenatal mice receiving intraventricular injections of serum or LPA showed schizophrenia-like behavior as adults and had altered expression profiles of genes relevant to schizophrenia; these effects were blocked by a $LPA_{1/3}$ antagonist [81]. Thus, LPA receptors blockage is a potential strategy to prevent infant and neonatal brain damage from various insults.

High levels of LPA_1 expression was also observed in NPCs in brain of adult mice, as revealed in a LPA_1-GFP reporter mouse line [82]. Intracerebroventricular infusion of LPA resulted in increased adult hippocampal neurogenesis [82, 83] and cell proliferation [83] in WT but not in *Lpar1*-KO mice nor in $LPA_{1/3}$ inhibitor-treated mice [83]. Hippocampal neurogenesis was also reduced in adult *Lpar1*-KO mice under normal conditions [84]. Possibly reflecting the developmental or adult neurological defects, *Lpar1*-KO mice displayed various behavioral defects closely related to psychiatric and mental diseases including reduced prepulse inhibition (PPI) (a marker of schizophrenic disorders) [85], spatial memory defects [86–88], reduced exploratory behavior [86, 87], increased anxiety-like behavior [86, 87, 89], impaired fear extinction [90], and altered substance abuse behavior [91, 92]. Since *Lpar1*-KO mice suffer from altered neurodevelopment, as described above, it is plausible that these phenotypes are due to altered neurodevelopment or neuroadaptation. However, some studies have shown LPA signaling manipulation also causes acute effects on neuronal function. In rodents, intracerebroventricular infusion of LPA potentiates hippocampal-dependent memory [93], increases anxiety-like responses [94, 95] and decreases cocaine-induced conditioned place preference [83]. Furthermore, administration of an $LPA_{1/3}$ antagonist decreased context-dependent fear extinction [90], and increased anxiety-like behavior [96] or alcohol consumption [92, 97], all of which mimic the *Lpar1*-KO mice phenotypes [86, 87, 89, 90, 92, 96]. Intracerebroventricular injection of an $LPA_{1/3}$ antagonist also led to progressive increases of cocaine-induced conditioned place preference [83]. These results demonstrate neurodevelopment-independent effects of the LPA-LPA_1 axis in memory, fear extinction, anxiety and substance abuse.

In addition to LPA_1, a neuropsychiatric role of LPA_2 has been demonstrated with relevance to the function of plasticity related genes (PRGs), neuron specific membrane proteins located at the postsynaptic density of glutamatergic synapses. PRGs are highly homologous to lipid phosphate phosphatases (LPPs), ecto-phosphatases that degrade LPA and S1P [98]. However, PRGs generally lack ecto-phosphatase activities, and rather are proposed to modify synaptic phospholipid levels by promoting their uptake into the intracellular compartment [98]. *Prg1*-KO mice display epileptic seizures caused by neuronal hyperexcitability, leading to reduced body weight and increased mortality [99]. Neuronal hyperactivity in these mice was rescued by *in utero* electroporation of WT PRG1 but not by PRG1 point-mutated to disrupt the putative interaction with lipid phosphates [99]. Of note, the phenotypes of *Prg1*-KO mice were rescued by global deletion of *Lpar2* [99], astrocyte-specific deletion of *Enpp2* [100], or ATX inhibitors [100]. Mono- or bi-allelic removal of *Prg1* also induces behavioral abnormalities including reduced PPI and social interaction, higher exploratory behavior, and higher motor activity with running and jumping (referred to as rodent "Forrest Gump") [101]; these abnormalities are reverted by genetic or pharmacological inhibition of LPA_2 or ATX [100–102]. Therefore, it is proposed that PRG1 normally functions to limit bioavailability of synaptic LPA which is produced by astrocytic ATX and induces neuronal activation via LPA_2. Notably, a loss-of-function single nucleotide polymorphism (SNP) was identified in the human *PRG1* gene that is associated with an endophenotype for mental disorders [102].

These studies indicate LPA signaling may have important roles in mental and neurological disorders; however, there are several controversial and unsolved issues. For example, reduced PPI is observed under conditions of both loss (in *Lpar1*-KO mice [85]) and gain (in *Prg1*-heterozygous mice [100]) of LPA signaling, while genetic or pharmacological inhibition of ATX under normal physiological conditions did not alter animal behavior, including PPI, of WT mice [100]. PRG1 and ATX display spatially restricted expressions at postsynaptic neurons [99] and peri-synaptic astrocytic processes [100], respectively, suggesting that the microenvironmental LPA levels may be tightly regulated. Therefore, it is plausible that spatially distinct pools of LPA have different roles in the regulation of neuronal activities by signaling through different LPA receptor subtypes. Furthermore, in addition to LPA_1 and LPA_2, other LPA receptors which are also enriched in brain and neuron [103] may participate in neuronal function. Indeed, *Lpar5*-KO mice display altered behaviors including anxiolytic effects, reduced social exploration, increased behavioral flexibility, and increased motivational behavior [104]. Thus, more comprehensive understanding of LPA signaling in the brain including functions of the various LPA receptors in their different microenvironments is required to enable development of LPA signaling-based therapies to treat mental and neurological disorders.

7.5.3 Pain

Neuropathic pain, intractable pain directly caused by damage or disease of the somatosensory nervous system, is challenging to treat with currently available therapies. LPA signaling plays a central role in the initiation of neuropathic pain and therefore represents a potential therapeutic target to treat this debilitating condition. In rodents, intrathecal injection of LPA itself induces behavioral allodynia and hyperalgesia accompanied by demyelination [105]. Importantly, *Lpar1*-KO mice were resistant to both the nociceptive response and demyelination not only in LPA-induced model of neuropathic pain but also in a partial sciatic nerve injury model [105]. Behavioral allodynia and demyelination were attenuated by pre- but not post-injury blockage of $LPA_{1/3}$ [106, 107] or its downstream signaling molecules Rho/ROCK [105, 107], highlighting the essential role of LPA_1 signaling in the initiation of neuropathic pain [105]. Mechanistically, LPA_1 signaling in Schwann cells at the dorsal root activates multiple cascades that decrease myelin gene mRNA expression and increase proteolysis of myelin proteins, leading to the subsequent demyelination [105, 108].

ATX mediates the production of LPA which acts as an initiator of neuropathic pain. Indeed, *Enpp2*-heterozygous mice displayed improved recovery from neuropathic pain [109]. It's been known for nearly fifty years that direct injection of LPC into the spinal cord induces demyelination [110], and this procedure is still commonly used in animal models for studying demyelination and remyelination processes. Notably, LPC did not exhibit demyelinating activity in *Lpar1*-KO mice, and *Enpp2*-heterozygout mice displayed much milder phenotypes in a LPC-induced demyelination model [111]. In WT mice, LPA levels showed drastic increases after injury in spinal dorsal horn and dorsal roots but not in spinal nerve or sciatic nerve, and these increases were attenuated both in *Enpp2*-heterozygous mice as well as in PLA_2 inhibitor-treated mice [112, 113]. These observations suggest that the demyelination initiator LPA and its precursor LPC are locally produced during nerve injury.

Lpar3-KO mice are also protected from LPC- or LPA-induced demyelination. Notably, these mice did not show an initial increase in LPA levels at the spinal dorsal horn following nerve injury [112]; this suggests that LPA_3 facilitates feed-forward amplification of LPA production following nerve injury, a mechanism known to involve microglial activation [114]. Genetic or pharmacological interventions have also demonstrated the involvement of the LPA-$LPA_{1/3}$ axis in models of neuropathic pain induced by chemotherapeutic agents [115], osteoarthritis [116], and stroke [117, 118]. Given the pivotal role of the ATX-$LPA_{1/3}$ axis in the initiation stage of neuropathic pain, blockage of this pathway represents a therapeutic target for prevention and treatment of neuropathic pain.

In addition to LPA_1 and LPA_3, the role of LPA_5 in neuropathic pain has been established. *Lpar5*-KO mice were protected from neuropathic pain development in peripheral nerve injury models [104, 119] as well as in a cuprizone-induced multiple sclerosis model [120]. Intrathecal injection of a selective LPA_5 agonist itself induced pain [121] while injection of a LPA_5 antagonist had an analgesic effect [121, 122] in rodent neuropathic pain models. Despite their protection from neuropathic pain development, *Lpar5*-KO mice did not show decreased myelin sheath aberrations following nerve ligation [119], unlike what had been observed in *Lpar1*-KO mice [105]. Instead, *Lpar5*-KO mice showed abolished neuronal cAMP response element-binding protein phosphorylation at dorsal horn after nerve ligation [119], indicating a role for LPA_5 in central neurons during neuropathic pain induction. The analgesic effects achieved by genetic deletion or pharmacological inhibition of LPA_5 are not limited to neuropathic pain models but has also been demonstrated for inflammatory pain [121], supporting the central involvement of LPA_5 in pain-signal transmission and responses. Therefore, pharmacological targeting of LPA_5 may represent an attractive strategy to manage pain for broad range of pain conditions including neuropathic and inflammatory pain.

7.5.4 Cardiovascular System

The importance of ATX-derived LPA in vascular development was demonstrated by multiple studies that showed that *Enpp2*-KO embryos die due to severe vascular defects [123–126]. Overexpression of *Enpp2* also caused embryonic lethality with severe defects in angiogenesis and vessel formation [127]; therefore, tight regulation of ATX expression and LPA production seems to be essential for proper embryonic angiogenesis. This is further supported by the fact that genetic removal of LPP3, an LPA degrading enzyme, causes embryonic lethality that is accompanied by vascular defects resembling those of *Enpp2*-overexpressing embryos [128]. Since the vascular defects in *Enpp2*-KO embryos are reminiscent of *Gna13* (encoding $G\alpha_{13}$ protein)-KO embryos [129–131], $G\alpha_{13}$-coupled receptor targets were postulated to mediate the actions of angiogenic LPA. One of the target receptors for angiogenic LPA is LPA_4 since embryos deficient for *Lpar4*, a $G\alpha_{13}$-coupled LPA receptor, partially phenocopied the vascular defects in *Enpp2*-KO embryos [132]. However, the penetrance of lethal vascular defect in *Lpar4*-KO embryos is ~30%, indicating the existence of at least one more angiogenic LPA receptor existed that links ATX-produced LPA to $G\alpha_{13}$ signaling [132]. Very recently, Yasuda et al. established that LPA_4 and LPA_6 synergistically function to contribute to sprouting angiogenesis, and *Lpar4*/*Lpar6*-double KO mice almost completely phenocopy the vascular defects observed in *Enpp2*- or *Gna13*-KO embryos [44]. The role of endothelial LPA_4 in vascular development has been further highlighted in studies of tumor vasculature. In murine lung carcinoma [133] or glioblastoma [134] cell implantation models of tumors, LPA application promoted intra-tumor vascular network formation and improved drug delivery into tumors, which effect was not observed in *Lpar4*-KO mice [133]. Therefore, activation of the LPA-LPA_4 axis in endothelial cells may be a promising target to normalize vasculature in tumors, which has potential to facilitate anti-tumor drug delivery, improve anti-tumor immunity, and also normalize hypoxic environments which may promote cancer progression. Abnormal or excessive angiogenesis plays major roles in the development or pathologies of numerous diseases including cancer, arthritis, and blindness; and vascularization/vascular architecture is often a critical factor affecting delivery of drugs to treat various conditions. Therefore, modification of the $LPA_{4/6}$ axis in endothelial cells holds great promise as a therapeutic strategy to treat various diseases associated with disturbed angiogenesis or vascular maturation.

Historically, the vasoactive role of LPA was first identified in a soybean extract in 1978 [2], and hypertension has long been noted as a major physiological effect of LPA. This LPA-induced hypertensive effect was not altered in *Lpar1*/*Lpar2*-double KO mice [135], and the molecular mechanisms including LPA receptor subtypes which mediate the hypertensive actions of LPA had long been elusive. Recently, Kano et al. have utilized a variety of LPA receptor-selective agonists and LPA receptor-KO mice to show that the hypertensive response to LPA is mediated mainly by LPA_4, while LPA_6 is also crucial for normal vasoactivity [136]. Of note, elevated levels of plasma LPA or ATX have been reported in acute coronary syndrome [137, 138] and pregnancy, especially in women facing preterm delivery [9]. Therefore, the ATX-$LPA_{4/6}$-$G\alpha_{13}$ axis may be a major contributor to hypertension in these clinical conditions, and this pathway will be a promising drug target for blood pressure control.

Neointimal formation is a pathological vascular remodeling process that occurs after arterial injury, and it is a major cause of restenosis after stenting or angioplasty. LPA has a major role in neointima formation, and infusion of LPA itself induced neointimal formation in uninjured rodent arteries [139, 140]. The LPA-induced vascular remodeling was blunted by either a $LPA_{1/3}$ antagonist [141] or genetic depletion of *Lpar1* but not *Lpar2* or *Lpar3* [142], indicating a major role of LPA_1 in this process. However, the roles reported for various LPA receptors in neointimal responses after injury may be complex and often varies among studies. An $LPA_{1/3}$ antagonist reduced carotid wire-induced neointima formation in *Apoe*-KO mice, possibly by inhibiting mobilization and recruitment of smooth muscle progenitor cells [141]. However, in a carotid artery ligation model, *Lpar1*-KO mice showed enhanced neointimal formation, possibly by upregulation of LPA_3 [135]. In the same study, *Lpar1*/*Lpar2*-double KO mice but not *Lpar2*-KO mice were partially protected from the development of intimal formation [135], suggesting that LPA promotes neointimal formation by signaling that is cooperatively regulated by LPA_1, LPA_2 and possibly LPA_3. Exaggerated vascular injury-neointima formation was also observed in mice with smooth muscle cell (SMC)-specific deletion of LPP3, indicating a biological role for LPP3 to limit LPA-induced responses following vascular injury [143]. Additionally, a non-GPCR-mediated role of LPA was postulated whereby PPARγ may promote neointima formation by working as an intracellular receptor target of LPA independently of LPA_1 or LPA_2 [140, 144]. The variable and inconsistent roles reported for several LPA receptors in different neointima formation models may reflect the complexity and diversity of LPA signaling responses that occur in various cell types and vascular beds during different phases of vascular remodeling. Furthermore, exaggerated hypoxia-induced pulmonary vascular remodeling was observed in *Lpar1*/*Lpar2*-double KO and *Enpp2* heterozygous mice [145], indicating additional roles for the ATX-$LPA_{1/2}$ axis in vascular remodeling in lung that are distinct and rather opposite from those in the systemic vasculature. To better understand these discrepancies, it may be necessary to examine the role of LPA receptors utilizing cell type-specific KO approaches for several different cell types including endothelium, SMCs, and inflammatory cells.

LPA is abundant in the lipid rich core of human atherosclerotic lesions [146]. Exogenous application of LPA accelerated disease progression in a murine model of atherosclerosis by enhancing recruitment of inflammatory cells, whereas administration of $LPA_{1/3}$ antagonist suppressed atherogenesis by reducing inflammatory cell recruitment [147, 148] as well as plasma cholesterol levels [148]. Beneficial effects of a $LPA_{1/3}$ antagonist were also observed in a murine model of calcific aortic valve stenosis, which shares common pathophysiological mechanisms with atherosclerosis [149]. In addition, a recent study has shown that *Lpar4*-KO mice are protected from development of atherosclerosis, probably by changes in macrophage populations that are involved in the process [150, 151]. An important role of LPP3 in atherosclerosis was indicated by a genome-wide association study (GWAS) which identified a heritable SNP in *PPAP2B*, the gene encoding LPP3, as a susceptibility locus for cardiovascular disease [152]. Indeed, SMC- or hepatocyte-specific deletion of LPP3 led to exacerbated plaque formation in murine models of atherosclerosis [151, 153]. Together with the fact that endothelium-specific deletion of LPP3 increases vascular inflammation [154], it is likely that both systemic levels of LPA as well as LPA produced locally in the vascular walls have important roles in progression of atherosclerosis.

LPA also has important functions in heart, that are not well characterized. Cardiomyocyte-specific deletion of LPP3 in mice resulted in shorter life span due to heart failure, indicating deteriorative roles for excessive LPA signaling on normal cardiac homeostasis [155]. However, exaggerated cardiac dysfunction and infarction size were observed in *Lpar3*-KO mice [156] or mice treated with a $LPA_{1/3}$-antagonist [157] in myocardial infarction models, indicating that cardio-protective roles of LPA_1 and/or LPA_3 may exist. Although the specific roles of LPA signaling related to myocardial infarction remains elusive, a six-fold increase in serum LPA levels was reported in patients following acute myocardial infarction [158]; therefore, LPA pathways represent a promising and emerging potential drug target for myocardial infarction and other heart diseases.

7.5.5 Respiratory System

LPA signaling is involved in multiple processes that impact both development and pathologies of the respiratory system. *Lpar1*-KO mice displayed alveolarization defects reminiscent of the alveolar dysplasia that occurs in bronchopulmonary dysplasia due to reduced alveolar septal elastogenesis [159], which may be related to the profibrotic action of LPA_1 observed in pulmonary fibrosis as discussed above (Sect. 7.5.1). Likewise, *LPAR1*-hypomorph (*LPAR1*$^{M318R/M318R}$) rats displayed enlarged alveoli [160], confirming the importance of LPA_1 in alveolar development.

Roles of LPA_1 have been implicated not only in chronic fibrogenic lung pathology (Sect. 7.5.1) but also in non-fibrogenic acute lung injury (ALI). In an endotoxin-induced ALI model, LPA levels in bronchial lavage (BAL) fluid were elevated [161], and *Lpar1*-KO and $LPA_{1/3}$ antagonist-treated mice [162] as well as *LPAR1*-hypomorph rats [163] showed reduced BAL cytokine levels and immune cell infiltration into alveolar spaces following ALI challenges. Unlike the lung fibrosis models, LPA_1 blockage did not affect vascular leak induction in ALI models, indicating that LPA_1 may promote ALI mainly by its pro-inflammatory properties. A pro-inflammatory role of LPA_1 is also implicated in rodent hyperoxia-induced ALI models [160, 164]; however, the involvement of ATX in ALI is not as clear as that of LPA_1. Although chronic systemic overexpression of *Enpp2* exacerbated lung inflammation in an endotoxin-induced ALI model [161], the contribution of endogenous ATX on the acute inflammatory phase appeared to be minor based on the results from genetic or pharmacological targeting of ATX in ALI models [161]. Rather, it is suggested that increased ATX may predispose or promote pulmonary inflammation differentially under acute or chronic inflammatory conditions, which may involve distinct LPA receptors expressed in various cell types during different phases of lung injury.

Increased LPA levels in BAL fluid is also observed in a murine model of allergic airway inflammation and humans with asthma [165]. In a *Schistosoma mansoni* egg-driven asthma model, *Lpar2*-heterozygous mice showed reduced airway goblet cell metaplasia, eosinophil infiltration into BAL, and prostaglandin (PG) synthesis compared to WT mice [166]. Reduction of allergic lung inflammation in *Lpar2*-KO mice was also observed in a triple-allergen-driven asthma model [165]. Pharmacological ATX blockage or haploinsufficiency of *Enpp2* also bestowed a protective effect, while *Enpp2* over-expression exacerbated airway inflammation in the same model [165], indicating a pro-inflammatory role of the ATX-LPA_2 axis in asthma. However, anti-inflammatory roles of LPA_2 have also been reported. In a murine model of ovalbumin-driven allergic asthmatic inflammation, *Lpar2*-KO mice displayed more severe airway inflammation, possibly due to increased dendritic cell (DC) activation [167]. Furthermore, pharmacological activation of LPA_2 showed protective effects in a house-dust mite-driven model of murine allergic airway inflammation [168], supporting a potential anti-inflammatory role of LPA_2. The reasons for the discrepancies are unknown but might be due to differences in nature of the antigens used, immunization protocols, or genetic backgrounds of the mice used. Although LPA_2 was shown to be a negative regulator of DC activation and allergic lung inflammation [167], LPA_2 has also been shown to enhance motility of T cells, which may promote their stimulation by antigen-bearing DCs in lymph node (LN) [169, 170] (discussed in Sect. 7.5.11). Together these studies suggest that LPA_2 may mediate either pro- or anti-inflammatory signaling during allergic reactions, which may vary between cell types and disease stage. A distinct neuronal mechanism is also reported to be important for LPA-mediated airway hypersensitivity in asthma, whereby direct activation of carotid body by LPA involving transient receptor potential vanilloid 1 and LPA-specific receptors was sufficient to induce vagal activity followed by acute bronchoconstriction [171].

7.5.6 Gastrointestinal Tract

LPA is involved in the homeostatic maintenance of gastrointestinal tissues. Detailed analysis on basal intestinal phenotype of *Lpar1*-KO mice revealed shortened intestinal villi and defective proliferation and migration of intestinal epithelial cells (IECs) [172]. *Lpar1*-KO mice also displayed epithelial barrier dysfunction accompanied by enhanced bacterial infiltration and cytokine gene induction in intestine [173]. Concordantly, inducible global deletion of *Enpp2* from adult mice lead to intestinal inflammation and aberrant IEC proliferation, which was accompanied by weight loss and increased mortality [174]. In addition, gastrointestinal application of LPA promoted gastric would repair [175] and reduced stress-induced gastric ulcer formation in rats [176]. These studies revealed a major role for ATX/LPA signaling in gastric wound healing, which may occur mainly via LPA_1 because the effect was absent in *Lpar1*-KO mice [172].

Inflammatory bowel disease (IBD), which includes ulcerative colitis and Crohn's disease, is a multifactorial intestinal disorder characterized by diffuse accumulation of lymphocytes in mucosa. Given the protective roles of ATX and LPA_1 reported in intestinal homeostasis including barrier maintenance, it follows that LPA signaling might also have protective roles in the context of IBD. Indeed, recovery from dextran sulfate sodium (DSS)-induced colitis was delayed in *Lpar1*-KO mice or LPA_1-antagonist treated mice [172]. However, other studies have indicated that detrimental roles of LPA signaling also exist in IBD. Global deletion of *Enpp2* [174] or pharmacological inhibition of ATX [177] protected mice from DSS-induced colitis. A beneficial effect by ATX inhibition was also reported in a murine model of Crohn's disease [178]. Notably, mucosal accumulation of lymphocytes, a characteristic of IBD inflammation, was diminished by ATX inhibition or gene deletion in these studies. As discussed later (Sect. 7.5.11), ATX has important roles in homing of lymphocytes to secondary lymphoid organs, thus it is plausible that the ATX-LPA receptors axis boosts mucosal lymphocyte accumulation to promote IBD progression. Therefore, immuno-modulation by ATX inhibitors or LPA receptors antagonists may hold clinical promise for treatment or control of IBD.

IECs, especially those in crypt compartment, abundantly express LPA_2 [179]. LPA administration displayed protective effects from cholera-toxin-induced diarrhea by inhibiting cystic fibrosis transmembrane conductance regulator (CFTR)-dependent Cl^- secretion [180]. This protective role of LPA was absent in *Lpar2*-KO mice, confirming the essential role of LPA_2 [179, 180]. LPA_2 has a unique carboxyl terminal PDZ-binding domain which is thought to allow formation of a macromolecular complex with the Na^+/H^+ exchanger regulator factor 2 (NHERF2) and other proteins to facilitate effective and compartmentalized inhibition of CFTR [180]. Indeed, *Lpar2*, *Nherf2* and *Cftr* are all more abundantly expressed in crypts rather than in villi compartments of intestine, and the protective role of the LPA-LPA_2 axis was absent in *Nherf2*-KO mice [179]. LPA_2 also has a unique protective role in radiation-induced intestinal injury. Oral administration of LPA or a stable analogue of LPA clearly reduced radiation-induced apoptosis of IECs in WT but not *Lpar2*-KO mice [181, 182]. In addition, *Lpar2*-KO mice showed increased rate of radiation-induced apoptosis and higher mortality, indicating that endogenous LPA also plays an essential role in radioprotection. Mechanistically, it is proposed that radiation increases systemic ATX activity, which facilitates LPA-dependent DNA damage repair via LPA_2 activation [183]. Therefore, LPA_2 agonists represent a promising therapeutic tool to promote recovery from radiation injuries due to environmental exposures or cancer therapy, and various chemically stable LPA_2 agonists are being developed as exemplified by Rx-100 and Radioprotectin-1 [184].

LPA_5 mRNA is also abundant in the intestinal tract. In contrast to LPA_2, which is enriched in crypts, LPA_5 expression is enriched in the villus compartment of the intestine. LPA acting on IECs in the villus compartment has a protective role in diarrhea by inducing fluid absorption via activation of Na^+/H^+ exchanger 3 (NHE3) [185]. This effect is absent in IEC-specific *Lpar5*-KO mice [186] but intact in *Lpar2*-KO mice [187], confirming the essential role of LPA_5 expressed on IECs for LPA-dependent enhancement of fluid absorption. Thus, LPA_2 and LPA_5 have protective roles in diarrhea via different mechanisms: inhibition of fluid secretion at crypt segment is LPA_2-dependent while activation of fluid absorption at villus segments occurs via LPA_5. Like LPA_2, LPA_5 also contains a PDZ-binding domain at its carboxyl terminus and interacts with NHERF2, which is necessary for the stimulation of NHE3 [187]. Indeed, the fluid absorption effect by LPA-LPA_5 axis was abolished in *Nherf2*-KO animals [187]. Of note, bioactive LPA and LPA precursors are present in a variety of herbal medicines and food stuffs such as cabbage where they have been suggested to have beneficial health effects that include promoting healing of gastric ulcers [188]. Thus, modulation of NHE3 by LPA_5 ligands including both drug-like compounds as well as natural sources enriched in certain foods hold promise for the treatment or prevention of diarrhea, as well as other gastric conditions which can be modulated by LPA signaling.

7.5.7 Metabolic Disorders

The involvement of ATX in the development of obesity and metabolic disorders has been extensively studied by many research groups. Notably, adipose tissue is a major source of systemic ATX, as revealed by adipocyte-specific *Enpp2*-KO mice which display a 50–70% reduction in blood ATX [189–192]. Therefore, ATX can be classified as an adipokine based on its LPA production. Adipocyte-specific *Enpp2*-KO mice [189–192] as well as global *Enpp2*-heterozygous mice [191, 193] display improved systemic metabolism including protection from obesity-induced hepatosteatosis, while systemic [194] or adipocyte-specific [191] overexpression of *Enpp2* increases adiposity in murine obesity models. Furthermore, obesity-related cardiac hypertrophy and dysfunction were also attenuated by pharmacological ATX inhibition [195] or *Enpp2* hemizygous deletion [193], indicating broad roles for ATX existing in several tissues that are impacted by metabolic disorders.

Important roles for LPA_1 and LPA_4 in promoting metabolic disorders have been reported. LPA administration acutely inhibited insulin secretion and lowered glucose tolerance in mice, which was inhibited by pre-injection of a $LPA_{1/3}$ antagonist [196]. Chronic inhibition of $LPA_{1/3}$ also decreased glucose tolerance in a murine model of obesity [196]. Although *Lpar1*-KO mice weighed less, they had higher adiposity than WT mice under low fat diet conditions [197, 198]. Under high fat diet (HFD) conditions, *Lpar1*-KO mice are protected from development of obesity, possibly due to decreased food intake [198]. In addition to LPA_1, the impact of LPA_4 on obesity-related metabolic disorders was recently evaluated [199]. *Lpar4*-KO mice showed enhanced expression of mitochondrial and adipogenic genes along with increased production of adiponectin when fed a standard diet. In a HFD-induced obesity model, *Lpar4*-KO mice displayed a metabolically healthy obese phenotype, including improved glucose homeostasis and reduced hepatosteatosis, which was mediated through continuous healthy expansion of adipose tissue. Nonalcoholic steatohepatitis is a progressive form of fatty liver disease that is characterized by inflammation and fibrosis, and there are currently no approved pharmacological therapies. Together with the beneficial effect of LPA_1 and ATX inhibition on liver fibrosis (Sect. 7.5.1), LPA signaling may have potential to solve multiple aspects of this unmet and urgent medical need.

7.5.8 Reproduction

The critical role of LPA_3 signaling in female reproduction has been established. *Lpar3*-KO female mice showed severe reproductive defects including delayed implantation and aberrant embryo spacing [200]. Similar phenotypes have been observed in mouse models where PG synthesis is genetically or pharmacologically blocked [201]. Indeed, reduced mRNA levels of cyclooxygenase-2 (COX-2), a key enzyme for PG synthesis, as well as reduced PG amounts are observed in uteri of *Lpar3*-KO mice [200]. Of note, exogenous PG rescued the delayed implantation but not the aberrant embryo spacing in *Lpar3*-KO mice, suggesting that LPA_3 regulates implantation timing and embryo spacing independently. An LPA_3-selective agonist induced a potent contractile response in isolated uteri from WT but not *Lpar3*-KO mice [202]. Therefore, it is possible that LPA_3-dependent uterine contraction may be needed for proper embryo spacing in a PG-independent manner. More detailed roles of LPA_3 in female reproductive tissue were further investigated by Aikawa et al. They studied the effects of an LPA_3-selective agonist and an ATX inhibitor on reproductive tissues of WT as well as *Lpar3*-KO mice [203]. In their study, the authors proposed a model whereby ATX-dependent production of LPA at the embryo-epithelial boundary leads to activation of epithelial LPA_3, which contributes to successful decidualization through induction of COX-2 and heparin-binding epidermal growth factor at the epithelial layer, followed by BMP2 and Wnt4 induction at the stromal layer. Notably, these downstream targets of LPA_3 have been implicated in the progression of endometriosis, and in fact *Lpar3*-KO mice displayed less developed endometrial tissues in a mouse endometriosis model [203]. Therefore, the ATX-LPA_3 axis is a potential therapeutic target not only for female infertility but also other female reproductive tissue conditions such as endometriosis.

The male reproductive system is also regulated by LPA signaling. In testis, *Lpar1*, *Lpar2* and *Lpar3* are highly expressed in germ cells. Male *Lpar1*-, *Lpar2*-, and *Lpar3*-single gene-KO mice all displayed mild reductions in sperm counts compared to age-matched WT mice; sperm reductions were even more pronounced in compound-KO mice lacking any two of these receptors [204]. The triple LPA receptor-KO males showed the most pronounced reductions, with over 60% of these males having azoospermia at 8 months age despite normal testosterone levels, suggesting that the reduced sperm counts were caused independent of systemic changes in endocrine function. Mechanistically, activation of these LPA receptors seems to protect germ cells from apoptosis and support sperm production in a partially redundant manner through multiple and convergent roles [204]. Male infertility accounts for 40–50% of all infertility and affects approximately 7% of all men. More precise elucidation of the molecular mechanisms by which LPA signaling promotes healthy sperm counts, including the sources of LPA and whether those LPA levels are altered in various conditions affecting sperm counts such as aging, may lead to new therapeutic tools to treat or prevent male infertility.

7.5.9 Bone

The first *in vivo* role of LPA signaling was demonstrated in *Lpar1*-KO mice, which display dwarfism as well as cranial deformities including shorter snouts and more widely spaced eyes [205]. In a follow up study, extensive analyses of the bone phenotypes in *Lpar1*-KO mice revealed additional abnormalities in ribs and vertebrae, which included fused rims and abnormal attachment of the ribs to the sternum [206]. *Lpar1*-KO mice showed decreased masses of trabecular and cortical bones without mineralization deficiency, indicative of true osteoporosis due to decreased osteoblast differentiation and osteogenesis. More recently, Nishioka et al. revealed further details about the molecular mechanisms underlying the cranial deformities in *Lpar1*-KO mice [207]. They found that *Lpar1* is highly expressed in chondrocytes, and *Lpar1*-KO mice displayed dyschondroplasia in skull, costa, and femur. Because these bones are largely formed by endochondral ossification, in which cartilage is replaced by bone, and disturbed chondrocyte and cartilage phenotypes preceded defective bone tissue formation in *Lpar1*-KO mice, the authors concluded that impaired bone development in *Lpar1*-KO mice is mainly caused by dyschondroplasia. *Enpp2* expression is also enriched in chondrocytes and typical cartilage, and bone abnormalities of *Lpar1*-KO mice are phenocopied by genetic reduction of *Enpp2*. Mechanistically, the ATX-LPA_1 axis in chondrocytes supports integrin-dependent fibronectin assembly which is essential for chondrocyte proliferation. Although the skeletal phenotypes of *Lpar1*-KO mice can be attributed to functional defects in chondrocytes and possibly osteoblasts, a role of osteoclastic LPA_1 is also possible. In fact, systemic treatment with a LPA_1 antagonist prevents osteoclast-mediated bone loss in an ovariectomy-induced osteoporosis mouse model [208].

LPA_4 is another LPA receptor involved in bone formation, and has been shown to regulate osteoblast function. *Lpar4*-KO mice displayed increased bone volume, trabecular thickness, and trabecular number [209], indicating that LPA_4

may have an opposing role to LPA_1 on osteoblast differentiation and development of bone mass. The inhibitory role of LPA_4 on bone formation was attributed to its $G\alpha_s$-coupled signaling [209]; however, $G\alpha_s$ stimulation generally enhances osteogenesis [210], in contrast to the anti-osteogenic role of LPA_4. LPA_4 has been shown to couple primarily to $G\alpha_{13}$ in several biological contexts including angiogenesis [44, 132], vascular normalization [133, 134], and metabolic disorders [199]; and it will be important to determine whether $G\alpha_{13}$-coupled signaling may also mediate actions of LPA_4 pertinent to bone formation. Elucidation of the mechanisms that coordinate the opposing actions of LPA_1 and LPA_4 on osteoblasts and determining whether these mechanisms are altered in disease condition such as osteoporosis are critical questions to be addressed in future studies. Determination of the sources of LPA that stimulate each receptor will enable a fuller understanding of the roles of LPA signaling pathway in bone pathophysiology, which may aid in the development of LPA signaling-based therapies for bone diseases.

7.5.10 Skin and Hair

In addition to the role in skin fibrosis (Sect. 7.5.1), LPA plays important roles in skin wound healing, itch response, and hair development. In rodents, topical application of LPA thickened the skin epidermis [211] and accelerated skin wound healing [212, 213]. LPA also promoted skin barrier function in a mouse model of dry skin [214]. Notably, LPLs containing high amount of LPA promoted skin moisturizing capacity of human skin [215]. Therefore, LPA may serve as a novel therapeutic target for skin barrier dysfunction. Although these observations reveal beneficial aspects of LPA application on skin, dermal injection of LPA can cause itchiness and scratching behavior in rodents [216–222]. Genetic and pharmacological approaches have identified LPA_1 [220], LPA_5 [221], and transient receptor potential channels [221] as molecular targets of LPA that mediate pruritic responses. Increased blood levels of both ATX and LPA have been reported in murine models of atopic dermatitis [220, 223] as well as in human atopic dermatitis patients [224, 225], indicating a possible role for LPA in initiation or exacerbation of pruritic skin disease. Furthermore, serum ATX levels were found to be markedly elevated in patients with cholestatic pruritus [226], and LPA was suggested to be a potential pruritogen in these patients [219].

Studies on human hereditary hair disorders and detailed studies utilizing KO mice have established a critical role for LPA signaling in hair development. In 2006, *LIPH*, encoding mPA-$PLA_{1\alpha}$, was identified as a causative gene for human autosomal recessive woolly hair/hypotricosis (ARWH/HT) [227]. This was followed by numerous reports on various *LIPH* mutations in ARWH/HT patients [228–256]. In 2008, *P2RY5* (former official gene name of *LPAR6*) was also identified as a causative gene for ARWH/HT [257, 258], and various mutations have been reported [253, 259–271]. In accordance with the findings from human genetics, LIPH has been identified as the causative genes for the rex hair coat phenotype in rabbits [272] and LPAR6 is causative for the curly coat phenotype of Cornish Rex cats [273], indicating a fundamental role for the LIPH-LPA_6 axis in hair development across several species. Using *Liph*-KO mice having wavy hairs, Inoue et al. revealed the detailed molecular mechanism underlying mPA-$PLA_{1\alpha}$-LPA_6 axis [43]. At hair follicles, the predominant LPA species are 2-acyl ones, and they are almost completely lacking in *Liph*-KO [43] but not in *Enpp*2-KO mice [274], indicating a critical role of mPA-$PLA_{1\alpha}$ rather than ATX in LPA production at hair follicles. At the inner root sheath of hair follicles, mPA-$PLA_{1\alpha}$-derived LPA activates LPA_6, an LPA receptor that prefers 2-acyl type LPA species, leading to $G\alpha_{13}$-dependent ectodomain shedding of TGFα, a key process for hair development. Together with the "ligand lateral diffusion" hypothesis suggested by the crystal structure of LPA_6 (Sect. 7.4.2), the mPA-$PLA_{1\alpha}$-LPA_6 axis-dependent model of hair development represents a more complete model of LPA-signaling in a specific biological context, connecting the production and function by describing how the LPA is produced, delivered to receptors, and exerts biology.

7.5.11 Hematopoiesis and Immune Function

The roles of LPA signaling in hematopoiesis were evaluated by pharmacological blockage of LPA_2 and LPA_3 in mice. LPA_2 agonists decreased while LPA_3 agonists increased red blood cell number in mice, indicating the opposing roles of these receptors in regulation of erythropoiesis [275]. However, a beneficial effect of LPA_2 activation for hematopoiesis has also been reported in a mouse model of irradiation [276]. Administration of a potent LPA_2 agonist after irradiation increased peripheral white blood cell and platelet counts and reduced mortality. LPA has critical roles in the maintenance and regulation of hematopoietic stem cells (HSCs), which include both cell-autonomous mechanisms as well as modulation of the HSC niche microenvironment. The importance of LPA signaling in the HSC niche was revealed by a study which found that *Lpar4*-KO mice have decreased numbers of HSCs in bone marrow and spleen [277]. The *Lpar4*-KO mice showed delayed hematopoietic recovery and increased mortality after chemical- or radiation-induced myelosuppression. The importance of LPA_4 in stromal cells that support HSC stem cell proliferation was elucidated by bone marrow transplantation experiments.

During their patrol through the body, lymphocytes continuously migrate from circulation into secondary lymphoid organs including LNs via specific type of blood vessels called high endothelial venules (HEVs). An extremely high amount of ATX is expressed in HEV endothelial cells (ECs) [278–280], and ATX plays essential role in lymphocyte transendothelial migration. Local [280] but not circulatory [279] ATX blockage attenuates lymphocyte entry into LNs, indicating the important role of LPA locally produced by ATX in the transendothelial migration of lymphocytes. The locally produced LPA seems to act in an autocrine manner on HEV ECs that express LPA_4 and LPA_6. Indeed, *Lpar4*- and *Lpar6*-KO mice displayed compromised lymphocyte transmigration across HEV EC layers [45]. T cells, after homing to LNs, dynamically move within the LNs which facilitates effective detection of antigens. Even here, LPA signaling is important for intranodal T cell movement via LPA_2-dependent signaling. Pharmacological inhibition of ATX slows down intranodal T cell motility [281, 282]. Fibroblastic reticular cells (FRCs) in HEVs express high amounts of ATX [281] and are proposed to be the source of the LPA responsible for the intranodal T cell movement regulation. Indeed, intranodal T cell motility was significantly decreased in FRC-specific *Enpp2*-KO mice [170]. Furthermore, *Lpar2*-KO T cells that have been adoptively transferred into WT mice display attenuated motility in LNs [169, 170]. An important role for LPA_2 in T cell trafficking and homing is also postulated to function in neuroinflammation. In a murine experimental autoimmune encephalomyelitis (EAE) model, *Lpar2*-KO mice displayed exaggerated disease progression which was accompanied by enhanced T cell infiltration into the spinal cord, suggesting defective lymphocyte homing in these mice [283]. Accordingly, an LPA_2 agonist reduced clinical signs of EAE, possibly by enhancing T cell homing [283]. Taken together, these studies suggest that LPA produced locally by ATX from several distinct cell types coordinately regulates lymphocyte movement into and within the LNs, via actions on specific LPA receptors on several cell types.

CD8 T cell cytotoxicity is often inhibited by prolonged antigen-specific stimulation associated with chronic infections and cancer. Recent studies have revealed that LPA, which is elevated in inflammation and cancer, inhibits activation of cytotoxic T cells via binding to LPA_5. LPA suppresses T cell receptor-mediated CD8 T cell signaling and cytotoxic activity *in vitro*, but this effect was abolished in *Lpar5*-KO cells [284, 285]. Genetic deletion of *Lpar5* in CD8 T cells enhanced their antigen-specific stimulation *in vivo*, while *in vivo* administration of a stable analogue of LPA dampened antigen-specific CD8 T cell responses [284, 285]. CD8 T cell cytotoxic activity was increased in *Enpp2*-heterozygous mice [285], further supporting that the ATX-LPA-LPA_5 axis has an inhibitory role on CD8 T cell immunity. LPA_5-dependent signaling also suppresses B cell activation in response to antigen receptor signaling and antibody responses [286]. Pharmacological suppression of the adaptive immune resistance is attracting attention due to recent successes in overcoming immunotolerances that limit the ability of the immune system to combat cancer. Notably, tumor-bearing mice transplanted with *Lpar5*-KO T cells displayed improved control over tumor progression [284, 285], and targeting of LPA_5 may be a key therapeutic strategy to unlock the potential of cancer immunotherapy.

7.6 LysoPS Receptors

The bioactivity of lysoPS was originally identified in 1979 as an activity which enhances mast cell degranulation [287]. Since then, several studies have elucidated the underlying mechanisms for the actions of lysoPS on mast cells. Similar to LPA, lysoPS was postulated to possess specific receptor(s), as indicated by biological responses to lysoPS that showed stereoselectivity [288]. However, it was not until 2006 that Sugo et al. identified the first cell-surface GPCR target of lysoPS, GPR34 (now called $lysoPS_1$) [289]. *Lysopsr1/GPR34* mRNA expression was found to be remarkably high in mast cells [289]; however, mast cell responses to lysoPS were not altered in *Lysopsr1/Gpr34*-KO mice, indicating the presence of additional lysoPS receptor(s) [290]. In 2012, Aoki, Inoue, and colleagues developed an excellent GPCR-ligand screening methodology that succeeded in identifying three additional lysoPS receptors: P2Y10 ($lysoPS_2$), A630033H20Rik ($lysoPS_{2L}$, a pseudogene in humans) and GPR174 ($lysoPS_3$) [291]. Their group further developed unique lysoPS analogues which differentially activate $lysoPS_{1/2/3}$ [292]. Notably, they identified an analogue with extremely high potency (>20-fold compared to authentic lysoPS) for mast cell activation, however this compound did not activate any of the known lysoPS receptors [293]. Thus, the receptor target of lysoPS responsible for mast cell degranulation is still enigmatic, and there may be additional unidentified lysoPS receptors, either GPCR or non-GPCR, in mast cells.

7.7 Putative Bioactive lysoPS Synthesis Pathways *In Vivo*

LysoPS can be produced from phosphatidylserine (PS) by enzymes with PLA activities. Currently, the identification of the specific enzyme(s) and synthesis pathway(s) to produce bioactive lysoPS *in vivo* are not firmly established; however one candidate enzyme is PS-specific PLA_1 (PS-PLA_1) [294]. PS-PLA_1 shows the highest homology with mPA-$PLA_{1\alpha}$, a bioactive LPA-producing enzyme [11]; and a positive correlation between lysoPS and PS-PLA_1 concentration has been observed in plasma from acute coronary syndrome patients [295] as well as in gastric cancerous ascites [296]. In *in vitro* experiments, exogenous application of PS-PLA_1 strongly enhanced mast cell degranulation, clearly mimicking the bioactive effects of lysoPS [297]. Another candidate enzyme for bioactive lysoPS production is α/β-hydrolase domain-containing 16A (ABHD16A), a principle PS lipase in mammalian cells and tissues. Indeed, reduced lysoPS levels are observed by genetic removal or pharmacological blockage of ABHD16A in tumor cell lines, macrophages, and mouse brain [298]. Although both PS-PLA_1 and ABHD16A produce lysoPS, further studies, perhaps linking the phenotypes of the specific enzyme- and receptor-KO mice, will be required to determine the enzymes that supply ligands for the lysoPS receptors *in vivo*.

7.8 Pathophysiology of lysoPS Signaling

Compared with LPA, much less is known about the roles of lysoPS signaling *in vivo*. However, recent studies with KO mice have started to unveil the biological significance of lysoPS, especially in the immune system.

Similar to the role for LPA, lysoPS signaling seems to be essential in the pathogenesis of neuropathic pain, likely through effects on microglial function. In nerve injury models of neuropathic pain, *Lysopsr1/Gpr34*-KO mice show reduced pain as well as microglia-associated pro-inflammatory gene expression [299]. Consistent with this, intrathecal administration of an antagonist for $lysoPS_1$/GPR34 suppresses neuropathic pain [299]. Notably, increased lysoPS levels accompanied by dramatically enhanced microglial activation had been observed in brains of mice gene-deficient for ABHD12, a principle lysoPS lipase in brain [300]. Although it is not clear whether the enhanced microglial activation is mediated through $lysoPS_1$/GPR34, it is plausible that lysoPS signaling plays pro-inflammatory roles in microglia. In addition, *Lysopsr1/Gpr34*-KO mice display altered microglial morphology reminiscent of activated microglia even under non-inflammatory conditions [301]. Furthermore, freshly isolated microglia from *Lysopsr1/Gpr34*-KO mice displayed defective phagocytosis activity [301], suggesting that basal activities of $lysoPS_1$/GPR34 may be necessary to maintain microglia in a homeostatic non-phagocytic state that is required to initiate phagocytosis upon stimuli [302]. This scenario may also explain the higher brain pathogen burden of *Lysopsr1/Gpr34*-KO mice in a murine model of pulmonary infection of fungus [290]. Considering that enhanced levels of brain lysoPS is suggested to contribute to neuroinflammatory disease caused by inactivating mutations of ABHD12 [298, 300], further understanding of lysoPS signaling in microglia will help to establish novel therapeutic approaches toward various inflammation-related neurological diseases.

Studies with *Lysopsr3/Gpr174*-KO mice have revealed important roles for lysoPS signaling in T cell function. Under homeostatic condition, increased frequency of *Lysopsr3/Gpr174*-KO regulatory T cells (Tregs) was observed in sites where thymus-derived and peripherally induced Tregs can accumulate [303]. In *in vitro* experiments, production of IL-2, which is necessary for the growth and differentiation of Tregs, was clearly suppressed by lysoPS in T cells from WT mice but not in T cells from *Lysopsr3/Gpr174*-KO or *Gnas* (gene encoding $G\alpha_s$ protein)-KO mice [304, 305]. Notably, basal levels of IL-2 production were increased in *Lysopsr3/Gpr174*-KO T cells even without addition of exogenous lysoPS [304, 305]. Furthermore, in *in vitro* culture experiments of CD4 T cells purified from WT mice, lysoPS was almost exclusively detected in cells and levels were further increased after antigen stimulation, while very little lysoPS was detected in culture medias [304]. These results suggest that a T cell intrinsic lysoPS-$lysoPS_3$/GPR174 axis restrains T cell proliferation partly by suppressing IL-2 production in a $G\alpha_s$-dependent manner. The absence of this "downer" signaling on Tregs likely contributes to protective effects observed for *Lysopsr3/Gpr174*-KO mice in EAE [303] and cytokine storm of sepsis [306]. Recent GWAS studies have linked human SNPs in *LYSOPSR3/GPR174* and *LYSOPSR2/P2RY10* (adjacent locus to *LYSOPSR3/GPR174*) to two different autoimmune diseases; Graves' disease [307–309] and rheumatoid arthritis [310], respectively. Considering the pivotal role of Tregs in limiting excessive immune responses and preventing autoimmunity, unlocking the Tregs "downer" signaling by inhibition of the lysoPS-$lysoPS_3$/GPR174 axis may be a promising therapeutic strategy to combat various autoimmune diseases.

7.9 LPI and Lysophosphatidylglucoside (LysoPtdGlc)

Dual LPL Ligands for LPI_1/GPR55 GPR55 was an orphan GPCR belonging to the P2Y cluster that was cloned from human brain [311]. In early 2000's, GPR55 was proposed as a new cannabinoid receptor by two patent reports (reviewed in [312]). Since then, numerous studies have sought to elucidate the biological roles of GPR55, especially its potential as an atypical cannabinoid pharmacological target. However, the identity of GPR55 as a cannabinoid receptor is still under debate by conflicting reports [313]. Firmly establishing the natural ligand specificity of GPR55 had been elusive for many years, which had hampered our understanding of pathophysiological roles of GPR55.

7.9.1 LPI

In 2009, Oka et al. identified LPI as a high affinity ligand for GPR55 [314]. In brain, two major LPI species are 1-stearoyl-LPI (18.9 nmol/g of tissue) and 2-arachidonoyl-LPI (8.29 nmol/g of tissue) [315]. In structure-activity relationship experiments, 2-arachidonoyl-LPI has much higher activity than 1-stearoyl-LPI and is proposed to be a natural ligand for GPR55 [315]. The identity of GPR55 as a LPI receptor has been confirmed by various studies from many independent groups, and GPR55 has now been assigned with the provisional name LPI_1 and the gene name *LPIR1* by a nomenclature review for LPL receptors [316]. Of note, LPI_1/GPR55-independent actions of LPI are also reported [317, 318], suggesting that additional receptor target(s) for LPI may exist.

LPI can be produced from phosphatidylinositol (PI) by various PLAs; however, a *bona fide* enzymatic pathway for synthesis of LPI which functions as a LPI_1/GPR55 ligand is not yet established. One promising candidate enzyme to produce bioactive signaling LPI is DDHD domain containing 1 (DDHD1), a brain-enriched intracellular enzyme that can produce 2-acyl type LPLs, including 2-arachidonoyl-LPI, *in vitro* [319]. Importantly, *Ddhd1*-KO brain displays a clear reduction of arachidonoyl- or docosahexaenoyl-LPI but not of other LPI species [320]. *Ddhd1*-KO mice show reduced mechanical nociception [320], which may be related to the impaired nociception that was also observed in *Lpir1/Gpr55*-KO mice in several pain models [321, 322] (Sect. 7.10.2). However, given that brains of *Ddhd1*-KO mice display reductions of additional LPLs including lysoPS [320], future analyses will be needed to conclude whether DDHD1 supplies LPI as a *bona fide* natural ligand for LPI_1/GPR55.

7.9.2 LysoPtdGlc: Repulsive Axon Guidance Cue

As developing dorsal root ganglion axons of sensory neurons enter the spinal cord, they become segregated to project to distinct positions of the spinal cord without mixing; nociceptive axons to lateral positions and proprioceptive axons to dorsomedial positions. This developmental process is thought to be facilitated by the coordinated chemoattraction and chemorepulsion of individual axons; however, the molecular mechanism including the guidance cues had been elusive. In 2015, Guy et al. identified that LysoPtdGlc, a glial cell-derived glycosylated LPL molecule, acts as a spatially repulsive signaling cue during pathfinding of nociceptive axons [318]. In a screening assay that examined over one hundred GPCRs, LPI_1/GPR55 was identified as a receptor target of LysoPtdGlc. Importantly, injection of a neutralizing antibody against LysoPtdGlc induced guidance errors of the nociceptive afferents, and the same phenotype was completely recapitulated in *Lpir1/Gpr55*-KO mice, validating the ligand-receptor interaction *in vivo*.

The LysoPtdGlc that acts as a spatial cue for nociceptive axons is likely to be produced from phosphatidylglucoside (PtdGlc) via an unidentified enzyme with PLA_2 activity. PtdGlc is a component of plasma membrane of radial glia, and specifically localized in the dorsomedial position of spinal cord [318]. This confined distribution of PtdGlc facilitates the confinement of LysoPtdGlc to the position where proprioceptive afferents will project; by repulsing LPI_1/GPR55-positive nociceptive afferents and thereby aiding their guidance to the lateral positions. Thus, fine spatial compartmentalization of LysoPtdGlc is a key feature for the accurate repulsive guidance of nociceptive afferents via LPI_1/GPR55, and may distinguish a role of LysoPtdGlc that is distinct from that of LPI, another LPI_1/GPR55 ligand. Further detailed analyses on the molecular mechanism underlying LysoPtdGlc (and PtdGlc) confinement as well as regulation of LPI synthesis is warranted to more fully understand the biological complexities that have shaped a "dual" LPL ligand system for LPI_1/GPR55 signaling.

7.10 Pathophysiology of LPI_1/GPR55 Signaling

After the first proposal of LPI_1/GPR55 as new cannabinoid receptor, numerous studies have assessed the *in vivo* roles of LPI_1/GPR55, especially in the context of pharmacology of cannabinoid ligands. However, considering the debates on the identity of LPI_1/GPR55 as a cannabinoid receptor [313], studies that utilize *Lpir1*/*Gpr55*-KO mice will provide more confident information on the specific roles of LPI_1/GPR55. Such studies have already validated important roles for LPI_1/GPR55 existing in the nervous system (Sects. 7.10.1 and 7.10.2), gastrointestinal tract [323–326], cardiovascular system [327–329], pancreas [330–332], cancer [333–335], bone [336], salivary gland [337], lymphocytes [338, 339], and metabolic disorders [340, 341]. Below, we discuss the role of LPI_1/GPR55 signaling in the nervous system as revealed by studies using *Lpir1*/*Gpr55*-KO mice.

7.10.1 Central Nervous System

Roles for LPI_1/GPR55 are implicated in neurogenesis and synaptic function. *In vivo* infusion of O-1602, an atypical cannabinoid agonist for LPI_1/GPR55, into hippocampus of adult WT mice increases NSC proliferation as well as immature neuron generation, but this effect is abolished by genetic deletion of *Lpir1*/*Gpr55* [342]. Notably, *Lpir1*/*Gpr55*-KO mice display reduced basal rates of NSC proliferation and immature neuron generation [342], indicating that endogenously activated LPI_1/GPR55 normally facilitates homeostatic and continuous neurogenesis in the adult mouse hippocampus. LPI_1/GPR55 is also involved in synaptic function. LPI_1/GPR55 is expressed within submicron proximity to glutamatergic synaptic vesicles, and application of LPI or O-1602 enhances presynaptic Ca^{2+} elevation and glutamate release in acute hippocampal slices from WT but not from *Lpir1*/*Gpr55*-KO mice [343]. Even without ligand application, *Lpir1*/*Gpr55*-KO slices show decreased post-burst synaptic potentiation that is thought to depend on presynaptic Ca^{2+} elevation [343]. Therefore, activation of LPI_1/GPR55 by an endogenous ligand seems to positively regulate a presynaptic Ca^{2+} elevation that leads to enhanced neurotransmitter release. In addition, LPI enhances long-term potentiation, likely by postsynaptic modification, in hippocampal slices from WT but not *Lpir1*/*Gpr55*-KO mice [344]. Furthermore, *Lpir1*/*Gpr55*-KO mice display neuronal projection abnormalities not only in nociceptive afferents [318] (Sect. 7.9.2), but also in retinal ganglion cells targeting towards the visual thalamus [345]. Therefore, it is likely that in addition to its roles in the peripheral nervous system, LPI_1/GPR55 also modulates anatomical neuronal circuits in the central nervous system during development. A major challenge for future research is to identify the endogenous ligand (LPI *versus* LysoPtdGlc) that signals through LPI_1/GPR55 to regulate these several processes in the central nervous system. Although behavioral tests with *Lpir1*/*Gpr55*-KO mice did not reveal any memory defects [344], increasing LPI_1/GPR55 signaling may be a promising therapeutic strategy for memory disorders such as dementia and Alzheimer's disease.

7.10.2 Pain

Like LPA and lysoPS receptors, LPI_1/GPR55 has important roles in pain. Intraplantar injection of LPI in hind paws of mice causes a marked mechanical hypersensitivity lasting for 6 days without causing local inflammation [322]. The nociceptive hypersensitivity triggered by LPI is completely abolished in *Lpir1*/*Gpr55*-KO mice as well as sensory neuron-specific *Gna13*-KO or *Gnaq*/*Gna11* (encoding $G\alpha_q$/$G\alpha_{11}$ proteins)-double KO mice [322]. Importantly, unlike LPA, LPI did not cause demyelination. Mechanistically, it is hypothesized that activation of the LPI_1/GPR55-$G\alpha_{13}$ and $G\alpha_q$/$_{11}$ axis in peripheral sensory neurons in dorsal root ganglia leads to hyperexcitability and sensitization of sensory neurons. Accordingly, *Lpir1*/*Gpr55*-KO failed to develop mechanical hyperalgesia in several different pain models including a partial nerve ligation model of neuropathic pain [321], an inflammatory pain model [321], and a tumor-associated pain model [322]. In addition, microinjection of LPI into intra-periaqueductal gray caused thermic hyperalgesia, which was blocked by pre-treatment with a LPI_1/GPR55 antagonist [346]. Therefore, LPI_1/GPR55 has pronociceptive roles not only at peripheral but also at central levels. Although these studies have revealed pronociceptive role of LPI_1/GPR55, *Lpir1/Gpr55*-KO mice display mild thermic hyperalgesia under normal conditions [321, 347, 348], suggesting an analgesic role of LPI_1/GPR55 signaling exists under homeostasis. Together with the pivotal role of LPI_1/GPR55 for nociceptive afferent guidance (Sect. 7.9.2), this indicates that LPI_1/GPR55 is involved in multiple levels of nociception, from neural network formation to central sensation. As mentioned above, some of the major constituents of cannabis, including cannabidiol and tetrahydrocannabinol, are reported to be ligands for LPI_1/GPR55 [313]. Considering the increasing focus and debates concerning the usage of cannabinoids for pain relief, further understanding the details of LPI_1/GPR55-dependent regulation of nociception may be essential to guide development and usage of novel cannabinoid-based pain relievers.

7.11 Conclusion

Here we have reviewed some of the many roles of LPLs, especially LPA, in physiology and pathophysiological states that hold potential as druggable signaling pathways. Targeting of bioactive LPA signaling pathways including the LPA-synthesizing enzymes and individual receptor subtypes holds potential to treat or ameliorate a wide range of disease of conditions, as exemplified by the current clinical trials of ATX inhibitors and LPA_1 antagonists for fibrotic diseases. Development of highly selective and effective compounds has been aided by the solved crystal structures of LPA_1, LPA_6 and ATX; and will be accelerated by future solving of additional LPA receptor structures. Besides for LPA signaling pathway molecules, more recently identified bioactive LPL receptor targets LPI_1 and $lysoPS_{1-3}$ are emerging as promising therapeutic targets, and other as yet unidentified receptors may also exist to mediate effects of various LPL signaling molecules. These receptors, along with the enzymes that regulate production and availability of their ligands, will be the next wave of druggable targets of LPL signaling pathways. Determining the *bona fide* LPL receptor subtype and ligand source mediating biological responses remains an ongoing challenge in many instances, and future studies utilizing various cell type-specific KO mice or receptor-subtype-compound KOs will be highly beneficial to unravel some of these complexities. This will aid not only in novel drug development, but also for avoidance of unwanted side effects and may allow potential drug repositioning. Furthermore, it is expected that where and how the LPL ligand is produced and how it is delivered to its cognate receptor are critical determinants of biological outputs. Elucidating the biological mechanisms that mediate precise regulation of LPL signaling required for physiological responses will reveal exciting new opportunities to pharmacologically target LPL signaling pathways to treat human diseases.

Acknowledgements The Lipid Signaling Project, National Center for Global Health and Medicine receives funding from ONO Pharmaceuticals.

Conflicts of Interest The authors have no conflicts of interest affecting the objectivity of this review.

References

1. Vogt W (1960) Intestine-stimulating activity of various phosphatides and glycolipids. Naunyn Schmiedebergs Arch Exp Pathol Pharmakol 240:134–139
2. Tokumura A, Fukuzawa K, Tsukatani H (1978) Effects of synthetic and natural lysophosphatidic acids on the arterial blood pressure of different animal species. Lipids 13:572–574
3. van Corven EJ, Groenink A, Jalink K, Eichholtz T, Moolenaar WH (1989) Lysophosphatidate-induced cell proliferation: identification and dissection of

signaling pathways mediated by G proteins. Cell 59:45–54
4. Hecht JH, Weiner JA, Post SR, Chun J (1996) Ventricular zone gene-1 (vzg-1) encodes a lysophosphatidic acid receptor expressed in neurogenic regions of the developing cerebral cortex. J Cell Biol 135:1071–1083
5. Yanagida K, Kurikawa Y, Shimizu T, Ishii S (2013) Current progress in non-Edg family LPA receptor research. Biochim Biophys Acta 1831:33–41
6. Bandoh K, Aoki J, Taira A, Tsujimoto M, Arai H, Inoue K (2000) Lysophosphatidic acid (LPA) receptors of the EDG family are differentially activated by LPA species. Structure-activity relationship of cloned LPA receptors. FEBS Lett 478:159–165
7. Yanagida K, Masago K, Nakanishi H, Kihara Y, Hamano F, Tajima Y, Taguchi R, Shimizu T, Ishii S (2009) Identification and characterization of a novel lysophosphatidic acid receptor, p2y5/LPA6. J Biol Chem 284:17731–17741
8. Aoki J, Inoue A, Okudaira S (2008) Two pathways for lysophosphatidic acid production. Biochim Biophys Acta 1781:513–518
9. Tokumura A, Majima E, Kariya Y, Tominaga K, Kogure K, Yasuda K, Fukuzawa K (2002) Identification of human plasma lysophospholipase D, a lysophosphatidic acid-producing enzyme, as autotaxin, a multifunctional phosphodiesterase. J Biol Chem 277:39436–39442
10. Umezu-Goto M, Kishi Y, Taira A, Hama K, Dohmae N, Takio K, Yamori T, Mills GB, Inoue K, Aoki J, Arai H (2002) Autotaxin has lysophospholipase D activity leading to tumor cell growth and motility by lysophosphatidic acid production. J Cell Biol 158:227–233
11. Sonoda H, Aoki J, Hiramatsu T, Ishida M, Bandoh K, Nagai Y, Taguchi R, Inoue K, Arai H (2002) A novel phosphatidic acid-selective phospholipase A1 that produces lysophosphatidic acid. J Biol Chem 277:34254–34263
12. Aoki J (2004) Mechanisms of lysophosphatidic acid production. Semin Cell Dev Biol 15:477–489
13. An S, Bleu T, Hallmark OG, Goetzl EJ (1998) Characterization of a novel subtype of human G protein-coupled receptor for lysophosphatidic acid. J Biol Chem 273:7906–7910
14. Bandoh K, Aoki J, Hosono H, Kobayashi S, Kobayashi T, Murakami-Murofushi K, Tsujimoto M, Arai H, Inoue K (1999) Molecular cloning and characterization of a novel human G-protein-coupled receptor, EDG7, for lysophosphatidic acid. J Biol Chem 274:27776–27785
15. Noguchi K, Ishii S, Shimizu T (2003) Identification of p2y9/GPR23 as a novel G protein-coupled receptor for lysophosphatidic acid, structurally distant from the Edg family. J Biol Chem 278:25600–25606
16. Lee CW, Rivera R, Gardell S, Dubin AE, Chun J (2006) GPR92 as a new G12/13- and Gq-coupled lysophosphatidic acid receptor that increases cAMP, LPA5. J Biol Chem 281:23589–23597
17. Hausmann J, Kamtekar S, Christodoulou E, Day JE, Wu T, Fulkerson Z, Albers HM, van Meeteren LA, Houben AJ, van Zeijl L, Jansen S, Andries M, Hall T, Pegg LE, Benson TE, Kasiem M, Harlos K, Kooi CW, Smyth SS, Ovaa H, Bollen M, Morris AJ, Moolenaar WH, Perrakis A (2011) Structural basis of substrate discrimination and integrin binding by autotaxin. Nat Struct Mol Biol 18:198–204
18. Nishimasu H, Okudaira S, Hama K, Mihara E, Dohmae N, Inoue A, Ishitani R, Takagi J, Aoki J, Nureki O (2011) Crystal structure of autotaxin and insight into GPCR activation by lipid mediators. Nat Struct Mol Biol 18:205–212
19. Albers HM, Hendrickx LJ, van Tol RJ, Hausmann J, Perrakis A, Ovaa H (2011) Structure-based design of novel boronic acid-based inhibitors of autotaxin. J Med Chem 54:4619–4626
20. Kawaguchi M, Okabe T, Okudaira S, Nishimasu H, Ishitani R, Kojima H, Nureki O, Aoki J, Nagano T (2013) Screening and X-ray crystal structure-based optimization of autotaxin (ENPP2) inhibitors, using a newly developed fluorescence probe. ACS Chem Biol 8:1713–1721
21. Stein AJ, Bain G, Prodanovich P, Santini AM, Darlington J, Stelzer NM, Sidhu RS, Schaub J, Goulet L, Lonergan D, Calderon I, Evans JF, Hutchinson JH (2015) Structural basis for inhibition of human autotaxin by four potent compounds with distinct modes of binding. Mol Pharmacol 88:982–992
22. Keune WJ, Hausmann J, Bolier R, Tolenaars D, Kremer A, Heidebrecht T, Joosten RP, Sunkara M, Morris AJ, Matas-Rico E, Moolenaar WH, Oude Elferink RP, Perrakis A (2016) Steroid binding to Autotaxin links bile salts and lysophosphatidic acid signalling. Nat Commun 7:11248
23. Kato K, Ikeda H, Miyakawa S, Futakawa S, Nonaka Y, Fujiwara M, Okudaira S, Kano K, Aoki J, Morita J, Ishitani R, Nishimasu H, Nakamura Y, Nureki O (2016) Structural basis for specific inhibition of Autotaxin by a DNA aptamer. Nat Struct Mol Biol 23:395–401
24. Hausmann J, Keune WJ, Hipgrave Ederveen AL, van Zeijl L, Joosten RP, Perrakis A (2016) Structural snapshots of the catalytic cycle of the phosphodiesterase Autotaxin. J Struct Biol 195:199–206
25. Jones SB, Pfeifer LA, Bleisch TJ, Beauchamp TJ, Durbin JD, Klimkowski VJ, Hughes NE, Rito CJ, Dao Y, Gruber JM, Bui H, Chambers MG, Chandrasekhar S, Lin C, McCann DJ, Mudra DR, Oskins JL, Swearingen CA, Thirunavukkarasu K, Norman BH (2016) Novel autotaxin inhibitors for the treatment of osteoarthritis pain: lead optimization via structure-based drug design. ACS Med Chem Lett 7:857–861
26. Shah P, Cheasty A, Foxton C, Raynham T, Farooq M, Gutierrez IF, Lejeune A, Pritchard M, Turnbull A, Pang L, Owen P, Boyd S, Stowell A, Jordan A, Hamilton NM, Hitchin JR, Stockley M, MacDonald E, Quesada MJ, Trivier E, Skeete J, Ovaa H,

Moolenaar WH, Ryder H (2016) Discovery of potent inhibitors of the lysophospholipase autotaxin. Bioorg Med Chem Lett 26:5403–5410
27. Bain G, Shannon KE, Huang F, Darlington J, Goulet L, Prodanovich P, Ma GL, Santini AM, Stein AJ, Lonergan D, King CD, Calderon I, Lai A, Hutchinson JH, Evans JF (2017) Selective inhibition of autotaxin is efficacious in mouse models of liver fibrosis. J Pharmacol Exp Ther 360:1–13
28. Miller LM, Keune WJ, Castagna D, Young LC, Duffy EL, Potjewyd F, Salgado-Polo F, Engel Garcia P, Semaan D, Pritchard JM, Perrakis A, Macdonald SJ, Jamieson C, Watson AJ (2017) Structure-activity relationships of small molecule autotaxin inhibitors with a discrete binding mode. J Med Chem 60:722–748
29. Keune WJ, Potjewyd F, Heidebrecht T, Salgado-Polo F, Macdonald SJ, Chelvarajan L, Abdel Latif A, Soman S, Morris AJ, Watson AJ, Jamieson C, Perrakis A (2017) Rational design of autotaxin inhibitors by structural evolution of endogenous modulators. J Med Chem 60:2006–2017
30. Desroy N, Housseman C, Bock X, Joncour A, Bienvenu N, Cherel L, Labeguere V, Rondet E, Peixoto C, Grassot JM, Picolet O, Annoot D, Triballeau N, Monjardet A, Wakselman E, Roncoroni V, Le Tallec S, Blanque R, Cottereaux C, Vandervoort N, Christophe T, Mollat P, Lamers M, Auberval M, Hrvacic B, Ralic J, Oste L, van der Aar E, Brys R, Heckmann B (2017) Discovery of 2-[[2-Ethyl-6-[4-[2-(3-hydroxyazetidin-1-yl)-2-oxoethyl]piperazin-1-yl]-8-methyli midazo[1,2-a] pyridin-3-yl]methylamino]-4-(4-fluorophenyl) thiazole-5-carbonitrile (GLPG1690), a first-in-class autotaxin inhibitor undergoing clinical evaluation for the treatment of idiopathic pulmonary fibrosis. J Med Chem 60:3580–3590
31. Joncour A, Desroy N, Housseman C, Bock X, Bienvenu N, Cherel L, Labeguere V, Peixoto C, Annoot D, Lepissier L, Heiermann J, Hengeveld WJ, Pilzak G, Monjardet A, Wakselman E, Roncoroni V, Le Tallec S, Galien R, David C, Vandervoort N, Christophe T, Conrath K, Jans M, Wohlkonig A, Soror S, Steyaert J, Touitou R, Fleury D, Vercheval L, Mollat P, Triballeau N, van der Aar E, Brys R, Heckmann B (2017) Discovery, structure-activity relationship, and binding mode of an imidazo[1,2-a] pyridine series of autotaxin inhibitors. J Med Chem 60:7371–7392
32. Kuttruff CA, Ferrara M, Bretschneider T, Hoerer S, Handschuh S, Nosse B, Romig H, Nicklin P, Roth GJ (2017) Discovery of BI-2545: a novel autotaxin inhibitor that significantly reduces LPA levels in vivo. ACS Med Chem Lett 8:1252–1257
33. Salgado-Polo F, Perrakis A (2019) The structural binding mode of the four autotaxin inhibitor types that differentially affect catalytic and non-catalytic functions. Cancers (Basel) 11:1577
34. Hanson MA, Roth CB, Jo E, Griffith MT, Scott FL, Reinhart G, Desale H, Clemons B, Cahalan SM, Schuerer SC, Sanna MG, Han GW, Kuhn P, Rosen H, Stevens RC (2012) Crystal structure of a lipid G protein-coupled receptor. Science 335:851–855
35. Audet M, Stevens RC (2019) Emerging structural biology of lipid G protein-coupled receptors. Protein Sci 28:292–304
36. Chrencik JE, Roth CB, Terakado M, Kurata H, Omi R, Kihara Y, Warshaviak D, Nakade S, Asmar-Rovira G, Mileni M, Mizuno H, Griffith MT, Rodgers C, Han GW, Velasquez J, Chun J, Stevens RC, Hanson MA (2015) Crystal structure of antagonist bound human lysophosphatidic acid receptor 1. Cell 161:1633–1643
37. Hori T, Okuno T, Hirata K, Yamashita K, Kawano Y, Yamamoto M, Hato M, Nakamura M, Shimizu T, Yokomizo T, Miyano M, Yokoyama S (2018) Na(+)-mimicking ligands stabilize the inactive state of leukotriene B4 receptor BLT1. Nat Chem Biol 14:262–269
38. Audet M, White KL, Breton B, Zarzycka B, Han GW, Lu Y, Gati C, Batyuk A, Popov P, Velasquez J, Manahan D, Hu H, Weierstall U, Liu W, Shui W, Katritch V, Cherezov V, Hanson MA, Stevens RC (2019) Crystal structure of misoprostol bound to the labor inducer prostaglandin E2 receptor. Nat Chem Biol 15:11–17
39. Hua T, Vemuri K, Nikas SP, Laprairie RB, Wu Y, Qu L, Pu M, Korde A, Jiang S, Ho JH, Han GW, Ding K, Li X, Liu H, Hanson MA, Zhao S, Bohn LM, Makriyannis A, Stevens RC, Liu ZJ (2017) Crystal structures of agonist-bound human cannabinoid receptor CB1. Nature 547:468–471
40. Hua T, Vemuri K, Pu M, Qu L, Han GW, Wu Y, Zhao S, Shui W, Li S, Korde A, Laprairie RB, Stahl EL, Ho JH, Zvonok N, Zhou H, Kufareva I, Wu B, Zhao Q, Hanson MA, Bohn LM, Makriyannis A, Stevens RC, Liu ZJ (2016) Crystal Structure of the Human Cannabinoid Receptor CB1. Cell 167:750–762.e714
41. Shao Z, Yin J, Chapman K, Grzemska M, Clark L, Wang J, Rosenbaum DM (2016) High-resolution crystal structure of the human CB1 cannabinoid receptor. Nature 540:602–606
42. Taniguchi R, Inoue A, Sayama M, Uwamizu A, Yamashita K, Hirata K, Yoshida M, Tanaka Y, Kato HE, Nakada-Nakura Y, Otani Y, Nishizawa T, Doi T, Ohwada T, Ishitani R, Aoki J, Nureki O (2017) Structural insights into ligand recognition by the lysophosphatidic acid receptor LPA6. Nature 548:356–360
43. Inoue A, Arima N, Ishiguro J, Prestwich GD, Arai H, Aoki J (2011) LPA-producing enzyme PA-PLA(1) alpha regulates hair follicle development by modulating EGFR signalling. EMBO J 30:4248–4260
44. Yasuda D, Kobayashi D, Akahoshi N, Ohto-Nakanishi T, Yoshioka K, Takuwa Y, Mizuno S, Takahashi S, Ishii S (2019) Lysophosphatidic acid-induced YAP/TAZ activation promotes developmental angiogenesis by repressing Notch ligand Dll4. J Clin Invest 130:4332–4349

45. Hata E, Sasaki N, Takeda A, Tohya K, Umemoto E, Akahoshi N, Ishii S, Bando K, Abe T, Kano K, Aoki J, Hayasaka H, Miyasaka M (2016) Lysophosphatidic acid receptors LPA4 and LPA6 differentially promote lymphocyte transmigration across high endothelial venules in lymph nodes. Int Immunol 28:283–292
46. Tager AM, LaCamera P, Shea BS, Campanella GS, Selman M, Zhao Z, Polosukhin V, Wain J, Karimi-Shah BA, Kim ND, Hart WK, Pardo A, Blackwell TS, Xu Y, Chun J, Luster AD (2008) The lysophosphatidic acid receptor LPA1 links pulmonary fibrosis to lung injury by mediating fibroblast recruitment and vascular leak. Nat Med 14:45–54
47. Pradere JP, Klein J, Gres S, Guigne C, Neau E, Valet P, Calise D, Chun J, Bascands JL, Saulnier-Blache JS, Schanstra JP (2007) LPA1 receptor activation promotes renal interstitial fibrosis. J Am Soc Nephrol 18:3110–3118
48. Funke M, Zhao Z, Xu Y, Chun J, Tager AM (2012) The lysophosphatidic acid receptor LPA1 promotes epithelial cell apoptosis after lung injury. Am J Respir Cell Mol Biol 46:355–364
49. Sakai N, Chun J, Duffield JS, Lagares D, Wada T, Luster AD, Tager AM (2017) Lysophosphatidic acid signaling through its receptor initiates profibrotic epithelial cell fibroblast communication mediated by epithelial cell derived connective tissue growth factor. Kidney Int 91:628–641
50. Sakai N, Chun J, Duffield JS, Wada T, Luster AD, Tager AM (2013) LPA1-induced cytoskeleton reorganization drives fibrosis through CTGF-dependent fibroblast proliferation. FASEB J 27:1830–1846
51. Knipe RS, Probst CK, Lagares D, Franklin A, Spinney JJ, Brazee PL, Grasberger P, Zhang L, Black KE, Sakai N, Shea BS, Liao JK, Medoff BD, Tager AM (2018) The rho kinase isoforms ROCK1 and ROCK2 each contribute to the development of experimental pulmonary fibrosis. Am J Respir Cell Mol Biol 58:471–481
52. Cao P, Aoki Y, Badri L, Walker NM, Manning CM, Lagstein A, Fearon ER, Lama VN (2017) Autocrine lysophosphatidic acid signaling activates beta-catenin and promotes lung allograft fibrosis. J Clin Invest 127:1517–1530
53. Castelino FV, Seiders J, Bain G, Brooks SF, King CD, Swaney JS, Lorrain DS, Chun J, Luster AD, Tager AM (2011) Amelioration of dermal fibrosis by genetic deletion or pharmacologic antagonism of lysophosphatidic acid receptor 1 in a mouse model of scleroderma. Arthritis Rheum 63:1405–1415
54. Swaney JS, Chapman C, Correa LD, Stebbins KJ, Bundey RA, Prodanovich PC, Fagan P, Baccei CS, Santini AM, Hutchinson JH, Seiders TJ, Parr TA, Prasit P, Evans JF, Lorrain DS (2010) A novel, orally active LPA(1) receptor antagonist inhibits lung fibrosis in the mouse bleomycin model. Br J Pharmacol 160:1699–1713
55. Swaney JS, Chapman C, Correa LD, Stebbins KJ, Broadhead AR, Bain G, Santini AM, Darlington J, King CD, Baccei CS, Lee C, Parr TA, Roppe JR, Seiders TJ, Ziff J, Prasit P, Hutchinson JH, Evans JF, Lorrain DS (2011) Pharmacokinetic and pharmacodynamic characterization of an oral lysophosphatidic acid type 1 receptor-selective antagonist. J Pharmacol Exp Ther 336:693–700
56. Gan L, Xue JX, Li X, Liu DS, Ge Y, Ni PY, Deng L, Lu Y, Jiang W (2011) Blockade of lysophosphatidic acid receptors LPAR1/3 ameliorates lung fibrosis induced by irradiation. Biochem Biophys Res Commun 409:7–13
57. Tang N, Zhao Y, Feng R, Liu Y, Wang S, Wei W, Ding Q, An MS, Wen J, Li L (2014) Lysophosphatidic acid accelerates lung fibrosis by inducing differentiation of mesenchymal stem cells into myofibroblasts. J Cell Mol Med 18:156–169
58. Cong C, Mao L, Zhang Y, Zhao Z, Xu X, Zhao J (2014) Regulation of silicosis formation by lysophosphatidic acid and its receptors. Exp Lung Res 40:317–326
59. Ninou I, Kaffe E, Muller S, Budd DC, Stevenson CS, Ullmer C, Aidinis V (2018) Pharmacologic targeting of the ATX/LPA axis attenuates bleomycin-induced pulmonary fibrosis. Pulm Pharmacol Ther 52:32–40
60. Sakai N, Bain G, Furuichi K, Iwata Y, Nakamura M, Hara A, Kitajima S, Sagara A, Miyake T, Toyama T, Sato K, Nakagawa S, Shimizu M, Kaneko S, Wada T (2019) The involvement of autotaxin in renal interstitial fibrosis through regulation of fibroblast functions and induction of vascular leakage. Sci Rep 9:7414
61. Ohashi T, Yamamoto T (2015) Antifibrotic effect of lysophosphatidic acid receptors LPA1 and LPA3 antagonist on experimental murine scleroderma induced by bleomycin. Exp Dermatol 24:698–702
62. Huang LS, Fu P, Patel P, Harijith A, Sun T, Zhao Y, Garcia JG, Chun J, Natarajan V (2013) Lysophosphatidic acid receptor-2 deficiency confers protection against bleomycin-induced lung injury and fibrosis in mice. Am J Respir Cell Mol Biol 49:912–922
63. Oikonomou N, Mouratis MA, Tzouvelekis A, Kaffe E, Valavanis C, Vilaras G, Karameris A, Prestwich GD, Bouros D, Aidinis V (2012) Pulmonary autotaxin expression contributes to the pathogenesis of pulmonary fibrosis. Am J Respir Cell Mol Biol 47:566–574
64. Kaffe E, Katsifa A, Xylourgidis N, Ninou I, Zannikou M, Harokopos V, Foka P, Dimitriadis A, Evangelou K, Moulas AN, Georgopoulou U, Gorgoulis VG, Dalekos GN, Aidinis V (2017) Hepatocyte autotaxin expression promotes liver fibrosis and cancer. Hepatology 65:1369–1383
65. Black KE, Berdyshev E, Bain G, Castelino FV, Shea BS, Probst CK, Fontaine BA, Bronova I, Goulet L, Lagares D, Ahluwalia N, Knipe RS, Natarajan V, Tager AM (2016) Autotaxin activity increases locally following lung injury, but is not required for pulmonary lysophosphatidic acid production or fibrosis. FASEB J 30:2435–2450

66. Nikolaou A, Ninou I, Kokotou MG, Kaffe E, Afantitis A, Aidinis V, Kokotos G (2018) Hydroxamic acids constitute a novel class of autotaxin inhibitors that exhibit in vivo efficacy in a pulmonary fibrosis model. J Med Chem 61:3697–3711
67. Cortinovis M, Aiello S, Mister M, Conde-Knape K, Noris M, Novelli R, Solini S, Rodriguez Ordonez PY, Benigni A, Remuzzi G (2020) Autotaxin inhibitor protects from chronic allograft injury in rat kidney allotransplantation. Nephron 144(1):38–48
68. Castelino FV, Bain G, Pace VA, Black KE, George L, Probst CK, Goulet L, Lafyatis R, Tager AM (2016) An autotaxin/lysophosphatidic acid/interleukin-6 amplification loop drives scleroderma fibrosis. Arthritis Rheumatol 68:2964–2974
69. Higuchi T, Takagi K, Tochimoto A, Ichimura Y, Norose T, Katsumata Y, Masuda I, Yamanaka H, Morohoshi T, Kawaguchi Y (2019) Antifibrotic effects of 2-carba cyclic phosphatidic acid (2ccPA) in systemic sclerosis: contribution to the novel treatment. Arthritis Res Ther 21:103
70. Wu L, Petrigliano FA, Ba K, Lee S, Bogdanov J, McAllister DR, Adams JS, Rosenthal AK, Van Handel B, Crooks GM, Lin Y, Evseenko D (2015) Lysophosphatidic acid mediates fibrosis in injured joints by regulating collagen type I biosynthesis. Osteoarthr Cartil 23:308–318
71. Palmer SM, Snyder L, Todd JL, Soule B, Christian R, Anstrom K, Luo Y, Gagnon R, Rosen G (2018) Randomized, double-blind, placebo-controlled, phase 2 trial of BMS-986020, a lysophosphatidic acid receptor antagonist for the treatment of idiopathic pulmonary fibrosis. Chest 154:1061–1069
72. Gill MW, Lakshmi S, Cheng PTW, Murphy BJ, Chadwick K, Lehman-McKeeman L, Graziano M (2019) BMS-986278, an LPA1 receptor antagonist for idiopathic pulmonary fibrosis: preclinical assessments of potential hepatobiliary toxicity. Am J Respir Crit Care Med 199:A5882
73. Allanore Y, Distler O, Jagerschmidt A, Illiano S, Ledein L, Boitier E, Agueusop I, Denton CP, Khanna D (2018) Lysophosphatidic acid receptor 1 antagonist SAR100842 for patients with diffuse cutaneous systemic sclerosis: a double-blind, randomized, eight-week placebo-controlled study followed by a sixteen-week open-label extension study. Arthritis Rheumatol 70:1634–1643
74. Maher TM, van der Aar EM, Van de Steen O, Allamassey L, Desrivot J, Dupont S, Fagard L, Ford P, Fieuw A, Wuyts W (2018) Safety, tolerability, pharmacokinetics, and pharmacodynamics of GLPG1690, a novel autotaxin inhibitor, to treat idiopathic pulmonary fibrosis (FLORA): a phase 2a randomised placebo-controlled trial. Lancet Respir Med 6:627–635
75. Maher TM, Kreuter M, Lederer DJ, Brown KK, Wuyts W, Verbruggen N, Stutvoet S, Fieuw A, Ford P, Abi-Saab W, Wijsenbeek M (2019) Rationale, design and objectives of two phase III, randomised, placebo-controlled studies of GLPG1690, a novel autotaxin inhibitor, in idiopathic pulmonary fibrosis (ISABELA 1 and 2). BMJ Open Respir Res 6:e000422
76. Lee G, Kang SU, Ryou JH, Lim JJ, Lee YH (2019) Late breaking abstract – BBT-877, a potent autotaxin inhibitor in clinical development to treat idiopathic pulmonary fibrosis. Eur Respir J 54:PA1293
77. Estivill-Torrús G, Llebrez-Zayas P, Matas-Rico E, Santín L, Pedraza C, De Diego I, Del Arco I, Fernández-Llebrez P, Chun J, De Fonseca FR (2008) Absence of LPA1 signaling results in defective cortical development. Cereb Cortex 18:938–950
78. Herr KJ, Herr DR, Lee CW, Noguchi K, Chun J (2011) Stereotyped fetal brain disorganization is induced by hypoxia and requires lysophosphatidic acid receptor 1 (LPA1) signaling. Proc Natl Acad Sci U S A 108:15444–15449
79. Yung YC, Mutoh T, Lin ME, Noguchi K, Rivera RR, Choi JW, Kingsbury MA, Chun J (2011) Lysophosphatidic acid signaling may initiate fetal hydrocephalus. Sci Transl Med 3:99ra87
80. Lummis NC, Sanchez-Pavon P, Kennedy G, Frantz AJ, Kihara Y, Blaho VA, Chun J (2019) LPA1/3 overactivation induces neonatal posthemorrhagic hydrocephalus through ependymal loss and ciliary dysfunction. Sci Adv 5:eaax2011
81. Mirendil H, Thomas EA, De Loera C, Okada K, Inomata Y, Chun J (2015) LPA signaling initiates schizophrenia-like brain and behavioral changes in a mouse model of prenatal brain hemorrhage. Transl Psychiatry 5:e541
82. Walker TL, Overall RW, Vogler S, Sykes AM, Ruhwald S, Lasse D, Ichwan M, Fabel K, Kempermann G (2016) Lysophosphatidic acid receptor is a functional marker of adult hippocampal precursor cells. Stem Cell Reports 6:552–565
83. Ladron de Guevara-Miranda D, Moreno-Fernandez RD, Gil-Rodriguez S, Rosell-Valle C, Estivill-Torrus G, Serrano A, Pavon FJ, Rodriguez de Fonseca F, Santin LJ, Castilla-Ortega E (2019) Lysophosphatidic acid-induced increase in adult hippocampal neurogenesis facilitates the forgetting of cocaine-contextual memory. Addict Biol 24:458–470
84. Matas-Rico E, Garcia-Diaz B, Llebrez-Zayas P, Lopez-Barroso D, Santin L, Pedraza C, Smith-Fernandez A, Fernandez-Llebrez P, Tellez T, Redondo M, Chun J, De Fonseca FR, Estivill-Torrus G (2008) Deletion of lysophosphatidic acid receptor LPA1 reduces neurogenesis in the mouse dentate gyrus. Mol Cell Neurosci 39:342–355
85. Harrison SM, Reavill C, Brown G, Brown JT, Cluderay JE, Crook B, Davies CH, Dawson LA, Grau E, Heidbreder C, Hemmati P, Hervieu G, Howarth A, Hughes ZA, Hunter AJ, Latcham J, Pickering S, Pugh P, Rogers DC, Shilliam CS, Maycox PR (2003) LPA1 receptor-deficient mice have phenotypic changes observed in psychiatric disease. Mol Cell Neurosci 24:1170–1179

86. Santin LJ, Bilbao A, Pedraza C, Matas-Rico E, Lopez-Barroso D, Castilla-Ortega E, Sanchez-Lopez J, Riquelme R, Varela-Nieto I, de la Villa P, Suardiaz M, Chun J, De Fonseca FR, Estivill-Torrus G (2009) Behavioral phenotype of maLPA1-null mice: increased anxiety-like behavior and spatial memory deficits. Genes Brain Behav 8:772–784
87. Castilla-Ortega E, Sanchez-Lopez J, Hoyo-Becerra C, Matas-Rico E, Zambrana-Infantes E, Chun J, De Fonseca FR, Pedraza C, Estivill-Torrus G, Santin LJ (2010) Exploratory, anxiety and spatial memory impairments are dissociated in mice lacking the LPA1 receptor. Neurobiol Learn Mem 94:73–82
88. Castilla-Ortega E, Pedraza C, Chun J, de Fonseca FR, Estivill-Torrus G, Santin LJ (2012) Hippocampal c-Fos activation in normal and LPA(1)-null mice after two object recognition tasks with different memory demands. Behav Brain Res 232:400–405
89. Moreno-Fernandez RD, Perez-Martin M, Castilla-Ortega E, Del Valle CR, Garcia-Fernandez MI, Chun J, Estivill-Torrus G, de Fonseca FR, Santin LJ, Pedraza C (2017) maLPA1-null mice as an endophenotype of anxious depression. Transl Psychiatry 7:e1077
90. Pedraza C, Sanchez-Lopez J, Castilla-Ortega E, Rosell-Valle C, Zambrana-Infantes E, Garcia-Fernandez M, Rodriguez de Fonseca F, Chun J, Santin LJ, Estivill-Torrus G (2014) Fear extinction and acute stress reactivity reveal a role of LPA(1) receptor in regulating emotional-like behaviors. Brain Struct Funct 219:1659–1672
91. Blanco E, Bilbao A, Luque-Rojas MJ, Palomino A, Bermudez-Silva FJ, Suarez J, Santin LJ, Estivill-Torrus G, Gutierrez A, Campos-Sandoval JA, Alonso-Carrion FJ, Marquez J, de Fonseca FR (2012) Attenuation of cocaine-induced conditioned locomotion is associated with altered expression of hippocampal glutamate receptors in mice lacking LPA1 receptors. Psychopharmacology (Berl) 220:27–42
92. Castilla-Ortega E, Pavon FJ, Sanchez-Marin L, Estivill-Torrus G, Pedraza C, Blanco E, Suarez J, Santin L, Rodriguez de Fonseca F, Serrano A (2016) Both genetic deletion and pharmacological blockade of lysophosphatidic acid LPA1 receptor results in increased alcohol consumption. Neuropharmacology 103:92–103
93. Dash PK, Orsi SA, Moody M, Moore AN (2004) A role for hippocampal Rho-ROCK pathway in long-term spatial memory. Biochem Biophys Res Commun 322:893–898
94. Castilla-Ortega E, Escuredo L, Bilbao A, Pedraza C, Orio L, Estivill-Torrus G, Santin LJ, de Fonseca FR, Pavon FJ (2014) 1-Oleoyl lysophosphatidic acid: a new mediator of emotional behavior in rats. PLoS One 9:e85348
95. Yamada M, Tsukagoshi M, Hashimoto T, Oka J, Saitoh A, Yamada M (2015) Lysophosphatidic acid induces anxiety-like behavior via its receptors in mice. J Neural Transm (Vienna) 122:487–494
96. Moreno-Fernandez RD, Nieto-Quero A, Gomez-Salas FJ, Chun J, Estivill-Torrus G, de Fonseca FR, Santin LJ, Perez-Martin M, Pedraza C (2018) Effects of genetic deletion versus pharmacological blockade of the LPA1 receptor on depression-like behaviour and related brain functional activity. Dis Model Mech 11:dmm035519
97. Sanchez-Marin L, Ladron de Guevara-Miranda D, Manas-Padilla MC, Alen F, Moreno-Fernandez RD, Diaz-Navarro C, Perez-Del Palacio J, Garcia-Fernandez M, Pedraza C, Pavon FJ, Rodriguez de Fonseca F, Santin LJ, Serrano A, Castilla-Ortega E (2018) Systemic blockade of LPA1/3 lysophosphatidic acid receptors by ki16425 modulates the effects of ethanol on the brain and behavior. Neuropharmacology 133:189–201
98. Strauss U, Brauer AU (2013) Current views on regulation and function of plasticity-related genes (PRGs/LPPRs) in the brain. Biochim Biophys Acta 1831:133–138
99. Trimbuch T, Beed P, Vogt J, Schuchmann S, Maier N, Kintscher M, Breustedt J, Schuelke M, Streu N, Kieselmann O, Brunk I, Laube G, Strauss U, Battefeld A, Wende H, Birchmeier C, Wiese S, Sendtner M, Kawabe H, Kishimoto-Suga M, Brose N, Baumgart J, Geist B, Aoki J, Savaskan NE, Brauer AU, Chun J, Ninnemann O, Schmitz D, Nitsch R (2009) Synaptic PRG-1 modulates excitatory transmission via lipid phosphate-mediated signaling. Cell 138:1222–1235
100. Thalman C, Horta G, Qiao L, Endle H, Tegeder I, Cheng H, Laube G, Sigurdsson T, Hauser MJ, Tenzer S, Distler U, Aoki J, Morris AJ, Geisslinger G, Roper J, Kirischuk S, Luhmann HJ, Radyushkin K, Nitsch R, Vogt J (2018) Synaptic phospholipids as a new target for cortical hyperexcitability and E/I balance in psychiatric disorders. Mol Psychiatry 23:1699–1710
101. Schneider P, Petzold S, Sommer A, Nitsch R, Schwegler H, Vogt J, Roskoden T (2018) Altered synaptic phospholipid signaling in PRG-1 deficient mice induces exploratory behavior and motor hyperactivity resembling psychiatric disorders. Behav Brain Res 336:1–7
102. Vogt J, Yang JW, Mobascher A, Cheng J, Li Y, Liu X, Baumgart J, Thalman C, Kirischuk S, Unichenko P, Horta G, Radyushkin K, Stroh A, Richers S, Sahragard N, Distler U, Tenzer S, Qiao L, Lieb K, Tuscher O, Binder H, Ferreiros N, Tegeder I, Morris AJ, Gropa S, Nurnberg P, Toliat MR, Winterer G, Luhmann HJ, Huai J, Nitsch R (2016) Molecular cause and functional impact of altered synaptic lipid signaling due to a prg-1 gene SNP. EMBO Mol Med 8:25–38
103. Gross I, Brauer AU (2020) Modulation of lysophosphatidic acid (LPA) receptor activity: the key to successful neural regeneration? Neural Regen Res 15:53–54
104. Callaerts-Vegh Z, Leo S, Vermaercke B, Meert T, D'Hooge R (2012) LPA5 receptor plays a role in

pain sensitivity, emotional exploration and reversal learning. Genes Brain Behav 11:1009–1019
105. Inoue M, Rashid MH, Fujita R, Contos JJ, Chun J, Ueda H (2004) Initiation of neuropathic pain requires lysophosphatidic acid receptor signaling. Nat Med 10:712–718
106. Ma L, Matsumoto M, Xie W, Inoue M, Ueda H (2009) Evidence for lysophosphatidic acid 1 receptor signaling in the early phase of neuropathic pain mechanisms in experiments using Ki-16425, a lysophosphatidic acid 1 receptor antagonist. J Neurochem 109:603–610
107. Ahn DK, Lee SY, Han SR, Ju JS, Yang GY, Lee MK, Youn DH, Bae YC (2009) Intratrigeminal ganglionic injection of LPA causes neuropathic pain-like behavior and demyelination in rats. Pain 146:114–120
108. Xie W, Uchida H, Nagai J, Ueda M, Chun J, Ueda H (2010) Calpain-mediated down-regulation of myelin-associated glycoprotein in lysophosphatidic acid-induced neuropathic pain. J Neurochem 113:1002–1011
109. Inoue M, Ma L, Aoki J, Chun J, Ueda H (2008) Autotaxin, a synthetic enzyme of lysophosphatidic acid (LPA), mediates the induction of nerve-injured neuropathic pain. Mol Pain 4:6
110. Hall SM (1972) The effect of injections of lysophosphatidyl choline into white matter of the adult mouse spinal cord. J Cell Sci 10:535–546
111. Inoue M, Xie W, Matsushita Y, Chun J, Aoki J, Ueda H (2008) Lysophosphatidylcholine induces neuropathic pain through an action of autotaxin to generate lysophosphatidic acid. Neuroscience 152:296–298
112. Ma L, Uchida H, Nagai J, Inoue M, Chun J, Aoki J, Ueda H (2009) Lysophosphatidic acid-3 receptor-mediated feed-forward production of lysophosphatidic acid: an initiator of nerve injury-induced neuropathic pain. Mol Pain 5:64
113. Nagai J, Uchida H, Matsushita Y, Yano R, Ueda M, Niwa M, Aoki J, Chun J, Ueda H (2010) Autotaxin and lysophosphatidic acid1 receptor-mediated demyelination of dorsal root fibers by sciatic nerve injury and intrathecal lysophosphatidylcholine. Mol Pain 6:78
114. Ma L, Nagai J, Ueda H (2010) Microglial activation mediates de novo lysophosphatidic acid production in a model of neuropathic pain. J Neurochem 115:643–653
115. Uchida H, Nagai J, Ueda H (2014) Lysophosphatidic acid and its receptors LPA1 and LPA3 mediate paclitaxel-induced neuropathic pain in mice. Mol Pain 10:71
116. McDougall JJ, Albacete S, Schuelert N, Mitchell PG, Lin C, Oskins JL, Bui HH, Chambers MG (2017) Lysophosphatidic acid provides a missing link between osteoarthritis and joint neuropathic pain. Osteoarthr Cartil 25:926–934
117. Halder SK, Yano R, Chun J, Ueda H (2013) Involvement of LPA1 receptor signaling in cerebral ischemia-induced neuropathic pain. Neuroscience 235:10–15
118. Ueda H, Neyama H, Sasaki K, Miyama C, Iwamoto R (2019) Lysophosphatidic acid LPA1 and LPA3 receptors play roles in the maintenance of late tissue plasminogen activator-induced central poststroke pain in mice. Neurobiol Pain 5:100020
119. Lin ME, Rivera RR, Chun J (2012) Targeted deletion of LPA5 identifies novel roles for lysophosphatidic acid signaling in development of neuropathic pain. J Biol Chem 287:17608–17617
120. Tsukahara R, Yamamoto S, Yoshikawa K, Gotoh M, Tsukahara T, Neyama H, Ishii S, Akahoshi N, Yanagida K, Sumida H, Araki M, Araki K, Yamamura KI, Murakami-Murofushi K, Ueda H (2018) LPA5 signaling is involved in multiple sclerosis-mediated neuropathic pain in the cuprizone mouse model. J Pharmacol Sci 136:93–96
121. Murai N, Hiyama H, Kiso T, Sekizawa T, Watabiki T, Oka H, Aoki T (2017) Analgesic effects of novel lysophosphatidic acid receptor 5 antagonist AS2717638 in rodents. Neuropharmacology 126:97–107
122. Kawamoto Y, Seo R, Murai N, Hiyama H, Oka H (2018) Identification of potent lysophosphatidic acid receptor 5 (LPA5) antagonists as potential analgesic agents. Bioorg Med Chem 26:257–265
123. van Meeteren LA, Ruurs P, Stortelers C, Bouwman P, van Rooijen MA, Pradere JP, Pettit TR, Wakelam MJ, Saulnier-Blache JS, Mummery CL, Moolenaar WH, Jonkers J (2006) Autotaxin, a secreted lysophospholipase D, is essential for blood vessel formation during development. Mol Cell Biol 26:5015–5022
124. Tanaka M, Okudaira S, Kishi Y, Ohkawa R, Iseki S, Ota M, Noji S, Yatomi Y, Aoki J, Arai H (2006) Autotaxin stabilizes blood vessels and is required for embryonic vasculature by producing lysophosphatidic acid. J Biol Chem 281:25822–25830
125. Koike S, Keino-Masu K, Ohto T, Sugiyama F, Takahashi S, Masu M (2009) Autotaxin/lysophospholipase D-mediated lysophosphatidic acid signaling is required to form distinctive large lysosomes in the visceral endoderm cells of the mouse yolk sac. J Biol Chem 284:33561–33570
126. Fotopoulou S, Oikonomou N, Grigorieva E, Nikitopoulou I, Paparountas T, Thanassopoulou A, Zhao Z, Xu Y, Kontoyiannis DL, Remboutsika E, Aidinis V (2010) ATX expression and LPA signalling are vital for the development of the nervous system. Dev Biol 339:451–464
127. Yukiura H, Kano K, Kise R, Inoue A, Aoki J (2015) Autotaxin overexpression causes embryonic lethality and vascular defects. PLoS One 10:e0126734
128. Escalante-Alcalde D, Hernandez L, Le Stunff H, Maeda R, Lee HS, Jr Gang C, Sciorra VA, Daar I, Spiegel S, Morris AJ, Stewart CL (2003) The lipid phosphatase LPP3 regulates extra-embryonic vasculogenesis and axis patterning. Development 130:4623–4637
129. Offermanns S, Mancino V, Revel JP, Simon MI (1997) Vascular system defects and impaired cell

chemokinesis as a result of Galpha13 deficiency. Science 275:533–536
130. Gu JL, Muller S, Mancino V, Offermanns S, Simon MI (2002) Interaction of G alpha(12) with G alpha(13) and G alpha(q) signaling pathways. Proc Natl Acad Sci U S A 99:9352–9357
131. Ruppel KM, Willison D, Kataoka H, Wang A, Zheng YW, Cornelissen I, Yin L, Xu SM, Coughlin SR (2005) Essential role for Galpha13 in endothelial cells during embryonic development. Proc Natl Acad Sci U S A 102:8281–8286
132. Sumida H, Noguchi K, Kihara Y, Abe M, Yanagida K, Hamano F, Sato S, Tamaki K, Morishita Y, Kano MR, Iwata C, Miyazono K, Sakimura K, Shimizu T, Ishii S (2010) LPA4 regulates blood and lymphatic vessel formation during mouse embryogenesis. Blood 116:5060–5070
133. Takara K, Eino D, Ando K, Yasuda D, Naito H, Tsukada Y, Iba T, Wakabayashi T, Muramatsu F, Kidoya H, Fukuhara S, Mochizuki N, Ishii S, Kishima H, Takakura N (2017) Lysophosphatidic acid receptor 4 activation augments drug delivery in tumors by tightening endothelial cell-cell contact. Cell Rep 20:2072–2086
134. Eino D, Tsukada Y, Naito H, Kanemura Y, Iba T, Wakabayashi T, Muramatsu F, Kidoya H, Arita H, Kagawa N, Fujimoto Y, Takara K, Kishima H, Takakura N (2018) LPA4-mediated vascular network formation increases the efficacy of anti-PD-1 therapy against brain tumors. Cancer Res 78:6607–6620
135. Panchatcharam M, Miriyala S, Yang F, Rojas M, End C, Vallant C, Dong A, Lynch K, Chun J, Morris AJ, Smyth SS (2008) Lysophosphatidic acid receptors 1 and 2 play roles in regulation of vascular injury responses but not blood pressure. Circ Res 103:662–670
136. Kano K, Matsumoto H, Inoue A, Yukiura H, Kanai M, Chun J, Ishii S, Shimizu T, Aoki J (2019) Molecular mechanism of lysophosphatidic acid-induced hypertensive response. Sci Rep 9:2662
137. Dohi T, Miyauchi K, Ohkawa R, Nakamura K, Kurano M, Kishimoto T, Yanagisawa N, Ogita M, Miyazaki T, Nishino A, Yaginuma K, Tamura H, Kojima T, Yokoyama K, Kurata T, Shimada K, Daida H, Yatomi Y (2013) Increased lysophosphatidic acid levels in culprit coronary arteries of patients with acute coronary syndrome. Atherosclerosis 229:192–197
138. Kurano M, Suzuki A, Inoue A, Tokuhara Y, Kano K, Matsumoto H, Igarashi K, Ohkawa R, Nakamura K, Dohi T, Miyauchi K, Daida H, Tsukamoto K, Ikeda H, Aoki J, Yatomi Y (2015) Possible involvement of minor lysophospholipids in the increase in plasma lysophosphatidic acid in acute coronary syndrome. Arterioscler Thromb Vasc Biol 35:463–470
139. Yoshida K, Nishida W, Hayashi K, Ohkawa Y, Ogawa A, Aoki J, Arai H, Sobue K (2003) Vascular remodeling induced by naturally occurring unsaturated lysophosphatidic acid in vivo. Circulation 108:1746–1752
140. Zhang C, Baker DL, Yasuda S, Makarova N, Balazs L, Johnson LR, Marathe GK, McIntyre TM, Xu Y, Prestwich GD, Byun HS, Bittman R, Tigyi G (2004) Lysophosphatidic acid induces neointima formation through PPARgamma activation. J Exp Med 199:763–774
141. Subramanian P, Karshovska E, Reinhard P, Megens RT, Zhou Z, Akhtar S, Schumann U, Li X, van Zandvoort M, Ludin C, Weber C, Schober A (2010) Lysophosphatidic acid receptors LPA1 and LPA3 promote CXCL12-mediated smooth muscle progenitor cell recruitment in neointima formation. Circ Res 107:96–105
142. Hao F, Zhang F, Wu DD, An D, Shi J, Li G, Xu X, Cui MZ (2016) Lysophosphatidic acid-induced vascular neointimal formation in mouse carotid arteries is mediated by the matricellular protein CCN1/Cyr61. Am J Physiol Cell Physiol 311:C975–c984
143. Panchatcharam M, Miriyala S, Salous A, Wheeler J, Dong A, Mueller P, Sunkara M, Escalante-Alcalde D, Morris AJ, Smyth SS (2013) Lipid phosphate phosphatase 3 negatively regulates smooth muscle cell phenotypic modulation to limit intimal hyperplasia. Arterioscler Thromb Vasc Biol 33:52–59
144. Cheng Y, Makarova N, Tsukahara R, Guo H, Shuyu E, Farrar P, Balazs L, Zhang C, Tigyi G (2009) Lysophosphatidic acid-induced arterial wall remodeling: requirement of PPARgamma but not LPA1 or LPA2 GPCR. Cell Signal 21:1874–1884
145. Cheng HY, Dong A, Panchatcharam M, Mueller P, Yang F, Li Z, Mills G, Chun J, Morris AJ, Smyth SS (2012) Lysophosphatidic acid signaling protects pulmonary vasculature from hypoxia-induced remodeling. Arterioscler Thromb Vasc Biol 32:24–32
146. Siess W, Zangl KJ, Essler M, Bauer M, Brandl R, Corrinth C, Bittman R, Tigyi G, Aepfelbacher M (1999) Lysophosphatidic acid mediates the rapid activation of platelets and endothelial cells by mildly oxidized low density lipoprotein and accumulates in human atherosclerotic lesions. Proc Natl Acad Sci U S A 96:6931–6936
147. Zhou Z, Subramanian P, Sevilmis G, Globke B, Soehnlein O, Karshovska E, Megens R, Heyll K, Chun J, Saulnier-Blache JS, Reinholz M, van Zandvoort M, Weber C, Schober A (2011) Lipoprotein-derived lysophosphatidic acid promotes atherosclerosis by releasing CXCL1 from the endothelium. Cell Metab 13:592–600
148. Kritikou E, van Puijvelde GH, van der Heijden T, van Santbrink PJ, Swart M, Schaftenaar FH, Kroner MJ, Kuiper J, Bot I (2016) Inhibition of lysophosphatidic acid receptors 1 and 3 attenuates atherosclerosis development in LDL-receptor deficient mice. Sci Rep 6:37585
149. Nsaibia MJ, Boulanger MC, Bouchareb R, Mkannez G, Le Quang K, Hadji F, Argaud D, Dahou A, Bosse Y, Koschinsky ML, Pibarot P, Arsenault BJ, Marette A, Mathieu P (2017) OxLDL-derived lysophosphatidic acid promotes the progression of aortic valve

stenosis through a LPAR1-RhoA-NF-kappaB pathway. Cardiovasc Res 113:1351–1363
150. Yang L, Kraemer M, Fang XF, Angel PM, Drake RR, Morris AJ, Smyth SS (2019) LPA receptor 4 deficiency attenuates experimental atherosclerosis. J Lipid Res 60:972–980
151. Mueller PA, Yang L, Ubele M, Mao G, Brandon J, Vandra J, Nichols TC, Escalante-Alcalde D, Morris AJ, Smyth SS (2019) Coronary artery disease risk-associated Plpp3 gene and its product lipid phosphate phosphatase 3 regulate experimental atherosclerosis. Arterioscler Thromb Vasc Biol 39:2261–2272
152. Schunkert H, Konig IR, Kathiresan S, Reilly MP, Assimes TL, Holm H, Preuss M, Stewart AF, Barbalic M, Gieger C, Absher D, Aherrahrou Z, Allayee H, Altshuler D, Anand SS, Andersen K, Anderson JL, Ardissino D, Ball SG, Balmforth AJ, Barnes TA, Becker DM, Becker LC, Berger K, Bis JC, Boekholdt SM, Boerwinkle E, Braund PS, Brown MJ, Burnett MS, Buysschaert I, Carlquist JF, Chen L, Cichon S, Codd V, Davies RW, Dedoussis G, Dehghan A, Demissie S, Devaney JM, Diemert P, Do R, Doering A, Eifert S, Mokhtari NE, Ellis SG, Elosua R, Engert JC, Epstein SE, de Faire U, Fischer M, Folsom AR, Freyer J, Gigante B, Girelli D, Gretarsdottir S, Gudnason V, Gulcher JR, Halperin E, Hammond N, Hazen SL, Hofman A, Horne BD, Illig T, Iribarren C, Jones GT, Jukema JW, Kaiser MA, Kaplan LM, Kastelein JJ, Khaw KT, Knowles JW, Kolovou G, Kong A, Laaksonen R, Lambrechts D, Leander K, Lettre G, Li M, Lieb W, Loley C, Lotery AJ, Mannucci PM, Maouche S, Martinelli N, McKeown PP, Meisinger C, Meitinger T, Melander O, Merlini PA, Mooser V, Morgan T, Muhleisen TW, Muhlestein JB, Munzel T, Musunuru K, Nahrstaedt J, Nelson CP, Nothen MM, Olivieri O, Patel RS, Patterson CC, Peters A, Peyvandi F, Qu L, Quyyumi AA, Rader DJ, Rallidis LS, Rice C, Rosendaal FR, Rubin D, Salomaa V, Sampietro ML, Sandhu MS, Schadt E, Schafer A, Schillert A, Schreiber S, Schrezenmeir J, Schwartz SM, Siscovick DS, Sivananthan M, Sivapalaratnam S, Smith A, Smith TB, Snoep JD, Soranzo N, Spertus JA, Stark K, Stirrups K, Stoll M, Tang WH, Tennstedt S, Thorgeirsson G, Thorleifsson G, Tomaszewski M, Uitterlinden AG, van Rij AM, Voight BF, Wareham NJ, Wells GA, Wichmann HE, Wild PS, Willenborg C, Witteman JC, Wright BJ, Ye S, Zeller T, Ziegler A, Cambien F, Goodall AH, Cupples LA, Quertermous T, Marz W, Hengstenberg C, Blankenberg S, Ouwehand WH, Hall AS, Deloukas P, Thompson JR, Stefansson K, Roberts R, Thorsteinsdottir U, O'Donnell CJ, McPherson R, Erdmann J, Samani NJ (2011) Large-scale association analysis identifies 13 new susceptibility loci for coronary artery disease. Nat Genet 43:333–338
153. Busnelli M, Manzini S, Hilvo M, Parolini C, Ganzetti GS, Dellera F, Ekroos K, Janis M, Escalante-Alcalde D, Sirtori CR, Laaksonen R, Chiesa G (2017) Liver-specific deletion of the Plpp3 gene alters plasma lipid composition and worsens atherosclerosis in apoE(-/-) mice. Sci Rep 7:44503
154. Panchatcharam M, Salous AK, Brandon J, Miriyala S, Wheeler J, Patil P, Sunkara M, Morris AJ, Escalante-Alcalde D, Smyth SS (2014) Mice with targeted inactivation of ppap2b in endothelial and hematopoietic cells display enhanced vascular inflammation and permeability. Arterioscler Thromb Vasc Biol 34:837–845
155. Chandra M, Escalante-Alcalde D, Bhuiyan MS, Orr AW, Kevil C, Morris AJ, Nam H, Dominic P, McCarthy KJ, Miriyala S, Panchatcharam M (2018) Cardiac-specific inactivation of LPP3 in mice leads to myocardial dysfunction and heart failure. Redox Biol 14:261–271
156. Cai L, Fan G, Wang F, Liu S, Li T, Cong X, Chun J, Chen X (2017) Protective role for LPA3 in cardiac hypertrophy induced by myocardial infarction but not by isoproterenol. Front Physiol 8:356
157. Chen H, Liu S, Liu X, Yang J, Wang F, Cong X, Chen X (2017) Lysophosphatidic acid pretreatment attenuates myocardial ischemia/reperfusion injury in the immature hearts of rats. Front Physiol 8:153
158. Chen X, Yang XY, Wang ND, Ding C, Yang YJ, You ZJ, Su Q, Chen JH (2003) Serum lysophosphatidic acid concentrations measured by dot immunogold filtration assay in patients with acute myocardial infarction. Scand J Clin Lab Invest 63:497–503
159. Funke M, Knudsen L, Lagares D, Ebener S, Probst CK, Fontaine BA, Franklin A, Kellner M, Kuhnel M, Matthieu S, Grothausmann R, Chun J, Roberts JD Jr, Ochs M, Tager AM (2016) Lysophosphatidic acid signaling through the lysophosphatidic acid-1 receptor is required for alveolarization. Am J Respir Cell Mol Biol 55:105–116
160. Chen X, Walther FJ, van Boxtel R, Laghmani EH, Sengers RM, Folkerts G, DeRuiter MC, Cuppen E, Wagenaar GT (2016) Deficiency or inhibition of lysophosphatidic acid receptor 1 protects against hyperoxia-induced lung injury in neonatal rats. Acta Physiol (Oxf) 216:358–375
161. Mouratis MA, Magkrioti C, Oikonomou N, Katsifa A, Prestwich GD, Kaffe E, Aidinis V (2015) Autotaxin and endotoxin-induced acute lung injury. PLoS One 10:e0133619
162. Zhao J, He D, Su Y, Berdyshev E, Chun J, Natarajan V, Zhao Y (2011) Lysophosphatidic acid receptor 1 modulates lipopolysaccharide-induced inflammation in alveolar epithelial cells and murine lungs. Am J Physiol Lung Cell Mol Physiol 301:L547–L556
163. Chen X, Walther FJ, Laghmani EH, Hoogeboom AM, Hogen-Esch AC, van Ark I, Folkerts G, Wagenaar GT (2017) Adult lysophosphatidic acid receptor 1-deficient rats with hyperoxia-induced neonatal chronic lung disease are protected against lipopolysaccharide-induced acute lung injury. Front Physiol 8:155
164. Nowak-Machen M, Lange M, Exley M, Wu S, Usheva A, Robson SC (2015) Lysophosphatidic acid generation by pulmonary NKT cell ENPP-2/auto-

taxin exacerbates hyperoxic lung injury. Purinergic Signal 11:455–461
165. Park GY, Lee YG, Berdyshev E, Nyenhuis S, Du J, Fu P, Gorshkova IA, Li Y, Chung S, Karpurapu M, Deng J, Ranjan R, Xiao L, Jaffe HA, Corbridge SJ, Kelly EA, Jarjour NN, Chun J, Prestwich GD, Kaffe E, Ninou I, Aidinis V, Morris AJ, Smyth SS, Ackerman SJ, Natarajan V, Christman JW (2013) Autotaxin production of lysophosphatidic acid mediates allergic asthmatic inflammation. Am J Respir Crit Care Med 188:928–940
166. Zhao Y, Tong J, He D, Pendyala S, Evgeny B, Chun J, Sperling AI, Natarajan V (2009) Role of lysophosphatidic acid receptor LPA2 in the development of allergic airway inflammation in a murine model of asthma. Respir Res 10:114
167. Emo J, Meednu N, Chapman TJ, Rezaee F, Balys M, Randall T, Rangasamy T, Georas SN (2012) Lpa2 is a negative regulator of both dendritic cell activation and murine models of allergic lung inflammation. J Immunol 188:3784–3790
168. Knowlden SA, Hillman SE, Chapman TJ, Patil R, Miller DD, Tigyi G, Georas SN (2016) Novel inhibitory effect of a lysophosphatidic acid 2 agonist on allergen-driven airway inflammation. Am J Respir Cell Mol Biol 54:402–409
169. Knowlden SA, Capece T, Popovic M, Chapman TJ, Rezaee F, Kim M, Georas SN (2014) Regulation of T cell motility in vitro and in vivo by LPA and LPA2. PLoS One 9:e101655
170. Takeda A, Kobayashi D, Aoi K, Sasaki N, Sugiura Y, Igarashi H, Tohya K, Inoue A, Hata E, Akahoshi N, Hayasaka H, Kikuta J, Scandella E, Ludewig B, Ishii S, Aoki J, Suematsu M, Ishii M, Takeda K, Jalkanen S, Miyasaka M, Umemoto E (2016) Fibroblastic reticular cell-derived lysophosphatidic acid regulates confined intranodal T-cell motility. Elife 5:e10561
171. Jendzjowsky NG, Roy A, Barioni NO, Kelly MM, Green FHY, Wyatt CN, Pye RL, Tenorio-Lopes L, Wilson RJA (2018) Preventing acute asthmatic symptoms by targeting a neuronal mechanism involving carotid body lysophosphatidic acid receptors. Nat Commun 9:4030
172. Lee SJ, Leoni G, Neumann PA, Chun J, Nusrat A, Yun CC (2013) Distinct phospholipase C-beta isozymes mediate lysophosphatidic acid receptor 1 effects on intestinal epithelial homeostasis and wound closure. Mol Cell Biol 33:2016–2028
173. Lin S, Han Y, Jenkin K, Lee SJ, Sasaki M, Klapproth JM, He P, Yun CC (2018) Lysophosphatidic acid receptor 1 is important for intestinal epithelial barrier function and susceptibility to colitis. Am J Pathol 188:353–366
174. Lin S, Haque A, Raeman R, Guo L, He P, Denning TL, El-Rayes B, Moolenaar WH, Yun CC (2019) Autotaxin determines colitis severity in mice and is secreted by B cells in the colon. FASEB J 33:3623–3635
175. Sturm A, Sudermann T, Schulte KM, Goebell H, Dignass AU (1999) Modulation of intestinal epithelial wound healing in vitro and in vivo by lysophosphatidic acid. Gastroenterology 117:368–377
176. Adachi M, Horiuchi G, Ikematsu N, Tanaka T, Terao J, Satouchi K, Tokumura A (2011) Intragastrically administered lysophosphatidic acids protect against gastric ulcer in rats under water-immersion restraint stress. Dig Dis Sci 56:2252–2261
177. Dong YL, Duan XY, Liu YJ, Fan H, Xu M, Chen QY, Nan Z, Wu H, Deng SJ (2019) Autotaxin-lysophosphatidic acid axis blockade improves inflammation by regulating Th17 cell differentiation in DSS-induced chronic colitis mice. Inflammation 42:1530–1541
178. He P, Haque A, Lin S, Cominelli F, Yun CC (2018) Inhibition of autotaxin alleviates inflammation and increases the expression of sodium-dependent glucose cotransporter 1 and Na(+)/H(+) exchanger 3 in SAMP1/Fc mice. Am J Physiol Gastrointest Liver Physiol 315:G762–g771
179. Singh AK, Riederer B, Krabbenhoft A, Rausch B, Bonhagen J, Lehmann U, de Jonge HR, Donowitz M, Yun C, Weinman EJ, Kocher O, Hogema BM, Seidler U (2009) Differential roles of NHERF1, NHERF2, and PDZK1 in regulating CFTR-mediated intestinal anion secretion in mice. J Clin Invest 119:540–550
180. Li C, Dandridge KS, Di A, Marrs KL, Harris EL, Roy K, Jackson JS, Makarova NV, Fujiwara Y, Farrar PL, Nelson DJ, Tigyi GJ, Naren AP (2005) Lysophosphatidic acid inhibits cholera toxin-induced secretory diarrhea through CFTR-dependent protein interactions. J Exp Med 202:975–986
181. Deng W, Shuyu E, Tsukahara R, Valentine WJ, Durgam G, Gududuru V, Balazs L, Manickam V, Arsura M, VanMiddlesworth L, Johnson LR, Parrill AL, Miller DD, Tigyi G (2007) The lysophosphatidic acid type 2 receptor is required for protection against radiation-induced intestinal injury. Gastroenterology 132:1834–1851
182. Patil R, Szabo E, Fells JI, Balogh A, Lim KG, Fujiwara Y, Norman DD, Lee SC, Balazs L, Thomas F, Patil S, Emmons-Thompson K, Boler A, Strobos J, McCool SW, Yates CR, Stabenow J, Byrne GI, Miller DD, Tigyi GJ (2015) Combined mitigation of the gastrointestinal and hematopoietic acute radiation syndromes by an LPA2 receptor-specific nonlipid agonist. Chem Biol 22:206–216
183. Balogh A, Shimizu Y, Lee SC, Norman DD, Gangwar R, Bavaria M, Moon C, Shukla P, Rao R, Ray R, Naren AP, Banerjee S, Miller DD, Balazs L, Pelus L, Tigyi G (2015) The autotaxin-LPA2 GPCR axis is modulated by gamma-irradiation and facilitates DNA damage repair. Cell Signal 27:1751–1762
184. Tigyi GJ, Johnson LR, Lee SC, Norman DD, Szabo E, Balogh A, Thompson K, Boler A, McCool WS (2019) Lysophosphatidic acid type 2 receptor agonists in targeted drug development offer broad therapeutic potential. J Lipid Res 60:464–474
185. He P, Zhao L, Zhu L, Weinman EJ, De Giorgio R, Koval M, Srinivasan S, Yun CC (2015) Restoration

of Na+/H+ exchanger NHE3-containing macrocomplexes ameliorates diabetes-associated fluid loss. J Clin Invest 125:3519–3531
186. Jenkin KA, He P, Yun CC (2018) Expression of lysophosphatidic acid receptor 5 is necessary for the regulation of intestinal Na(+)/H(+) exchanger 3 by lysophosphatidic acid in vivo. Am J Physiol Gastrointest Liver Physiol 315:G433–G442
187. Lin S, Yeruva S, He P, Singh AK, Zhang H, Chen M, Lamprecht G, de Jonge HR, Tse M, Donowitz M, Hogema BM, Chun J, Seidler U, Yun CC (2010) Lysophosphatidic acid stimulates the intestinal brush border Na(+)/H(+) exchanger 3 and fluid absorption via LPA(5) and NHERF2. Gastroenterology 138:649–658
188. Lee BH, Choi SH, Kim HJ, Jung SW, Kim HK, Nah SY (2016) Plant lysophosphatidic acids: a rich source for bioactive lysophosphatidic acids and their pharmacological applications. Biol Pharm Bull 39:156–162
189. Dusaulcy R, Rancoule C, Gres S, Wanecq E, Colom A, Guigne C, van Meeteren LA, Moolenaar WH, Valet P, Saulnier-Blache JS (2011) Adipose-specific disruption of autotaxin enhances nutritional fattening and reduces plasma lysophosphatidic acid. J Lipid Res 52:1247–1255
190. Rancoule C, Dusaulcy R, Treguer K, Gres S, Attane C, Saulnier-Blache JS (2014) Involvement of autotaxin/lysophosphatidic acid signaling in obesity and impaired glucose homeostasis. Biochimie 96:140–143
191. Nishimura S, Nagasaki M, Okudaira S, Aoki J, Ohmori T, Ohkawa R, Nakamura K, Igarashi K, Yamashita H, Eto K, Uno K, Hayashi N, Kadowaki T, Komuro I, Yatomi Y, Nagai R (2014) ENPP2 contributes to adipose tissue expansion and insulin resistance in diet-induced obesity. Diabetes 63:4154–4164
192. Brandon JA, Kraemer M, Vandra J, Halder S, Ubele M, Morris AJ, Smyth SS (2019) Adipose-derived autotaxin regulates inflammation and steatosis associated with diet-induced obesity. PLoS One 14:e0208099
193. D'Souza K, Nzirorera C, Cowie AM, Varghese GP, Trivedi P, Eichmann TO, Biswas D, Touaibia M, Morris AJ, Aidinis V, Kane DA, Pulinilkunnil T, Kienesberger PC (2018) Autotaxin-LPA signaling contributes to obesity-induced insulin resistance in muscle and impairs mitochondrial metabolism. J Lipid Res 59:1805–1817
194. Federico L, Ren H, Mueller PA, Wu T, Liu S, Popovic J, Blalock EM, Sunkara M, Ovaa H, Albers HM, Mills GB, Morris AJ, Smyth SS (2012) Autotaxin and its product lysophosphatidic acid suppress brown adipose differentiation and promote diet-induced obesity in mice. Mol Endocrinol 26:786–797
195. Weng J, Jiang S, Ding L, Xu Y, Zhu X, Jin P (2019) Autotaxin/lysophosphatidic acid signaling mediates obesity-related cardiomyopathy in mice and human subjects. J Cell Mol Med 23:1050–1058
196. Rancoule C, Attane C, Gres S, Fournel A, Dusaulcy R, Bertrand C, Vinel C, Treguer K, Prentki M, Valet P, Saulnier-Blache JS (2013) Lysophosphatidic acid impairs glucose homeostasis and inhibits insulin secretion in high-fat diet obese mice. Diabetologia 56:1394–1402
197. Simon MF, Daviaud D, Pradere JP, Gres S, Guigne C, Wabitsch M, Chun J, Valet P, Saulnier-Blache JS (2005) Lysophosphatidic acid inhibits adipocyte differentiation via lysophosphatidic acid 1 receptor-dependent down-regulation of peroxisome proliferator-activated receptor gamma2. J Biol Chem 280:14656–14662
198. Dusaulcy R, Daviaud D, Pradere JP, Gres S, Valet P, Saulnier-Blache JS (2009) Altered food consumption in mice lacking lysophosphatidic acid receptor-1. J Physiol Biochem 65:345–350
199. Yanagida K, Igarashi H, Yasuda D, Kobayashi D, Ohto-Nakanishi T, Akahoshi N, Sekiba A, Toyoda T, Ishijima T, Nakai Y, Shojima N, Kubota N, Abe K, Kadowaki T, Ishii S, Shimizu T (2018) The Galpha12/13-coupled receptor LPA4 limits proper adipose tissue expansion and remodeling in diet-induced obesity. JCI Insight 3:e97293
200. Ye X, Hama K, Contos JJ, Anliker B, Inoue A, Skinner MK, Suzuki H, Amano T, Kennedy G, Arai H, Aoki J, Chun J (2005) LPA3-mediated lysophosphatidic acid signalling in embryo implantation and spacing. Nature 435:104–108
201. Sugimoto Y, Inazumi T, Tsuchiya S (2015) Roles of prostaglandin receptors in female reproduction. J Biochem 157:73–80
202. Hama K, Aoki J, Inoue A, Endo T, Amano T, Motoki R, Kanai M, Ye X, Chun J, Matsuki N, Suzuki H, Shibasaki M, Arai H (2007) Embryo spacing and implantation timing are differentially regulated by LPA3-mediated lysophosphatidic acid signaling in mice. Biol Reprod 77:954–959
203. Aikawa S, Kano K, Inoue A, Wang J, Saigusa D, Nagamatsu T, Hirota Y, Fujii T, Tsuchiya S, Taketomi Y, Sugimoto Y, Murakami M, Arita M, Kurano M, Ikeda H, Yatomi Y, Chun J, Aoki J (2017) Autotaxin-lysophosphatidic acid-LPA3 signaling at the embryo-epithelial boundary controls decidualization pathways. EMBO J 36:2146–2160
204. Ye X, Skinner MK, Kennedy G, Chun J (2008) Age-dependent loss of sperm production in mice via impaired lysophosphatidic acid signaling. Biol Reprod 79:328–336
205. Contos JJ, Fukushima N, Weiner JA, Kaushal D, Chun J (2000) Requirement for the lpA1 lysophosphatidic acid receptor gene in normal suckling behavior. Proc Natl Acad Sci U S A 97:13384–13389
206. Gennero I, Laurencin-Dalicieux S, Conte-Auriol F, Briand-Mesange F, Laurencin D, Rue J, Beton N, Malet N, Mus M, Tokumura A, Bourin P, Vico L, Brunel G, Oreffo RO, Chun J, Salles JP (2011) Absence of the lysophosphatidic acid receptor LPA1

results in abnormal bone development and decreased bone mass. Bone 49:395–403

207. Nishioka T, Arima N, Kano K, Hama K, Itai E, Yukiura H, Kise R, Inoue A, Kim SH, Solnica-Krezel L, Moolenaar WH, Chun J, Aoki J (2016) ATX-LPA1 axis contributes to proliferation of chondrocytes by regulating fibronectin assembly leading to proper cartilage formation. Sci Rep 6:23433
208. David M, Machuca-Gayet I, Kikuta J, Ottewell P, Mima F, Leblanc R, Bonnelye E, Ribeiro J, Holen I, Lopez Vales R, Jurdic P, Chun J, Clezardin P, Ishii M, Peyruchaud O (2014) Lysophosphatidic acid receptor type 1 (LPA1) plays a functional role in osteoclast differentiation and bone resorption activity. J Biol Chem 289:6551–6564
209. Liu YB, Kharode Y, Bodine PV, Yaworsky PJ, Robinson JA, Billiard J (2010) LPA induces osteoblast differentiation through interplay of two receptors: LPA1 and LPA4. J Cell Biochem 109:794–800
210. Cong Q, Xu R, Yang Y (2019) Galphas signaling in skeletal development, homeostasis and diseases. Curr Top Dev Biol 133:281–307
211. Piazza GA, Ritter JL, Baracka CA (1995) Lysophosphatidic acid induction of transforming growth factors alpha and beta: modulation of proliferation and differentiation in cultured human keratinocytes and mouse skin. Exp Cell Res 216:51–64
212. Demoyer JS, Skalak TC, Durieux ME (2000) Lysophosphatidic acid enhances healing of acute cutaneous wounds in the mouse. Wound Repair Regen 8:530–537
213. Balazs L, Okolicany J, Ferrebee M, Tolley B, Tigyi G (2001) Topical application of the phospholipid growth factor lysophosphatidic acid promotes wound healing in vivo. Am J Physiol Regul Integr Comp Physiol 280:R466–R472
214. Sumitomo A, Siriwach R, Thumkeo D, Ito K, Nakagawa R, Tanaka N, Tanabe K, Watanabe A, Kishibe M, Ishida-Yamamoto A, Honda T, Kabashima K, Aoki J, Narumiya S (2019) LPA induces keratinocyte differentiation and promotes skin barrier function through the LPAR1/LPAR5-RHO-ROCK-SRF axis. J Invest Dermatol 139:1010–1022
215. Yahagi S, Koike M, Okano Y, Masaki H (2011) Lysophospholipids improve skin moisturization by modulating of calcium-dependent cell differentiation pathway. Int J Cosmet Sci 33:251–256
216. Hashimoto T, Ohata H, Momose K (2004) Itch-scratch responses induced by lysophosphatidic acid in mice. Pharmacology 72:51–56
217. Hashimoto T, Ohata H, Momose K, Honda K (2005) Lysophosphatidic acid induces histamine release from mast cells and skin fragments. Pharmacology 75:13–20
218. Rhim JH, Jang IS, Kwon ST, Song KY, Yeo EJ, Park SC (2010) Activation of wound healing in aged rats by altering the cellular mitogenic potential. J Gerontol A Biol Sci Med Sci 65:704–711
219. Kremer AE, Martens JJ, Kulik W, Rueff F, Kuiper EM, van Buuren HR, van Erpecum KJ, Kondrackiene J, Prieto J, Rust C, Geenes VL, Williamson C, Moolenaar WH, Beuers U, Oude Elferink RP (2010) Lysophosphatidic acid is a potential mediator of cholestatic pruritus. Gastroenterology 139:1008–1018. 1018.e1001
220. Shimizu Y, Morikawa Y, Okudaira S, Kimoto S, Tanaka T, Aoki J, Tokumura A (2014) Potentials of the circulating pruritogenic mediator lysophosphatidic acid in development of allergic skin inflammation in mice: role of blood cell-associated lysophospholipase D activity of autotaxin. Am J Pathol 184:1593–1603
221. Kittaka H, Uchida K, Fukuta N, Tominaga M (2017) Lysophosphatidic acid-induced itch is mediated by signalling of LPA5 receptor, phospholipase D and TRPA1/TRPV1. J Physiol 595:2681–2698
222. Yamanoi Y, Kittaka H, Tominaga M (2019) Cheek injection model for simultaneous measurement of pain and itch-related behaviors. J Vis Exp 151:e58943
223. Lee BH, Kim HK, Jang M, Kim HJ, Choi SH, Hwang SH, Kim HC, Rhim H, Cho IH, Nah SY (2017) Effects of Gintonin-Enriched Fraction in an Atopic Dermatitis Animal Model: Involvement of Autotaxin Regulation. Biol Pharm Bull 40:1063–1070
224. Nakao M, Sugaya M, Suga H, Kawaguchi M, Morimura S, Kai H, Ohmatsu H, Fujita H, Asano Y, Tada Y, Kadono T, Sato S (2014) Serum autotaxin levels correlate with pruritus in patients with atopic dermatitis. J Invest Dermatol 134:1745–1747
225. Shimizu Y, Murao K, Tanaka T, Kubo Y, Tokumura A (2014) Increased lysophospholipase D activity of autotaxin in sera of patients with atopic dermatitis. J Dermatol Sci 74:162–165
226. Kremer AE, van Dijk R, Leckie P, Schaap FG, Kuiper EM, Mettang T, Reiners KS, Raap U, van Buuren HR, van Erpecum KJ, Davies NA, Rust C, Engert A, Jalan R, Oude Elferink RP, Beuers U (2012) Serum autotaxin is increased in pruritus of cholestasis, but not of other origin, and responds to therapeutic interventions. Hepatology 56:1391–1400
227. Kazantseva A, Goltsov A, Zinchenko R, Grigorenko AP, Abrukova AV, Moliaka YK, Kirillov AG, Guo Z, Lyle S, Ginter EK, Rogaev EI (2006) Human hair growth deficiency is linked to a genetic defect in the phospholipase gene LIPH. Science 314:982–985
228. Ali G, Chishti MS, Raza SI, John P, Ahmad W (2007) A mutation in the lipase H (LIPH) gene underlie autosomal recessive hypotrichosis. Hum Genet 121:319–325
229. Jelani M, Wasif N, Ali G, Chishti M, Ahmad W (2008) A novel deletion mutation in LIPH gene causes autosomal recessive hypotrichosis (LAH2). Clin Genet 74:184–188
230. Nahum S, Pasternack SM, Pforr J, Indelman M, Wollnik B, Bergman R, Nothen MM, Konig A, Khamaysi Z, Betz RC, Sprecher E (2009) A large duplication in LIPH underlies autosomal recessive

hypotrichosis simplex in four Middle Eastern families. Arch Dermatol Res 301:391–393

231. Shimomura Y, Wajid M, Petukhova L, Shapiro L, Christiano AM (2009) Mutations in the lipase H gene underlie autosomal recessive woolly hair/hypotrichosis. J Invest Dermatol 129:622–628
232. Shinkuma S, Akiyama M, Inoue A, Aoki J, Natsuga K, Nomura T, Arita K, Abe R, Ito K, Nakamura H, Ujiie H, Shibaki A, Suga H, Tsunemi Y, Nishie W, Shimizu H (2010) Prevalent LIPH founder mutations lead to loss of P2Y5 activation ability of PA-PLA1alpha in autosomal recessive hypotrichosis. Hum Mutat 31:602–610
233. Yoshimasu T, Kanazawa N, Kambe N, Nakamura M, Furukawa F (2011) Identification of 736T>A mutation of lipase H in Japanese siblings with autosomal recessive woolly hair. J Dermatol 38:900–904
234. Tanahashi K, Sugiura K, Takeichi T, Takama H, Shinkuma S, Shimizu H, Akiyama M (2013) Prevalent founder mutation c.736T>A of LIPH in autosomal recessive woolly hair of Japanese leads to variable severity of hypotrichosis in adulthood. J Eur Acad Dermatol Venereol 27:1182–1184
235. Shinkuma S, Inoue A, Aoki J, Nishie W, Natsuga K, Ujiie H, Nomura T, Abe R, Akiyama M, Shimizu H (2012) The beta9 loop domain of PA-PLA1alpha has a crucial role in autosomal recessive woolly hair/hypotrichosis. J Invest Dermatol 132:2093–2095
236. Tariq M, Azhar A, Baig SM, Dahl N, Klar J (2012) A novel mutation in the Lipase H gene underlies autosomal recessive hypotrichosis and woolly hair. Sci Rep 2:730
237. Harada K, Inozume T, Kawamura T, Shibagaki N, Kinoshita T, Deguchi N, Shimada S (2013) Two cases of autosomal recessive woolly hair with LIPH gene mutations. Int J Dermatol 52:572–574
238. Yoshizawa M, Nakamura M, Farooq M, Inoue A, Aoki J, Shimomura Y (2013) A novel mutation, c.699C>G (p.C233W), in the LIPH gene leads to a loss of the hydrolytic activity and the LPA6 activation ability of PA-PLA1alpha in autosomal recessive wooly hair/hypotrichosis. J Dermatol Sci 72:61–64
239. Matsuno N, Kunisada M, Kanki H, Simomura Y, Nishigori C (2013) A case of autosomal recessive woolly hair/hypotrichosis with alternation in severity: deterioration and improvement with age. Case Rep Dermatol 5:363–367
240. Tanahashi K, Sugiura K, Kono M, Takama H, Hamajima N, Akiyama M (2014) Highly prevalent LIPH founder mutations causing autosomal recessive woolly hair/hypotrichosis in Japan and the genotype/phenotype correlations. PLoS One 9:e89261
241. Mehmood S, Jan A, Muhammad D, Ahmad F, Mir H, Younus M, Ali G, Ayub M, Ansar M, Ahmad W (2015) Mutations in the lipase-H gene causing autosomal recessive hypotrichosis and woolly hair. Australas J Dermatol 56:e66–e70
242. Hamada K, Kubo R, Nishio D, Nakamura M (2014) A homozygous mutation, c.736T>A (p.C246S), in LIPH gene in a patient manifesting woolly hair, hypotrichosis, hearing difficulty, cleft palate and amblyopia. Eur J Dermatol 24:272–273
243. Hayashi R, Akasaka T, Ito M, Shimomura Y (2014) Compound heterozygous mutations in two distinct catalytic residues of the LIPH gene underlie autosomal recessive woolly hair in a Japanese family. J Dermatol 41:937–938
244. Hayashi R, Inui S, Farooq M, Ito M, Shimomura Y (2014) Expression studies of a novel splice site mutation in the LIPH gene identified in a Japanese patient with autosomal recessive woolly hair. J Dermatol 41:890–894
245. Hayashi R, Inoue A, Suga Y, Aoki J, Shimomura Y (2015) Analysis of unique mutations in the LPAR6 gene identified in a Japanese family with autosomal recessive woolly hair/hypotrichosis: Establishment of a useful assay system for LPA6. J Dermatol Sci 78:197–205
246. Ito T, Shimomura Y, Hayashi R, Tokura Y (2015) Identification of a novel mutation, c.686delAins18 (p.Asp229Glyfs*22), in the LIPH gene as a compound heterozygote with c.736T>A (p.Cys246Ser) in autosomal recessive woolly hair/hypotrichosis. J Dermatol 42:752–753
247. Sleiman MB, Sleiman MB, Abbas O, Btadini W, Najjar T, Tofaili M, Chedraoui A, Khalil S, Kibbi AG, Kurban M (2015) Novel mutation in LIPH in a Lebanese patient with autosomal recessive woolly hair/hypotrichosis. J Dermatol 42:822–824
248. Mehmood S, Shah SH, Jan A, Younus M, Ahmad F, Ayub M, Ahmad W (2016) Frameshift sequence variants in the human lipase-H gene causing hypotrichosis. Pediatr Dermatol 33:e40–e42
249. Kinoshita-Ise M, Kubo A, Sasaki T, Umegaki-Arao N, Amagai M, Ohyama M (2017) Identification of factors contributing to phenotypic divergence via quantitative image analyses of autosomal recessive woolly hair/hypotrichosis with homozygous c.736T>A LIPH mutation. Br J Dermatol 176:138–144
250. Takeichi T, Tanahashi K, Taki T, Kono M, Sugiura K, Akiyama M (2017) Mutational analysis of 29 patients with autosomal-recessive woolly hair and hypotrichosis: LIPH mutations are extremely predominant in autosomal-recessive woolly hair and hypotrichosis in Japan. Br J Dermatol 177:290–292
251. Itoh E, Nakahara T, Furumura M, Furue M, Shimomura Y (2017) Case of autosomal recessive woolly hair/hypotrichosis with atopic dermatitis. J Dermatol 44:1185–1186
252. Matsuo Y, Tanaka A, Shimomura Y, Hide M (2016) Novel splice site mutation in LIPH identified in a Japanese patient with autosomal recessive woolly hair. J Dermatol 43:1384–1385
253. Ahmad F, Sharif S, Furqan Ubaid M, Shah K, Khan MN, Umair M, Azeem Z, Ahmad W (2018) Novel sequence variants in the LIPH and LPAR6 genes underlies autosomal recessive woolly hair/hypotrichosis in consanguineous families. Congenit Anom (Kyoto) 58:24–28

254. Chang XD, Gu YJ, Dai S, Chen XR, Zhang CL, Zhao HS, Song QH (2017) Novel mutations in the lipase H gene lead to secretion defects of LIPH in Chinese patients with autosomal recessive woolly hair/hypotrichosis (ARWH/HT). Mutagenesis 32:599–606
255. Mizukami Y, Hayashi R, Tsuruta D, Shimomura Y, Sugawara K (2018) Novel splice site mutation in the LIPH gene in a patient with autosomal recessive woolly hair/hypotrichosis: case report and published work review. J Dermatol 45:613–617
256. Asano N, Okita T, Yasuno S, Yamaguchi M, Kashiwagi K, Kanekura T, Shimomura Y (2019) Identification of a novel splice site mutation in the LIPH gene in a Japanese family with autosomal recessive woolly hair. J Dermatol 46:e19–e20
257. Pasternack SM, von Kugelgen I, Al Aboud K, Lee YA, Ruschendorf F, Voss K, Hillmer AM, Molderings GJ, Franz T, Ramirez A, Nurnberg P, Nothen MM, Betz RC (2008) G protein-coupled receptor P2Y5 and its ligand LPA are involved in maintenance of human hair growth. Nat Genet 40:329–334
258. Shimomura Y, Wajid M, Ishii Y, Shapiro L, Petukhova L, Gordon D, Christiano AM (2008) Disruption of P2RY5, an orphan G protein-coupled receptor, underlies autosomal recessive woolly hair. Nat Genet 40:335–339
259. Azeem Z, Jelani M, Naz G, Tariq M, Wasif N, Kamran-Ul-Hassan Naqvi S, Ayub M, Yasinzai M, Amin-Ud-Din M, Wali A, Ali G, Chishti MS, Ahmad W (2008) Novel mutations in G protein-coupled receptor gene (P2RY5) in families with autosomal recessive hypotrichosis (LAH3). Hum Genet 123:515–519
260. Petukhova L, Sousa EC Jr, Martinez-Mir A, Vitebsky A, Dos Santos LG, Shapiro L, Haynes C, Gordon D, Shimomura Y, Christiano AM (2008) Genome-wide linkage analysis of an autosomal recessive hypotrichosis identifies a novel P2RY5 mutation. Genomics 92:273–278
261. Shimomura Y, Garzon MC, Kristal L, Shapiro L, Christiano AM (2009) Autosomal recessive woolly hair with hypotrichosis caused by a novel homozygous mutation in the P2RY5 gene. Exp Dermatol 18:218–221
262. Tariq M, Ayub M, Jelani M, Basit S, Naz G, Wasif N, Raza SI, Naveed AK, Khan SU, Azeem Z, Yasinzai M, Wali A, Ali G, Chishti MS, Ahmad W (2009) Mutations in the P2RY5 gene underlie autosomal recessive hypotrichosis in 13 Pakistani families. Br J Dermatol 160:1006–1010
263. Pasternack SM, Murugusundram S, Eigelshoven S, Muller M, Kruse R, Lehmann P, Betz RC (2009) Novel mutations in the P2RY5 gene in one Turkish and two Indian patients presenting with hypotrichosis and woolly hair. Arch Dermatol Res 301:621–624
264. Kurban M, Ghosn S, Abbas O, Shimomura Y, Christiano A (2010) A missense mutation in the P2RY5 gene leading to autosomal recessive woolly hair in a Syrian patient. J Dermatol Sci 57:132–134
265. Horev L, Saad-Edin B, Ingber A, Zlotogorski A (2010) A novel deletion mutation in P2RY5/LPA(6) gene cause autosomal recessive woolly hair with hypotrichosis. J Eur Acad Dermatol Venereol 24:858–859
266. Nahum S, Morice-Picard F, Taieb A, Sprecher E (2011) A novel mutation in LPAR6 causes autosomal recessive hypotrichosis of the scalp. Clin Exp Dermatol 36:188–194
267. Khan S, Habib R, Mir H, Umme K, Naz G, Ayub M, Shafique S, Yamin T, Ali N, Basit S, Wasif N, Naqvi SK-U-H, Ali G, Wali A, Ansar M, Ahmad W (2011) Mutations in the LPAR6 and LIPH genes underlie autosomal recessive hypotrichosis/woolly hair in 17 consanguineous families from Pakistan. Clin Exp Dermatol 36:652–654
268. Kurban M, Wajid M, Shimomura Y, Christiano AM (2013) Mutations in LPAR6/P2RY5 and LIPH are associated with woolly hair and/or hypotrichosis. J Eur Acad Dermatol Venereol 27:545–549
269. Mahmoudi H, Tug E, Parlak AH, Atasoy HI, Ludwig M, Polat M, Pasternack SM, Betz RC (2012) Identification of an Alu-mediated 12.2-kb deletion of the complete LPAR6 (P2RY5) gene in a Turkish family with hypotrichosis and woolly hair. Exp Dermatol 21:469–471
270. Liu LH, Chen G, Wang JW, Liu SX, Wang JB, Zhou FS, Zhu J, Sun LD, Gao M, Wang PG, Yang S, Zhang XJ (2013) A novel deletion mutation in the LPAR6 gene underlies autosomal recessive woolly hair with hypotrichosis. Clin Exp Dermatol 38:796–798
271. Khan GM, Hassan N, Khan N, Humayun M, Khan K, Khaliq S, Rehman FU, Ahmed S, Shah K, Khan SA, Muhammad N, Wali A, Khan S, Basit S, Ayub M (2019) Biallelic mutations in the LPAR6 gene causing autosomal recessive wooly hair/hypotrichosis phenotype in five Pakistani families. Int J Dermatol 58:946–952
272. Diribarne M, Mata X, Chantry-Darmon C, Vaiman A, Auvinet G, Bouet S, Deretz S, Cribiu EP, de Rochambeau H, Allain D, Guerin G (2011) A deletion in exon 9 of the LIPH gene is responsible for the rex hair coat phenotype in rabbits (Oryctolagus cuniculus). PLoS One 6:e19281
273. Gandolfi B, Alhaddad H, Affolter VK, Brockman J, Haggstrom J, Joslin SE, Koehne AL, Mullikin JC, Outerbridge CA, Warren WC, Lyons LA (2013) To the root of the curl: a signature of a recent selective sweep identifies a mutation that defines the Cornish Rex cat breed. PLoS One 8:e67105
274. Grisanti L, Rezza A, Clavel C, Sennett R, Rendl M (2013) Enpp2/Autotaxin in dermal papilla precursors is dispensable for hair follicle morphogenesis. J Invest Dermatol 133:2332–2339
275. Lin KH, Ho YH, Chiang JC, Li MW, Lin SH, Chen WM, Chiang CL, Lin YN, Yang YJ, Chen CN, Lu J, Huang CJ, Tigyi G, Yao CL, Lee H (2016) Pharmacological activation of lysophosphatidic acid receptors regulates erythropoiesis. Sci Rep 6:27050

276. Deng W, Kimura Y, Gududuru V, Wu W, Balogh A, Szabo E, Thompson KE, Yates CR, Balazs L, Johnson LR, Miller DD, Strobos J, McCool WS, Tigyi GJ (2015) Mitigation of the hematopoietic and gastrointestinal acute radiation syndrome by octadecenyl thiophosphate, a small molecule mimic of lysophosphatidic acid. Radiat Res 183:465–475
277. Igarashi H, Akahoshi N, Ohto-Nakanishi T, Yasuda D, Ishii S (2015) The lysophosphatidic acid receptor LPA4 regulates hematopoiesis-supporting activity of bone marrow stromal cells. Sci Rep 5:11410
278. Kanda H, Newton R, Klein R, Morita Y, Gunn MD, Rosen SD (2008) Autotaxin, an ectoenzyme that produces lysophosphatidic acid, promotes the entry of lymphocytes into secondary lymphoid organs. Nat Immunol 9:415–423
279. Nakasaki T, Tanaka T, Okudaira S, Hirosawa M, Umemoto E, Otani K, Jin S, Bai Z, Hayasaka H, Fukui Y, Aozasa K, Fujita N, Tsuruo T, Ozono K, Aoki J, Miyasaka M (2008) Involvement of the lysophosphatidic acid-generating enzyme autotaxin in lymphocyte-endothelial cell interactions. Am J Pathol 173:1566–1576
280. Bai Z, Cai L, Umemoto E, Takeda A, Tohya K, Komai Y, Veeraveedu PT, Hata E, Sugiura Y, Kubo A, Suematsu M, Hayasaka H, Okudaira S, Aoki J, Tanaka T, Albers HM, Ovaa H, Miyasaka M (2013) Constitutive lymphocyte transmigration across the basal lamina of high endothelial venules is regulated by the autotaxin/lysophosphatidic acid axis. J Immunol 190:2036–2048
281. Katakai T, Kondo N, Ueda Y, Kinashi T (2014) Autotaxin produced by stromal cells promotes LFA-1-independent and Rho-dependent interstitial T cell motility in the lymph node paracortex. J Immunol 193:617–626
282. Katakai T, Kinashi T (2016) Microenvironmental control of high-speed interstitial T cell migration in the lymph node. Front Immunol 7:194
283. Schmitz K, Brunkhorst R, de Bruin N, Mayer CA, Haussler A, Ferreiros N, Schiffmann S, Parnham MJ, Tunaru S, Chun J, Offermanns S, Foerch C, Scholich K, Vogt J, Wicker S, Lotsch J, Geisslinger G, Tegeder I (2017) Dysregulation of lysophosphatidic acids in multiple sclerosis and autoimmune encephalomyelitis. Acta Neuropathol Commun 5:42
284. Oda SK, Strauch P, Fujiwara Y, Al-Shami A, Oravecz T, Tigyi G, Pelanda R, Torres RM (2013) Lysophosphatidic acid inhibits CD8 T cell activation and control of tumor progression. Cancer Immunol Res 1:245–255
285. Mathew D, Kremer KN, Strauch P, Tigyi G, Pelanda R, Torres RM (2019) LPA5 is an inhibitory receptor that suppresses CD8 T-Cell cytotoxic function via disruption of early TCR signaling. Front Immunol 10:1159
286. Hu J, Oda SK, Shotts K, Donovan EE, Strauch P, Pujanauski LM, Victorino F, Al-Shami A, Fujiwara Y, Tigyi G, Oravecz T, Pelanda R, Torres RM (2014) Lysophosphatidic acid receptor 5 inhibits B cell antigen receptor signaling and antibody response. J Immunol 193:85–95
287. Martin TW, Lagunoff D (1979) Interactions of lysophospholipids and mast cells. Nature 279:250–252
288. Chang HW, Inoue K, Bruni A, Boarato E, Toffano G (1988) Stereoselective effects of lysophosphatidylserine in rodents. Br J Pharmacol 93:647–653
289. Sugo T, Tachimoto H, Chikatsu T, Murakami Y, Kikukawa Y, Sato S, Kikuchi K, Nagi T, Harada M, Ogi K, Ebisawa M, Mori M (2006) Identification of a lysophosphatidylserine receptor on mast cells. Biochem Biophys Res Commun 341:1078–1087
290. Liebscher I, Muller U, Teupser D, Engemaier E, Engel KM, Ritscher L, Thor D, Sangkuhl K, Ricken A, Wurm A, Piehler D, Schmutzler S, Fuhrmann H, Albert FW, Reichenbach A, Thiery J, Schoneberg T, Schulz A (2011) Altered immune response in mice deficient for the G protein-coupled receptor GPR34. J Biol Chem 286:2101–2110
291. Inoue A, Ishiguro J, Kitamura H, Arima N, Okutani M, Shuto A, Higashiyama S, Ohwada T, Arai H, Makide K, Aoki J (2012) TGFalpha shedding assay: an accurate and versatile method for detecting GPCR activation. Nat Methods 9:1021–1029
292. Uwamizu A, Inoue A, Suzuki K, Okudaira M, Shuto A, Shinjo Y, Ishiguro J, Makide K, Ikubo M, Nakamura S, Jung S, Sayama M, Otani Y, Ohwada T, Aoki J (2015) Lysophosphatidylserine analogues differentially activate three LysoPS receptors. J Biochem 157:151–160
293. Kishi T, Kawana H, Sayama M, Makide K, Inoue A, Otani Y, Ohwada T, Aoki J (2016) Identification of lysophosphatidylthreonine with an aromatic fatty acid surrogate as a potent inducer of mast cell degranulation. Biochem Biophys Rep 8:346–351
294. Sato T, Aoki J, Nagai Y, Dohmae N, Takio K, Doi T, Arai H, Inoue K (1997) Serine phospholipid-specific phospholipase A that is secreted from activated platelets. A new member of the lipase family. J Biol Chem 272:2192–2198
295. Kurano M, Dohi T, Nojiri T, Kobayashi T, Hirowatari Y, Inoue A, Kano K, Matsumoto H, Igarashi K, Nishikawa M, Miyauchi K, Daida H, Ikeda H, Aoki J, Yatomi Y (2015) Blood levels of serotonin are specifically correlated with plasma lysophosphatidylserine among the glycero-lysophospholipids. BBA Clin 4:92–98
296. Emoto S, Kurano M, Kano K, Matsusaki K, Yamashita H, Nishikawa M, Igarashi K, Ikeda H, Aoki J, Kitayama J, Yatomi Y (2017) Analysis of glycero-lysophospholipids in gastric cancerous ascites. J Lipid Res 58:763–771
297. Hosono H, Aoki J, Nagai Y, Bandoh K, Ishida M, Taguchi R, Arai H, Inoue K (2001) Phosphatidylserine-specific phospholipase A1 stimulates histamine release from rat peritoneal mast cells through production of 2-acyl-1-lysophosphatidylserine. J Biol Chem 276:29664–29670

298. Kamat SS, Camara K, Parsons WH, Chen DH, Dix MM, Bird TD, Howell AR, Cravatt BF (2015) Immunomodulatory lysophosphatidylserines are regulated by ABHD16A and ABHD12 interplay. Nat Chem Biol 11:164–171
299. Sayo A, Konishi H, Kobayashi M, Kano K, Kobayashi H, Hibi H, Aoki J, Kiyama H (2019) GPR34 in spinal microglia exacerbates neuropathic pain in mice. J Neuroinflammation 16:82
300. Blankman JL, Long JZ, Trauger SA, Siuzdak G, Cravatt BF (2013) ABHD12 controls brain lysophosphatidylserine pathways that are deregulated in a murine model of the neurodegenerative disease PHARC. Proc Natl Acad Sci U S A 110:1500–1505
301. Preissler J, Grosche A, Lede V, Le Duc D, Krugel K, Matyash V, Szulzewsky F, Kallendrusch S, Immig K, Kettenmann H, Bechmann I, Schoneberg T, Schulz A (2015) Altered microglial phagocytosis in GPR34-deficient mice. Glia 63:206–215
302. Schoneberg T, Meister J, Knierim AB, Schulz A (2018) The G protein-coupled receptor GPR34 – The past 20 years of a grownup. Pharmacol Ther 189:71–88
303. Barnes MJ, Li CM, Xu Y, An J, Huang Y, Cyster JG (2015) The lysophosphatidylserine receptor GPR174 constrains regulatory T cell development and function. J Exp Med 212:1011–1020
304. Shinjo Y, Makide K, Satoh K, Fukami F, Inoue A, Kano K, Otani Y, Ohwada T, Aoki J (2017) Lysophosphatidylserine suppresses IL-2 production in CD4 T cells through LPS3/GPR174. Biochem Biophys Res Commun 494:332–338
305. Barnes MJ, Cyster JG (2018) Lysophosphatidylserine suppression of T-cell activation via GPR174 requires Galphas proteins. Immunol Cell Biol 96:439–445
306. Qiu D, Chu X, Hua L, Yang Y, Li K, Han Y, Yin J, Zhu M, Mu S, Sun Z, Tong C, Song Z (2019) Gpr174-deficient regulatory T cells decrease cytokine storm in septic mice. Cell Death Dis 10:233
307. Chu X, Shen M, Xie F, Miao XJ, Shou WH, Liu L, Yang PP, Bai YN, Zhang KY, Yang L, Hua Q, Liu WD, Dong Y, Wang HF, Shi JX, Wang Y, Song HD, Chen SJ, Chen Z, Huang W (2013) An X chromosome-wide association analysis identifies variants in GPR174 as a risk factor for Graves' disease. J Med Genet 50:479–485
308. Szymanski K, Miskiewicz P, Pirko K, Jurecka-Lubieniecka B, Kula D, Hasse-Lazar K, Krajewski P, Bednarczuk T, Ploski R (2014) rs3827440, a nonsynonymous single nucleotide polymorphism within GPR174 gene in X chromosome, is associated with Graves' disease in Polish Caucasian population. Tissue Antigens 83:41–44
309. Zhao SX, Xue LQ, Liu W, Gu ZH, Pan CM, Yang SY, Zhan M, Wang HN, Liang J, Gao GQ, Zhang XM, Yuan GY, Li CG, Du WH, Liu BL, Liu LB, Chen G, Su Q, Peng YD, Zhao JJ, Ning G, Huang W, Liang L, Qi L, Chen SJ, Chen Z, Chen JL, Song HD (2013) Robust evidence for five new Graves' disease risk loci from a staged genome-wide association analysis. Hum Mol Genet 22:3347–3362
310. Okada Y, Wu D, Trynka G, Raj T, Terao C, Ikari K, Kochi Y, Ohmura K, Suzuki A, Yoshida S, Graham RR, Manoharan A, Ortmann W, Bhangale T, Denny JC, Carroll RJ, Eyler AE, Greenberg JD, Kremer JM, Pappas DA, Jiang L, Yin J, Ye L, Su DF, Yang J, Xie G, Keystone E, Westra HJ, Esko T, Metspalu A, Zhou X, Gupta N, Mirel D, Stahl EA, Diogo D, Cui J, Liao K, Guo MH, Myouzen K, Kawaguchi T, Coenen MJ, van Riel PL, van de Laar MA, Guchelaar HJ, Huizinga TW, Dieude P, Mariette X, Bridges SL Jr, Zhernakova A, Toes RE, Tak PP, Miceli-Richard C, Bang SY, Lee HS, Martin J, Gonzalez-Gay MA, Rodriguez-Rodriguez L, Rantapaa-Dahlqvist S, Arlestig L, Choi HK, Kamatani Y, Galan P, Lathrop M, Eyre S, Bowes J, Barton A, de Vries N, Moreland LW, Criswell LA, Karlson EW, Taniguchi A, Yamada R, Kubo M, Liu JS, Bae SC, Worthington J, Padyukov L, Klareskog L, Gregersen PK, Raychaudhuri S, Stranger BE, De Jager PL, Franke L, Visscher PM, Brown MA, Yamanaka H, Mimori T, Takahashi A, Xu H, Behrens TW, Siminovitch KA, Momohara S, Matsuda F, Yamamoto K, Plenge RM (2014) Genetics of rheumatoid arthritis contributes to biology and drug discovery. Nature 506:376–381
311. Sawzdargo M, Nguyen T, Lee DK, Lynch KR, Cheng R, Heng HH, George SR, O'Dowd BF (1999) Identification and cloning of three novel human G protein-coupled receptor genes GPR52, PsiGPR53 and GPR55: GPR55 is extensively expressed in human brain. Brain Res Mol Brain Res 64:193–198
312. Petitet F, Donlan M, Michel A (2006) GPR55 as a new cannabinoid receptor: still a long way to prove it. Chem Biol Drug Des 67:252–253
313. Yang H, Zhou J, Lehmann C (2016) GPR55 – a putative "type 3" cannabinoid receptor in inflammation. J Basic Clin Physiol Pharmacol 27:297–302
314. Oka S, Nakajima K, Yamashita A, Kishimoto S, Sugiura T (2007) Identification of GPR55 as a lysophosphatidylinositol receptor. Biochem Biophys Res Commun 362:928–934
315. Oka S, Toshida T, Maruyama K, Nakajima K, Yamashita A, Sugiura T (2009) 2-Arachidonoyl-sn-glycero-3-phosphoinositol: a possible natural ligand for GPR55. J Biochem 145:13–20
316. Kihara Y, Maceyka M, Spiegel S, Chun J (2014) Lysophospholipid receptor nomenclature review: IUPHAR Review 8. Br J Pharmacol 171:3575–3594
317. Bondarenko A, Waldeck-Weiermair M, Naghdi S, Poteser M, Malli R, Graier WF (2010) GPR55-dependent and -independent ion signalling in response to lysophosphatidylinositol in endothelial cells. Br J Pharmacol 161:308–320
318. Guy AT, Nagatsuka Y, Ooashi N, Inoue M, Nakata A, Greimel P, Inoue A, Nabetani T, Murayama A, Ohta K, Ito Y, Aoki J, Hirabayashi Y, Kamiguchi H (2015) Neuronal development. Glycerophospholipid regulation of modality-specific sensory axon guidance in the spinal cord. Science 349:974–977

319. Yamashita A, Kumazawa T, Koga H, Suzuki N, Oka S, Sugiura T (2010) Generation of lysophosphatidylinositol by DDHD domain containing 1 (DDHD1): possible involvement of phospholipase D/phosphatidic acid in the activation of DDHD1. Biochim Biophys Acta 1801:711–720
320. Inloes JM, Jing H, Cravatt BF (2018) The spastic paraplegia-associated phospholipase DDHD1 is a primary brain phosphatidylinositol lipase. Biochemistry 57:5759–5767
321. Staton PC, Hatcher JP, Walker DJ, Morrison AD, Shapland EM, Hughes JP, Chong E, Mander PK, Green PJ, Billinton A, Fulleylove M, Lancaster HC, Smith JC, Bailey LT, Wise A, Brown AJ, Richardson JC, Chessell IP (2008) The putative cannabinoid receptor GPR55 plays a role in mechanical hyperalgesia associated with inflammatory and neuropathic pain. Pain 139:225–236
322. Gangadharan V, Selvaraj D, Kurejova M, Njoo C, Gritsch S, Skoricova D, Horstmann H, Offermanns S, Brown AJ, Kuner T, Tappe-Theodor A, Kuner R (2013) A novel biological role for the phospholipid lysophosphatidylinositol in nociceptive sensitization via activation of diverse G-protein signalling pathways in sensory nerves in vivo. Pain 154:2801–2812
323. Schicho R, Bashashati M, Bawa M, McHugh D, Saur D, Hu HM, Zimmer A, Lutz B, Mackie K, Bradshaw HB, McCafferty DM, Sharkey KA, Storr M (2011) The atypical cannabinoid O-1602 protects against experimental colitis and inhibits neutrophil recruitment. Inflamm Bowel Dis 17:1651–1664
324. Ross GR, Lichtman A, Dewey WL, Akbarali HI (2012) Evidence for the putative cannabinoid receptor (GPR55)-mediated inhibitory effects on intestinal contractility in mice. Pharmacology 90:55–65
325. Li K, Fichna J, Schicho R, Saur D, Bashashati M, Mackie K, Li Y, Zimmer A, Goke B, Sharkey KA, Storr M (2013) A role for O-1602 and G protein-coupled receptor GPR55 in the control of colonic motility in mice. Neuropharmacology 71:255–263
326. Stancic A, Jandl K, Hasenohrl C, Reichmann F, Marsche G, Schuligoi R, Heinemann A, Storr M, Schicho R (2015) The GPR55 antagonist CID16020046 protects against intestinal inflammation. Neurogastroenterol Motil 27:1432–1445
327. Walsh SK, Hector EE, Andreasson AC, Jonsson-Rylander AC, Wainwright CL (2014) GPR55 deletion in mice leads to age-related ventricular dysfunction and impaired adrenoceptor-mediated inotropic responses. PLoS One 9:e108999
328. Rinne P, Guillamat-Prats R, Rami M, Bindila L, Ring L, Lyytikainen LP, Raitoharju E, Oksala N, Lehtimaki T, Weber C, van der Vorst EPC, Steffens S (2018) Palmitoylethanolamide promotes a proresolving macrophage phenotype and attenuates atherosclerotic plaque formation. Arterioscler Thromb Vasc Biol 38:2562–2575
329. Robertson-Gray OJ, Walsh SK, Ryberg E, Jonsson-Rylander AC, Lipina C, Wainwright CL (2019) l-alpha-Lysophosphatidylinositol (LPI) aggravates myocardial ischemia/reperfusion injury via a GPR55/ROCK-dependent pathway. Pharmacol Res Perspect 7:e00487
330. Romero-Zerbo SY, Rafacho A, Diaz-Arteaga A, Suarez J, Quesada I, Imbernon M, Ross RA, Dieguez C, Rodriguez de Fonseca F, Nogueiras R, Nadal A, Bermudez-Silva FJ (2011) A role for the putative cannabinoid receptor GPR55 in the islets of Langerhans. J Endocrinol 211:177–185
331. Liu B, Song S, Ruz-Maldonado I, Pingitore A, Huang GC, Baker D, Jones PM, Persaud SJ (2016) GPR55-dependent stimulation of insulin secretion from isolated mouse and human islets of Langerhans. Diabetes Obes Metab 18:1263–1273
332. Ruz-Maldonado I, Pingitore A, Liu B, Atanes P, Huang GC, Baker D, Alonso FJ, Bermudez-Silva FJ, Persaud SJ (2018) LH-21 and abnormal cannabidiol improve beta-cell function in isolated human and mouse islets through GPR55-dependent and -independent signalling. Diabetes Obes Metab 20:930–942
333. Perez-Gomez E, Andradas C, Flores JM, Quintanilla M, Paramio JM, Guzman M, Sanchez C (2013) The orphan receptor GPR55 drives skin carcinogenesis and is upregulated in human squamous cell carcinomas. Oncogene 32:2534–2542
334. Hasenoehrl C, Feuersinger D, Sturm EM, Barnthaler T, Heitzer E, Graf R, Grill M, Pichler M, Beck S, Butcher L, Thomas D, Ferreiros N, Schuligoi R, Schweiger C, Haybaeck J, Schicho R (2018) G protein-coupled receptor GPR55 promotes colorectal cancer and has opposing effects to cannabinoid receptor 1. Int J Cancer 142:121–132
335. Ferro R, Adamska A, Lattanzio R, Mavrommati I, Edling CE, Arifin SA, Fyffe CA, Sala G, Sacchetto L, Chiorino G, De Laurenzi V, Piantelli M, Sansom OJ, Maffucci T, Falasca M (2018) GPR55 signalling promotes proliferation of pancreatic cancer cells and tumour growth in mice, and its inhibition increases effects of gemcitabine. Oncogene 37:6368–6382
336. Whyte LS, Ryberg E, Sims NA, Ridge SA, Mackie K, Greasley PJ, Ross RA, Rogers MJ (2009) The putative cannabinoid receptor GPR55 affects osteoclast function in vitro and bone mass in vivo. Proc Natl Acad Sci U S A 106:16511–16516
337. Korchynska S, Lutz MI, Borok E, Pammer J, Cinquina V, Fedirko N, Irving AJ, Mackie K, Harkany T, Keimpema E (2019) GPR55 controls functional differentiation of self-renewing epithelial progenitors for salivation. JCI Insight 4:e122947
338. Sisay S, Pryce G, Jackson SJ, Tanner C, Ross RA, Michael GJ, Selwood DL, Giovannoni G, Baker D (2013) Genetic background can result in a marked or minimal effect of gene knockout (GPR55 and CB2 receptor) in experimental autoimmune encephalomyelitis models of multiple sclerosis. PLoS One 8:e76907
339. Sumida H, Lu E, Chen H, Yang Q, Mackie K, Cyster JG (2017) GPR55 regulates intraepithelial lympho-

cyte migration dynamics and susceptibility to intestinal damage. Sci Immunol 2:eaao1135
340. Meadows A, Lee JH, Wu CS, Wei Q, Pradhan G, Yafi M, Lu HC, Sun Y (2016) Deletion of G-protein-coupled receptor 55 promotes obesity by reducing physical activity. Int J Obes (Lond) 40:417–424
341. Lipina C, Walsh SK, Mitchell SE, Speakman JR, Wainwright CL, Hundal HS (2019) GPR55 deficiency is associated with increased adiposity and impaired insulin signaling in peripheral metabolic tissues. FASEB J 33:1299–1312
342. Hill JD, Zuluaga-Ramirez V, Gajghate S, Winfield M, Persidsky Y (2018) Activation of GPR55 increases neural stem cell proliferation and promotes early adult hippocampal neurogenesis. Br J Pharmacol 175:3407–3421
343. Sylantyev S, Jensen TP, Ross RA, Rusakov DA (2013) Cannabinoid- and lysophosphatidylinositol-sensitive receptor GPR55 boosts neurotransmitter release at central synapses. Proc Natl Acad Sci U S A 110:5193–5198
344. Hurst K, Badgley C, Ellsworth T, Bell S, Friend L, Prince B, Welch J, Cowan Z, Williamson R, Lyon C, Anderson B, Poole B, Christensen M, McNeil M, Call J, Edwards JG (2017) A putative lysophosphatidylinositol receptor GPR55 modulates hippocampal synaptic plasticity. Hippocampus 27:985–998
345. Cherif H, Argaw A, Cecyre B, Bouchard A, Gagnon J, Javadi P, Desgent S, Mackie K, Bouchard JF (2015) Role of GPR55 during Axon Growth and Target Innervation. eNeuro 2:11–15
346. Deliu E, Sperow M, Console-Bram L, Carter RL, Tilley DG, Kalamarides DJ, Kirby LG, Brailoiu GC, Brailoiu E, Benamar K, Abood ME (2015) The lysophosphatidylinositol receptor GPR55 modulates pain perception in the periaqueductal gray. Mol Pharmacol 88:265–272
347. Wu CS, Chen H, Sun H, Zhu J, Jew CP, Wager-Miller J, Straiker A, Spencer C, Bradshaw H, Mackie K, Lu HC (2013) GPR55, a G-protein coupled receptor for lysophosphatidylinositol, plays a role in motor coordination. PLoS One 8:e60314
348. Bjursell M, Ryberg E, Wu T, Greasley PJ, Bohlooly YM, Hjorth S (2016) Deletion of Gpr55 results in subtle effects on energy metabolism, motor activity and thermal pain sensation. PLoS One 11:e0167965

Druggable Targets in Endocannabinoid Signaling

8

Ann M. Gregus and Matthew W. Buczynski

Abstract

Cannabis and cannabinoid-based extracts have long been utilized for their perceived therapeutic value, and support for the legalization of cannabis for medicinal purposes continues to increase worldwide. Since the discovery of Δ^9-tetrahydrocannabinol (THC) as the primary psychoactive component of cannabis over 50 years ago, substantial effort has been directed toward detection of endogenous mediators of cannabinoid activity. The discovery of anandamide and 2-arachidonoylglycerol as two endogenous lipid mediators of cannabinoid-like effects (endocannabinoids) has inspired exponential growth in our understanding of this essential pathway, as well as the pathological conditions that result from dysregulated endocannabinoid signaling. This review examines current knowledge of the endocannabinoid system including metabolic enzymes involved in biosynthesis and degradation and their receptors, and evaluates potential druggable targets for therapeutic intervention.

Keywords

Endocannabinoid · GPCR · CB_1 · CB_2 · FAAH

8.1 Cannabinoids as Therapeutics

For centuries, cannabis and cannabinoid-based extracts were thought to possess therapeutic value. In recent years, the use of medical marijuana has increased for a wide variety of disorders in the United States, and changes in the legal landscape and public opinion support expanding its recreational availability nationwide. The 2010, a resolution adopted by the American Medical Association advocated reconsideration of marijuana as a Schedule I controlled substance given the potential therapeutic value of marijuana and cannabis-based products. As of January 1, 2020, 33 states have legalized the sale of medical marijuana, with additional states considering similar legislation with possible enaction in the near future.

The primary psychoactive component of cannabis was identified as Δ^9-tetrahydrocannabinol (THC) by Yechiel Gaoni and Raphael Mechoulam in the 1960s [1, 2]. While hundreds of bioactive molecules have been identified in cannabis thus far [3], THC recapitulates many of the pharmacological properties attributed to marijuana in both rodent models and in humans [4, 5]. Subsequent

A. M. Gregus · M. W. Buczynski (✉)
School of Neuroscience, Virginia Polytechnic Institute and State University, Blacksburg, VA, USA
e-mail: mwb@vt.edu

Y. Kihara (ed.), *Druggable Lipid Signaling Pathways*, Advances in Experimental Medicine and Biology 1274, https://doi.org/10.1007/978-3-030-50621-6_8

worldwide efforts were aimed at the discovery of both synthetic and semi-synthetic cannabinoids capable of producing cannabinoid-like effects *in vivo* for the eventual development of patentable drugs with verifiable therapeutic value. While many hundreds of cannabinoid compounds were created in subsequent years, the pharmacological properties of these compounds often retained or exacerbated psychoactive effects when compared with THC [6], and thus many of these early cannabinoid-mimetics were not heavily pursued in clinical trials. Alternatively, synthetic cannabinoids began to reemerge as a recreational alternative to traditional cannabis in convenience stores and online marketplaces during the mid-2000s under pseudonyms such as "Spice" and "K2". Synthetic cannabinoids were generally consumed by inhalation via cigarettes containing herbal substances along with these synthetic molecules to obtain euphoric, anxiolytic, and antidepressant-like effects. Whereas traditional cannabis products generally have been considered safe across a wide dose range, numerous case reports illustrate that synthetic cannabinoids produce deleterious effects including paranoia, tachycardia, panic, convulsions, psychosis, visual/auditory hallucinations, vomiting, and seizures [6].

Thus far, two cannabinoid-based therapeutics have obtained FDA approval: Marinol® (dronabinol or THC) and Cesamet® (nabilone), a synthetic cannabinoid [7, 8], for the treatment of chemotherapy-induced nausea and emesis. Marinol also has been indicated as an appetite stimulant to treat cachexia in AIDS patients. A third medication, formulated with equivalent concentrations of THC and cannabidiol (CBD) known as Sativex®, has been approved in several countries outside the United States for the relief of spasticity in multiple sclerosis (MS) patients [9]. While the widespread use of medical marijuana suggests potential therapeutic value for a number of diseases, psychoactive effects and addictive potential of cannabinoids with chronic usage may limit widespread use in clinical practice. Additionally, their CNS effects complicate interpretation of efficacy in clinical trials as patients can easily determine whether or not they are receiving the drug or a placebo. Thus, substantial efforts are directed toward evaluating alternative targets in the cannabinoid signaling pathway for the development of safe and effective therapeutics.

8.2 Selective Modulation of Cannabinoid Receptors

While significant adverse effects and lack of efficacy have hampered the development of cannabinoid receptor antagonists for clinical use [10–14], these compounds served as important tools for the discovery of endogenous cannabinoid receptors 1 (CB_1) and 2 (CB_2) and their classification as G protein-coupled receptors [15]. Binding studies conducted using radiolabeled versions of potent synthetic cannabinoids such as CP-55,940 revealed high-affinity cannabinoid-specific binding sites via radioactive displacement by THC or other synthetic cannabinoids [16, 17]. Subsequent efforts harnessed these approaches to discover CB_1 [18, 19] and CB_2 receptors [20], respectively. Both CB_1 and CB_2 receptors couple to $G\alpha_{i/o}$ proteins to inhibit adenylate cyclase activity and reduce production of cyclic AMP [21, 22]. While CB_1 receptors are enriched in neuronal synapses (where they inhibit neurotransmitter release), CB_2 is strongly expressed in immune cells and glia [23–25]. Many of the psychoactive effects of THC and other cannabinoids can be attributed to actions on the CB_1 receptor [26], yet mounting evidence paints a more complex picture of the cell-type specific expression patterns of cannabinoid receptors *in vivo*.

Following the discovery of endogenously expressed cannabinoid receptors, investigators raced to develop the first potent and selective CB_1 and CB_2 receptor modulators. From the many compounds identified, SR141716A (rimonabant) represents the most well-characterized drug in this class [27]. Subsequently, it was demonstrated that CB_1-selective rimonabant blocks acute cannabinoid-induced tetrad behaviors in mice [27], alters dopamine release in rats [28, 29], precipitates withdrawal in THC-dependent rats [30–32], and inhibits long-term potentiation in rodent brain slices [33, 34]. It has been shown in multiple preclinical models of nociception that rimonabant exacerbates hyperalgesia [35–37] and attenuates cannabinoid-induced analgesic effects [35, 37] suggesting an opportunity for newer generation therapeutics that modify endocannabinoid signaling in the treatment of chronic pain. Rimonabant approved in 2006 as an anti-obesity medication in Europe (Acomplia, Zimulti), but was later associated with increased incidence of severe adverse psychiatric consequences during Phase III clinical trials to examine its efficacy as an obesity treatment and smoking cessation therapy [38, 39]. As a result, rimonabant did not garner FDA approval in the United States, and was subsequently pulled from the market worldwide in 2008.

The localization of CB_2 primarily in immune cells with limited expression in neurons may underlie its implication in several diseases with an inflammatory component including neurodegenerative and autoimmune diseases [40–42]. While a suite of CB_2 agonists have been synthesized to date, a collaborative effort between multiple academic and industry laboratories identified substantial differences in their mechanisms of action, pharmacokinetic properties, and off-target effects *in vivo* [43]. Based on their collective findings on a wide range of compounds, this research team recommends using HU910, HU308, or JWH133 as potent and *in vivo* active agonists of the CB_2 receptor for subsequent drug discovery efforts of clinically useful CB_2-based therapeutics. This endeavor has proven to be more challenging than initially expected, as only a few synthetic CB_2 agonists have reached clinical trials (GW842166X, CP-55,940, S-777469, and JTE-907), with none completing phase II for chronic pain indications [42]. Currently, the CB_2 agonist JBT-101 is undergoing Phase II testing for efficacy in autoimmune diseases including systemic lupus erythematosus (NCT03093402) and diffuse scleroderma, where it has shown some beneficial effects (NCT02465437).

8.3 Two Primary Endogenous Cannabinoids: Anandamide and 2-Arachidonoylglycerol

Nearly 30 years ago, two derivatives of arachidonic acid were identified as the endogenous cannabinoid receptor ligands. Anandamide (AEA) was the first endocannabinoid (eCB) to be discovered [44], closely followed by identification of a second endogenous molecule, 2-arachido noylglycerol (2-AG), signaling via CB_1 and CB_2 receptors [45, 46]. AEA and 2-AG retain an arachidonoyl moiety that imparts a significant amount of their bioactivity. While endocannabinoid-related lipids generated from other fatty acids substrates, including palmitoylethanolamide [47] and oleoylethanolamide [48] have described as eCBs, these molecules do not interact with cannabinoid receptors [49, 50]. Thus, AEA and 2-AG are still viewed as the primary endogenous mediators of cannabinoid signaling.

Historically, evidence for an eCB mechanism *in vivo* was determined indirectly using cannabinoid receptor antagonists, without certainty of the identity of the signaling molecule(s). Most studies quantified eCB content primarily by lipid extraction and purification from bulk tissue, followed by subsequent analysis with liquid chromatography coupled with mass spectrometry. Several excellent articles outline this process [51–54]. However, there is significant debate regarding the physiological range of eCB concentrations in various regions, as there is considerable variability in estimates of brain AEA and 2-AG content. Notably, a significant pool of 2-AG serves as an intracellular substrate for triacylglycerol formation in energy metabolism and may not participate in cannabinoid signaling [55]. An alternate approach utilizing *in vivo* microdialysis samples of interstitial, signaling-competent eCBs from awake, behaving animals with exquisite sensitivity [55, 56]. Using this method, basal interstitial AEA and 2-AG in the brain are estimated at low to mid-nanomolar levels, physiologically relevant concentrations for activating cannabinoid receptors *in vivo* [55, 57].

8.4 Selective Inhibitor Development Using Activity-Based Protein Profiling (ABPP)

Drug selectivity presented a major challenge in early efforts in the discovery of drug candidates that modulate endocannabinoid metabolism. Initial pharmacology studies suggested that these enzymes were not rate-limiting, and thus near-complete inhibition is required to produce therapeutic effects [58–61]. Moreover, many of these enzymes utilized the same mechanism of action (serine hydrolase), so compounds used at doses needed for complete inhibition were more likely to exhibit off-target effects. Thus, it was particularly challenging to develop inhibitors for the serine hydrolases, a class of over 200 enzymes with a wide range of biological functions [62]. For example, partial inhibition of acetylcholinesterase by donepezil can improve cognitive function in patients with Alzheimer's disease [63]. However, acetylcholinesterase knockout mice typically do not survive to adulthood [64], and complete inhibition by non-selective nerve agents such as Sarin produce lethal neurotoxicity [65]. Given the large number of unannotated serine hydrolases in the human body [62], it is plausible that inhibition of additional off-target serine hydrolases may have similar safety and toxicity issues. Thus, the development of a safe and effective therapeutic targeting endocannabinoid metabolism requires substantial preclinical pharmacokinetic and pharmacodynamic validation prior to entering clinical trials.

The use of activity-based protein profiling (ABPP) approaches has greatly facilitated the development of many of the selective inhibitors currently used in academic research and clinical trials [66]. While non-selective serine hydrolase inhibitors such as organophosphates would raise safety concerns as chronically administered therapeutics, their broad-spectrum capacity to covalently capture a wide range of endogenous serine hydrolases render them an excellent tool for evaluating potency and selectivity of potential drug candidates *in vivo*. The first broad-spectrum fluorophosphonate probes contained a rhodamine or biotin tag [67], which allowed any serine hydrolases captured by these probes to be visualized by in-gel fluorescence or identified using mass spectrometry. In competition experiments, any serine hydrolase inhibited by a drug would fail to be captured by the fluorophosphonate probe under those treatment conditions and the corresponding fluorescence or mass spectra would be diminished. In addition to broad-spectrum probes that capture high abundance serine hydrolases, a number of more selective ABPP probes have been synthesized for discovery of inhibitors for difficult targets, such as diacylglycerol lipases [59, 68–71], with procedural details outlined in several excellent reviews [71, 72]. It follows that potential drug-like molecules can be modified to contain alkyne moieties which have minimal effect on their selectivity but allow their direct targets to then be bound by an azide-functionalized rhodamine or biotin using click chemistry techniques. Collectively, this approach can provide an *in vivo* readout of both potency and selectivity while significantly facilitating the identification of off-targets for novel inhibitors.

These techniques have been employed to great effect in the development of a selective fatty acid amide hydrolase (FAAH) inhibitor by Pfizer. Systemic delivery of the initial lead compound (PF-3845) exhibits minimal serine hydrolase off-target effects in both brain and liver tissue at doses that abolish FAAH activity [58]. Moreover, an alkyne-functionalized PF-3845 provides direct evidence for the selectivity of this compound, as minimal off-target binding was identified. While minor modifications of the lead compound were made to improve efficacy and reduce interactions with liver cytochrome P450s (CYPs) that cause unwanted drug-drug interactions [73], the final candidate to enter clinical trials (PF-04457845) is based on the chemotype PF-3845, retaining the selectivity profile of the original lead compound [74].

8.5 Targeting Endocannabinoid Degradation

Increasing endogenous cannabinoid signaling represents an alternative therapeutic approach to using cannabinoid receptor agonists such as THC. By inhibiting the natural breakdown of endogenous AEA and/or 2-AG, this approach would ideally recapitulate some of the beneficial therapeutic effects of cannabinoids while reducing undesirable side effects. In support of this approach, Long and colleagues used a combination of chemical inhibitors and genetic approaches to show that inhibiting all of the primary endocannabinoid degradative enzymes (FAAH, MGLL, ABHD6) in mice produces similar cannabinoid-appropriate responding in a drug discrimination test to THC [75]. Importantly, selective inhibition of either AEA or 2-AG degradation failed to recapitulate THC-like responsivity. This study provides clear support for the therapeutic viability of selective inhibition of specific endocannabinoid pathways with potential for reduced side-effect profile. Accordingly, the following sections will evaluate current clinical and preclinical studies evaluating inhibitors fatty acid amide hydrolase (FAAH), monoacylglycerol lipase (MGLL), and α/β hydrolase domain 6 (ABHD6).

Fatty Acid Amide Hydrolase (FAAH) Following the discovery of AEA as an endogenous ligand for the CB_1 receptor [44], the search began for the enzyme(s) regulating its metabolism. In 1996, the fatty acid amide hydrolase (FAAH) was identified as the primary enzyme responsible for AEA degradation [76, 77]. Perturbation of FAAH activity via genetic deletion or pharmacological inhibition markedly reduces AEA hydrolysis, thereby elevating AEA levels in multiple organ systems in rodents [78]. In addition to AEA, FAAH metabolizes a number of other fatty acid amide substrates such as oleoylethanolamine and palmitoylethanoamine [58, 73], resulting in a battery of CB_1-dependent and CB_1-independent behavioral changes in rodents. These effects include, but are not limited to, decreased anxiety-like [79–81] and depression-like behaviors [82, 83], gastrointestinal function [84–88], altered expression of drug and alcohol withdrawal [89–93], as well as diminished inflammatory and neuropathic pain states [74, 94–97]. This substantial body of preclinical evidence inspired clinical development of FAAH inhibitors by several pharmaceutical companies including PF-04457845 (Pfizer), JNJ-42165279 (Janssen), ASP3652 (Astellas), V158866 (Vernalis) and BIA 10–2474 (Bial). Likewise, FAAH inhibitors have been evaluated for treatment of several disease indications, including cannabis use disorder [98], fear memory extinction [99], Tourette Syndrome (NCT02134080 – terminated for lack of funding), chronic pain due to spinal cord injury (NCT01748695), osteoarthritis [100], and prostatitis [101].

The most notable failure in FAAH drug development to date is BIA 10-2474, a potent and long-acting CNS-active inhibitor of FAAH which produced acute neurotoxicity in 5 patients, one of which resulted in death in the multiple ascending dose part of the study in Phase I [102]. The expression of severe adverse events had not yet been observed at such a late stage of a first-in-human study [103]. It was later revealed that BIA 10-2474 inhibits several lipases that are not targeted by PF04457845, and produces substantial changes in lipid networks in human cortical neurons that may lead to metabolic dysregulation [104]. One of these off-targets is Aldehyde Dehydrogenase 2, which has been implicated in neuroprotection from oxidative stress-related damage [105]. While BIAL has since conducted a series of toxicity studies in animals [106], in-depth analysis of the trial PK/PD parameters and study periods may help determine conclusively the drug and metabolite concentrations underlying these adverse events.

In contrast to BIA 10-2474, most clinical candidates were generally safe and well-tolerated [101, 107–109], and some report efficacy of FAAH inhibition in primary outcome measurements for reducing stress reactivity [99] and cannabis withdrawal symptoms [98]. However, PF-04457845 and ASP3652 failed to attenuate osteoarthritis or prostatitis pain, respectively [101, 110] despite considerable data demonstrating antihyperalgesic effects of FAAH inhibition in several preclinical models of chronic pain states. It has been hypothesized that premature termination of the osteoarthritis study due to lack of efficacy, while necessary, may have prevented the adequate assessment of contribution of non-responders and placebo effects [110]. Furthermore, the human trial focused on assessment of affective measures of pain that generally were not evaluated in rodent models, with exception of one study reporting no effect of FAAH inhibition on osteoarthritis-induced burrowing behavior [111]. Thus, it may be advantageous to include spontaneous and functional output measures of pain-like behaviors (e.g., locomotor activity, grip force, nesting, sucrose preference) in preclinical studies in order to help inform selection of disease indication for clinical trials.

Monoacylglycerol Lipase (MGLL) The discovery of 2-AG as a bonafide endocannabinoid led to the search for enzymes regulating its metabolism *in vivo*. While FAAH was briefly considered a potential candidate, both chemical and genetic inhibition of FAAH failed to substantially elevate 2-AG compared with AEA. Monoacylglycerol lipase was identified as the primary metabolic driver of cannabinergic 2-AG breakdown [112, 113], and genetic deletion of MGLL in mice confirmed these findings [114]. Genetic and pharmacological analyses demonstrate ubiquitous expression of MGLL across most tissues including brain, liver, kidney, lung, heart, muscle, intestines, and adipose tissue [60, 115, 116]. These mice exhibited decreased body weight as adults, hastened increased latency to inflammatory thermal hyperalgesia, and alterations in basal pain sensitivity [117]. Genetic inactivation of MGLL enhanced extinction and reversal learning [118] and facilitates anxiety-like behavior [119], neurophysiological analyses reveals cannabinoid-dependent changes in excitatory and inhibitor synaptic plasticity in multiple brain regions in mice [119–121]. Accordingly, selective chemical inhibitors would be necessary to distinguish between role of 2-AG signaling in adults with the critical role of this pathway in neuronal development.

The early chemical inhibitors developed in academic labs suggested a prominent role for MGLL in 2-AG metabolism and neuronal signaling [122–127]. The first ABPP-validated selective murine MGLL inhibitor JZL-184 confirmed many of these findings, including enhancement of depolarization-induced 2-AG release and reduction of cannabinoid-sensitive pain states [128]. The development of JZL-184 greatly accelerated the evaluation of the molecular and behavioral role of MGLL in mice, with over 200 publications using this compound to date. A number of potential physiological roles have been discovered using JZL-184 including, but not limited to, drug withdrawal [91, 129, 130], pain states [131–133], stress [134, 135], immune function [136–138], cancer [139–141], gastrointestinal function [137, 142, 143], and neurodegeneration [144–147]. Importantly, chronic administration of JZL-184 produces CB_1 receptor desensitization and functional antagonism [148], suggesting that pharmacological tolerance of an MGLL inhibitor may produce a therapeutic profile more similar to a CB_1 antagonist than with cannabinoid-based therapeutics such as THC. However, limitations in JZL-184 pharmacology including partial inhibition of ABHD6 and FAAH during chronic dosing procedures [149] and limited efficacy in rats [60] have helped open the door for next-generation compounds that address these concerns [149–152]. Thus, future studies should validate the pharmacological selectivity of these compounds and dosing procedures for proper interpretation of results.

Recently, Abide Therapeutics discovered ABX-1431 (acquired by Lundbeck, and rebranded as Lu AG06466) as the potent first-in-class, orally bioavailable and selective inhibitor of MGLL, now under development for as a therapeutic for movement disorders, neurodegenerative diseases and pain [153]. A Phase I Experimental Hyperalgesia study of ABX-1431 may yield insight into its viability in this therapeutic space (NCT02929264), as this model is generally highly predictive of clinical success for Neuropathic Pain [154]. At present, this molecule is undergoing a Phase IIa trial for the treatment of Tourette syndrome (NCT03625453) and a Phase I trial for neuropathic pain (NCT03138421). Likewise, Pfizer has developed a selective covalent MGLL inhibitor [155] and a corresponding ^{11}C-PET tracer [156] for evaluating the pharmacokinetics of this compound as part of a Phase 1 clinical trial (NCT03100136). Additionally, both Takeda Pharmaceutical Co. [157] and Janssen Research [158] recently have developed non-covalent MGLL inhibitors that show preclinical viability for target engagement and increased 2-AG levels in CNS and peripheral tissue, but these candidates have not yet entered clinical trials.

α/β Hydrolase Domain 6 (ABHD6) While MGLL drives the majority of 2-AG breakdown *in vivo*, emerging evidence suggests that other enzymes play a supportive role in this process. Initial studies using immortalized BV-2 microglial cells, which do not express MGLL, provide clear evidence of alternative 2-AG metabolism pathways [159]. While compensation by FAAH accounted for about half of the 2-AG metabolism, the use of selective inhibitors revealed that the remaining activity could be attributed to unknown enzyme(s). Using a functional proteomic approach, two additional serine hydrolases (ABHD6 and ABHD12) were identified as potential 2-AG metabolic enzymes [160]. Ultimately, the unknown 2-AG activity in BV-2 microglia was attributed to ABHD6, which regulates endocannabinergic signals to alter excitatory [161] and inhibitory [162] synapses in the brain.

Our understanding of the significance of ABHD6 in regulating endocannabinoid signaling and other lipid pathways has emerged from studies of targeted inactivation using genetic and chemical tools [163]. Constitutive deletion of ABHD6 reveals its critical role in energy metabolism [164, 165] but not in lysosomal storage disorders [166]. ABHD6 knockout mice are protected from high-fat diet-induced obesity [164] and displayed increased decreased body weight, increased energy expenditure, improved glucose tolerance and insulin sensitivity, and changes in white and brown adipose tissue composition [165]. Many of these metabolic effects can be recapitulated by antisense oligonucleotides or treatment with WWL70, the first chemical inhibitor designed to target ABHD6. Early studies using WWL70 implicate ABHD6 function in traumatic brain injury [167], obesity and type II diabetes [165, 168], seizure activity [169, 170], inflammation and pain [171, 172]. Selective blockade of ABHD6 with the newer generation peripherally-restricted inhibitor KT-203 decreases pancreatic cancer cell metastasis [173], while KT-182 modestly attenuates autoimmune demyelination [174, 175], in contrast with a previous study reporting significant reduction of clinical signs in the experimental autoimmune encephalitis model of Multiple Sclerosis with WWL70 [176]. At present, it is not clear if ABHD6 drives metabolic changes by enzymatic regulation of 2-AG signaling [177] or other lipid pathways [164]. Since ABHD6 exhibits promiscuity in its acceptance of lipid substrates [164], the process of determining the specific lipid mediators responsible for its effects presents a substantial challenge. Thus, studies utilizing inhibitors with off-target effects should be interpreted with caution for future drug development efforts.

α/β Hydrolase Domain 12 (ABHD12) Since its discovery as a serine hydrolase with in vitro 2-AG metabolic activity [160], the physiological role of ABHD12 has remained poorly understood. Loss-of-function mutations in ABHD12 cause the rare neurodegenerative disorder PHARC (polyneuropathy, hearing loss, retinosis pigmentosa, and cataract) in humans [178–182], which is phenocopied in ABHD12 knockout mice [178]. While ABHD12 knockout mice have increased levels of 2-AG in multiple brain regions [183], the primary function of this enzyme *in vivo* is likely a lysophosphatidylserine lipase. Indeed, both genetic deletion [178, 184] and selective inhibition [184, 185] of ABHD12 generate substantially elevated levels of very long chain lysophosphatidylserine levels *in vivo* [186]. ABHD12 and downstream target lysophosphatidylserine receptors are coexpressed in glia and in immune cells, and treatment of macrophages with the selective ABHD12 inhibitor DO264 exacerbated immune responsivity [178, 184,

187]. Although acute chemical inactivation of ABHD12 did not produce the severe behavioral deficits found in knockout mice [184, 187], it is presently unclear if the PHARC phenotype would emerge following chronic, long-term drug treatment. Given its restricted contribution to endocannabinoid signaling and potential role in neuroprotection, ABHD12 may offer limited opportunity for therapeutic development when compared with other targets in the endocannabinoid metabolic pathway.

8.6 Targeting Endocannabinoid Biosynthesis

Due to concerns regarding safety and tolerability with cannabinoid receptor antagonism, efforts have emerged to selectively decrease cannabinoid signaling as an alternative therapeutic approach. While CB_1 receptor antagonists or inverse agonists such as rimonabant and taranabant demonstrated efficacy in the treatment of obesity, type II diabetes and nicotine dependence [188–192], serious adverse psychiatric consequences significantly limit clinical utility [10–14], ultimately precluding their approval by the FDA. Instead, by inhibiting the natural production of endogenous AEA and/or 2-AG, targeting endocannabinoid biosynthesis might recapitulate some of the beneficial therapeutic effects of receptor antagonists while mitigating undesirable side effects.

The FDA-approved drug Orlistat (tetrahydrolipstatin, sold over-the-counter as Alli®) indicated for the treatment of obesity was designed as a pancreatic and gastric lipase inhibitor [193, 194], but nonetheless has a number of potential off-target effects including inhibition of the two diacylglycerol lipases α (DAGLα) and β (DAGLβ) that generate 2-AG *in vivo*. Unlike rimonabant, Orlistat does not produce serious psychiatric events, suggesting that inhibition of endocannabinoid biosynthesis may represent a viable alternative to CB_1 receptor antagonists. While the study of endocannabinoid biosynthetic pathways and the subsequent discovery of corresponding selective chemical inhibitors is in its nascent stage, a number of recent advances have bolstered this field and may yield unique candidates for future drug development. Several selective ABPP probes have been developed to facilitate inhibitor development against DAGLs and other potential endocannabinoid biosynthases, including some based on the tetrahydrolipstatin structure [195]. Accordingly, the following sections will evaluate current clinical and preclinical studies evaluating inhibitors of DAGLα and β (DAGLβ, n-acyl phosphatidylethanolamine phospholipase D (NAPE-PLD), and other enzymes involved in endocannabinoid biosynthesis.

Diacylglycerol Lipase α (DAGLα) The endogenous biosynthesis of signaling competent 2-AG is driven by the enzymes diacylglycerol lipase α (DAGLα) and diacylglycerol lipase β (DAGLβ). While both of these serine hydrolases convert diacylglycerol into 2-AG, they exhibit unique cellular and tissue specific expression. Specifically, DAGLα is predominantly expressed in neuronal tissue [196] and DAGLα knockout mice exhibit approximately 80% lower levels of 2-AG in the brain and spinal cord, compared with a 50% reduction of 2-AG in liver and adipose tissue [197, 198]. Within the central nervous system, neurons contain the predominant amount of DAGLα as compared with glial cells [196, 199–201]. In addition, many of the metabolic and behavioral phenotypes found in CB_1 receptor knockout mice are recapitulated by genetic deletion of DAGLα. Mice with genetic inactivation of either CB_1 or DAGLα both exhibit signs of enhanced metabolic function including lower body weight and decreased body fat, as well as reduced fasting insulin release and blood lipid levels [197, 202]. However, both genotypes also showed signs of psychiatric dysfunction including less marble-burying and shorter latencies in the forced swim test [202–204]. DAGLα mice also exhibited increased mortality beginning around 8–10 weeks of age [202]. Mounting evidence indicates that DAGLα plays a critical role in neuronal developmental processes [205–210], suggesting that psychiatric behavioral phenotypes may results from neurodevelopmental deficiencies as opposed to the direct signaling actions of DAGLα. Collectively, these findings highlight the necessity for rigorous behavioral evaluation of a selective DAGLα inhibitor prior to initiation of clinical development.

Early DAGLα inhibitors such as tetrahydrolipstatin were utilized mainly in electrophysiological studies, yet the recent development of *in vivo*-active diacylglycerol lipase inhibitors has offered important insight into the consequences of global chemical inactivation of these enzymes [211]. Ogasawara and colleagues created a suite of CNS-active covalent inhibitors of diacylglycerol lipases, most notably DO34 and DH376 [61]. Compared with DH376, DO34 exhibits enhanced selectivity for DAGLα, however both compounds block DAGLα and DAGLβ following systemic administration and are blood-brain barrier permeable. Despite their off-target effects, each molecule targets a unique group of serine hydrolases. Importantly, DO53 blocks all spurious targets of DO34 while sparing DAGLα and DAGLβ, thus serving as a critical negative control compound for *in vivo* pharmacology studies. Furthermore, Baggelaar et al. report LEI105 as a reversible inhibitor of diacylglycerol lipases, with potent inactivation of DAGLα and DAGLβ and minimal cross-reactivity with other endocannabinoid metabolic enzymes in mice [69]. Surprisingly, intracerebroventricular administration in rats of the DAGLβ inhibitor KT172 (originally validated in mice) actually reduces activity of DAGLα (~80%) and DAGLβ (~50%) in the brain [162], suggesting this approach as a potential alternative to rimonabant antagonism of CB_1 for smoking cessation. While the selectivity of these compounds generally has been evaluated against the human and mouse orthologs, there is precedent for species-specific selectivity when targeting enzymes in the eCB pathway [60]. Thus, optimization is needed in order to fully validate effects of these compounds in rats for behavioral studies and in multiple species for future clinical development.

Diacylglycerol Lipase β (DAGLβ) With the initial cloning in 2003 of DAGLα and DAGLβ [196], it was demonstrated that DAGL expression pattern shifts from axonal tracts to dendritic fields, consistent with later reports that DAGL activity is required for synaptic plasticity [125, 126, 212, 213], axonal growth and guidance [214], adult neurogenesis [215, 216] and oligodendrocyte differentiation [217]. Several studies suggest that while DAGLα is the predominant 2-AG synthesizing enzyme for endocannabinoid-mediated modulation of neurotransmission in adults [197, 198, 218], DAGLβ is more abundantly expressed in the developing CNS [196, 219]. DAGLβ contributes to depolarization-induced suppression of excitation in early postnatal hippocampal autaptic neurons [209] and neurite outgrowth in culture models [208]. Outside of the CNS, DAGLβ is widely expressed in multiple sites including white blood cells [220], liver [197] and adipose tissue [221], where it may be correlated with serum high-density lipoprotein cholesterol levels. Most notably, while LPS-induced eCB-eicosanoid crosstalk is dependent on DAGLβ in microglia [201], macrophages [59] and dendritic cells [222]. These observations indicate a crucial role for DAGLβ in immune function and inflammation, consistent with the implication of this enzyme in pathologies associated with alcoholic fatty liver disease [223], Alzheimer's disease [224] as well as inflammatory, neuropathic and post-surgical pain [225–227].

The development of *in vivo*-active DAGLβ inhibitors has facilitated significantly our understanding of the role of this enzyme. Systemic administration of currently available blood-brain barrier-permeable DAGLβ inhibitors exhibit cross-reactivity with both DAGLα and DAGLβ [211], however Hsu and colleagues have developed peripherally-restricted inhibitors KT109 and KT172, which exhibit ~60-fold selectivity for DAGLβ versus DAGLα in mice [59]. To account for its limited off-target effects on ABHD6 and to determine DAGLβ-specific biology, the negative control compound KT195 was utilized as a selective ABHD6 inhibitor. Acute treatment with KT109 or KT172 reveals a role for DAGLβ in 2-AG metabolism, as well as in downstream eicosanoid production and inflammatory signaling in peripheral macrophages [59]. Delivery of liposome-encapsulated KT109 produces macrophage-specific targeted inhibition of DAGLβ, with no apparent activity of other tissues *in vivo* [228]. This mode of administration substantially enhances anti-nociceptive potency of KT109 compared with traditional systemic treatment, thereby demonstrating potential for DAGLβ as a novel druggable target with potential indications in inflammatory diseases.

N-Acyl Phosphatidylethanolamine Phospholipase D (NAPE-PLD) While multiple enzymatic pathways have been implicated in the endogenous biosynthesis of AEA, none has been definitively nominated as the "AEA synthase" to date. The initial evaluation of n-acyl phosphatidylethanolamine phospholipase D (NAPE-PLD) as a zinc hydrolase capable of producing n-acylethanolamines (including AEA) offered the first example of a potential anandamide synthase [229]. Consistent with this purported role of NAPE-PLD *in vivo*, overexpression of this enzyme in mammalian culture systems increases AEA and other n-acylethanolamines [230]. However, while NAPE-PLD knockout mice express lower levels of saturated and monounsaturated n-acylethanolamines, reduction of AEA is inconsistent between reports [231–233].

Recent studies utilizing global and spatiotemporal genetic inactivation of NAPE-PLD have attempted to elucidate the physiological role of this enzyme. Mice expressing a constitutive knockout of NAPE-PLD exhibit no changes in body mass composition or glucose tolerance on a normal chow diet, yet inactivation of NAPE-PLD selectively in either adipose [234] or intestinal tissue [235] leads to an exaggerated obese phenotype on a high-fat diet. Spatio-temporal deletion of NAPE-PLD produces significant decreases in AEA and other n-acylethanolamines in the targeted tissues and alters the composition of gut microbiota in these animals [234, 235]. NAPE-PLD may limit development of obesity through non-cannabinergic pathways, as other n-acylethanolamines such as oleoylethanolamine produce robust anorexic effects [48] through multiple receptors including GPR119 [236] and PPARα [237]. However, it should be noted that given the potential for compensatory lipid metabolic pathways in constitutive knockout models, selective chemical inhibitors are critical in order to clarify the function of NAPE-PLD in activity-dependent AEA signaling. Unfortunately, potent CNS-active inhibitors are currently not available as existing compounds lack the necessary pharmacological properties to inhibit NAPE-PLD *in vivo* [230, 238, 239]. Future work will evaluate if these metabolic changes can be recapitulated and treated using pharmacological tools and subsequently translated into a potential therapeutic for obesity.

Other Potential Anandamide Biosynthetic Enzymes In addition to NAPE-PLD, multiple enzymatic pathways leading to the biosynthesis of n-acylethanolamines, a family of lipid species that includes AEA, have been discovered. However, current evidence highlights several obstacles to therapeutic drug discovery in this arena. For example, the serine hydrolase ABHD4 can act in concert with glycerophosphodiesterase GDE1 to produce anandamide and other n-acylethanolamines *in vitro* [240], however levels of AEA in brain tissue are unaltered in GDE1 knockout mice [241]. Likewise, ABHD4 regulates multiple lipid classes, and genetic inactivation of this enzyme elicits comparatively greater changes in lysophosphatidylserine levels [242]. An additional pathway for AEA production utilizing sequential activity of phospholipase C and tyrosine phosphatases such as lymphoid-specific tyrosine phosphatase (PTPN22) was discovered in macrophages [243, 244]. However, targeting either phospholipases C or PTPN22 alone is expected to exert substantial non-cannabinergic effects, as these enzymes are broadly involved in lipid metabolism and in responsiveness of B and T cells, respectively. Collectively, research on AEA biosynthesis suggests that multiple redundant pathways likely exist *in vivo* [241, 244], as illustrated by biological compensation when one of these enzymes is inactivated. It follows that the physiological source(s) of AEA and thus the therapeutic utility of inhibitors for these enzymes remain to be fully elucidated.

8.7 The Interaction Between Endocannabinoids and Eicosanoid Production

Precursors for Eicosanoid Production The primary focus of research on endocannabinoid metabolism centers on changes in 2-AG signaling via cannabinoid receptors. However, recent work has uncovered an important role for prostaglandins and other arachidonic acid metabolites that are derived from endocannabinoid precursors. Although it was previously thought that prostaglandins arise mainly from actions of cytosolic phospholipase A_2, genetic inactivation of this enzyme exerts only minimal effect on arachidonic acid levels in the brain [245]. Instead, metabolic breakdown of 2-AG by MGLL supplies arachidonic acid for production of proinflammatory prostaglandins by cyclooxygenases during neuroinflammation [138]. Both genetic and chemical inactivation of MGLL attenuate lipopolysaccharide-induced cytokine release and protect against neurodegeneration of dopamine neurons through a CB_1-independent mechanism of reduced prostaglandin synthesis. Accordingly, MGLL activity liberates prostaglandins in the brain to facilitate neurodegeneration in mouse models of Alzheimer's disease [246–248], while deletion or pharmacological inhibition of MGLL facilitates cannabinoid receptor-independent blunting of disease progression [246]. MGLL-dependent prostaglandins also mediate the fever response in mice, as genetic or pharmacological inactivation of MGLL blunts LPS-induced elevations in body temperature without altering core body temperature in control mice [249, 250].

Upstream of MGLL, DAGLα and DAGLβ both contribute to release of arachidonic acid [197, 201] and subsequent production of prostaglandins [201], with a prominent role of DAGLα in the brain [59, 197, 201] and DAGLβ in microglia and other immune cells [59, 201], respectively. For example, inhibition of DAGLβ following injury decreases local PGE_2 production and attenuates chronic pain-like behaviors in mouse models of neuropathic and inflammatory pain [225]. Furthermore, targeted delivery of the DAGLβ inhibitor KT-109 using liposomes produces 80% inactivation in macrophages without altering activity in other tissues such as brain and heart, and reduces LPS-induced allodynia [228], thus providing more direct evidence for peripheral immune DAGLβ in the inflammatory response. Alternatively, pharmacological or genetic inactivation of DAGLα abrogates production of prostaglandins in brain tissue, blocks central LPS-induced prostaglandin release and blunts the fever response in these mice [225]. While the role of eCBs in prostaglandin signaling has become clear, our understanding of their contribution(s) to other eicosanoid pathways remains limited, and future studies will establish potential links with lipoxygenases and cytochrome P450s.

Eicosanoid-Like 2-AG Metabolites Endocannabinoids also may interact directly with cyclooxygenases as substrates to produce prostaglandin-like compounds with unique biological effects [251]. While the (S)-enantiomers of nonsteroidal anti-inflammatory drugs ($NSAID_S$) such as ibuprofen and naproxen inhibit cyclooxygenase to prevent the formation of proinflammatory prostaglandins [252], the (R)-enantiomers of these compounds accomplish only minimal inhibition of enzymatic activity against arachidonic acid substrate. However, (R)-NSAIDs attenuate cyclooxygenase-dependent activity with AEA and 2-AG, thereby acting as potent substrate-selective inhibitors of this class of lipid signals [253]. Importantly, (R)-NSAIDS exhibit antihyperalgesic activity in models of neuropathic pain that is superior to that of traditionally prescribed (S)-$NSAID_S$ [254, 255]. Furthermore, levels of endogenous prostaglandin E_2 glycerol ester are elevated in the carrageenan model of inflammatory pain, contributing to thermal hyperalgesia that is not fully reversed by prostaglandin receptor antagonists [256]. Accordingly, prostaglandin

E_2 glycerol ester functions as an agonist for the g-protein coupled receptor P_2Y_6, with almost four orders of magnitude more potency than the prototypical agonist uridine diphosphate [257]. In a model of colon inflammation, prostaglandin D_2 glycerol ester derived from 2-AG, but not related metabolites arising from arachidonic acid or AEA, reduced dextran sulfate sodium-induced colitis in mice and are blocked by antagonists of traditional prostaglandin receptors, DP_1 and PPARy [258].

Eicosanoid-Like Anandamide Metabolites While serving only a limited role in colon inflammation, prostaglandin D_2 ethanolamine induces skin cancer apoptosis independent of the putative DP receptors, but the precise mechanism remains unclear [259]. Perhaps the most well-established endogenous prostamide, prostaglandin $F_2\alpha$ ethanolamine exerts minimal activity through the prostaglandin $F_{2\alpha}$ receptor and instead likely acts through an FP receptor variant [260]. A structural analogue of prostaglandin $F_{2\alpha}$ ethanolamine, Bimatoprost is an FDA-approved drug marketed under the name Lumigan® (Allergan) for reducing intraocular pressure as a treatment for glaucoma [261], and a sustained-release formulation of Bimatoprost is currently undergoing testing in Phase I/II clinical trials [262]. Patients reported longer and fuller eyelashes during administration of Bimatoprost, so the drug was repurposed as Latisse® and received FDA approval for the treatment of eyelash hypotrichosis [263] In contrast, an antagonist of the prostamide $F_{2\alpha}$ receptor AGN211336 reduces inflammatory pain in mice [264]. Substrate-selective inhibitors affect a number of different pathologies in mice including stress and anxiety-like behaviors (23912944), and future research likely will uncover additional biological roles for these lipids as signaling molecules.

While it has become clear that cyclooxygenases utilize 2-AG and AEA as substrates, research investigating the biological activity of corresponding lipoxygenase and cytochrome P450 enzymes on these lipid species is limited [251] and future studies will help clarify their role(s) in endocannabinoid biology. Collectively, these results suggest a number of novel potential therapeutic avenues for endocannabinoid metabolism inhibitors.

8.8 Conclusions and Future Directions

Considerable efforts have been concentrated on targeting endocannabinoid biosynthetic and degradative enzymes as alternatives to CB_1-receptor-based therapeutics that can produce serious adverse effects associated with a number of failures in clinical trials. In general, potent and selective inhibitors of FAAH or MGLL that are devoid of off-target effects have demonstrated safety and tolerability in human volunteers. These drug candidates could be met with success in the treatment of neurological diseases and pain if caution is exercised in interpreting preclinical data as well as selection of clinical indication and output measures. Small molecule inhibitors of other eCB enzymes remain in the preclinical discovery stage, but current research suggests some potential druggable targets. DAGLα serves an important physiological role in metabolism and brain function, yet its temporal inhibition may improve smoking cessation. Chemical inactivation of DAGLβ by liposome-mediated delivery of a peripherally-restricted inhibitor reduces inflammation and pain-like behaviors in mice. DAGL inhibitors undoubtedly will benefit from further chemical optimization in order to improve both selectivity and brain-barrier permeability for CNS indications. In addition, future studies elucidating pathways of anandamide synthesis and exploring substrate-specific inhibitors of cyclooxygenases, lipoxygenases and cytochrome P450s may yield additional promising targets for drug discovery in the cannabinoid therapeutic space.

Acknowledgements This work was supported by the National Institute on Drug Abuse (MWB, R00-DA03586) and National Institute of Arthritis and Musculoskeletal and Skin Diseases (AMG, R01-AR075241).

Declaration of Competing Interest The authors declare that they have no competing interests.

References

1. Mechoulam R, Gaoni Y (1965) A total synthesis of Dl-Delta-1-tetrahydrocannabinol, the active constituent of hashish. J Am Chem Soc 87:3273–3275
2. Pertwee RG (2006) Cannabinoid pharmacology: the first 66 years. Br J Pharmacol 147(Suppl 1):S163–S171
3. Mechoulam R (2005) Plant cannabinoids: a neglected pharmacological treasure trove. Br J Pharmacol 146:913–915
4. Perez-Reyes M, Timmons MC, Lipton MA, Davis KH, Wall ME (1972) Intravenous injection in man of 9 -tetrahydrocannabinol and 11-OH- 9 -tetrahydrocannabinol. Science 177:633–635
5. Little PJ, Compton DR, Johnson MR, Melvin LS, Martin BR (1988) Pharmacology and stereoselectivity of structurally novel cannabinoids in mice. J Pharmacol Exp Ther 247:1046–1051
6. Miliano C, Serpelloni G, Rimondo C, Mereu M, Marti M, De Luca MA (2016) Neuropharmacology of new psychoactive substances (NPS): focus on the rewarding and reinforcing properties of cannabimimetics and amphetamine-like stimulants. Front Neurosci 10:153
7. Pertwee RG (2009) Emerging strategies for exploiting cannabinoid receptor agonists as medicines. Br J Pharmacol 156:397–411
8. Rahn EJ, Hohmann AG (2009) Cannabinoids as pharmacotherapies for neuropathic pain: from the bench to the bedside. Neurotherapeutics 6:713–737
9. Syed YY, McKeage K, Scott LJ (2014) Delta-9-tetrahydrocannabinol/cannabidiol (Sativex(R)): a review of its use in patients with moderate to severe spasticity due to multiple sclerosis. Drugs 74:563–578
10. Horder J, Browning M, Di Simplicio M, Cowen PJ, Harmer CJ (2012) Effects of 7 days of treatment with the cannabinoid type 1 receptor antagonist, rimonabant, on emotional processing. J Psychopharmacol 26:125–132
11. Horder J, Cowen PJ, Di Simplicio M, Browning M, Harmer CJ (2009) Acute administration of the cannabinoid CB1 antagonist rimonabant impairs positive affective memory in healthy volunteers. Psychopharmacology 205:85–91
12. Kipnes MS, Hollander P, Fujioka K, Gantz I, Seck T, Erondu N, Shentu Y, Lu K, Suryawanshi S, Chou M, Johnson-Levonas AO, Heymsfield SB, Shapiro D, Kaufman KD, Amatruda JM (2010) A one-year study to assess the safety and efficacy of the CB1R inverse agonist taranabant in overweight and obese patients with type 2 diabetes. Diabetes Obes Metab 12:517–531
13. Morrison MF, Ceesay P, Gantz I, Kaufman KD, Lines CR (2010) Randomized, controlled, double-blind trial of taranabant for smoking cessation. Psychopharmacology 209:245–253
14. Wadden TA, Fujioka K, Toubro S, Gantz I, Erondu NE, Chen M, Suryawanshi S, Carofano W, Johnson-Levonas AO, Shapiro DR, Kaufman KD, Heymsfield SB, Amatruda JM (2010) A randomized trial of lifestyle modification and taranabant for maintaining weight loss achieved with a low-calorie diet. Obesity (Silver Spring) 18:2301–2310
15. Pertwee RG (1997) Pharmacology of cannabinoid CB1 and CB2 receptors. Pharmacol Ther 74:129–180
16. Devane WA, Dysarz FA 3rd, Johnson MR, Melvin LS, Howlett AC (1988) Determination and characterization of a cannabinoid receptor in rat brain. Mol Pharmacol 34:605–613
17. Herkenham M, Lynn AB, Little MD, Johnson MR, Melvin LS, de Costa BR, Rice KC (1990) Cannabinoid receptor localization in brain. Proc Natl Acad Sci U S A 87:1932–1936
18. Gerard CM, Mollereau C, Vassart G, Parmentier M (1991) Molecular cloning of a human cannabinoid receptor which is also expressed in testis. Biochem J 279(Pt 1):129–134
19. Matsuda LA, Lolait SJ, Brownstein MJ, Young AC, Bonner TI (1990) Structure of a cannabinoid receptor and functional expression of the cloned cDNA. Nature 346:561–564
20. Munro S, Thomas KL, Abu-Shaar M (1993) Molecular characterization of a peripheral receptor for cannabinoids. Nature 365:61–65
21. Bayewitch M, Avidor-Reiss T, Levy R, Barg J, Mechoulam R, Vogel Z (1995) The peripheral cannabinoid receptor: adenylate cyclase inhibition and G protein coupling. FEBS Lett 375:143–147
22. Felder CC, Briley EM, Axelrod J, Simpson JT, Mackie K, Devane WA (1993) Anandamide, an endogenous cannabimimetic eicosanoid, binds to the cloned human cannabinoid receptor and stimulates receptor-mediated signal transduction. Proc Natl Acad Sci U S A 90:7656–7660
23. Matsuda LA (1997) Molecular aspects of cannabinoid receptors. Crit Rev Neurobiol 11:143–166
24. Maresz K, Carrier EJ, Ponomarev ED, Hillard CJ, Dittel BN (2005) Modulation of the cannabinoid CB2 receptor in microglial cells in response to inflammatory stimuli. J Neurochem 95:437–445
25. Cabral GA, Marciano-Cabral F (2005) Cannabinoid receptors in microglia of the central nervous system: immune functional relevance. J Leukoc Biol 78:1192–1197

26. Herkenham M, Lynn AB, Johnson MR, Melvin LS, de Costa BR, Rice KC (1991) Characterization and localization of cannabinoid receptors in rat brain: a quantitative in vitro autoradiographic study. J Neurosci 11:563–583
27. Rinaldi-Carmona M, Barth F, Heaulme M, Shire D, Calandra B, Congy C, Martinez S, Maruani J, Neliat G, Caput D et al (1994) SR141716A, a potent and selective antagonist of the brain cannabinoid receptor. FEBS Lett 350:240–244
28. Gueudet C, Santucci V, Rinaldi-Carmona M, Soubrie P, Le Fur G (1995) The CB1 cannabinoid receptor antagonist SR 141716A affects A9 dopamine neuronal activity in the rat. Neuroreport 6:1421–1425
29. French ED (1997) delta9-tetrahydrocannabinol excites rat VTA dopamine neurons through activation of cannabinoid CB1 but not opioid receptors. Neurosci Lett 226:159–162
30. Aceto MD, Scates SM, Lowe JA, Martin BR (1995) Cannabinoid precipitated withdrawal by the selective cannabinoid receptor antagonist, SR 141716A. Eur J Pharmacol 282:R1–R2
31. Tsou K, Patrick SL, Walker JM (1995) Physical withdrawal in rats tolerant to delta 9-tetrahydrocannabinol precipitated by a cannabinoid receptor antagonist. Eur J Pharmacol 280:R13–R15
32. Aceto MD, Scates SM, Lowe JA, Martin BR (1996) Dependence on delta 9-tetrahydrocannabinol: studies on precipitated and abrupt withdrawal. J Pharmacol Exp Ther 278:1290–1295
33. Collins DR, Pertwee RG, Davies SN (1995) Prevention by the cannabinoid antagonist, SR141716A, of cannabinoid-mediated blockade of long-term potentiation in the rat hippocampal slice. Br J Pharmacol 115:869–870
34. Terranova JP, Michaud JC, Le Fur G, Soubrie P (1995) Inhibition of long-term potentiation in rat hippocampal slices by anandamide and WIN55212-2: reversal by SR141716 A, a selective antagonist of CB1 cannabinoid receptors. Naunyn Schmiedeberg's Arch Pharmacol 352:576–579
35. Reche I, Fuentes JA, Ruiz-Gayo M (1996) A role for central cannabinoid and opioid systems in peripheral delta 9-tetrahydrocannabinol-induced analgesia in mice. Eur J Pharmacol 301:75–81
36. Richardson JD, Aanonsen L, Hargreaves KM (1997) SR 141716A, a cannabinoid receptor antagonist, produces hyperalgesia in untreated mice. Eur J Pharmacol 319:R3–R4
37. Herzberg U, Eliav E, Bennett GJ, Kopin IJ (1997) The analgesic effects of R(+)-WIN 55,212-2 mesylate, a high affinity cannabinoid agonist, in a rat model of neuropathic pain. Neurosci Lett 221:157–160
38. Christensen R, Kristensen PK, Bartels EM, Bliddal H, Astrup A (2007) Efficacy and safety of the weight-loss drug rimonabant: a meta-analysis of randomised trials. Lancet 370:1706–1713
39. Sam AH, Salem V, Ghatei MA (2011) Rimonabant: from RIO to ban. J Obes 2011:432607
40. Ashton JC, Glass M (2007) The cannabinoid CB2 receptor as a target for inflammation-dependent neurodegeneration. Curr Neuropharmacol 5:73–80
41. Katchan V, David P, Shoenfeld Y (2016) Cannabinoids and autoimmune diseases: a systematic review. Autoimmun Rev 15:513–528
42. Morales P, Goya P, Jagerovic N (2018) Emerging strategies targeting CB2 cannabinoid receptor: biased agonism and allosterism. Biochem Pharmacol 157:8–17
43. Soethoudt M, Grether U, Fingerle J, Grim TW, Fezza F, de Petrocellis L, Ullmer C, Rothenhausler B, Perret C, van Gils N, Finlay D, MacDonald C, Chicca A, Gens MD, Stuart J, de Vries H, Mastrangelo N, Xia L, Alachouzos G, Baggelaar MP, Martella A, Mock ED, Deng H, Heitman LH, Connor M, Di Marzo V, Gertsch J, Lichtman AH, Maccarrone M, Pacher P, Glass M, van der Stelt M (2017) Cannabinoid CB2 receptor ligand profiling reveals biased signalling and off-target activity. Nat Commun 8:13958
44. Devane WA, Hanus L, Breuer A, Pertwee RG, Stevenson LA, Griffin G, Gibson D, Mandelbaum A, Etinger A, Mechoulam R (1992) Isolation and structure of a brain constituent that binds to the cannabinoid receptor. Science 258:1946–1949
45. Sugiura T, Kondo S, Sukagawa A, Nakane S, Shinoda A, Itoh K, Yamashita A, Waku K (1995) 2-Arachidonoylglycerol: a possible endogenous cannabinoid receptor ligand in brain. Biochem Biophys Res Commun 215:89–97
46. Mechoulam R, Ben-Shabat S, Hanus L, Ligumsky M, Kaminski NE, Schatz AR, Gopher A, Almog S, Martin BR, Compton DR et al (1995) Identification of an endogenous 2-monoglyceride, present in canine gut, that binds to cannabinoid receptors. Biochem Pharmacol 50:83–90
47. Re G, Barbero R, Miolo A, Di Marzo V (2007) Palmitoylethanolamide, endocannabinoids and related cannabimimetic compounds in protection against tissue inflammation and pain: potential use in companion animals. Vet J 173:21–30
48. Rodriguez de Fonseca F, Navarro M, Gomez R, Escuredo L, Nava F, Fu J, Murillo-Rodriguez E, Giuffrida A, LoVerme J, Gaetani S, Kathuria S, Gall C, Piomelli D (2001) An anorexic lipid mediator regulated by feeding. Nature 414:209–212
49. Lambert DM, Di Marzo V (1999) The palmitoylethanolamide and oleamide enigmas: are these two fatty acid amides cannabimimetic? Curr Med Chem 6:757–773
50. Thabuis C, Tissot-Favre D, Bezelgues JB, Martin JC, Cruz-Hernandez C, Dionisi F, Destaillats F (2008) Biological functions and metabolism of oleoylethanolamide. Lipids 43:887–894
51. Hardison S, Weintraub ST, Giuffrida A (2006) Quantification of endocannabinoids in rat biological samples by GC/MS: technical and theoretical considerations. Prostaglandins Other Lipid Mediat 81:106–112

52. Richardson D, Ortori CA, Chapman V, Kendall DA, Barrett DA (2007) Quantitative profiling of endocannabinoids and related compounds in rat brain using liquid chromatography-tandem electrospray ionization mass spectrometry. Anal Biochem 360:216–226
53. Kingsley PJ, Marnett LJ (2007) LC-MS-MS analysis of neutral eicosanoids. Methods Enzymol 433:91–112
54. Kingsley PJ, Marnett LJ (2009) Analysis of endocannabinoids, their congeners and COX-2 metabolites. J Chromatogr B Analyt Technol Biomed Life Sci 877:2746–2754
55. Buczynski MW, Parsons LH (2010) Quantification of brain endocannabinoid levels: methods, interpretations and pitfalls. Br J Pharmacol 160:423–442
56. Giuffrida A, Parsons LH, Kerr TM, Rodriguez de Fonseca F, Navarro M, Piomelli D (1999) Dopamine activation of endogenous cannabinoid signaling in dorsal striatum. Nat Neurosci 2:358–363
57. Wiskerke J, Irimia C, Cravatt BF, De Vries TJ, Schoffelmeer AN, Pattij T, Parsons LH (2012) Characterization of the effects of reuptake and hydrolysis inhibition on interstitial endocannabinoid levels in the brain: an in vivo microdialysis study. ACS Chem Neurosci 3:407–417
58. Ahn K, Johnson DS, Mileni M, Beidler D, Long JZ, McKinney MK, Weerapana E, Sadagopan N, Liimatta M, Smith SE, Lazerwith S, Stiff C, Kamtekar S, Bhattacharya K, Zhang Y, Swaney S, Van Becelaere K, Stevens RC, Cravatt BF (2009) Discovery and characterization of a highly selective FAAH inhibitor that reduces inflammatory pain. Chem Biol 16:411–420
59. Hsu KL, Tsuboi K, Adibekian A, Pugh H, Masuda K, Cravatt BF (2012) DAGLbeta inhibition perturbs a lipid network involved in macrophage inflammatory responses. Nat Chem Biol 8:999–1007
60. Long JZ, Nomura DK, Cravatt BF (2009) Characterization of monoacylglycerol lipase inhibition reveals differences in central and peripheral endocannabinoid metabolism. Chem Biol 16:744–753
61. Ogasawara D, Deng H, Viader A, Baggelaar MP, Breman A, den Dulk H, van den Nieuwendijk AM, Soethoudt M, van der Wel T, Zhou J, Overkleeft HS, Sanchez-Alavez M, Mori S, Nguyen W, Conti B, Liu X, Chen Y, Liu QS, Cravatt BF, van der Stelt M (2016) Rapid and profound rewiring of brain lipid signaling networks by acute diacylglycerol lipase inhibition. Proc Natl Acad Sci U S A 113:26–33
62. Bachovchin DA, Cravatt BF (2012) The pharmacological landscape and therapeutic potential of serine hydrolases. Nat Rev Drug Discov 11:52–68
63. Rogers SL, Doody RS, Mohs RC, Friedhoff LT (1998) Donepezil improves cognition and global function in Alzheimer disease: a 15-week, double-blind, placebo-controlled study. Donepezil Study Group. Arch Intern Med 158:1021–1031
64. Duysen EG, Stribley JA, Fry DL, Hinrichs SH, Lockridge O (2002) Rescue of the acetylcholinesterase knockout mouse by feeding a liquid diet; phenotype of the adult acetylcholinesterase deficient mouse. Brain Res Dev Brain Res 137:43–54
65. Tuin AW, Mol MA, van den Berg RM, Fidder A, van der Marel GA, Overkleeft HS, Noort D (2009) Activity-based protein profiling reveals broad reactivity of the nerve agent sarin. Chem Res Toxicol 22:683–689
66. Schreiber SL, Kotz JD, Li M, Aube J, Austin CP, Reed JC, Rosen H, White EL, Sklar LA, Lindsley CW, Alexander BR, Bittker JA, Clemons PA, de Souza A, Foley MA, Palmer M, Shamji AF, Wawer MJ, McManus O, Wu M, Zou B, Yu H, Golden JE, Schoenen FJ, Simeonov A, Jadhav A, Jackson MR, Pinkerton AB, Chung TD, Griffin PR, Cravatt BF, Hodder PS, Roush WR, Roberts E, Chung DH, Jonsson CB, Noah JW, Severson WE, Ananthan S, Edwards B, Oprea TI, Conn PJ, Hopkins CR, Wood MR, Stauffer SR, Emmitte KA, Team NIHMLP (2015) Advancing biological understanding and therapeutics discovery with small-molecule probes. Cell 161:1252–1265
67. Liu Y, Patricelli MP, Cravatt BF (1999) Activity-based protein profiling: the serine hydrolases. Proc Natl Acad Sci U S A 96:14694–14699
68. Janssen FJ, Deng H, Baggelaar MP, Allara M, van der Wel T, den Dulk H, Ligresti A, van Esbroeck AC, McGuire R, Di Marzo V, Overkleeft HS, van der Stelt M (2014) Discovery of glycine sulfonamides as dual inhibitors of sn-1-diacylglycerol lipase alpha and alpha/beta-hydrolase domain 6. J Med Chem 57:6610–6622
69. Baggelaar MP, Chameau PJ, Kantae V, Hummel J, Hsu KL, Janssen F, van der Wel T, Soethoudt M, Deng H, den Dulk H, Allara M, Florea BI, Di Marzo V, Wadman WJ, Kruse CG, Overkleeft HS, Hankemeier T, Werkman TR, Cravatt BF, van der Stelt M (2015) Highly selective, reversible inhibitor identified by comparative chemoproteomics modulates diacylglycerol lipase activity in neurons. J Am Chem Soc 137:8851–8857
70. Baggelaar MP, Van der Stelt M (2017) Competitive ABPP of serine hydrolases: a case study on DAGL-alpha. Methods Mol Biol 1491:161–169
71. van Rooden EJ, Florea BI, Deng H, Baggelaar MP, van Esbroeck ACM, Zhou J, Overkleeft HS, van der Stelt M (2018) Mapping in vivo target interaction profiles of covalent inhibitors using chemical proteomics with label-free quantification. Nat Protoc 13:752–767
72. Niphakis MJ, Cravatt BF (2014) Enzyme inhibitor discovery by activity-based protein profiling. Annu Rev Biochem 83:341–377
73. Johnson DS, Stiff C, Lazerwith SE, Kesten SR, Fay LK, Morris M, Beidler D, Liimatta MB, Smith SE, Dudley DT, Sadagopan N, Bhattachar SN, Kesten SJ, Nomanbhoy TK, Cravatt BF, Ahn K (2011) Discovery of PF-04457845: a highly potent, orally bioavailable, and selective urea FAAH inhibitor. ACS Med Chem Lett 2:91–96

74. Ahn K, Smith SE, Liimatta MB, Beidler D, Sadagopan N, Dudley DT, Young T, Wren P, Zhang Y, Swaney S, Van Becelaere K, Blankman JL, Nomura DK, Bhattachar SN, Stiff C, Nomanbhoy TK, Weerapana E, Johnson DS, Cravatt BF (2011) Mechanistic and pharmacological characterization of PF-04457845: a highly potent and selective fatty acid amide hydrolase inhibitor that reduces inflammatory and noninflammatory pain. J Pharmacol Exp Ther 338:114–124
75. Long JZ, Nomura DK, Vann RE, Walentiny DM, Booker L, Jin X, Burston JJ, Sim-Selley LJ, Lichtman AH, Wiley JL, Cravatt BF (2009) Dual blockade of FAAH and MAGL identifies behavioral processes regulated by endocannabinoid crosstalk in vivo. Proc Natl Acad Sci U S A 106:20270–20275
76. Cravatt BF, Giang DK, Mayfield SP, Boger DL, Lerner RA, Gilula NB (1996) Molecular characterization of an enzyme that degrades neuromodulatory fatty-acid amides. Nature 384:83–87
77. Giang DK, Cravatt BF (1997) Molecular characterization of human and mouse fatty acid amide hydrolases. Proc Natl Acad Sci U S A 94:2238–2242
78. Long JZ, LaCava M, Jin X, Cravatt BF (2011) An anatomical and temporal portrait of physiological substrates for fatty acid amide hydrolase. J Lipid Res 52:337–344
79. Doenni VM, Gray JM, Song CM, Patel S, Hill MN, Pittman QJ (2016) Deficient adolescent social behavior following early-life inflammation is ameliorated by augmentation of anandamide signaling. Brain Behav Immun 58:237–247
80. Natividad LA, Buczynski MW, Herman MA, Kirson D, Oleata CS, Irimia C, Polis I, Ciccocioppo R, Roberto M, Parsons LH (2017) Constitutive increases in amygdalar corticotropin-releasing factor and fatty acid amide hydrolase drive an anxious phenotype. Biol Psychiatry 82:500–510
81. Bedse G, Bluett RJ, Patrick TA, Romness NK, Gaulden AD, Kingsley PJ, Plath N, Marnett LJ, Patel S (2018) Therapeutic endocannabinoid augmentation for mood and anxiety disorders: comparative profiling of FAAH, MAGL and dual inhibitors. Transl Psychiatry 8:92
82. Gobbi G, Bambico FR, Mangieri R, Bortolato M, Campolongo P, Solinas M, Cassano T, Morgese MG, Debonnel G, Duranti A, Tontini A, Tarzia G, Mor M, Trezza V, Goldberg SR, Cuomo V, Piomelli D (2005) Antidepressant-like activity and modulation of brain monoaminergic transmission by blockade of anandamide hydrolysis. Proc Natl Acad Sci U S A 102:18620–18625
83. Naidu PS, Varvel SA, Ahn K, Cravatt BF, Martin BR, Lichtman AH (2007) Evaluation of fatty acid amide hydrolase inhibition in murine models of emotionality. Psychopharmacology 192:61–70
84. Fichna J, Salaga M, Stuart J, Saur D, Sobczak M, Zatorski H, Timmermans JP, Bradshaw HB, Ahn K, Storr MA (2014) Selective inhibition of FAAH produces antidiarrheal and antinociceptive effect mediated by endocannabinoids and cannabinoid-like fatty acid amides. Neurogastroenterol Motil 26:470–481
85. Salaga M, Mokrowiecka A, Zakrzewski PK, Cygankiewicz A, Leishman E, Sobczak M, Zatorski H, Malecka-Panas E, Kordek R, Storr M, Krajewska WM, Bradshaw HB, Fichna J (2014) Experimental colitis in mice is attenuated by changes in the levels of endocannabinoid metabolites induced by selective inhibition of fatty acid amide hydrolase (FAAH). J Crohns Colitis 8:998–1009
86. Sakin YS, Dogrul A, Ilkaya F, Seyrek M, Ulas UH, Gulsen M, Bagci S (2015) The effect of FAAH, MAGL, and dual FAAH/MAGL inhibition on inflammatory and colorectal distension-induced visceral pain models in rodents. Neurogastroenterol Motil 27:936–944
87. Rock EM, Limebeer CL, Ward JM, Cohen A, Grove K, Niphakis MJ, Cravatt BF, Parker LA (2015) Interference with acute nausea and anticipatory nausea in rats by fatty acid amide hydrolase (FAAH) inhibition through a PPARalpha and CB1 receptor mechanism, respectively: a double dissociation. Psychopharmacology 232:3841–3848
88. Wasilewski A, Misicka A, Sacharczuk M, Fichna J (2017) Modulation of the endocannabinoid system by the fatty acid amide hydrolase, monoacylglycerol and diacylglycerol lipase inhibitors as an attractive target for secretory diarrhoea therapy. J Physiol Pharmacol 68:591–596
89. Cippitelli A, Cannella N, Braconi S, Duranti A, Tontini A, Bilbao A, Defonseca FR, Piomelli D, Ciccocioppo R (2008) Increase of brain endocannabinoid anandamide levels by FAAH inhibition and alcohol abuse behaviours in the rat. Psychopharmacology 198:449–460
90. Merritt LL, Martin BR, Walters C, Lichtman AH, Damaj MI (2008) The endogenous cannabinoid system modulates nicotine reward and dependence. J Pharmacol Exp Ther 326:483–492
91. Schlosburg JE, Carlson BL, Ramesh D, Abdullah RA, Long JZ, Cravatt BF, Lichtman AH (2009) Inhibitors of endocannabinoid-metabolizing enzymes reduce precipitated withdrawal responses in THC-dependent mice. AAPS J 11:342–352
92. Cippitelli A, Astarita G, Duranti A, Caprioli G, Ubaldi M, Stopponi S, Kallupi M, Sagratini G, Rodriguez de Fonseca F, Piomelli D, Ciccocioppo R (2011) Endocannabinoid regulation of acute and protracted nicotine withdrawal: effect of FAAH inhibition. PLoS One 6:e28142
93. Muldoon PP, Lichtman AH, Parsons LH, Damaj MI (2013) The role of fatty acid amide hydrolase inhibition in nicotine reward and dependence. Life Sci 92:458–462
94. Kinsey SG, Long JZ, Cravatt BF, Lichtman AH (2010) Fatty acid amide hydrolase and monoacylglycerol lipase inhibitors produce anti-allodynic effects in mice through distinct cannabinoid receptor mechanisms. J Pain 11:1420–1428

95. Booker L, Kinsey SG, Abdullah RA, Blankman JL, Long JZ, Ezzili C, Boger DL, Cravatt BF, Lichtman AH (2012) The fatty acid amide hydrolase (FAAH) inhibitor PF-3845 acts in the nervous system to reverse LPS-induced tactile allodynia in mice. Br J Pharmacol 165:2485–2496
96. Nasirinezhad F, Jergova S, Pearson JP, Sagen J (2015) Attenuation of persistent pain-related behavior by fatty acid amide hydrolase (FAAH) inhibitors in a rat model of HIV sensory neuropathy. Neuropharmacology 95:100–109
97. Bhuniya D, Kharul RK, Hajare A, Shaikh N, Bhosale S, Balwe S, Begum F, De S, Athavankar S, Joshi D, Madgula V, Joshi K, Raje AA, Meru AV, Magdum A, Mookhtiar KA, Barbhaiya R (2019) Discovery and evaluation of novel FAAH inhibitors in neuropathic pain model. Bioorg Med Chem Lett 29:238–243
98. D'Souza DC, Cortes-Briones J, Creatura G, Bluez G, Thurnauer H, Deaso E, Bielen K, Surti T, Radhakrishnan R, Gupta A, Gupta S, Cahill J, Sherif MA, Makriyannis A, Morgan PT, Ranganathan M, Skosnik PD (2019) Efficacy and safety of a fatty acid amide hydrolase inhibitor (PF-04457845) in the treatment of cannabis withdrawal and dependence in men: a double-blind, placebo-controlled, parallel group, phase 2a single-site randomised controlled trial. Lancet Psychiatry 6:35–45
99. Mayo LM, Asratian A, Linde J, Morena M, Haataja R, Hammar V, Augier G, Hill MN, Heilig M (2019) Elevated anandamide, enhanced recall of fear extinction, and attenuated stress responses following inhibition of fatty acid amide hydrolase: a randomized, controlled experimental medicine trial. Biol Psychiatry 87(6):538–547
100. Huggins JP, Smart TS, Langman S, Taylor L, Young T (2012) An efficient randomised, placebo-controlled clinical trial with the irreversible fatty acid amide hydrolase-1 inhibitor PF-04457845, which modulates endocannabinoids but fails to induce effective analgesia in patients with pain due to osteoarthritis of the knee. Pain 153:1837–1846
101. Wagenlehner FME, van Till JWO, Houbiers JGA, Martina RV, Cerneus DP, Melis J, Majek A, Vjaters E, Urban M, Ramonas H, Shoskes DA, Nickel JC (2017) Fatty acid amide hydrolase inhibitor treatment in men with chronic prostatitis/chronic pelvic pain syndrome: an adaptive double-blind, randomized controlled trial. Urology 103:191–197
102. Kerbrat A, Ferre JC, Fillatre P, Ronziere T, Vannier S, Carsin-Nicol B, Lavoue S, Verin M, Gauvrit JY, Le Tulzo Y, Edan G (2016) Acute neurologic disorder from an inhibitor of fatty acid amide hydrolase. N Engl J Med 375:1717–1725
103. Brosen K, Funck-Brentano C, Kroemer HK, Pirmohamed M, Schwab M (2017) Open letter on access to the BIA 10-2474 clinical trial data. Lancet 389:156
104. van Esbroeck ACM, Janssen APA, Cognetta AB 3rd, Ogasawara D, Shpak G, van der Kroeg M, Kantae V, Baggelaar MP, de Vrij FMS, Deng H, Allara M, Fezza F, Lin Z, van der Wel T, Soethoudt M, Mock ED, den Dulk H, Baak IL, Florea BI, Hendriks G, De Petrocellis L, Overkleeft HS, Hankemeier T, De Zeeuw CI, Di Marzo V, Maccarrone M, Cravatt BF, Kushner SA, van der Stelt M (2017) Activity-based protein profiling reveals off-target proteins of the FAAH inhibitor BIA 10–2474. Science 356:1084–1087
105. Huang Z, Ogasawara D, Seneviratne UI, Cognetta AB 3rd, Am Ende CW, Nason DM, Lapham K, Litchfield J, Johnson DS, Cravatt BF (2019) Global portrait of protein targets of metabolites of the neurotoxic compound BIA 10-2474. ACS Chem Biol 14:192–197
106. Hayes AW (2019) Commentary on BIA 10-2474. Regul Toxicol Pharmacol 104541
107. Li GL, Winter H, Arends R, Jay GW, Le V, Young T, Huggins JP (2012) Assessment of the pharmacology and tolerability of PF-04457845, an irreversible inhibitor of fatty acid amide hydrolase-1, in healthy subjects. Br J Clin Pharmacol 73:706–716
108. Pawsey S, Wood M, Browne H, Donaldson K, Christie M, Warrington S (2016) Safety, tolerability and pharmacokinetics of FAAH inhibitor V158866: a double-blind, randomised, placebo-controlled phase I study in healthy volunteers. Drugs R D 16:181–191
109. Postnov A, Schmidt ME, Pemberton DJ, de Hoon J, van Hecken A, van den Boer M, Zannikos P, van der Ark P, Palmer JA, Rassnick S, Celen S, Bormans G, van Laere K (2018) Fatty acid amide hydrolase inhibition by JNJ-42165279: a multiple-ascending dose and a positron emission tomography study in healthy volunteers. Clin Transl Sci 11:397–404
110. Di Marzo V (2012) Inhibitors of endocannabinoid breakdown for pain: not so FA(AH)cile, after all. Pain 153:1785–1786
111. Bryden LA, Nicholson JR, Doods H, Pekcec A (2015) Deficits in spontaneous burrowing behavior in the rat bilateral monosodium iodoacetate model of osteoarthritis: an objective measure of pain-related behavior and analgesic efficacy. Osteoarthr Cartil 23:1605–1612
112. Goparaju SK, Ueda N, Taniguchi K, Yamamoto S (1999) Enzymes of porcine brain hydrolyzing 2-arachidonoylglycerol, an endogenous ligand of cannabinoid receptors. Biochem Pharmacol 57:417–423
113. Dinh TP, Carpenter D, Leslie FM, Freund TF, Katona I, Sensi SL, Kathuria S, Piomelli D (2002) Brain monoglyceride lipase participating in endocannabinoid inactivation. Proc Natl Acad Sci U S A 99:10819–10824
114. Chanda PK, Gao Y, Mark L, Btesh J, Strassle BW, Lu P, Piesla MJ, Zhang MY, Bingham B, Uveges A, Kowal D, Garbe D, Kouranova EV, Ring RH, Bates B, Pangalos MN, Kennedy JD, Whiteside GT, Samad TA (2010) Monoacylglycerol lipase activity is a critical modulator of the tone and integrity

of the endocannabinoid system. Mol Pharmacol 78:996–1003
115. Pasquarelli N, Porazik C, Hanselmann J, Weydt P, Ferger B, Witting A (2015) Comparative biochemical characterization of the monoacylglycerol lipase inhibitor KML29 in brain, spinal cord, liver, spleen, fat and muscle tissue. Neuropharmacology 91:148–156
116. Taschler U, Eichmann TO, Radner FP, Grabner GF, Wolinski H, Storr M, Lass A, Schicho R, Zimmermann R (2015) Monoglyceride lipase deficiency causes desensitization of intestinal cannabinoid receptor type 1 and increased colonic mu-opioid receptor sensitivity. Br J Pharmacol 172:4419–4429
117. Petrenko AB, Yamazaki M, Sakimura K, Kano M, Baba H (2014) Augmented tonic pain-related behavior in knockout mice lacking monoacylglycerol lipase, a major degrading enzyme for the endocannabinoid 2-arachidonoylglycerol. Behav Brain Res 271:51–58
118. Kishimoto Y, Cagniard B, Yamazaki M, Nakayama J, Sakimura K, Kirino Y, Kano M (2015) Task-specific enhancement of hippocampus-dependent learning in mice deficient in monoacylglycerol lipase, the major hydrolyzing enzyme of the endocannabinoid 2-arachidonoylglycerol. Front Behav Neurosci 9:134
119. Imperatore R, Morello G, Luongo L, Taschler U, Romano R, De Gregorio D, Belardo C, Maione S, Di Marzo V, Cristino L (2015) Genetic deletion of monoacylglycerol lipase leads to impaired cannabinoid receptor CB(1)R signaling and anxiety-like behavior. J Neurochem 135:799–813
120. Zhong P, Pan B, Gao XP, Blankman JL, Cravatt BF, Liu QS (2011) Genetic deletion of monoacylglycerol lipase alters endocannabinoid-mediated retrograde synaptic depression in the cerebellum. J Physiol 589:4847–4855
121. Pan B, Wang W, Zhong P, Blankman JL, Cravatt BF, Liu QS (2011) Alterations of endocannabinoid signaling, synaptic plasticity, learning, and memory in monoacylglycerol lipase knock-out mice. J Neurosci 31:13420–13430
122. Hohmann AG, Suplita RL, Bolton NM, Neely MH, Fegley D, Mangieri R, Krey JF, Walker JM, Holmes PV, Crystal JD, Duranti A, Tontini A, Mor M, Tarzia G, Piomelli D (2005) An endocannabinoid mechanism for stress-induced analgesia. Nature 435:1108–1112
123. Makara JK, Mor M, Fegley D, Szabo SI, Kathuria S, Astarita G, Duranti A, Tontini A, Tarzia G, Rivara S, Freund TF, Piomelli D (2005) Selective inhibition of 2-AG hydrolysis enhances endocannabinoid signaling in hippocampus. Nat Neurosci 8:1139–1141
124. Quistad GB, Klintenberg R, Caboni P, Liang SN, Casida JE (2006) Monoacylglycerol lipase inhibition by organophosphorus compounds leads to elevation of brain 2-arachidonoylglycerol and the associated hypomotility in mice. Toxicol Appl Pharmacol 211:78–83
125. Szabo B, Urbanski MJ, Bisogno T, Di Marzo V, Mendiguren A, Baer WU, Freiman I (2006) Depolarization-induced retrograde synaptic inhibition in the mouse cerebellar cortex is mediated by 2-arachidonoylglycerol. J Physiol 577:263–280
126. Hashimotodani Y, Ohno-Shosaku T, Kano M (2007) Presynaptic monoacylglycerol lipase activity determines basal endocannabinoid tone and terminates retrograde endocannabinoid signaling in the hippocampus. J Neurosci 27:1211–1219
127. Nomura DK, Hudak CS, Ward AM, Burston JJ, Issa RS, Fisher KJ, Abood ME, Wiley JL, Lichtman AH, Casida JE (2008) Monoacylglycerol lipase regulates 2-arachidonoylglycerol action and arachidonic acid levels. Bioorg Med Chem Lett 18:5875–5878
128. Long JZ, Li W, Booker L, Burston JJ, Kinsey SG, Schlosburg JE, Pavon FJ, Serrano AM, Selley DE, Parsons LH, Lichtman AH, Cravatt BF (2009) Selective blockade of 2-arachidonoylglycerol hydrolysis produces cannabinoid behavioral effects. Nat Chem Biol 5:37–44
129. Ramesh D, Ross GR, Schlosburg JE, Owens RA, Abdullah RA, Kinsey SG, Long JZ, Nomura DK, Sim-Selley LJ, Cravatt BF, Akbarali HI, Lichtman AH (2011) Blockade of endocannabinoid hydrolytic enzymes attenuates precipitated opioid withdrawal symptoms in mice. J Pharmacol Exp Ther 339:173–185
130. Muldoon PP, Chen J, Harenza JL, Abdullah RA, Sim-Selley LJ, Cravatt BF, Miles MF, Chen X, Lichtman AH, Damaj MI (2015) Inhibition of monoacylglycerol lipase reduces nicotine withdrawal. Br J Pharmacol 172:869–882
131. Spradley JM, Guindon J, Hohmann AG (2010) Inhibitors of monoacylglycerol lipase, fatty-acid amide hydrolase and endocannabinoid transport differentially suppress capsaicin-induced behavioral sensitization through peripheral endocannabinoid mechanisms. Pharmacol Res 62:249–258
132. Khasabova IA, Chandiramani A, Harding-Rose C, Simone DA, Seybold VS (2011) Increasing 2-arachidonoyl glycerol signaling in the periphery attenuates mechanical hyperalgesia in a model of bone cancer pain. Pharmacol Res 64:60–67
133. Woodhams SG, Wong A, Barrett DA, Bennett AJ, Chapman V, Alexander SP (2012) Spinal administration of the monoacylglycerol lipase inhibitor JZL184 produces robust inhibitory effects on nociceptive processing and the development of central sensitization in the rat. Br J Pharmacol 167:1609–1619
134. Kinsey SG, O'Neal ST, Long JZ, Cravatt BF, Lichtman AH (2011) Inhibition of endocannabinoid catabolic enzymes elicits anxiolytic-like effects in the marble burying assay. Pharmacol Biochem Behav 98:21–27
135. Sumislawski JJ, Ramikie TS, Patel S (2011) Reversible gating of endocannabinoid plasticity in the amygdala by chronic stress: a potential role for monoacylglycerol lipase inhibition in the pre-

vention of stress-induced behavioral adaptation. Neuropsychopharmacology 36:2750–2761
136. Guindon J, Guijarro A, Piomelli D, Hohmann AG (2011) Peripheral antinociceptive effects of inhibitors of monoacylglycerol lipase in a rat model of inflammatory pain. Br J Pharmacol 163:1464–1478
137. Alhouayek M, Lambert DM, Delzenne NM, Cani PD, Muccioli GG (2011) Increasing endogenous 2-arachidonoylglycerol levels counteracts colitis and related systemic inflammation. FASEB J 25:2711–2721
138. Nomura DK, Morrison BE, Blankman JL, Long JZ, Kinsey SG, Marcondes MC, Ward AM, Hahn YK, Lichtman AH, Conti B, Cravatt BF (2011) Endocannabinoid hydrolysis generates brain prostaglandins that promote neuroinflammation. Science 334:809–813
139. Ye L, Zhang B, Seviour EG, Tao KX, Liu XH, Ling Y, Chen JY, Wang GB (2011) Monoacylglycerol lipase (MAGL) knockdown inhibits tumor cells growth in colorectal cancer. Cancer Lett 307:6–17
140. Zhang J, Liu Z, Lian Z, Liao R, Chen Y, Qin Y, Wang J, Jiang Q, Wang X, Gong J (2016) Monoacylglycerol lipase: a novel potential therapeutic target and prognostic Indicator for hepatocellular carcinoma. Sci Rep 6:35784
141. Xiang W, Shi R, Kang X, Zhang X, Chen P, Zhang L, Hou A, Wang R, Zhao Y, Zhao K, Liu Y, Ma Y, Luo H, Shang S, Zhang J, He F, Yu S, Gan L, Shi C, Li Y, Yang W, Liang H, Miao H (2018) Monoacylglycerol lipase regulates cannabinoid receptor 2-dependent macrophage activation and cancer progression. Nat Commun 9:2574
142. Kinsey SG, Nomura DK, O'Neal ST, Long JZ, Mahadevan A, Cravatt BF, Grider JR, Lichtman AH (2011) Inhibition of monoacylglycerol lipase attenuates nonsteroidal anti-inflammatory drug-induced gastric hemorrhages in mice. J Pharmacol Exp Ther 338:795–802
143. Crowe MS, Kinsey SG (2017) MAGL inhibition modulates gastric secretion and motility following NSAID exposure in mice. Eur J Pharmacol 807:198–204
144. Chen R, Zhang J, Wu Y, Wang D, Feng G, Tang YP, Teng Z, Chen C (2012) Monoacylglycerol lipase is a therapeutic target for Alzheimer's disease. Cell Rep 2:1329–1339
145. Valdeolivas S, Pazos MR, Bisogno T, Piscitelli F, Iannotti FA, Allara M, Sagredo O, Di Marzo V, Fernandez-Ruiz J (2013) The inhibition of 2-arachidonoyl-glycerol (2-AG) biosynthesis, rather than enhancing striatal damage, protects striatal neurons from malonate-induced death: a potential role of cyclooxygenase-2-dependent metabolism of 2-AG. Cell Death Dis 4:e862
146. Nader J, Rapino C, Gennequin B, Chavant F, Francheteau M, Makriyannis A, Duranti A, Maccarrone M, Solinas M, Thiriet N (2014) Prior stimulation of the endocannabinoid system prevents methamphetamine-induced dopaminergic neurotoxicity in the striatum through activation of CB2 receptors. Neuropharmacology 87:214–221
147. Fernandez-Suarez D, Celorrio M, Riezu-Boj JI, Ugarte A, Pacheco R, Gonzalez H, Oyarzabal J, Hillard CJ, Franco R, Aymerich MS (2014) Monoacylglycerol lipase inhibitor JZL184 is neuroprotective and alters glial cell phenotype in the chronic MPTP mouse model. Neurobiol Aging 35:2603–2616
148. Schlosburg JE, Blankman JL, Long JZ, Nomura DK, Pan B, Kinsey SG, Nguyen PT, Ramesh D, Booker L, Burston JJ, Thomas EA, Selley DE, Sim-Selley LJ, Liu QS, Lichtman AH, Cravatt BF (2010) Chronic monoacylglycerol lipase blockade causes functional antagonism of the endocannabinoid system. Nat Neurosci 13:1113–1119
149. Chang JW, Niphakis MJ, Lum KM, Cognetta AB 3rd, Wang C, Matthews ML, Niessen S, Buczynski MW, Parsons LH, Cravatt BF (2012) Highly selective inhibitors of monoacylglycerol lipase bearing a reactive group that is bioisosteric with endocannabinoid substrates. Chem Biol 19:579–588
150. Niphakis MJ, Cognetta AB 3rd, Chang JW, Buczynski MW, Parsons LH, Byrne F, Burston JJ, Chapman V, Cravatt BF (2013) Evaluation of NHS carbamates as a potent and selective class of endocannabinoid hydrolase inhibitors. ACS Chem Neurosci 4:1322–1332
151. Aaltonen N, Kedzierska E, Orzelska-Gorka J, Lehtonen M, Navia-Paldanius D, Jakupovic H, Savinainen JR, Nevalainen T, Laitinen JT, Parkkari T, Gynther M (2016) In vivo characterization of the ultrapotent monoacylglycerol lipase inhibitor {4-[bis-(benzo[d][1,3]dioxol-5-yl)methyl]-piperidin-1-yl}(1H-1,2,4-triazol-1-yl)m ethanone (JJKK-048). J Pharmacol Exp Ther 359:62–72
152. Granchi C, Lapillo M, Glasmacher S, Bononi G, Licari C, Poli G, El Boustani M, Caligiuri I, Rizzolio F, Gertsch J, Macchia M, Minutolo F, Tuccinardi T, Chicca A (2019) Optimization of a benzoylpiperidine class identifies a highly potent and selective reversible monoacylglycerol lipase (MAGL) inhibitor. J Med Chem 62:1932–1958
153. Cisar JS, Weber OD, Clapper JR, Blankman JL, Henry CL, Simon GM, Alexander JP, Jones TK, Ezekowitz RAB, O'Neill GP, Grice CA (2018) Identification of ABX-1431, a selective inhibitor of monoacylglycerol lipase and clinical candidate for treatment of neurological disorders. J Med Chem 61:9062–9084
154. Lotsch J, Oertel BG, Ultsch A (2014) Human models of pain for the prediction of clinical analgesia. Pain 155:2014–2021
155. McAllister LA, Butler CR, Mente S, O'Neil SV, Fonseca KR, Piro JR, Cianfrogna JA, Foley TL, Gilbert AM, Harris AR, Helal CJ, Johnson DS, Montgomery JI, Nason DM, Noell S, Pandit J, Rogers BN, Samad TA, Shaffer CL, da Silva RG, Uccello DP, Webb D, Brodney MA (2018) Discovery of Trifluoromethyl glycol Carbamates as potent and

selective covalent monoacylglycerol lipase (MAGL) inhibitors for treatment of neuroinflammation. J Med Chem 61:3008–3026
156. Zhang L, Butler CR, Maresca KP, Takano A, Nag S, Jia Z, Arakawa R, Piro JR, Samad T, Smith DL, Nason DM, O'Neil S, McAllister L, Schildknegt K, Trapa P, McCarthy TJ, Villalobos A, Halldin C (2019) Identification and development of an irreversible monoacylglycerol lipase (MAGL) positron emission tomography (PET) radioligand with high specificity. J Med Chem 62:8532–8543
157. Aida J, Fushimi M, Kusumoto T, Sugiyama H, Arimura N, Ikeda S, Sasaki M, Sogabe S, Aoyama K, Koike T (2018) Design, synthesis, and evaluation of piperazinyl pyrrolidin-2-ones as a novel series of reversible monoacylglycerol lipase inhibitors. J Med Chem 61:9205–9217
158. Wyatt RM, Fraser I, Welty N, Lord B, Wennerholm M, Sutton SW, Ameriks MK, Dugovic C, Yun S, White A, Nguyen L, Koudriakova T, Tian G, Suarez J, Szewczuk L, Bonnette W, Ahn K, Ghosh B, Flores CM, Connolly PJ, Zhu B, Macielag MJ, Brandt MR, Chevalier K, Zhang SP, Lovenberg TW, Bonaventure P (2019) Pharmacologic characterization of JNJ-42226314, [1-(4-fluorophenyl)indol-5-yl]-[3-[4-(thiazole-2-carbonyl)piperazin-1-yl]azetidin −1-yl] methanone, a reversible, selective and potent monoacylglycerol lipase inhibitor. J Pharmacol Exp Ther
159. Muccioli GG, Xu C, Odah E, Cudaback E, Cisneros JA, Lambert DM, Lopez Rodriguez ML, Bajjalieh S, Stella N (2007) Identification of a novel endocannabinoid-hydrolyzing enzyme expressed by microglial cells. J Neurosci 27:2883–2889
160. Blankman JL, Simon GM, Cravatt BF (2007) A comprehensive profile of brain enzymes that hydrolyze the endocannabinoid 2-arachidonoylglycerol. Chem Biol 14:1347–1356
161. Marrs WR, Blankman JL, Horne EA, Thomazeau A, Lin YH, Coy J, Bodor AL, Muccioli GG, Hu SS, Woodruff G, Fung S, Lafourcade M, Alexander JP, Long JZ, Li W, Xu C, Moller T, Mackie K, Manzoni OJ, Cravatt BF, Stella N (2010) The serine hydrolase ABHD6 controls the accumulation and efficacy of 2-AG at cannabinoid receptors. Nat Neurosci 13:951–957
162. Buczynski MW, Herman MA, Hsu KL, Natividad LA, Irimia C, Polis IY, Pugh H, Chang JW, Niphakis MJ, Cravatt BF, Roberto M, Parsons LH (2016) Diacylglycerol lipase disinhibits VTA dopamine neurons during chronic nicotine exposure. Proc Natl Acad Sci U S A 113:1086–1091
163. Cao JK, Kaplan J, Stella N (2019) ABHD6: its place in endocannabinoid Signaling and beyond. Trends Pharmacol Sci 40:267–277
164. Thomas G, Betters JL, Lord CC, Brown AL, Marshall S, Ferguson D, Sawyer J, Davis MA, Melchior JT, Blume LC, Howlett AC, Ivanova PT, Milne SB, Myers DS, Mrak I, Leber V, Heier C, Taschler U, Blankman JL, Cravatt BF, Lee RG, Crooke RM, Graham MJ, Zimmermann R, Brown HA, Brown JM (2013) The serine hydrolase ABHD6 is a critical regulator of the metabolic syndrome. Cell Rep 5:508–520
165. Zhao S, Mugabo Y, Ballentine G, Attane C, Iglesias J, Poursharifi P, Zhang D, Nguyen TA, Erb H, Prentki R, Peyot ML, Joly E, Tobin S, Fulton S, Brown JM, Madiraju SR, Prentki M (2016) Alpha/beta-hydrolase domain 6 deletion induces adipose browning and prevents obesity and type 2 diabetes. Cell Rep 14:2872–2888
166. Grabner GF, Fawzy N, Pribasnig MA, Trieb M, Taschler U, Holzer M, Schweiger M, Wolinski H, Kolb D, Horvath A, Breinbauer R, Rulicke T, Rabl R, Lass A, Stadlbauer V, Hutter-Paier B, Stauber RE, Fickert P, Zechner R, Marsche G, Eichmann TO, Zimmermann R (2019) Metabolic disease and ABHD6 alter the circulating bis(monoacylglycerol) phosphate profile in mice and humans. J Lipid Res 60:1020–1031
167. Tchantchou F, Zhang Y (2013) Selective inhibition of alpha/beta-hydrolase domain 6 attenuates neurodegeneration, alleviates blood brain barrier breakdown, and improves functional recovery in a mouse model of traumatic brain injury. J Neurotrauma 30:565–579
168. Patel JZ, Nevalainen TJ, Savinainen JR, Adams Y, Laitinen T, Runyon RS, Vaara M, Ahenkorah S, Kaczor AA, Navia-Paldanius D, Gynther M, Aaltonen N, Joharapurkar AA, Jain MR, Haka AS, Maxfield FR, Laitinen JT, Parkkari T (2015) Optimization of 1,2,5-thiadiazole carbamates as potent and selective ABHD6 inhibitors. ChemMedChem 10:253–265
169. Naydenov AV, Horne EA, Cheah CS, Swinney K, Hsu KL, Cao JK, Marrs W, Blankman JL, Tu S, Cherry AE, Fung S, Wen A, Li W, Saporito MS, Selley DE, Cravatt BF, Oakley JC, Stella N (2014) ABHD6 blockade exerts antiepileptic activity in PTZ-induced seizures and in spontaneous seizures in R6/2 mice. Neuron 83:361–371
170. Zareie P, Sadegh M, Palizvan MR, Moradi-Chameh H (2018) Anticonvulsive effects of endocannabinoids; an investigation to determine the role of regulatory components of endocannabinoid metabolism in the pentylenetetrazol induced tonic- clonic seizures. Metab Brain Dis 33:939–948
171. Alhouayek M, Masquelier J, Cani PD, Lambert DM, Muccioli GG (2013) Implication of the anti-inflammatory bioactive lipid prostaglandin D2-glycerol ester in the control of macrophage activation and inflammation by ABHD6. Proc Natl Acad Sci U S A 110:17558–17563
172. Wen J, Jones M, Tanaka M, Selvaraj P, Symes AJ, Cox B, Zhang Y (2018) WWL70 protects against chronic constriction injury-induced neuropathic pain in mice by cannabinoid receptor-independent mechanisms. J Neuroinflammation 15:9
173. Gruner BM, Schulze CJ, Yang D, Ogasawara D, Dix MM, Rogers ZN, Chuang CH, McFarland CD, Chiou SH, Brown JM, Cravatt BF, Bogyo M, Winslow

MM (2016) An in vivo multiplexed small-molecule screening platform. Nat Methods 13:883–889
174. Manterola A, Bernal-Chico A, Cipriani R, Canedo-Antelo M, Moreno-Garcia A, Martin-Fontecha M, Perez-Cerda F, Sanchez-Gomez MV, Ortega-Gutierrez S, Brown JM, Hsu KL, Cravatt B, Matute C, Mato S (2018) Deregulation of the endocannabinoid system and therapeutic potential of ABHD6 blockade in the cuprizone model of demyelination. Biochem Pharmacol 157:189–201
175. Manterola A, Bernal-Chico A, Cipriani R, Ruiz A, Perez-Samartin A, Moreno-Rodriguez M, Hsu KL, Cravatt BF, Brown JM, Rodriguez-Puertas R, Matute C, Mato S (2018) Re-examining the potential of targeting ABHD6 in multiple sclerosis: efficacy of systemic and peripherally restricted inhibitors in experimental autoimmune encephalomyelitis. Neuropharmacology 141:181–191
176. Wen J, Ribeiro R, Tanaka M, Zhang Y (2015) Activation of CB2 receptor is required for the therapeutic effect of ABHD6 inhibition in experimental autoimmune encephalomyelitis. Neuropharmacology 99:196–209
177. Fisette A, Tobin S, Decarie-Spain L, Bouyakdan K, Peyot ML, Madiraju SRM, Prentki M, Fulton S, Alquier T (2016) Alpha/beta-hydrolase domain 6 in the ventromedial hypothalamus controls energy metabolism flexibility. Cell Rep 17:1217–1226
178. Blankman JL, Long JZ, Trauger SA, Siuzdak G, Cravatt BF (2013) ABHD12 controls brain lysophosphatidylserine pathways that are deregulated in a murine model of the neurodegenerative disease PHARC. Proc Natl Acad Sci U S A 110:1500–1505
179. Chen DH, Naydenov A, Blankman JL, Mefford HC, Davis M, Sul Y, Barloon AS, Bonkowski E, Wolff J, Matsushita M, Smith C, Cravatt BF, Mackie K, Raskind WH, Stella N, Bird TD (2013) Two novel mutations in ABHD12: expansion of the mutation spectrum in PHARC and assessment of their functional effects. Hum Mutat 34:1672–1678
180. Yoshimura H, Hashimoto T, Murata T, Fukushima K, Sugaya A, Nishio SY, Usami S (2015) Novel ABHD12 mutations in PHARC patients: the differential diagnosis of deaf-blindness. Ann Otol Rhinol Laryngol 124(Suppl 1):77S–83S
181. Tingaud-Sequeira A, Raldua D, Lavie J, Mathieu G, Bordier M, Knoll-Gellida A, Rambeau P, Coupry I, Andre M, Malm E, Moller C, Andreasson S, Rendtorff ND, Tranebjaerg L, Koenig M, Lacombe D, Goizet C, Babin PJ (2017) Functional validation of ABHD12 mutations in the neurodegenerative disease PHARC. Neurobiol Dis 98:36–51
182. Frasquet M, Lupo V, Chumillas MJ, Vazquez-Costa JF, Espinos C, Sevilla T (2018) Phenotypical features of two patients diagnosed with PHARC syndrome and carriers of a new homozygous mutation in the ABHD12 gene. J Neurol Sci 387:134–138
183. Leishman E, Mackie K, Bradshaw HB (2019) Elevated levels of arachidonic acid-derived lipids including prostaglandins and endocannabinoids are present throughout ABHD12 knockout brains: novel insights into the neurodegenerative phenotype. Front Mol Neurosci 12:142
184. Ogasawara D, Ichu TA, Vartabedian VF, Benthuysen J, Jing H, Reed A, Ulanovskaya OA, Hulce JJ, Roberts A, Brown S, Rosen H, Teijaro JR, Cravatt BF (2018) Selective blockade of the lyso-PS lipase ABHD12 stimulates immune responses in vivo. Nat Chem Biol 14:1099–1108
185. Kamat SS, Camara K, Parsons WH, Chen DH, Dix MM, Bird TD, Howell AR, Cravatt BF (2015) Immunomodulatory lysophosphatidylserines are regulated by ABHD16A and ABHD12 interplay. Nat Chem Biol 11:164–171
186. Joshi A, Shaikh M, Singh S, Rajendran A, Mhetre A, Kamat SS (2018) Biochemical characterization of the PHARC-associated serine hydrolase ABHD12 reveals its preference for very-long-chain lipids. J Biol Chem 293:16953–16963
187. Ogasawara D, Ichu TA, Jing H, Hulce JJ, Reed A, Ulanovskaya OA, Cravatt BF (2019) Discovery and optimization of selective and in vivo active inhibitors of the lysophosphatidylserine lipase alpha/beta-hydrolase domain-containing 12 (ABHD12). J Med Chem 62:1643–1656
188. Addy C, Wright H, Van Laere K, Gantz I, Erondu N, Musser BJ, Lu K, Yuan J, Sanabria-Bohorquez SM, Stoch A, Stevens C, Fong TM, De Lepeleire I, Cilissen C, Cote J, Rosko K, Gendrano IN 3rd, Nguyen AM, Gumbiner B, Rothenberg P, de Hoon J, Bormans G, Depre M, Eng WS, Ravussin E, Klein S, Blundell J, Herman GA, Burns HD, Hargreaves RJ, Wagner J, Gottesdiener K, Amatruda JM, Heymsfield SB (2008) The acyclic CB1R inverse agonist taranabant mediates weight loss by increasing energy expenditure and decreasing caloric intake. Cell Metab 7:68–78
189. Rosenstock J, Hollander P, Chevalier S, Iranmanesh A, S. S. Group (2008) SERENADE: the study evaluating rimonabant efficacy in drug-naive diabetic patients: effects of monotherapy with rimonabant, the first selective CB1 receptor antagonist, on glycemic control, body weight, and lipid profile in drug-naive type 2 diabetes. Diabetes Care 31:2169–2176
190. Rigotti NA, Gonzales D, Dale LC, Lawrence D, Chang Y, C. S. Group (2009) A randomized controlled trial of adding the nicotine patch to rimonabant for smoking cessation: efficacy, safety and weight gain. Addiction 104:266–276
191. Hollander PA, Amod A, Litwak LE, Chaudhari U, A. S. Group (2010) Effect of rimonabant on glycemic control in insulin-treated type 2 diabetes: the ARPEGGIO trial. Diabetes Care 33:605–607
192. Proietto J, Rissanen A, Harp JB, Erondu N, Yu Q, Suryawanshi S, Jones ME, Johnson-Levonas AO, Heymsfield SB, Kaufman KD, Amatruda JM (2010) A clinical trial assessing the safety and efficacy of the CB1R inverse agonist taranabant in obese and overweight patients: low-dose study. Int J Obes 34:1243–1254

193. Sjostrom L, Rissanen A, Andersen T, Boldrin M, Golay A, Koppeschaar HP, Krempf M (1998) Randomised placebo-controlled trial of orlistat for weight loss and prevention of weight regain in obese patients. European Multicentre Orlistat Study Group. Lancet 352:167–172
194. Padwal RS, Majumdar SR (2007) Drug treatments for obesity: orlistat, sibutramine, and rimonabant. Lancet 369:71–77
195. Hoover HS, Blankman JL, Niessen S, Cravatt BF (2008) Selectivity of inhibitors of endocannabinoid biosynthesis evaluated by activity-based protein profiling. Bioorg Med Chem Lett 18:5838–5841
196. Bisogno T, Howell F, Williams G, Minassi A, Cascio MG, Ligresti A, Matias I, Schiano-Moriello A, Paul P, Williams EJ, Gangadharan U, Hobbs C, Di Marzo V, Doherty P (2003) Cloning of the first sn1-DAG lipases points to the spatial and temporal regulation of endocannabinoid signaling in the brain. J Cell Biol 163:463–468
197. Gao Y, Vasilyev DV, Goncalves MB, Howell FV, Hobbs C, Reisenberg M, Shen R, Zhang MY, Strassle BW, Lu P, Mark L, Piesla MJ, Deng K, Kouranova EV, Ring RH, Whiteside GT, Bates B, Walsh FS, Williams G, Pangalos MN, Samad TA, Doherty P (2010) Loss of retrograde endocannabinoid signaling and reduced adult neurogenesis in diacylglycerol lipase knock-out mice. J Neurosci 30:2017–2024
198. Tanimura A, Yamazaki M, Hashimotodani Y, Uchigashima M, Kawata S, Abe M, Kita Y, Hashimoto K, Shimizu T, Watanabe M, Sakimura K, Kano M (2010) The endocannabinoid 2-arachidonoylglycerol produced by diacylglycerol lipase alpha mediates retrograde suppression of synaptic transmission. Neuron 65:320–327
199. Yoshida T, Fukaya M, Uchigashima M, Miura E, Kamiya H, Kano M, Watanabe M (2006) Localization of diacylglycerol lipase-alpha around postsynaptic spine suggests close proximity between production site of an endocannabinoid, 2-arachidonoyl-glycerol, and presynaptic cannabinoid CB1 receptor. J Neurosci 26:4740–4751
200. Katona I, Urban GM, Wallace M, Ledent C, Jung KM, Piomelli D, Mackie K, Freund TF (2006) Molecular composition of the endocannabinoid system at glutamatergic synapses. J Neurosci 26:5628–5637
201. Viader A, Ogasawara D, Joslyn CM, Sanchez-Alavez M, Mori S, Nguyen W, Conti B, Cravatt BF (2016) A chemical proteomic atlas of brain serine hydrolases identifies cell type-specific pathways regulating neuroinflammation. Elife 5:e12345
202. Powell DR, Gay JP, Wilganowski N, Doree D, Savelieva KV, Lanthorn TH, Read R, Vogel P, Hansen GM, Brommage R, Ding ZM, Desai U, Zambrowicz B (2015) Diacylglycerol lipase alpha knockout mice demonstrate metabolic and behavioral phenotypes similar to those of cannabinoid receptor 1 knockout mice. Front Endocrinol (Lausanne) 6:86
203. Shonesy BC, Bluett RJ, Ramikie TS, Baldi R, Hermanson DJ, Kingsley PJ, Marnett LJ, Winder DG, Colbran RJ, Patel S (2014) Genetic disruption of 2-arachidonoylglycerol synthesis reveals a key role for endocannabinoid signaling in anxiety modulation. Cell Rep 9:1644–1653
204. Jenniches I, Ternes S, Albayram O, Otte DM, Bach K, Bindila L, Michel K, Lutz B, Bilkei-Gorzo A, Zimmer A (2016) Anxiety, stress, and fear response in mice with reduced endocannabinoid levels. Biol Psychiatry 79:858–868
205. Begbie J, Doherty P, Graham A (2004) Cannabinoid receptor, CB1, expression follows neuronal differentiation in the early chick embryo. J Anat 205:213–218
206. Mulder J, Aguado T, Keimpema E, Barabas K, Ballester Rosado CJ, Nguyen L, Monory K, Marsicano G, Di Marzo V, Hurd YL, Guillemot F, Mackie K, Lutz B, Guzman M, Lu HC, Galve-Roperh I, Harkany T (2008) Endocannabinoid signaling controls pyramidal cell specification and long-range axon patterning. Proc Natl Acad Sci U S A 105:8760–8765
207. Keimpema E, Barabas K, Morozov YM, Tortoriello G, Torii M, Cameron G, Yanagawa Y, Watanabe M, Mackie K, Harkany T (2010) Differential subcellular recruitment of monoacylglycerol lipase generates spatial specificity of 2-arachidonoyl glycerol signaling during axonal pathfinding. J Neurosci 30:13992–14007
208. Jung KM, Astarita G, Thongkham D, Piomelli D (2011) Diacylglycerol lipase-alpha and -beta control neurite outgrowth in neuro-2a cells through distinct molecular mechanisms. Mol Pharmacol 80:60–67
209. Jain T, Wager-Miller J, Mackie K, Straiker A (2013) Diacylglycerol lipasealpha (DAGLalpha) and DAGLbeta cooperatively regulate the production of 2-arachidonoyl glycerol in autaptic hippocampal neurons. Mol Pharmacol 84:296–302
210. Keimpema E, Alpar A, Howell F, Malenczyk K, Hobbs C, Hurd YL, Watanabe M, Sakimura K, Kano M, Doherty P, Harkany T (2013) Diacylglycerol lipase alpha manipulation reveals developmental roles for intercellular endocannabinoid signaling. Sci Rep 3:2093
211. Deng H, van der Stelt M (2018) Chemical tools to modulate 2-arachidonoylglycerol biosynthesis. Biotechnol Appl Biochem 65:9–15
212. Chevaleyre V, Castillo PE (2003) Heterosynaptic LTD of hippocampal GABAergic synapses: a novel role of endocannabinoids in regulating excitability. Neuron 38:461–472
213. Hashimotodani Y, Ohno-Shosaku T, Maejima T, Fukami K, Kano M (2008) Pharmacological evidence for the involvement of diacylglycerol lipase in depolarization-induced endocanabinoid release. Neuropharmacology 54:58–67
214. Williams EJ, Walsh FS, Doherty P (2003) The FGF receptor uses the endocannabinoid signaling system

to couple to an axonal growth response. J Cell Biol 160:481–486
215. Goncalves MB, Suetterlin P, Yip P, Molina-Holgado F, Walker DJ, Oudin MJ, Zentar MP, Pollard S, Yanez-Munoz RJ, Williams G, Walsh FS, Pangalos MN, Doherty P (2008) A diacylglycerol lipase-CB2 cannabinoid pathway regulates adult subventricular zone neurogenesis in an age-dependent manner. Mol Cell Neurosci 38:526–536
216. Oudin MJ, Gajendra S, Williams G, Hobbs C, Lalli G, Doherty P (2011) Endocannabinoids regulate the migration of subventricular zone-derived neuroblasts in the postnatal brain. J Neurosci 31:4000–4011
217. Gomez O, Arevalo-Martin A, Garcia-Ovejero D, Ortega-Gutierrez S, Cisneros JA, Almazan G, Sanchez-Rodriguez MA, Molina-Holgado F, Molina-Holgado E (2010) The constitutive production of the endocannabinoid 2-arachidonoylglycerol participates in oligodendrocyte differentiation. Glia 58:1913–1927
218. Yoshino H, Miyamae T, Hansen G, Zambrowicz B, Flynn M, Pedicord D, Blat Y, Westphal RS, Zaczek R, Lewis DA, Gonzalez-Burgos G (2011) Postsynaptic diacylglycerol lipase mediates retrograde endocannabinoid suppression of inhibition in mouse prefrontal cortex. J Physiol 589:4857–4884
219. Wu CS, Zhu J, Wager-Miller J, Wang S, O'Leary D, Monory K, Lutz B, Mackie K, Lu HC (2010) Requirement of cannabinoid CB(1) receptors in cortical pyramidal neurons for appropriate development of corticothalamic and thalamocortical projections. Eur J Neurosci 32:693–706
220. Consortium GT (2013) The genotype-tissue expression (GTEx) project. Nat Genet 45:580–585
221. Zhou D, Zhang D, Sun X, Li Z, Ni Y, Shan Z, Li H, Liu C, Zhang S, Liu Y, Zheng R, Pan F, Zhu Y, Shi Y, Lai M (2018) A novel variant associated with HDL-C levels by modifying DAGLB expression levels: an annotation-based genome-wide association study. Eur J Hum Genet 26:838–847
222. Shin M, Buckner A, Prince J, Bullock TNJ, Hsu KL (2019) Diacylglycerol lipase-beta is required for TNF-alpha response but not CD8(+) T cell priming capacity of dendritic cells. Cell Chem Biol 26(1036–1041):e1033
223. Jeong WI, Osei-Hyiaman D, Park O, Liu J, Batkai S, Mukhopadhyay P, Horiguchi N, Harvey-White J, Marsicano G, Lutz B, Gao B, Kunos G (2008) Paracrine activation of hepatic CB1 receptors by stellate cell-derived endocannabinoids mediates alcoholic fatty liver. Cell Metab 7:227–235
224. Mulder J, Zilberter M, Pasquare SJ, Alpar A, Schulte G, Ferreira SG, Kofalvi A, Martin-Moreno AM, Keimpema E, Tanila H, Watanabe M, Mackie K, Hortobagyi T, de Ceballos ML, Harkany T (2011) Molecular reorganization of endocannabinoid signalling in Alzheimer's disease. Brain 134:1041–1060
225. Wilkerson JL, Ghosh S, Bagdas D, Mason BL, Crowe MS, Hsu KL, Wise LE, Kinsey SG, Damaj MI, Cravatt BF, Lichtman AH (2016) Diacylglycerol lipase beta inhibition reverses nociceptive behaviour in mouse models of inflammatory and neuropathic pain. Br J Pharmacol 173:1678–1692
226. Wilkerson JL, Donvito G, Grim TW, Abdullah RA, Ogasawara D, Cravatt BF, Lichtman AH (2017) Investigation of Diacylglycerol lipase alpha inhibition in the mouse lipopolysaccharide inflammatory pain model. J Pharmacol Exp Ther 363:394–401
227. Luk J, Lu Y, Ackermann A, Peng X, Bogdan D, Puopolo M, Komatsu DE, Tong S, Ojima I, Rebecchi MJ, Kaczocha M (2018) Contribution of diacylglycerol lipase beta to pain after surgery. J Pain Res 11:473–482
228. Shin M, Snyder HW, Donvito G, Schurman LD, Fox TE, Lichtman AH, Kester M, Hsu KL (2018) Liposomal delivery of diacylglycerol lipase-beta inhibitors to macrophages dramatically enhances selectivity and efficacy in vivo. Mol Pharm 15:721–728
229. Okamoto Y, Morishita J, Tsuboi K, Tonai T, Ueda N (2004) Molecular characterization of a phospholipase D generating anandamide and its congeners. J Biol Chem 279:5298–5305
230. Scott SA, Spencer CT, O'Reilly MC, Brown KA, Lavieri RR, Cho CH, Jung DI, Larock RC, Brown HA, Lindsley CW (2015) Discovery of desketoraloxifene analogues as inhibitors of mammalian, Pseudomonas aeruginosa, and NAPE phospholipase D enzymes. ACS Chem Biol 10:421–432
231. Leung D, Saghatelian A, Simon GM, Cravatt BF (2006) Inactivation of N-acyl phosphatidylethanolamine phospholipase D reveals multiple mechanisms for the biosynthesis of endocannabinoids. Biochemistry 45:4720–4726
232. Tsuboi K, Okamoto Y, Ikematsu N, Inoue M, Shimizu Y, Uyama T, Wang J, Deutsch DG, Burns MP, Ulloa NM, Tokumura A, Ueda N (2011) Enzymatic formation of N-acylethanolamines from N-acylethanolamine plasmalogen through N-acylphosphatidylethanolamine-hydrolyzing phospholipase D-dependent and -independent pathways. Biochim Biophys Acta 1811:565–577
233. Leishman E, Mackie K, Luquet S, Bradshaw HB (2016) Lipidomics profile of a NAPE-PLD KO mouse provides evidence of a broader role of this enzyme in lipid metabolism in the brain. Biochim Biophys Acta 1861:491–500
234. Geurts L, Everard A, Van Hul M, Essaghir A, Duparc T, Matamoros S, Plovier H, Castel J, Denis RG, Bergiers M, Druart C, Alhouayek M, Delzenne NM, Muccioli GG, Demoulin JB, Luquet S, Cani PD (2015) Adipose tissue NAPE-PLD controls fat mass development by altering the browning process and gut microbiota. Nat Commun 6:6495
235. Everard A, Plovier H, Rastelli M, Van Hul M, de Wouters d'Oplinter A, Geurts L, Druart C, Robine S, Delzenne NM, Muccioli GG, de Vos WM, Luquet S, Flamand N, Di Marzo V, Cani PD (2019) Intestinal epithelial N-acylphosphatidylethanolamine phos-

pholipase D links dietary fat to metabolic adaptations in obesity and steatosis. Nat Commun 10:457
236. Overton HA, Fyfe MC, Reynet C (2008) GPR119, a novel G protein-coupled receptor target for the treatment of type 2 diabetes and obesity. Br J Pharmacol 153(Suppl 1):S76–S81
237. Fu J, Gaetani S, Oveisi F, Lo Verme J, Serrano A, Rodriguez De Fonseca F, Rosengarth A, Luecke H, Di Giacomo B, Tarzia G, Piomelli D (2003) Oleylethanolamide regulates feeding and body weight through activation of the nuclear receptor PPAR-alpha. Nature 425:90–93
238. Petersen G, Pedersen AH, Pickering DS, Begtrup M, Hansen HS (2009) Effect of synthetic and natural phospholipids on N-acylphosphatidylethanolamine-hydrolyzing phospholipase D activity. Chem Phys Lipids 162:53–61
239. Castellani B, Diamanti E, Pizzirani D, Tardia P, Maccesi M, Realini N, Magotti P, Garau G, Bakkum T, Rivara S, Mor M, Piomelli D (2017) Synthesis and characterization of the first inhibitor of N-acylphosphatidylethanolamine phospholipase D (NAPE-PLD). Chem Commun (Camb) 53:12814–12817
240. Simon GM, Cravatt BF (2008) Anandamide biosynthesis catalyzed by the phosphodiesterase GDE1 and detection of glycerophospho-N-acyl ethanolamine precursors in mouse brain. J Biol Chem 283:9341–9349
241. Simon GM, Cravatt BF (2010) Characterization of mice lacking candidate N-acyl ethanolamine biosynthetic enzymes provides evidence for multiple pathways that contribute to endocannabinoid production in vivo. Mol BioSyst 6:1411–1418
242. Lee HC, Simon GM, Cravatt BF (2015) ABHD4 regulates multiple classes of N-acyl phospholipids in the mammalian central nervous system. Biochemistry 54:2539–2549
243. Liu J, Wang L, Harvey-White J, Osei-Hyiaman D, Razdan R, Gong Q, Chan AC, Zhou Z, Huang BX, Kim HY, Kunos G (2006) A biosynthetic pathway for anandamide. Proc Natl Acad Sci U S A 103:13345–13350
244. Liu J, Wang L, Harvey-White J, Huang BX, Kim HY, Luquet S, Palmiter RD, Krystal G, Rai R, Mahadevan A, Razdan RK, Kunos G (2008) Multiple pathways involved in the biosynthesis of anandamide. Neuropharmacology 54:1–7
245. Rosenberger TA, Villacreses NE, Contreras MA, Bonventre JV, Rapoport SI (2003) Brain lipid metabolism in the cPLA2 knockout mouse. J Lipid Res 44:109–117
246. Piro JR, Benjamin DI, Duerr JM, Pi Y, Gonzales C, Wood KM, Schwartz JW, Nomura DK, Samad TA (2012) A dysregulated endocannabinoid-eicosanoid network supports pathogenesis in a mouse model of Alzheimer's disease. Cell Rep 1:617–623
247. Pihlaja R, Takkinen J, Eskola O, Vasara J, Lopez-Picon FR, Haaparanta-Solin M, Rinne JO (2015) Monoacylglycerol lipase inhibitor JZL184 reduces neuroinflammatory response in APdE9 mice and in adult mouse glial cells. J Neuroinflammation 12:81
248. Yan W, Yun Y, Ku T, Li G, Sang N (2016) NO2 inhalation promotes Alzheimer's disease-like progression: cyclooxygenase-2-derived prostaglandin E2 modulation and monoacylglycerol lipase inhibition-targeted medication. Sci Rep 6:22429
249. Kita Y, Yoshida K, Tokuoka SM, Hamano F, Yamazaki M, Sakimura K, Kano M, Shimizu T (2015) Fever is mediated by conversion of endocannabinoid 2-arachidonoylglycerol to prostaglandin E2. PLoS One 10:e0133663
250. Sanchez-Alavez M, Nguyen W, Mori S, Moroncini G, Viader A, Nomura DK, Cravatt BF, Conti B (2015) Monoacylglycerol lipase regulates fever response. PLoS One 10:e0134437
251. Urquhart P, Nicolaou A, Woodward DF (2015) Endocannabinoids and their oxygenation by cyclo-oxygenases, lipoxygenases and other oxygenases. Biochim Biophys Acta 1851:366–376
252. Davies NM (1998) Clinical pharmacokinetics of ibuprofen. The first 30 years. Clin Pharmacokinet 34:101–154
253. Duggan KC, Hermanson DJ, Musee J, Prusakiewicz JJ, Scheib JL, Carter BD, Banerjee S, Oates JA, Marnett LJ (2011) (R)-Profens are substrate-selective inhibitors of endocannabinoid oxygenation by COX-2. Nat Chem Biol 7:803–809
254. Lotsch J, Geisslinger G, Mohammadian P, Brune K, Kobal G (1995) Effects of flurbiprofen enantiomers on pain-related chemo-somatosensory evoked potentials in human subjects. Br J Clin Pharmacol 40:339–346
255. Bishay P, Schmidt H, Marian C, Haussler A, Wijnvoord N, Ziebell S, Metzner J, Koch M, Myrczek T, Bechmann I, Kuner R, Costigan M, Dehghani F, Geisslinger G, Tegeder I (2010) R-flurbiprofen reduces neuropathic pain in rodents by restoring endogenous cannabinoids. PLoS One 5:e10628
256. Hu SS, Bradshaw HB, Chen JS, Tan B, Walker JM (2008) Prostaglandin E2 glycerol ester, an endogenous COX-2 metabolite of 2-arachidonoylglycerol, induces hyperalgesia and modulates NFkappaB activity. Br J Pharmacol 153:1538–1549
257. Bruser A, Zimmermann A, Crews BC, Sliwoski G, Meiler J, Konig GM, Kostenis E, Lede V, Marnett LJ, Schoneberg T (2017) Prostaglandin E2 glyceryl ester is an endogenous agonist of the nucleotide receptor P2Y6. Sci Rep 7:2380
258. Alhouayek M, Buisseret B, Paquot A, Guillemot-Legris O, Muccioli GG (2018) The endogenous bioactive lipid prostaglandin D2-glycerol ester reduces murine colitis via DP1 and PPARgamma receptors. FASEB J 32:5000–5011
259. Elhassanny AEM, Ladin DA, Soliman E, Albassam H, Morris A, Kobet R, Thayne K, Burns C, Danell AS, Van Dross R (2019) Prostaglandin

D2-ethanolamide induces skin cancer apoptosis by suppressing the activity of cellular antioxidants. Prostaglandins Other Lipid Mediat 142:9–23

260. Liang Y, Woodward DF, Guzman VM, Li C, Scott DF, Wang JW, Wheeler LA, Garst ME, Landsverk K, Sachs G, Krauss AH, Cornell C, Martos J, Pettit S, Fliri H (2008) Identification and pharmacological characterization of the prostaglandin FP receptor and FP receptor variant complexes. Br J Pharmacol 154:1079–1093
261. Woodward DF, Phelps RL, Krauss AH, Weber A, Short B, Chen J, Liang Y, Wheeler LA (2004) Bimatoprost: a novel antiglaucoma agent. Cardiovasc Drug Rev 22:103–120
262. Craven ER, Walters T, Christie WC, Day DG, Lewis RA, Goodkin ML, Chen M, Wangsadipura V, Robinson MR, Bejanian M, Bimatoprost SRSG (2019) 24-month phase I/II clinical trial of Bimatoprost sustained-release implant (Bimatoprost SR) in Glaucoma patients. Drugs
263. Jha AK, Sarkar R, Udayan UK, Roy PK, Jha AK, Chaudhary RKP (2018) Bimatoprost in dermatology. Indian Dermatol Online J 9:224–228
264. Gatta L, Piscitelli F, Giordano C, Boccella S, Lichtman A, Maione S, Di Marzo V (2012) Discovery of prostamide F2alpha and its role in inflammatory pain and dorsal horn nociceptive neuron hyperexcitability. PLoS One 7:e31111

9 Drugging the Phosphoinositide 3-Kinase (PI3K) and Phosphatidylinositol 4-Kinase (PI4K) Family of Enzymes for Treatment of Cancer, Immune Disorders, and Viral/Parasitic Infections

Jacob A. McPhail and John E. Burke

Abstract

The lipid kinases that generate the lipid signalling phosphoinositides have been established as fundamental signalling enzymes that control numerous aspects of how cells respond to their extracellular environment. In addition, they play critical roles in regulating membrane trafficking and lipid transport within the cell. The class I phosphoinositide kinases which generate the critical lipid signal PIP_3 are hyperactivated in numerous human pathologies including cancer, overgrowth syndromes, and primary immunodeficiencies. The type III phosphatidylinositol 4-kinase beta isoform (PI4KB), which are evolutionarily similar to the class I PI3Ks, have been found to be essential host factors mediating the replication of numerous devastating pathogenic viruses. Finally, targeting the parasite variant of PI4KB has been established as one of the most promising strategies for the development of anti-malarial and anti-cryptosporidium strategies. Therefore, the development of targeted isoform selective inhibitors for these enzymes are of paramount importance. The first generation of PI3K inhibitors have recently been clinically approved for a number of different cancers, highlighting their therapeutic value. This review will examine the history of the class I PI3Ks, and the type III PI4Ks, their relevance to human disease, and the structural basis for their regulation and inhibition by potent and selective inhibitors.

Keywords

Phosphoinositides · Phosphatidylinositol · PI3K · p110 · p85 · PIK3CA · PIK3R1 · PI4K · PI4KB · PI4KA · Oncogenes · Cancer · Malaria · Cryptosporidiosis · Enteroviruses

9.1 Phosphoinositide Kinases and Their Role in Signalling/ Human Disease

9.1.1 Phosphoinositides

Phosphoinositides are essential membrane signalling molecules that regulate a multitude of cellular processes, from membrane identity

J. A. McPhail · J. E. Burke (✉)
Department of Biochemistry and Microbiology, University of Victoria, Victoria, BC, Canada
e-mail: jeburke@uvic.ca

Y. Kihara (ed.), *Druggable Lipid Signaling Pathways*, Advances in Experimental Medicine and Biology 1274, https://doi.org/10.1007/978-3-030-50621-6_9

and compartmentalization to growth and cell division. While they represent a small percentage of total cellular lipid composition, their correct spatiotemporal location in a cell is an essential mechanism maintaining organelle identify and membrane trafficking [1, 2]. All phosphoinositides are generated from phosphatidylinositol (PI), and the inositol headgroup can be phosphorylated at the hydroxyls present at the D3, D4 and D5 positions, generating a total of seven different phosphoinositides: phosphatidylinositol 3-phosphate (PI3P), phosphatidylinositol 4-phosphate (PI4P), phosphatidylinositol 5-phosphate (PI5P), phosphatidylinositol 3,4-bisphosphate (PI(3,4) P_2), phosphatidylinositol 3,5-bisphosphate (PI(3,5)P_2), phosphatidylinositol 4,5-bisphosphate (PI(4,5)P_2) and phosphatidylinositol 3,4,5-triphosphate (PIP_3). While numerous lipid kinases and phosphatases regulate the location and concentration of these phosphoinositide pools, this chapter will focus on the most clinically promising targets: the class I phosphoinositide 3-kinases (PI3Ks) which function downstream of cell surface receptors at the plasma membrane and are critically involved in cancer, and phosphatidylinositol 4-kinase type III beta (PI4KB), which functions at the Golgi, is involved in viral infection, and is a major target for treatment of parasitic infections (Fig. 9.1a).

9.1.2 Class I PI3Ks

The class I PI3Ks are the enzymes that mediate the phosphorylation of PI(3,4,5)P_3 from the substrate PI(4,5)P_2. They were originally discovered as oncoproteins derived from viral proteins associated with an unknown lipid kinase activity [3, 4]. It was quite rapidly determined that the generation of PIP_3 was critical in growth factor signalling [5, 6], and in immune cell regulation [7]. One of the most important discoveries was on the key role of PI3Ks in regulating signalling downstream of the insulin receptor [8, 9]. Excellent reviews recently have described the history of the discovery of the PI3K field, and readers are suggested to consult these for full details [1, 10].

The class I PI3Ks are divided into three distinct classes based upon their regulatory binding partners and lipid substrate specificity. Only the class I PI3Ks can generate PIP_3 from PI(4,5)P_2 at the plasma membrane *in vivo* [11]. These enzymes are major components of intracellular signalling networks downstream of receptor tyrosine kinases (RTKs) and G-protein coupled receptors (GPCRs) [12, 13]. Upon activation, the generation of PIP_3 by class I PI3Ks drives the recruitment of PIP_3 effectors to the plasma membrane, including protein kinases (i.e. PDK1, AKT) [14], and regulators of Ras superfamily GTPases, with signal cascades downstream of these effectors playing key roles in cell proliferation, growth, survival and tumorigenesis [12, 13, 15]. Class I enzymes form heterodimers with regulatory subunits, and are further subdivided into the class IA PI3Ks (p110α, p110β, p110δ) which can bind any of five different p85 like regulatory subunits (p85α, p55α, p50α, p85β, and p55γ), and the class IB PI3Ks (p110γ) which bind either a p84 or p101 regulatory subunit. The p110 subunits will be referred to by their gene names, PIK3CA (p110α), PIK3CB (p110β), PIK3CD (p110δ), and PIK3CG (p110γ) for the remainder of this review. Both PIK3CA and PIK3CB are ubiquitously expressed, with PIK3CA primarily responsible for insulin signalling [16], and PIK3CB playing key roles in platelet function and blood clotting [17]. The PIK3CG and PIK3CD isoforms are primarily expressed in immune cells and are important in lymphocyte activation, mast cell degranulation and leukocyte chemotaxis [18–23].

The first insight into the structural basis of regulation of these kinases was from the crystal structure of class IB PIK3CG, which revealed marked homology with protein kinases [24]. The kinase domain (also referred to as the catalytic domain) exhibits a bi-lobal organization consisting of an N-terminal-lobe and C-terminal lobe that form a cleft for ATP-binding and the machinery to facilitate phosphorylation of the PI headgroup (Fig. 9.1b) [24, 25]. This kinase domain packs against a helical domain, a conserved feature of all PI3Ks and type III PI4Ks [26–30]. The kinase hinge region, located at the cleft between the two lobes, is a key feature of the ATP binding

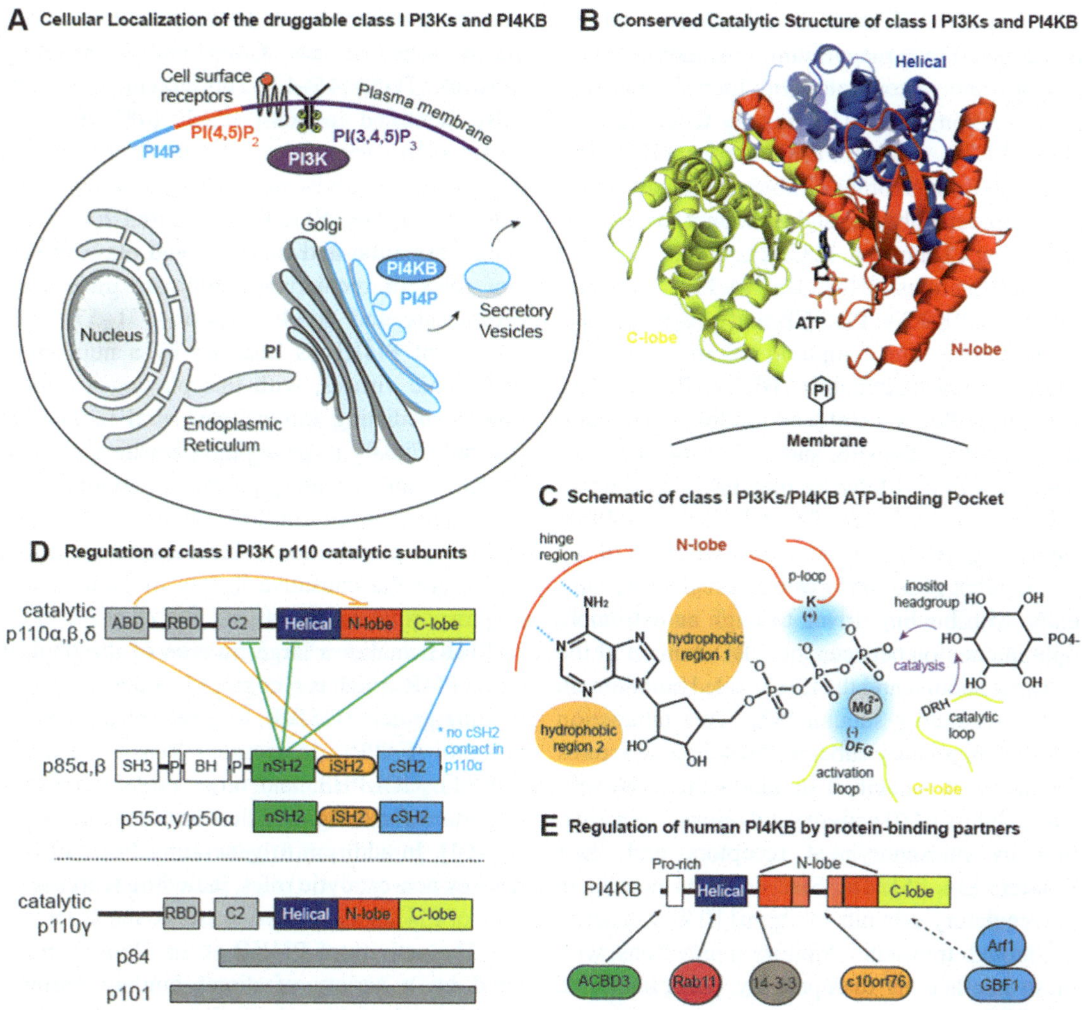

Fig. 9.1 (**a**) Cellular localizations of druggable class I PI3Ks and PI4KB. Class I PI3Ks are activated downstream of activated cell surface receptors, driving PIP_3 production at the plasma membrane. PI4KB generates PI4P from PI at the Golgi. (**b**) Conserved catalytic structure of class I PI3Ks and PI4KB. The kinase domain consists of a well conserved N-terminal lobe and C-terminal lobe which fold to form the active site. The kinase domain packs against the helical domain, a conserved feature of all PI3K's and class III PI4K's; PDB: 1E8X. (**c**) Schematic of the class I PI3Ks/PI4KB ATP-binding pocket. Key features of ATP binding include hydrogen bonds between the kinase hinge region and adenine moiety, accommodation of the nucleoside by hydrophobic regions, and charged interactions between ATP phosphates and lysine in the kinase P-loop. (**d**) Domain architecture of class IA PI3Ks (p110α,β,δ refer to PIK3CA, PIK3CB, PIK3CD) include a canonical helical and bi-lobal kinase domain, lipid binding domain (C2), Ras-binding domain (RBD) and an adaptor binding domain (ABD) which drives obligate hetero-dimerization with the p85, p55 or p50 regulatory subunits. Inhibitory contact sites between regulatory subunit SH2 domains and p110 are shown as colored lines. Class IB PI3K (p110γ refers to PIK3CG) associates with p84 and p101 regulatory subunits. (**e**) Domain architecture of the smaller PI4KB lacks the ABD, RBD and C2 domains of class I PI3Ks, and contains a long, disordered linker region within the N-lobe of the kinase domain. Solid arrows and dashed arrows indicate interactions with key regulatory PI4KB binding partners, and putative activators, respectively

pocket and makes key hydrogen bonds with the adenine moiety of ATP. The nucleoside is further accommodated by hydrophobic regions and charged interactions between ATP phosphates and lysine(s) in the kinase P-loop (Fig. 9.1c). Phosphorylation of the inositol headgroup is cat-

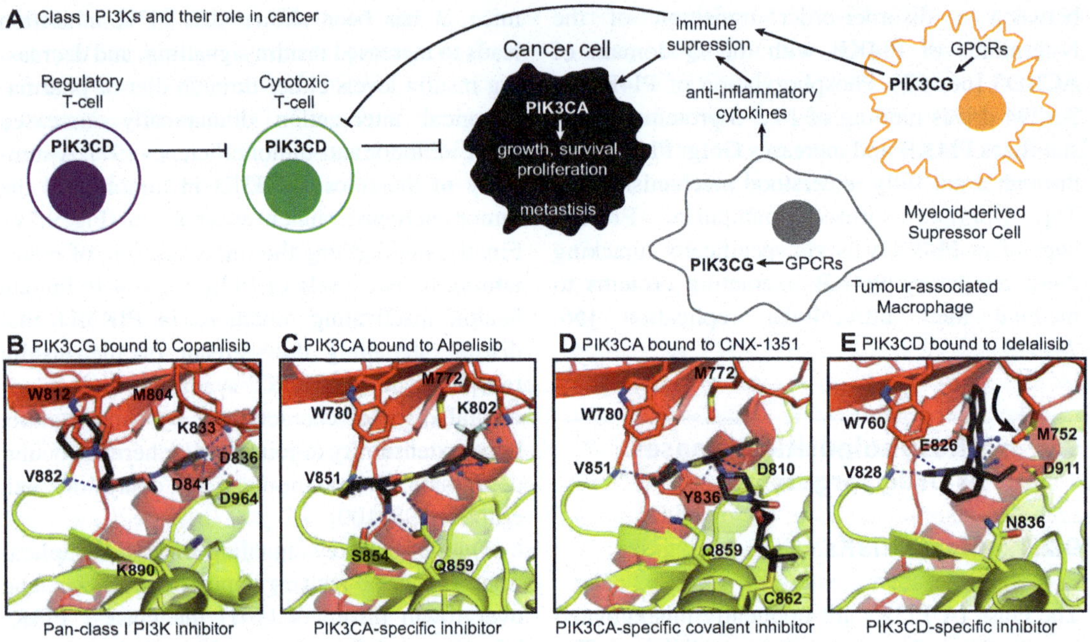

Fig. 9.2 (**a**) Class I PI3Ks are key signalling molecules involved in cancer growth, survival and progression. PIK3CA is ubiquitously expressed in all cell types and functions as a key driver of cancer growth and metastasis, while PIK3CD and PIK3CG are mainly expressed in immune cells that modulate the tumour environment. (**b**) Active site of PIK3CG bound to the Pan-PI3K, FDA-approved inhibitor copanlisib (PDB: 5G2N). (**c**) Active site of PIK3CA bound to the PIK3CA-specific, FDA-approved inhibitor alpelisib (PDB: 4JPS). (**d**) Active site of human PIK3CA bound to the PIK3CA-specific, covalent inhibitor CNX-1351 (PDB:3ZIM). (**e**) Active site of human PIK3CD bound to the PIK3CD-specific, FDA-approved inhibitor Idelalisib (PDB: 4XE0). Arrow indicates outward shift of M752 in presence of idelalisib. Key hydrogen bonds indicated as blue dotted lines; N-lobe colored red, C-lobe colored yellow; water molecules involved in hydrogen-bond networks indicated as blue spheres

lar lymphoma in 2017. While copanlisib predominantly inhibits PIK3CA and PIK3CD in the sub-nanomolar range, it still potently inhibits PIK3CB and PIK3CG in the low nanomolar range, and to a lesser extent mTOR, thus is considered a pan-PI3K selective inhibitor (Table 9.1) [104]. Characteristic of most kinase inhibitors, copanlisib forms the critical hydrogen bond to the hinge V882 (PIK3CG numbering) backbone amide with its imidazoline group, while the aminopyrimidine group fills the affinity pocket (a well conserved class I PI3K feature lined by Y867, I879, I963 and D841 in PIK3CG), forming hydrogen bonds with K833, D836, D841 and a pi-stacking interaction with D964 of the DFG motif (Fig. 9.2b) [105].

Additional pan-PI3K compounds from Adlai Noryte/Novartis (buparlisib) and Genentech (pictilisib) both recently entered phase II-phase III clinical trials for hormone-receptor positive, human epidermal growth factor receptor 2 (HER2)-negative breast cancer (Table 9.1) [106, 107]. Both compounds form a hydrogen bond with the hinge V882 and additional hydrogen bonds within the affinity pocket, while pictilisib also forms a hydrogen bond with K802 and the backbone amide of A805. Interestingly, buparlisib was identified to have an off-target effect as a tubulin inhibitor, with its antiproliferative activity possibly due to microtubule-dependent binding [108]. The compound bimiralisib, which shares a very similar molecular structure to buparlisib, has been designed with the replacement of a single carbon atom in the core of bupar-

Table 9.1 Representative class I PI3K inhibitors in advanced clinical trials

Target	Inhibitor	Trail phase	IC_{50} (nM)					Company (reference)
			PIK3CA	PIK3CB	PIK3CG	PIK3CD	mTOR	
Pan-class I	Buparlisib (BKM120)	III	52	166	262	116	4600	Adlai Noryte, Novartis [106]
Pan-class I	Pictilisib (GDC-0941)	II	3	33	75	3	580	Genentech [107]
Pan-class I & mTOR	Bimiralisib (PQR309)	II	33	661	708	451	89	PIQUR [109]
Pan-class I	Aliqopa™ (Copanlisib, BAY 80–6946)	FDA approved	0.5	3.7	6.4	0.7	45	Bayer [104]
PIK3CA	Piqray® (Alpelisib, BYL719)	FDA approved	5	1200	250	290	>9100	Novartis [113]
PIK3CG	IPI-549	II	3200	3500	16	8400	N.D.	Infinity Pharmaceuticals [125]
PIK3CG & PIK3CD	Copiktra® (Duvelisib, IPI-145)	FDA approved	1600	85	27	2.5	N.D.	Verastem Oncology & Infinity Pharmaceuticals [124]
PIK3CD	Zydelig (Idelalisib, CAL-101)	FDA approved	820	565	89	2.5	N.D.	Gilead [118]
PIK3CD	Leniolisib (CDZ173)	III	244	424	2230	11	N.D.	Novartis [122]

N.D. not determined

lisib that leads to disruption of microtubule binding [109], with this molecule also currently in clinical trials. The major issue with the majority of pan-PI3K inhibitors has been dose limiting toxicity [13, 110], with intensive efforts focused on the development of isoform selective inhibitors.

PIK3CA Specific Inhibitors Novartis' Piqray® (alpelisib, BYL719) was approved for HER2-negative, PI3KCA mutated breast cancer in 2019 [111]. Alpelisib represents the first FDA-approved PIK3CA-specific inhibitor, and continues to show promise in clinical trials for various other types of cancer. Intriguingly, this compound also showed promise in the treatment of PIK3CA dependent overgrowth conditions [112]. Apelisib potently inhibits PIK3CA with an IC_{50} of 50 nM, and displays a greater than 50-fold selectivity over other PI3K isoforms (Table 9.1) [113]. Apelisib satisfies the hinge hydrogen bond in which the thiazole nitrogen and 2-NH group form hydrogen bonds with the backbone amide and carbonyl of V851 (Fig. 9.2c). The pyrimidine moiety sits in the less solvent accessible affinity pocket which includes residues Y836, K802 and D933. The pyridine nitrogen atom is part of a hydrogen-bond network involving three water molecules and the side chains of residues Y836, D810, D933 and K802. One of the fluorine atoms of the trifluoromethyl group also makes a putative hydrogen bond with K802. The basis for PIK3CA specificity resides in the donor-acceptor interactions between alpelisib with the PIK3CA side chain Q859. Residue Q859 is unique to PIK3CA (D, N and K in PIK3CB, PIK3CD and PIK3CG respectively) with the other isoforms unable to form this particular set of hydrogen bonds, revealing the likely molecular mechanism of selectivity. Additional PI3K inhibitors of different chemotypes [114–116], also take advantage of the unique Q859 in PIK3CA to drive specificity. Another mechanism to target PIK3CA in an isoform specific manner, has been in the use of covalent inhibitors that modify C862, another residue unique to PIK3CA (L, L and Q in PIK3CB, PIK3CD and PIK3CG, respectively) to drive specificity. CNX-1351 potently inhibits PIK3CA (IC_{50} = 6.8 nM) with >20-fold selectivity over other class I PI3Ks due to hydrogen bonds between the hinge V851 amide backbone, the Y836 side chain, D810 side chain, and a covalent modification of PIK3CA C862 (Fig. 9.2d) [117].

PIK3CD & PIK3G Specific Inhibitors While PIK3CA and PIK3CB are ubiquitously expressed, PIK3CD and PIK3CG are expressed in immune cells. Both PIK3CD and PIK3CG play key roles in the immune-suppressive tumor microenvironment [92, 93]. Gilead's Zydelig® (idelalisib, CAL-101) was the first PI3K inhibitor to successfully progress through clinical trials, gaining FDA approval in 2014. Idelalisib specifically and potently targets PIK3CD over other class I PI3Ks, and was approved for relapsed follicular lymphoma (FL), chronic lymphocytic leukemia (CLL) and small lymphocytic lymphoma (SLL) (Table 9.1) [118, 119]. The purine group of idelalisib forms hydrogen bonds with backbone amide of hinge V828, the backbone carbonyl of E826, and a water-mediated hydrogen bond network involving the D911 sidechain. Idelalisib is a propeller-shaped inhibitor, a shape previously utilized to gain PIK3CD selectivity [16, 29]. In an apo state, M752 of the PIK3CD P-loop packs "in" against Y760, however binding of idelalisib and other propeller inhibitors pushes M752 "out", forming a hydrophobic specificity pocket [120]. While these two residues are conserved throughout class I PI3Ks, differences in P-loop residues and native hydrogen bond networks involving the tryptophan residues are thought to make the opening of this pocket more energetically unfavorable for PIK3CA, PIK3CB. Additionally, the phenyl group of idelalisib packs against the side chains of a wall capped by N836, which is unique to PIK3CD (Q859, D856 and K890 in PIK3CA, PIK3CB and PIK3CG, respectively). Studies on another propeller-shaped series has suggested modifications at this position can affect potency and selectivity of propeller compounds [121]. The PIK3CD selective inhibitor leniolisib (CDZ173; Table 9.1) exploits a different molecular mechanism to mediate PIK3CD selectivity, with a acetyl-piperazine group from the inhibitor forming a pi-stacking interaction with W760 in PIK3CD, with this corresponding interaction not possible with W780 in PIK3CA [122]. This inhibitor has shown initial promise for the treatment of APDS patients [123].

Verastem Oncology and Infinity Pharmaceuticals' Copiktra® (duvelisib, IPI-145) was approved for relapsed or refractory chronic lymphocytic leukemia/small lymphocytic lymphoma (CLL/SLL) in 2018. Duvelisib potently inhibits PIK3CG and PIK3CD in the low-nanomolar range (Table 9.1) [124]. While no co-crystal structure is available, duvelisib is very similar in structure to idelalisib and likely exerts inhibition through a similar propeller mechanism. In terms of specific PIK3CG inhibition, IPI-549 maintains the same 8-Cl isoquinolinone core as duvelisib, but achieves greater than 200-fold selectivity over PIK3CD and other class I isoforms (Table 9.1) [125]. The increase in potency is thought to be due to nonfavorable interactions between the alkyne moiety of IPI-549 and nonconserved residues adjacent to the specificity pocket such as PIK3CD T750 (PIK3CG K802). IPI-549 is currently in phase II clinical trials for several malignancies. Recently AstraZeneca has reported on a novel strategy to generate PIK3CG inhibitors that induce large allosteric conformational changes in PIK3CG, with these changes not possible in other class I PI3Ks [126].

9.2.2 PI4KB Inhibitors as Tool Compounds to Define the Role of PI4KB in Viral Infection

Picornaviruses hijack host PI4KB during infection to generate replication organelles which are essential for viral replication, and direct inhibition of PI4KB has been shown to disrupt viral infection progression [56]. Picornaviruses are nonenveloped, positive-sense, single-stranded RNA viruses that cause diverse human diseases including gastroenteritis, poliomyelitis, aseptic meningitis, hand-foot-and-mouth disease, respiratory illness and acute flaccid paralysis, and particularily affect young children [127]. Viruses that have been shown to specifically manipulate PI4KB include Aichivirus, Poliovirus, Rhinovirus, Coxsackievirus, Enterovirus D68 and Enterovirus A71 [77]. These viruses hijack host PI4KB during infection to generate replication organelles, which are essential for viral replication (Fig. 9.3a). These replication organelles are abnormal, PI4P-enriched membranes upon which multiple specific lipids are enriched, along with recruitment of viral replication machinery (i.e. RNA polymerase, etc.) [74]. PI4KB recruitment to replication organelles appears to be dependent on viral 3A proteins [56]. Viral 3A proteins do not directly interact with PI4KB, instead they recruit PI4KB-regulatory proteins such as ACBD3 and the Arf1 GEF GBF1 (Fig. 9.3a) [55, 66, 76]. Specific enteroviruses have also shown a dependence on the PI4KB-regulatory c10orf76 protein [128]. As infection progresses, the hijacking of PI4KB via regulatory proteins and subsequent redistribution of PI4P pools disrupts normal trafficking and eventually leads to collapse of the Golgi (Fig. 9.3a) [56, 129].

Pharmacologically or genetically inhibiting PI4KB prevents the formation of replication organelles and halts viral infection progression. When considering the potential value of PI4KB inhibitors as anti-viral therapeutics it is important to note that while these inhibitors are not generally cytotoxic and effectively disrupt viral replication, some PI4KB inhibitors have been shown to have immunosuppressive effects, specifically an antiproliferative effect on lymphocytes in mice [130, 131]. These currently are most useful as tool compounds for the study of the role of PI4KB in viral replication, with the discussion focused on the molecular mechanisms of inhibitor specificity.

Potent and Selective PI4KB Inhibitors The compound PIK-93 represents the original PI4KB "specific" inhibitor used extensively to study the cellular roles of PI4KB and its roles in viral replication [16]. While PIK93 is highly selective for

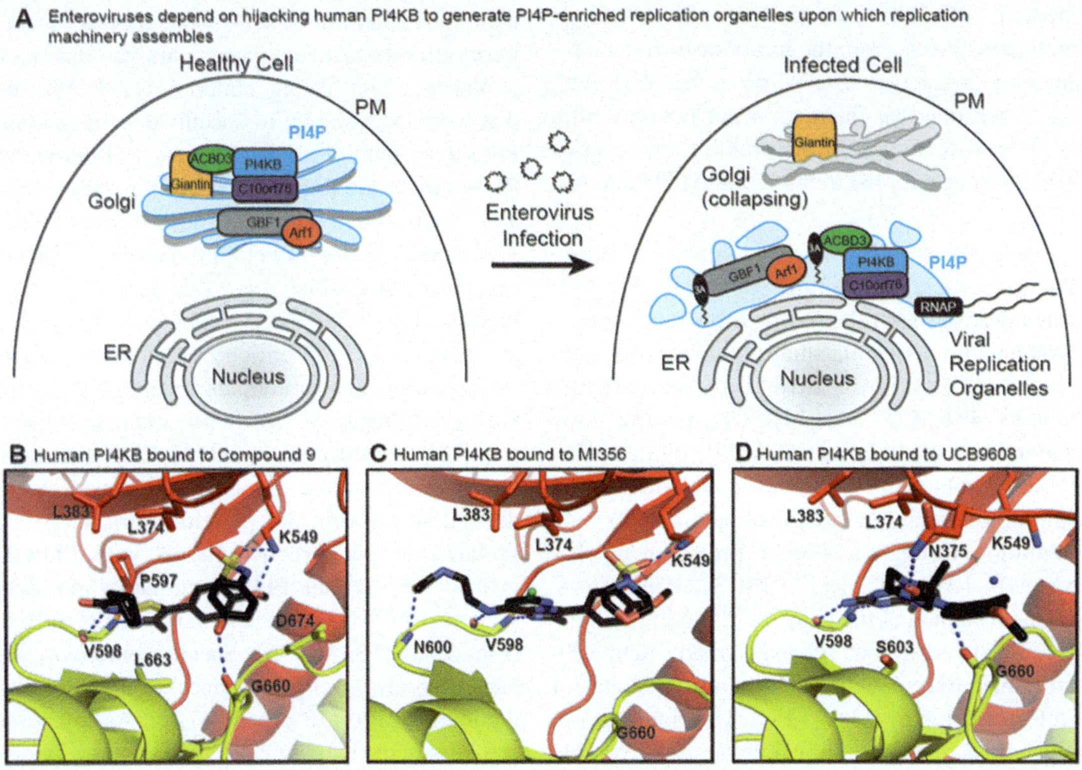

Fig. 9.3 (**a**) Human PI4KB is hijacked by enteroviruses to facilitate viral replication. The virus-encoded 3A protein indirectly hijacks PI4KB through its regulatory proteins and utilizes it to build PI4P-enriched replication organelles, upon which viral replication machinery assembles. PM = plasma membrane. (**b**) Active site of human PI4KB bound to the anti-viral Compound 9 (PDB: 5EUQ). (**c**) Active site of human PI4KB bound to the anti-viral MI356 (PDB: 5FBL). (**d**) Active site of human PI4KB bound to the immunosuppressant UCB9608 (PDB:6GL3). Key hydrogen bonds indicated as blue dotted lines; N-lobe colored red, C-lobe colored yellow

PI4KB over PI4KA, it has a high cross reactivity for PIK3CG, PIK3CD and the class III PI3K vps34. Derivatives of PIK-93 were optimized to specifically decrease reactivity with the class I PI3Ks [132]. This led to the identification of compound 9, which takes advantage of the less sterically crowded pocket of PI4KB (Fig. 9.3b). Compound 9 makes two key hydrogen bonds between the thiazole and acetamide groups with the hinge V598 backbone amide and carbonyl. A hydrogen bond is also present between the sulfonamide group and K549. These hydrogen bonds are conserved with PIK93, yet the N-phenol sulfonamide para-hydroxyl group of compound 9 forms an additional hydrogen bond with the G660 carbonyl, which is important for the increased potency (PI4KB IC_{50} 7 nM vs. 17 nM). The presence of this hydroxyl group also increases potency for PIK3CG and vps34, suggesting this hydrogen bond is conserved. The acetamide and methoxy substituent of compound 9 mediate selectivity, as the interacting residues in PI4KB (L383 and L663) are significantly bulkier in the class I PI3Ks.

The most potent human PI4KB inhibitor to date, MI356 (IC_{50} = 0.98 nM) [133], is a derivatization of the amino heterocyclic core of T-00127-HEV1, an anti-enterovirus PI4KB inhibitor [134]. MI356 forms an extensive hydrogen bond network with the active site of PI4KB (Fig. 9.3c). Amines from the imidazole of the bicyclic core and the (2-acetamidoethyl)amino moiety of MI356 form hydrogen bonds with the hinge V598 backbone amide and carbonyl, respectively (Fig. 9.3c). The carbonyl of the (2-acetamidoethyl)amino group forms a hydrogen bond with hinge N600 amide, and a hydrogen bond is also present between the sulfonamide moiety and K549. The sulfonamide moiety also likely coordinates hydrogen bonds to the backbone of G660 and N661, however these were not modelled in the crystal structure due to low resolution. While MI356 exhibits remarkable selectivity over PI4KA (IC_{50} 48 μM), it has not been tested against class I PI3Ks, however the (2-acetamidoethyl)amino group would likely cause a steric clash, similar to the PIK93 derivatives discussed above. When considering potential value as a tool in cells, MI356 and similar derivatives with the (2-acetamidoethyl) amino moiety appear to be poorly permeable through cell membranes. However, compound 10, a derivative of the MI356 series, substitutes a 2-ethylpyridine moiety for the (2-acetamidoethyl) amino group and still displays potent inhibition of PI4KB (IC_{50} = 6.1 nM) and great selectivity over class I PI3Ks (no IC_{50} below 10 μM), with improved cell penetration and anti-viral activity against HRV1 and HCV 1b (EC_{50} = 75 and 29 nM, respectively).

The pyrazolopyrimidine UCB9608, a potent PI4KB inhibitor (IC_{50} = 11 nM) was recently identified as a potent immunosuppressive agent capable of prolonging heterotopic allograft retention in mice [131]. The amino pyrimidine of UCB908 makes two hydrogen bonds with the backbone amide and carbonyl of PI4KB hinge V598 (Fig. 9.3d). There is no contact with K549 as seen in Compound 9 and MI356, however this residue appears to be coordinating a water molecule that may participate in a hydrogen bonding network. Another water molecule is present within hydrogen bonding range of the urea carbonyl and S618 side-chain. The urea NH and carbonyl of UCB9608 form hydrogen bonds with the G660 carbonyl and N375 side chain, respectively. The N375 interaction shifts the P-loop inwards, with the side-chain methylene of N375 and main-chain methylene of G660 available to make putative C − H/π interactions with the electron-rich aromatic ring of the UCB9608 urea group. UCB9608 displays greater than 1000-fold selectivity for PI4KB over class I and III PI3Ks, and 80-fold selectivity over class II PI3Ks. Selectivity for PI4KB over PIK3CG is likely due to the lack of the N375 to hydrogen bond (A805 in PIK3CG) and potentially steric hindrance by PIK3CG D950 and T887 residues (G660 and S603 in PI4KB, respectively). Selectivity over vps34 may be due to the presence of D747 in place of G660, with steric hindrance again preventing hydrogen bonding with the backbone carbonyl.

9.2.3 PI4KB Inhibitors as Anti-parasitics

PI4K activity is required for successful replication of the malaria parasite, which encodes a larger variant of PI4KB [60]. *Plasmodium falciparum* and *Plasmodium vivax* cause the vast majority of malaria cases worldwide, which are transferred to humans through the bites of infected female *Anopheles* mosquitoes. The parasite first infects hepatocytes in the liver, then eventually progresses to a symptomatic erythrocyte infection in the bloodstream. Artemisinin-based combination therapies have been recommended by the WHO since 2005, and have caused significant reductions in the global malaria burden and mortality – however, elimination of malaria has been threatened by the emergence of artemisinin resistance in *P. falciparum* across mainland Southeast Asia [135]. Complicating matters, *P. vivax* is able to persist in the liver for years after treatment before relapsing and reinitiating a blood stage infection [136]. Considering these challenges, to eradicate malaria, medicines are needed that cure the symptomatic asexual blood stage and clear the preceding liver stage that can cause relapses. *Plasmodium* PI4K inhibitors target all life stages of the parasite, in both mammalian hosts and the mosquito vector, kill drug-resistant parasites, and protect monkeys from malaria infection [60]. Inhibition of *Plasmodium* PI4K alters intracellular distribution of PI4P, which disrupts PI4P effector (including Rab11a) recruitment and regulation of transport vesicles destined for the ingressing plasma membrane, eventually causing failure of merozoite cytokinesis within asexual blood stage schizonts (Fig. 9.4a). These parasite PI4K inhibitors also display potent activity against another related parasite, *Cryptosporidium*, which is a leading cause of death in the developing world. Inhibition of *Cryptosporidium* PI4K leads to potent reduction in intestinal infection of immunocompromised mice, and rapid resolution of diarrhea and dehydration in neonatal calves [57]. Finally, a third parasite, *Trypanosoma brucei,* the causative agent of African Sleeping Sickness *Trypanosoma brucei*, also depends on activity of its PI4K homolog for growth, proper Golgi formation and cytokinesis [137]. While parasite PI4K inhibitors have only recently entered clinical trials, preclinical studies have been very promising, and the need for their development is clear.

Plasmodium and Cryptosporidium PI4KB Inhibitors The *Plasmodium* PI4K Imidazopyrazines and quinoxaline inhibitors (examples KDU691 and BQR695) were found to be active against all liver and blood-stage forms of the parasite. Whole genome sequencing of parasites resistant to imidazopyrazines and BQR695 identified mutations in *Plasmodium* PI4K were conferring resistance. These compounds potently inhibited recombinant *P. vivax* PI4K, and disrupt cytokinesis in the parasite blood-stage schizont, when membranes are built up around multiple nuclei of developing daughter merozoites (Fig. 9.4a) [60]. KDU691 is a potent inhibitor of *Plasmodium* PI4K (IC_{50} = 1.5 nM) with great selectivity over human PI4KB and class I PI3Ks (>1000-fold]). BQR695 is also a potent anti-malarial *Plasmodium* PI4K inhibitor (IC_{50} 3.5 nM) with sub-micromolar potency against human PI4KB (IC_{50} 88 nM) and good selectivity over other human kinases. MMV390048, a 2-aminopyridine [138], another *Plasmodium* PI4K inhibitor (IC_{50} = 3.4 nM) [58], showed single dose cures of malaria in mice (PMID:22390538). MMV390048 also exhibits a prophylactic effect against malaria in monkeys, blocking all life cycle stages of the *Plasmodium* parasite and is currently in phase II clinical trials for the treatment of malaria. Additional plasmodium selective PI4K inhibitors have been discovered including BRD73842 (IC_{50} = 21 nM) [59], and the MMV390048 derivative UCT943 (IC_{50} = 23 nM) [139]. In the absence of the *Plasmodium* PI4K structure, the structural basis for specificity of anti-malarial PI4K inhibitors over human PI4KB remains poorly defined. *P. vivax* PI4K (1528 residues) is much larger than the human PI4KB homolog (801 residues for human isoform 2) although the parasite variant maintains the canonical class I PI3K/class III PI4K helical and kinase domain (Fig. 9.4b). Crystallization and structure determination of human PI4KB is so far only possible through the removal of disordered regions throughout the protein that impede crystallization [28]. The larger *Plasmodium* PI4K homolog is predicted to have many additional such disordered regions, which have hindered efforts towards crystallization and structure determination. The structure of the anti-malarial BQR695 bound to human PI4KB provides limited insight into inhibition of *Plasmodium* PI4KB (Fig. 9.4c) [140]. BQR695 makes a key hydrogen bond in the human PI4KB active site between the central quinoxaline and V598 and another between the amino group off of the central quinoxaline and the A601 backbone carbonyl (Fig. 9.4c).

Comparing *P. vivax* PI4K to human PI4KB, a region of the *P. vivax* PI4K kinase domain spanning the C-terminal region of the N-lobe and entirety of the C-lobe (residues 1245–1525) is well conserved with the human PI4KB kinase domain (residues 520–801) with 43% of residues identical and 62% similar (Fig. 9.4d). While the C-lobe active site residues show good conservation, the N-terminal half of the *P. vivax* N-lobe and the N-lobe linker shows poor sequence similarity to human PI4KB, including a significantly divergent P-loop. These key differences are likely contributing to observed specificity observed for *Plasmodium* PI4K inhibitors over human PI4KB.

Cryptosporidium PI4K was identified as a therapeutic target of pyrazolopyridines, including the previously mentioned KDU691 [57]. Pyrazolopyridine compounds inhibit recombinant *Cryptosporidum* PI4K (KDU-691 IC_{50} = 25 nM, KDU-731 IC_{50} = 17 nM) with

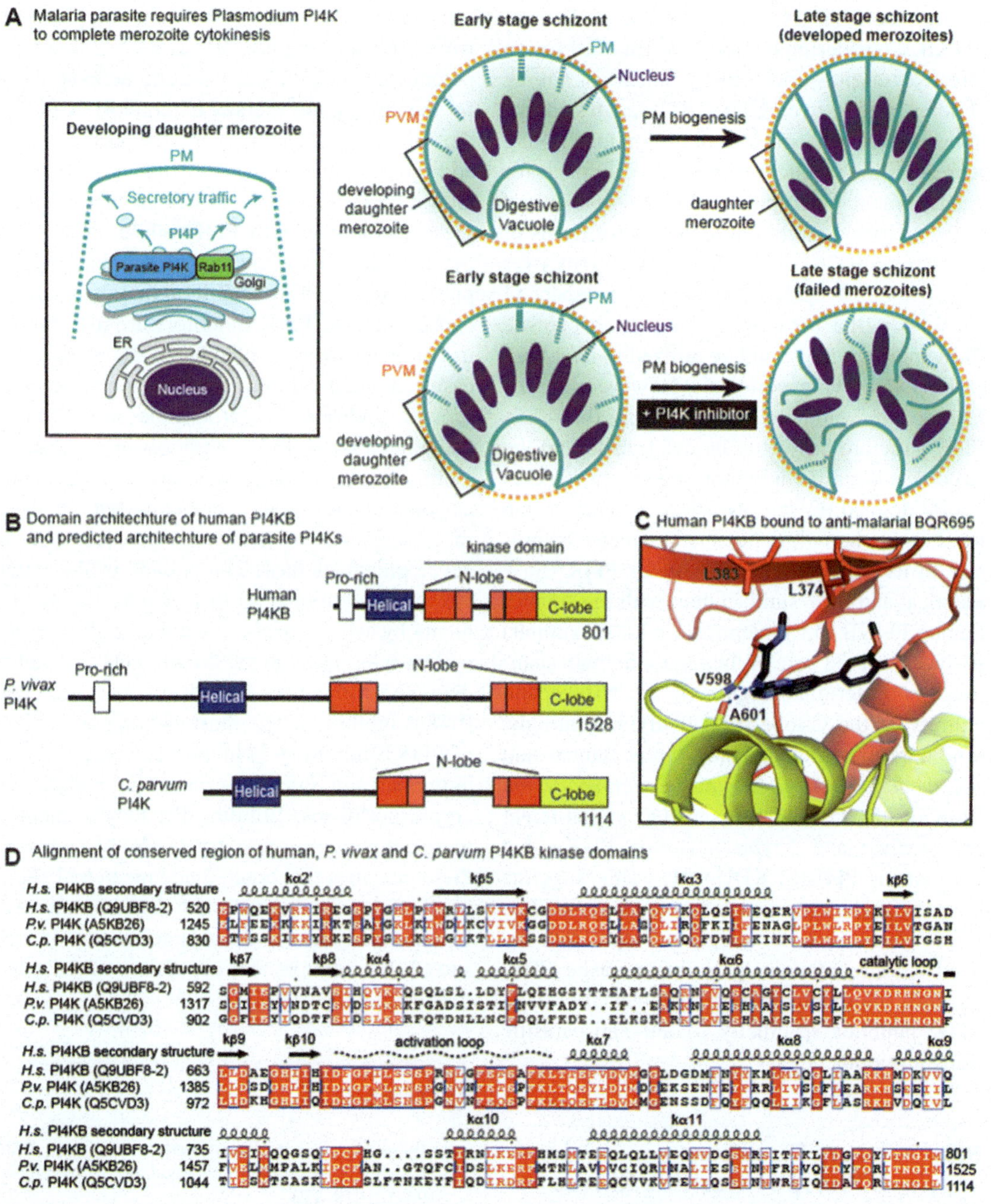

Fig. 9.4 (**a**) The malaria parasite requires the function of *Plasmodium* PI4K and Rab11a for proper plasma membrane biogenesis and cytokinesis of daughter merozoites, which are disrupted upon treatment with small molecule PI4K inhibitors. PM = plasma membrane, PVM = parisitophorous vacuolar membrane. (**b**) Active site of human PI4KB bound to the human and *Plasmodium* inhibitor BQR695 (PDB: 5C4G). Key hydrogen bonds indicated as blue dotted lines; N-lobe colored red, C-lobe colored yellow. (**c**) Domain architecture of human PI4KB (89 kDa) and predicted architecture of related *P. vivax* (172 kDa) and *C. parvum* (128 kda) PI4Ks. (**d**) Alignment of well-conserved region of human (residues 520–801), *Plasmodium vivax* (1245–1525) and *Cryptosporidium Parvum* (830–114) PI4KB kinase domains, consisting of the C-term of the N-lobe and entire C-lobe. Alignment generated using clustal omega and Espript 3.0 [141]

greater than 50-fold selectivity over human PI4KB. Considering the most urgent need for an effective cryptospiridiosis treatment is among children under two, KDU-731 was tested for numerous pharmacology safety parameters and represents an excellent preclinical candidate for the treatment of cryptospiridiosis – it displays low toxicity, no inhibition of any major forms of cytochrome P450, and > 20-fold selectivity over class I and III PI3Ks. An exciting prospect to consider is the potential to develop a parasite PI4K inhibitor that could be utilized to treat both malaria and cryptosporidiosis. However, like the *Plasmodium* homolog, the *Cryptosporidium* PI4K structure has yet to be determined and the structural basis for inhibitor specificity remains poorly defined. *C. parvum* PI4K is smaller than the *P. vivax* homolog, but shares greater homology in its active site with *P. vivax* PI4K (56% identical and 72% similar) than with the smaller human PI4KB (43% identical and 60% similar) in the conserved core of the kinase domain spanning the C-terminal region of the N-lobe and entirety of the C-lobe (Fig. 9.4d). Intriguingly, inhibitors exist that potently target human and *Plasmodium* PI4Ks, but not *Cryptosporidium* PI4K (BQR695). Inhibitors also exist that target *Plasmodium* and *Cryptosporidium* PI4Ks, but not human PI4KB (KDU-691). This suggests there are both conserved and non-conserved features within each homolog, and highlights the need for structural insights into parasite PI4K enzymes in order to define the molecular basis of PI4K inhibitor specificity to aid anti-parasitic drug development.

9.3 Future Directions of Phosphoinositide Kinase Drug Discovery

Recent years has seen an explosion in the development of potent and selective small molecules targeting the class I PI3Ks and type III PI4KBs. The class I PI3K inhibitors are now quite advanced in the clinic, with multiple being approved for treatment of blood cancers. There have been a number of failed trials of PI3K inhibitors in solid tumors, with many complications from dose-limiting toxicity. The development of potent and isoform selective compounds allows for the tailoring of therapy to target PI3K both in the tumor and the tumor microenvironment. In addition, new data on the role of insulin feedback allows for the design of specific metabolic strategies to combine with PI3K inhibition. Overall there are myriad exciting new combination therapies for which PI3K inhibition may be useful in cancer therapy. PI3K inhibitors are also finding use in the treatment of other PI3K related diseases, including immunodeficiencies and overgrowth syndromes. Exciting times are ahead for the application of PI3K inhibitors in the clinic.

Inhibitors for PI4KB were originally designed as potential anti-viral therapeutics, but it is likely that there will be numerous problems in the development of these compounds in the clinic. However, additional work revealing the mechanisms by which viruses manipulate PI4KB signalling may reveal novel drug targets that can be disrupted without the side effects that come with PI4KB inhibition. The most exciting application of PI4KB inhibitors is in the treatment of devastating parasitic infections, including malaria and cryptosporidiosis. Limiting the development of these compounds is any structural information on these enzymes that could guide medicinal chemistry efforts. Continued study on this approach will hopefully lead to the next generation of parasite PI4K inhibitors that hopefully will be effective, single-dose cures.

References

1. Balla T (2013) Phosphoinositides: tiny lipids with giant impact on cell regulation. Physiol Rev 93(3):1019–1137
2. Di Paolo G, De Camilli P (2006) Phosphoinositides in cell regulation and membrane dynamics. Nature 443(7112):651–657
3. Sugimoto Y, Whitman M, Cantley LC, Erikson RL (1984) Evidence that the Rous sarcoma virus transforming gene product phosphorylates phosphatidylinositol and diacylglycerol. Proc Natl Acad Sci U S A 81(7):2117–2121
4. Whitman M, Kaplan DR, Schaffhausen B, Cantley L, Roberts TM (1985) Association of phosphatidylino-

sitol kinase activity with polyoma middle-T competent for transformation. Nature 315(6016):239–242
5. Auger KR, Serunian LA, Soltoff SP, Libby P, Cantley LC (1989) PDGF-dependent tyrosine phosphorylation stimulates production of novel polyphosphoinositides in intact cells. Cell 57(1):167–175
6. Hawkins PT, Jackson TR, Stephens LR (1992) Platelet-derived growth factor stimulates synthesis of PtdIns(3,4,5)P3 by activating a PtdIns(4,5)P2 3-OH kinase. Nature 358(6382):157–159
7. Stephens LR, Hughes KT, Irvine RF (1991) Pathway of phosphatidylinositol(3,4,5)-trisphosphate synthesis in activated neutrophils. Nature 351(6321):33–39
8. Ruderman NB, Kapeller R, White MF, Cantley LC (1990) Activation of phosphatidylinositol 3-kinase by insulin. Proc Natl Acad Sci U S A 87(4):1411–1415
9. Backer JM, Myers MG, Shoelson SE, Chin DJ, Sun XJ, Miralpeix M et al (1992) Phosphatidylinositol 3′-kinase is activated by association with IRS-1 during insulin stimulation. EMBO J 11(9):3469–3479
10. Vanhaesebroeck B, Stephens L, Hawkins P (2012) PI3K signalling: the path to discovery and understanding. Nat Rev Mol Cell Biol [Internet] 13(3):195–203
11. Cantley L (2002) The phosphoinositide 3-kinase pathway. Science 296(5573):1655–1657
12. Burke JE, Williams RL (2015) Synergy in activating class I PI3Ks. Trends Biochem Sci 40(2):88–100
13. Fruman DA, Chiu H, Hopkins BD, Bagrodia S, Cantley LC, Abraham RT (2017) The PI3K pathway in human disease. Cell 170(4):605–635
14. Ebner M, Lučić I, Leonard TA, Yudushkin I (2017) PI(3,4,5)P3 engagement restricts Akt activity to cellular membranes. Mol Cell 65(3):416–416
15. Bilanges B, Posor Y, Vanhaesebroeck B (2019) PI3K isoforms in cell signalling and vesicle trafficking. Nat Rev Mol Cell Biol 10:6742
16. Knight Z, Gonzalez B, Feldman M, Zunder E, Goldenberg D, Williams O et al (2006) A pharmacological map of the PI3-K family defines a role for p110alpha in insulin signaling. Cell 125(4):733–747
17. Jackson S, Schoenwaelder S, Goncalves I, Nesbitt W, Yap C, Wright C et al (2005) PI 3-kinase p110beta: a new target for antithrombotic therapy. Nat Med 11(5):507–514
18. Okkenhaug K, Bilancio A, Farjot G, Priddle H, Sancho S, Peskett E et al (2002) Impaired B and T cell antigen receptor signaling in p110delta PI 3-kinase mutant mice. Science [Internet] 297(5583):1031–1034
19. Ji H, Rintelen F, Waltzinger C, Bertschy Meier D, Bilancio A, Pearce W et al (2007) Inactivation of PI3Kgamma and PI3Kdelta distorts T-cell development and causes multiple organ inflammation. Blood 110(8):2940–2947
20. Kim M-S, Rådinger M, Gilfillan AM (2008) The multiple roles of phosphoinositide 3-kinase in mast cell biology. Trends Immunol 29(10):493–501
21. Li Z, Jiang H, Xie W, Zhang Z, Smrcka AV, Wu D (2000) Roles of PLC-beta2 and -beta3 and PI3Kgamma in chemoattractant-mediated signal transduction. Science 287(5455):1046–1049
22. Ferguson G, Milne L, Kulkarni S, Sasaki T, Walker S, Andrews S et al (2007) PI(3)Kgamma has an important context-dependent role in neutrophil chemokinesis. Nat Cell Biol 9(1):86–91
23. Okkenhaug K (2013) Signaling by the phosphoinositide 3-kinase family in immune cells. Annu Rev Immunol 31:675–704
24. Walker EH, Perisic O, Ried C, Stephens L, Williams RL (1999) Structural insights into phosphoinositide 3-kinase catalysis and signalling. Nature 402(6759):313–320
25. Walker EH, Pacold ME, Perisic O, Stephens L, Hawkins PT, Wymann MP et al (2000) Structural determinants of phosphoinositide 3-kinase inhibition by wortmannin, LY294002, quercetin, myricetin, and staurosporine. Mol Cell [Internet] 6(4):909–919
26. Huang C, Mandelker D, Schmidt-Kittler O, Samuels Y, Velculescu V, Kinzler K et al (2007) The structure of a human p110alpha/p85alpha complex elucidates the effects of oncogenic PI3Kalpha mutations. Science 318(5857):1744–1748
27. Zhang X, Vadas O, Perisic O, Anderson KE, Clark J, Hawkins PT et al (2011) Structure of lipid kinase p110b/p85b elucidates an unusual SH2-domain-mediated inhibitory mechanism. Mol Cell [Internet] 41(5):567–578
28. Burke JE, Inglis AJ, Perisic O, Masson GR, McLaughlin SH, Rutaganira F et al (2014) Structures of PI4KIIIβ complexes show simultaneous recruitment of Rab11 and its effectors. Science 344(6187):1035–1038
29. Berndt A, Miller S, Williams O, Le DD, Houseman BT, Pacold JI et al (2010) The p110 delta structure: mechanisms for selectivity and potency of new PI(3)K inhibitors. Nat Chem Biol 6(2):117–124
30. Lees JA, Zhang Y, Oh MS, Schauder CM, Yu X, Baskin JM et al (2017) Architecture of the human PI4KIIIα lipid kinase complex. Proc Natl Acad Sci U S A 114(52):13720–13725
31. Vadas O, Burke JE, Zhang X, Berndt A, Williams RL (2011) Structural basis for activation and inhibition of class I phosphoinositide 3-kinases. Sci Signal [Internet] 4(195):1–13
32. Miller S, Tavshanjian B, Oleksy A, Perisic O, Houseman B, Shokat K et al (2010) Shaping development of autophagy inhibitors with the structure of the lipid kinase Vps34. Science 327(5973):1638–1642
33. Miled N, Yan Y, Hon W-C, Perisic O, Zvelebil M, Inbar Y et al (2007) Mechanism of two classes of cancer mutations in the phosphoinositide 3-kinase catalytic subunit. Science 317(5835):239–242
34. Yu J, Wjasow C, Backer JM (1998) Regulation of the p85/p110alpha phosphatidylinositol 3′-kinase. Distinct roles for the n-terminal and c-terminal SH2 domains. J Biol Chem 273(46):30199–30203

35. Yu J, Zhang Y, McIlroy J, Rordorf-Nikolic T, Orr GA, Backer JM (1998) Regulation of the p85/p110 phosphatidylinositol 3′-kinase: stabilization and inhibition of the p110alpha catalytic subunit by the p85 regulatory subunit. Mol Cell Biol 18(3):1379–1387
36. Songyang Z, Shoelson SE, Chaudhuri M, Gish G, Pawson T, Haser WG et al (1993) SH2 domains recognize specific phosphopeptide sequences. Cell 72(5):767–778
37. Mandelker D, Gabelli SB, Schmidt-Kittler O, Zhu J, Cheong I, Huang C-H et al (2009) A frequent kinase domain mutation that changes the interaction between PI3Kalpha and the membrane. Proc Natl Acad Sci U S A 106(40):16996–17001
38. Suire S, Coadwell J, Ferguson GJ, Davidson K, Hawkins P, Stephens L (2005) p84, a new Gbetagamma-activated regulatory subunit of the type IB phosphoinositide 3-kinase p110gamma. Curr Biol [Internet] 15(6):566–570
39. Stephens LR, Eguinoa A, Erdjument-Bromage H, Lui M, Cooke F, Coadwell J et al (1997) The G beta gamma sensitivity of a PI3K is dependent upon a tightly associated adaptor, p101. Cell 89(1):105–114
40. Kurig B, Shymanets A, Bohnacker T, Prajwal, Brock C, Ahmadian MR et al (2009) Ras is an indispensable coregulator of the class IB phosphoinositide 3-kinase p87/p110gamma. Proc Natl Acad Sci U S A [Internet] 106(48):20312–20317
41. Maier U, Babich A, Nurnberg B (1999) Roles of non-catalytic subunits in gbetagamma-induced activation of class I phosphoinositide 3-kinase isoforms beta and gamma. J Biol Chem 274(41):29311–29317
42. Pacold ME, Suire S, Perisic O, Lara-Gonzalez S, Davis CT, Walker EH et al (2000) Crystal structure and functional analysis of Ras binding to its effector phosphoinositide 3-kinase gamma. Cell [Internet] 103(6):931–943
43. Siempelkamp BD, Rathinaswamy MK, Jenkins ML, Burke JE (2017) Molecular mechanism of activation of class IA phosphoinositide 3-kinases (PI3Ks) by membrane-localized HRas. J Biol Chem 292(29):12256–12266
44. Fritsch R, de Krijger I, Fritsch K, George R, Reason B, Kumar MS et al (2013) RAS and RHO families of GTPases directly regulate distinct phosphoinositide 3-kinase isoforms. Cell 153(5):1050–1063
45. Burke JE (2018) Structural basis for regulation of phosphoinositide kinases and their involvement in human disease. Mol Cell 71(5):653–673
46. Dornan GL, McPhail JA, Burke JE (2016) Type III phosphatidylinositol 4 kinases: structure, function, regulation, signalling and involvement in disease. Biochem Soc Trans 44(1):260–266
47. Boura E, Nencka R (2015) Phosphatidylinositol 4-kinases: function, structure, and inhibition. Exp Cell Res 14
48. Flanagan CA, Schnieders EA, Emerick AW, Kunisawa R, Admon A, Thorner J (1993) Phosphatidylinositol 4-kinase – gene structure and requirement for yeast-cell viability. Science 262(5138):1444–1448
49. Balla T, Downing GJ, Jaffe H, Kim S, Zolyomi A, Catt KJ (1997) Isolation and molecular cloning of wortmannin-sensitive bovine type III phosphatidylinositol 4-kinases. J Biol Chem 272:18358–18366
50. Meyers R, Cantley LC (1997) Cloning and characterization of a wortmannin-sensitive human phosphatidylinositol 4-kinase. J Biol Chem 272(7):4384–4390
51. Polevoy G, Wei H-C, Wong R, Szentpetery Z, Kim YJ, Goldbach P et al (2009) Dual roles for the Drosophila PI 4-kinase four wheel drive in localizing Rab11 during cytokinesis. J Cell Biol 187(6):847–858
52. Godi A, Pertile P, Meyers R, Marra P, Di Tullio G, Iurisci C et al (1999) ARF mediates recruitment of PtdIns-4-OH kinase-beta and stimulates synthesis of PtdIns(4,5)P2 on the Golgi complex. Nat Cell Biol 1(5):280–287
53. Tóth B, Balla A, Ma H, Knight ZA, Shokat KM, Balla T (2006) Phosphatidylinositol 4-kinase IIIbeta regulates the transport of ceramide between the endoplasmic reticulum and Golgi. J Biol Chem 281(47):36369–36377
54. Greninger AL, Knudsen GM, Betegon M, Burlingame AL, DeRisi JL (2013) ACBD3 interaction with TBC1 domain 22 protein is differentially affected by enteroviral and kobuviral 3A protein binding. MBio 4(2):e00098–e00013
55. Sasaki J, Ishikawa K, Arita M, Taniguchi K (2012) ACBD3-mediated recruitment of PI4KB to picornavirus RNA replication sites. EMBO J 31(3):754–766
56. Hsu N-Y, Ilnytska O, Belov G, Santiana M, Chen Y-H, Takvorian PM et al (2010) Viral reorganization of the secretory pathway generates distinct organelles for RNA replication. Cell 141(5):799–811
57. Manjunatha UH, Vinayak S, Zambriski JA, Chao AT, Sy T, Noble CG et al (2017) A Cryptosporidium PI(4)K inhibitor is a drug candidate for cryptosporidiosis. Nature 546(7658):376–380
58. Paquet T, Le Manach C, Cabrera DG, Younis Y, Henrich PP, Abraham TS et al (2017) Antimalarial efficacy of MMV390048, an inhibitor of Plasmodium phosphatidylinositol 4-kinase. Sci Transl Med 9(387):eaad9735
59. Kato N, Comer E, Sakata-Kato T, Sharma A, Sharma M, Maetani M et al (2016) Diversity-oriented synthesis yields novel multistage antimalarial inhibitors. Nature 538(7625):344–349
60. McNamara CW, Lee MCS, Lim CS, Lim SH, Roland J, Nagle A et al (2013) Targeting Plasmodium PI(4)K to eliminate malaria. Nature 504(7479):248–253
61. D'Angelo G, Uemura T, Chuang C-C, Polishchuk E, Santoro M, Ohvo-Rekilä H et al (2013) Vesicular and non-vesicular transport feed distinct glycosylation pathways in the Golgi. Nature 501(7465):116–120
62. Mesmin B, Bigay J, Moser von Filseck J, Lacas-Gervais S, Drin G, Antonny B (2013) A four-step cycle driven by PI(4)P hydrolysis directs sterol/

PI(4)P exchange by the ER-Golgi tether OSBP. Cell 155(4):830–843
63. Rahajeng J, Kuna RS, Makowski SL, Tran TTT, Buschman MD, Li S et al (2019) Efficient Golgi forward trafficking requires GOLPH3-driven, PI4P-dependent membrane curvature. Dev Cell 50(5):573–575
64. Dippold HC, Ng MM, Farber-Katz SE, Lee S-K, Kerr ML, Peterman MC et al (2009) GOLPH3 bridges phosphatidylinositol-4- phosphate and actomyosin to stretch and shape the Golgi to promote budding. Cell 139(2):337–351
65. de Graaf P, Zwart WT, van Dijken RAJ, Deneka M, Schulz TKF, Geijsen N et al (2004) Phosphatidylinositol 4-kinasebeta is critical for functional association of rab11 with the Golgi complex. Mol Biol Cell 15(4):2038–2047
66. McPhail JA, Ottosen EH, Jenkins ML, Burke JE (2017) The molecular basis of Aichi virus 3A protein activation of phosphatidylinositol 4 kinase IIIβ, PI4KB, through ACBD3. Structure 25(1):121–131
67. Klima M, Tóth DJ, Hexnerova R, Baumlová A, Chalupská D, Tykvart J et al (2016) Structural insights and in vitro reconstitution of membrane targeting and activation of human PI4KB by the ACBD3 protein. Sci Rep 6(1):23641
68. Jovic M, Kean MJ, Szentpetery Z, Polevoy G, Gingras A-C, Brill JA et al (2012) Two phosphatidylinositol 4-kinases control lysosomal delivery of the Gaucher disease enzyme, β-glucocerebrosidase. Mol Biol Cell 23(8):1533–1545
69. Chalupská D, Eisenreichova A, Rózycki B, Rezabkova L, Humpolickova J, Klima M et al (2017) Structural analysis of phosphatidylinositol 4-kinase IIIβ (PI4KB) – 14-3-3 protein complex reveals internal flexibility and explains 14-3-3 mediated protection from degradation in vitro. J Struct Biol 31
70. Valente C, Turacchio G, Mariggiò S, Pagliuso A, Gaibisso R, Di Tullio G et al (2012) A 14-3-3γ dimer-based scaffold bridges CtBP1-S/BARS to PI(4)KIIIβ to regulate post-Golgi carrier formation. Nat Cell Biol 14(4):343–354
71. Hausser A, Storz P, Märtens S, Link G, Toker A, Pfizenmaier K (2005) Protein kinase D regulates vesicular transport by phosphorylating and activating phosphatidylinositol-4 kinase IIIbeta at the Golgi complex. Nat Cell Biol 7(9):880–886
72. Ile KE, Schaaf G, Bankaitis VA (2006) Phosphatidylinositol transfer proteins and cellular nanoreactors for lipid signaling. Nat Chem Biol 2(11):576–583
73. Schaaf G, Ortlund EA, Tyeryar KR, Mousley CJ, Ile KE, Garrett TA et al (2008) Functional anatomy of phospholipid binding and regulation of phosphoinositide homeostasis by proteins of the sec14 superfamily. Mol Cell 29(2):191–206
74. Altan-Bonnet N, Balla T (2012) Phosphatidylinositol 4-kinases: hostages harnessed to build panviral replication platforms. Trends Biochem Sci 37(7):293–302
75. Lyoo H, van der Schaar HM, Dorobantu CM, Rabouw HH, Strating JRPM, van Kuppeveld FJM (2019) ACBD3 is an essential pan-enterovirus host factor that mediates the interaction between viral 3A protein and cellular protein PI4KB. Racaniello VR, Racaniello VR (eds). MBio 10(1):282
76. Klima M, Chalupská D, Rózycki B, Humpolickova J, Rezabkova L, Silhan J et al (2017) Kobuviral non-structural 3A proteins act as molecular harnesses to Hijack the host ACBD3 protein. Structure 25(2):219–230
77. van der Schaar HM, Dorobantu CM, Albulescu L, Strating JRPM, van Kuppeveld FJM (2016) Fat(al) attraction: picornaviruses usurp lipid transfer at membrane contact sites to create replication organelles. Trends Microbiol 24(7):535–546
78. Ishikawa-Sasaki K, Sasaki J, Taniguchi K (2014) A complex comprising phosphatidylinositol 4-kinase IIIβ, ACBD3, and Aichi virus proteins enhances phosphatidylinositol 4-phosphate synthesis and is critical for formation of the viral replication complex. J Virol 88(12):6586–6598
79. Téoulé F, Brisac C, Pelletier I, Vidalain P-O, Jégouic S, Mirabelli C et al (2013) The Golgi protein ACBD3, an interactor for poliovirus protein 3A, modulates poliovirus replication. J Virol 87(20):11031–11046
80. Samuels Y, Wang Z, Bardelli A, Silliman N, Ptak J, Szabo S et al (2004) High frequency of mutations of the PIK3CA gene in human cancers. Science 304(5670):554
81. Urick ME, Rudd ML, Godwin AK, Sgroi D, Merino M, Bell DW (2011) PIK3R1 (p85α) is somatically mutated at high frequency in primary endometrial cancer. Cancer Res 71(12):4061–4067
82. Jaiswal BS, Janakiraman V, Kljavin NM, Chaudhuri S, Stern HM, Wang W et al (2009) Somatic mutations in p85a promote tumorigenesis through class IA PI3K activation. Cancer Cell [Internet] 16(6):463–474
83. Lindhurst MJ, Parker VER, Payne F, Sapp JC, Rudge S, Harris J et al (2012) Mosaic overgrowth with fibroadipose hyperplasia is caused by somatic activating mutations in PIK3CA. Nat Genet 44(8):928–933
84. Rivière J-B, Mirzaa GM, O'Roak BJ, Beddaoui M, Alcantara D, Conway RL et al (2012) De novo germline and postzygotic mutations in AKT3, PIK3R2 and PIK3CA cause a spectrum of related megalencephaly syndromes. Nat Genet 44(8):934–940
85. Takeda AJ, Zhang Y, Dornan GL, Siempelkamp BD, Jenkins ML, Matthews HF et al (2017) Novel PIK3CD mutations affecting N-terminal residues of p110δ cause activated PI3Kδ syndrome (APDS) in humans. J Allergy Clin Immunol 140(4):1152.e10–1156.e10
86. Dornan GL, Siempelkamp BD, Jenkins ML, Vadas O, Lucas CL, Burke JE (2017) Conformational disruption of PI3Kδ regulation by immunodeficiency mutations in PIK3CD and PIK3R1. Proc Natl Acad Sci U S A 114(8):1982–1987

87. Lucas CL, Chandra A, Nejentsev S, Condliffe AM, Okkenhaug K (2016) PI3Kδ and primary immunodeficiencies. Nat Rev Immunol 16(11):702–714
88. Deau M-C, Heurtier L, Frange P, Suarez F, Bole-Feysot C, Nitschke P et al (2014) A human immunodeficiency caused by mutations in the PIK3R1 gene. J Clin Invest 124(9):3923–3928
89. Angulo I, Vadas O, Garçon F, Banham-Hall E, Plagnol V, Leahy TR et al (2013) Phosphoinositide 3-kinase δ gene mutation predisposes to respiratory infection and airway damage. Science 342(6160):866–871
90. Lucas CL, Zhang Y, Venida A, Wang Y, Hughes J, McElwee J et al (2014) Heterozygous splice mutation in PIK3R1 causes human immunodeficiency with lymphoproliferation due to dominant activation of PI3K. J Exp Med 211(13):2537–2547
91. Lucas CL, Kuehn HS, Zhao F, Niemela JE, Deenick EK, Palendira U et al (2014) Dominant-activating germline mutations in the gene encoding the PI(3)K catalytic subunit p110δ result in T cell senescence and human immunodeficiency. Nat Immunol 15(1):88–97
92. Ali K, Soond DR, Piñeiro R, Hagemann T, Pearce W, Lim EL et al (2014) Inactivation of PI(3)K p110δ breaks regulatory T-cell-mediated immune tolerance to cancer. Nature 11
93. Kaneda MM, Messer KS, Ralainirina N, Li H, Leem CJ, Gorjestani S et al (2016) PI3Kγ is a molecular switch that controls immune suppression. Nature 539(7629):437–442
94. De Henau O, Rausch M, Winkler D, Campesato LF, Liu C, Cymerman DH et al (2016) Overcoming resistance to checkpoint blockade therapy by targeting PI3Kγ in myeloid cells. Nature 539(7629):443–447
95. Breasson L, Becattini B, Sardi C, Molinaro A, Zani F, Marone R et al (2017) PI3Kγ activity in leukocytes promotes adipose tissue inflammation and early-onset insulin resistance during obesity. Sci Signal 10(488):eaaf2969
96. Camps M, Rückle T, Ji H, Ardissone V, Rintelen F, Shaw J et al (2005) Blockade of PI3Kgamma suppresses joint inflammation and damage in mouse models of rheumatoid arthritis. Nat Med 11(9):936–943
97. Hopkins BD, Pauli C, Du X, Wang DG, Li X, Wu D et al (2018) Suppression of insulin feedback enhances the efficacy of PI3K inhibitors. Nature 560(7719):499–503
98. Thauvin-Robinet C, Auclair M, Duplomb L, Caron-Debarle M, Avila M, St-Onge J et al (2013) PIK3R1 mutations cause syndromic insulin resistance with lipoatrophy. Am J Hum Genet 93(1):141–149
99. Dyment DA, Smith AC, Alcantara D, Schwartzentruber JA, Basel-Vanagaite L, Curry CJ et al (2013) Mutations in PIK3R1 cause SHORT syndrome. Am J Hum Genet 93(1):158–166
100. Chudasama KK, Winnay J, Johansson S, Claudi T, König R, Haldorsen I et al (2013) SHORT syndrome with partial lipodystrophy due to impaired phosphatidylinositol 3 kinase signaling. Am J Hum Genet 93(1):150–157
101. Goncalves MD, Cantley LC (2018) Phosphatidylinositol 3-kinase, growth disorders, and cancer. N Engl J Med 379:2052–2062
102. Garces AE, Stocks MJ (2019) Class 1 PI3K clinical candidates and recent inhibitor design strategies: a medicinal chemistry perspective. J Med Chem 62(10):4815–4850
103. Miller MS, Thompson PE, Gabelli SB (2019) Structural determinants of isoform selectivity in PI3K inhibitors. Biomolecules 9(3):82
104. Liu N, Rowley BR, Bull CO, Schneider C, Haegebarth A, Schatz CA et al (2013) BAY 80–6946 is a highly selective intravenous PI3K inhibitor with potent p110α and p110δ activities in tumor cell lines and xenograft models. Mol Cancer Ther 12(11):2319–2330
105. Scott WJ, Hentemann MF, Rowley RB, Bull CO, Jenkins S, Bullion AM et al (2016) Discovery and SAR of novel 2,3-Dihydroimidazo[1,2-c]quinazoline PI3K inhibitors: identification of Copanlisib (BAY 80-6946). ChemMedChem 11(14):1517–1530
106. Burger MT, Pecchi S, Wagman A, Ni Z-J, Knapp M, Hendrickson T et al (2011) Identification of NVP-BKM120 as a potent, selective, orally bioavailable class I PI3 kinase inhibitor for treating cancer. ACS Med Chem Lett 2(10):774–779
107. Folkes AJ, Ahmadi K, Alderton WK, Alix S, Baker SJ, Box G et al (2008) The identification of 2-(1H-indazol-4-yl)-6-(4-methanesulfonyl-piperazin-1-ylmethyl)-4-morpholin-4-yl-thieno[3,2-d]pyrimidine (GDC-0941) as a potent, selective, orally bioavailable inhibitor of class I PI3 kinase for the treatment of cancer. J Med Chem 51(18):5522–5532
108. Bohnacker T, Prota AE, Beaufils F, Burke JE, Melone A, Inglis AJ et al (2017) Deconvolution of Buparlisib's mechanism of action defines specific PI3K and tubulin inhibitors for therapeutic intervention. Nat Commun 8:14683
109. Beaufils F, Cmiljanovic N, Cmiljanovic V, Bohnacker T, Melone A, Marone R et al (2017) 5-(4,6-Dimorpholino-1,3,5-triazin-2-yl)-4-(trifluoromethyl)pyridin-2-amine (PQR309), a potent, brain-penetrant, orally bioavailable, Pan-class I PI3K/mTOR inhibitor as clinical candidate in oncology. J Med Chem 60(17):7524–7538
110. Janku F, Yap TA, Meric-Bernstam F (2018) Targeting the PI3K pathway in cancer: are we making headway? Nat Rev Clin Oncol 26:1346
111. André F, Ciruelos E, Rubovszky G, Campone M, Loibl S, Rugo HS et al (2019) Alpelisib for PIK3CA-mutated, hormone receptor-positive advanced breast cancer. N Engl J Med 380(20):1929–1940
112. Venot Q, Blanc T, Rabia SH, Berteloot L, Ladraa S, Duong J-P et al (2018) Targeted therapy in patients with PIK3CA-related overgrowth syndrome. Nature 13:195

113. Furet P, Guagnano V, Fairhurst RA, Imbach-Weese P, Bruce I, Knapp M et al (2013) Discovery of NVP-BYL719 a potent and selective phosphatidylinositol-3 kinase alpha inhibitor selected for clinical evaluation. Bioorg Med Chem Lett 23(13):3741–3748
114. Barlaam B, Cosulich S, Fitzek M, Germain H, Green S, Hanson LL et al (2017) Discovery of a novel aminopyrazine series as selective PI3Kα inhibitors. Bioorg Med Chem Lett 27(13):3030–3035
115. Fan Y-H, Ding H-W, Liu D-D, Song H-R, Xu Y-N, Wang J (2018) Novel 4-aminoquinazoline derivatives induce growth inhibition, cell cycle arrest and apoptosis via PI3Kα inhibition. Bioorg Med Chem 26(8):1675–1685
116. Heffron TP, Heald RA, Ndubaku C, Wei B, Augistin M, Do S et al (2016) The rational design of selective benzoxazepin inhibitors of the α-isoform of phosphoinositide 3-kinase culminating in the identification of (S)-2-((2-(1-isopropyl-1H-1,2,4-triazol-5-yl)-5,6-dihydrobenzo[f]imidazo[1,2-d][1,4]oxazepin-9-yl)oxy)propanamide (GDC-0326). J Med Chem 59(3):985–1002
117. Nacht M, Qiao L, Sheets MP, St Martin T, Labenski M, Mazdiyasni H et al (2013) Discovery of a potent and isoform-selective targeted covalent inhibitor of the lipid kinase PI3Kα. J Med Chem 56(3):712–721
118. Lannutti BJ, Meadows SA, Herman SEM, Kashishian A, Steiner B, Johnson AJ et al (2011) CAL-101, a p110delta selective phosphatidylinositol-3-kinase inhibitor for the treatment of B-cell malignancies, inhibits PI3K signaling and cellular viability. Blood 117(2):591–594
119. Gopal AK, Kahl BS, de Vos S, Wagner-Johnston ND, Schuster SJ, Jurczak WJ et al (2014) PI3Kδ inhibition by idelalisib in patients with relapsed indolent lymphoma. N Engl J Med 370(11):1008–1018
120. Somoza JR, Koditek D, Villaseñor AG, Novikov N, Wong MH, Liclican A et al (2015) Structural, biochemical, and biophysical characterization of idelalisib binding to phosphoinositide 3-kinase δ. J Biol Chem 290(13):8439–8446
121. Cushing TD, Hao X, Shin Y, Andrews K, Brown M, Cardozo M et al (2015) Discovery and in vivo evaluation of (S)-N-(1-(7-fluoro-2-(pyridin-2-yl)quinolin-3-yl)ethyl)-9H-purin-6-amine (AMG319) and related PI3Kδ inhibitors for inflammation and autoimmune disease. J Med Chem 58(1):480–511
122. Hoegenauer K, Soldermann N, Zécri F, Strang RS, Graveleau N, Wolf RM et al (2017) Discovery of CDZ173 (Leniolisib), representing a structurally novel class of PI3K delta-selective inhibitors. ACS Med Chem Lett 8(9):975–980
123. Rao VK, Webster S, Dalm VASH, Sediva A, van Hagen PM, Holland S et al (2017) Effective "Activated PI3Kδ syndrome-"targeted therapy with the PI3Kδ inhibitor leniolisib. Blood 130(21):2307–2316
124. Winkler DG, Faia KL, DiNitto JP, Ali JA, White KF, Brophy EE et al (2013) PI3K-δ and PI3K-γ inhibition by IPI-145 abrogates immune responses and suppresses activity in autoimmune and inflammatory disease models. Chem Biol 20(11):1364–1374
125. Evans CA, Liu T, Lescarbeau A, Nair SJ, Grenier L, Pradeilles JA et al (2016) Discovery of a selective phosphoinositide-3-kinase (PI3K)-γ inhibitor (IPI-549) as an immuno-oncology clinical candidate. ACS Med Chem Lett 7(9):862–867
126. Gangadhara G, Dahl G, Bohnacker T, Rae R, Gunnarsson J, Blaho S et al (2019) A class of highly selective inhibitors bind to an active state of PI3Kγ. Nat Chem Biol 15(4):348–357
127. Tapparel C, Siegrist F, Petty TJ, Kaiser L (2013) Picornavirus and enterovirus diversity with associated human diseases. Infect Genet Evol 14:282–293
128. Blomen VA, Májek P, Jae LT, Bigenzahn JW, Nieuwenhuis J, Staring J et al (2015) Gene essentiality and synthetic lethality in haploid human cells. Science 350(6264):1092–1096
129. Ilnytska O, Santiana M, Hsu N-Y, Du W-L, Chen Y-H, Viktorova EG et al (2013) Enteroviruses harness the cellular endocytic machinery to remodel the host cell cholesterol landscape for effective viral replication. Cell Host Microbe 14(3):281–293
130. Lamarche MJ, Borawski J, Bose A, Capacci-Daniel C, Colvin R, Dennehy M et al (2012) Anti-hepatitis C virus activity and toxicity of type III phosphatidylinositol-4-kinase beta inhibitors. Antimicrob Agents Chemother 56(10):5149–5156
131. Reuberson J, Horsley H, Franklin RJ, Ford D, Neuss J, Brookings D et al (2018) Discovery of a potent, orally bioavailable PI4KIIIβ inhibitor (UCB9608) able to significantly prolong allogeneic organ engraftment in vivo. J Med Chem 61(15):6705–6723
132. Rutaganira FU, Fowler ML, McPhail JA, Gelman MA, Nguyen K, Xiong A et al (2016) Design and structural characterization of potent and selective inhibitors of phosphatidylinositol 4 kinase IIIβ. J Med Chem 59(5):1830–1839
133. Mejdrová I, Chalupská D, Plačková P, Müller C, Šála M, Klima M et al (2017) Rational design of novel highly potent and selective phosphatidylinositol 4-kinase IIIβ (PI4KB) inhibitors as broad-spectrum antiviral agents and tools for chemical biology. J Med Chem 60(1):100–118
134. Arita M, Kojima H, Nagano T, Okabe T, Wakita T, Shimizu H (2011) Phosphatidylinositol 4-kinase III beta is a target of enviroxime-like compounds for antipoliovirus activity. J Virol 85(5):2364–2372
135. Ashley EA, Dhorda M, Fairhurst RM, Amaratunga C, Lim P, Suon S et al (2014) Spread of artemisinin resistance in Plasmodium falciparum malaria. N Engl J Med 371(5):411–423
136. Vale N, Moreira R, Gomes P (2009) Primaquine revisited six decades after its discovery. Eur J Med Chem 44(3):937–953
137. Rodgers MJ, Albanesi JP, Phillips MA (2007) Phosphatidylinositol 4-kinase III-beta is required for Golgi maintenance and cytokinesis in Trypanosoma brucei. Eukaryot Cell 6(7):1108–1118

138. Younis Y, Douelle F, Feng T-S, González Cabrera D, Le Manach C, Nchinda AT et al (2012) 3,5-Diaryl-2-aminopyridines as a novel class of orally active antimalarials demonstrating single dose cure in mice and clinical candidate potential. J Med Chem 55(7):3479–3487
139. Brunschwig C, Lawrence N, Taylor D, Abay E, Njoroge M, Basarab GS et al (2018) UCT943, a next-generation Plasmodium falciparum PI4K inhibitor preclinical candidate for the treatment of malaria. Antimicrob Agents Chemother 62(9):62
140. Fowler ML, McPhail JA, Jenkins ML, Masson GR, Rutaganira FU, Shokat KM et al (2016) Using hydrogen deuterium exchange mass spectrometry to engineer optimized constructs for crystallization of protein complexes: case study of PI4KIIIβ with Rab11. Protein Sci 25(4):826–839
141. Robert X, Gouet P (2014) Deciphering key features in protein structures with the new ENDscript server. Nucleic Acids Res 42(W1):W320–W324

Druggable Lipid GPCRs: Past, Present, and Prospects

10

Hirotaka Mizuno and Yasuyuki Kihara

Abstract

G protein-coupled receptors (GPCRs) have seven transmembrane spanning domains and comprise the largest superfamily with ~800 receptors in humans. GPCRs are attractive targets for drug discovery because they transduce intracellular signaling in response to endogenous ligands via heterotrimeric G proteins or arrestins, resulting in a wide variety of physiological and pathophysiological responses. The endogenous ligands for GPCRs are highly chemically diverse and include ions, biogenic amines, nucleotides, peptides, and lipids. In this review, we follow the *KonMari* method to better understand druggable lipid GPCRs. First, we have a comprehensive *tidying up* of lipid GPCRs including receptors for prostanoids, leukotrienes, specialized pro-resolving mediators (SPMs), lysophospholipids, sphingosine 1-phosphate (S1P), cannabinoids, platelet-activating factor (PAF), free fatty acids (FFAs), and sterols. This *tidying up* consolidates 46 lipid GPCRs and *declutters* several perplexing lipid GPCRs. Then, we further *tidy up* the lipid GPCR-directed drugs from the literature and databases, which identified 24 clinical drugs targeting 16 unique lipid GPCRs available in the market and 44 drugs under evaluation in more than 100 clinical trials as of 2019. Finally, we introduce drug designs for GPCRs that *spark joy*, such as positive or negative allosteric modulators (PAM or NAM), biased agonism, functional antagonism like fingolimod, and monoclonal antibodies (MAbs). These strategic drug designs may increase the efficacy and specificity of drugs and reduce side effects. Technological advances will help to discover more endogenous lipid ligands from the vast number of remaining orphan GPCRs and will also lead to the development novel lipid GPCR drugs to treat various diseases.

Keywords

Lipid mediators · Bioactive lipids · Drug discovery · Mechanism of action

10.1 Introduction

G protein-coupled receptors (GPCRs) are the largest membrane receptor family in the human genome. Nearly 800 GPCRs have been identified in humans [1], about half of which are olfactory receptors [2], and the other half are classified into

H. Mizuno
ONO Pharmaceutical Co., Ltd, Osaka, Japan

Y. Kihara (✉)
Sanford Burnham Prebys Medical Discovery Institute, La Jolla, CA, USA
e-mail: kihara-yasuyuki@umin.net

Y. Kihara (ed.), *Druggable Lipid Signaling Pathways*, Advances in Experimental Medicine and Biology 1274, https://doi.org/10.1007/978-3-030-50621-6_10

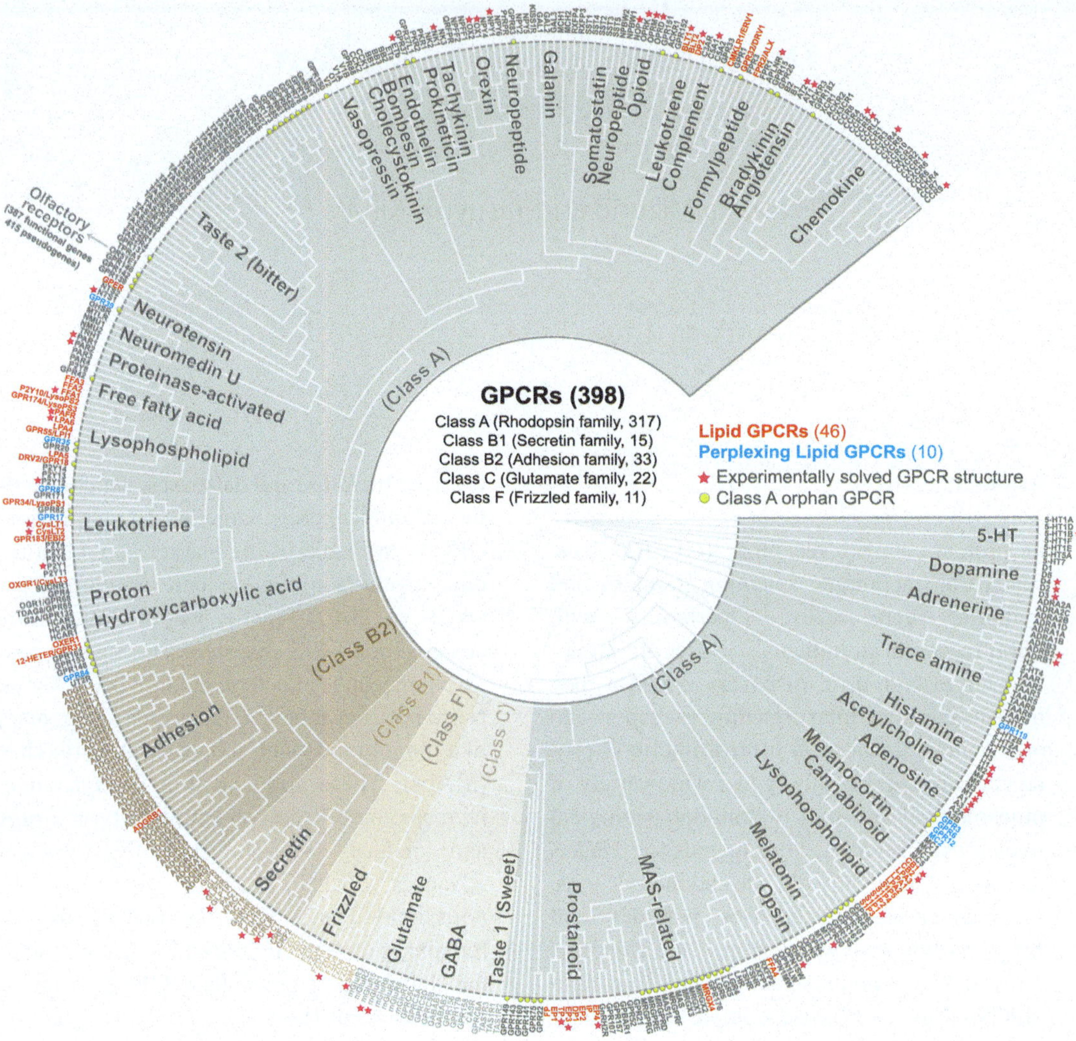

Fig. 10.1 A GPCR pyrogenetic tree

four major families including the Rhodopsin family (Class A), the Secretin and Adhesion family (Class B), the Glutamate family (Class C), and the Frizzled family (Class F) (Fig. 10.1). A list of GPCRs that includes their classification, structures, ligands, and other features is available in the Guide to Pharmacology, International Union of Basic and Clinical Pharmacology/ British Pharmacological Society (IUPHAR/BPS) [1, 3] and the GPCR database (GPCRdb) [4, 5]. GPCRs have seven transmembrane (TM) helices (TM1 ~ 7) that form an orthosteric ligand-binding pocket for small organic agonists, while some ligands such as glycoprotein hormones and glutamate bind to the amino terminus and extracellular domains [6]. Endogenous ligand binding to the orthosteric site causes a conformational change of the receptor that can interact with a heterotrimeric G protein (Gα, Gβ and Gγ), which promotes the guanine nucleotide exchange (i.e., releasing guanosine diphosphate, GDP, and guanosine triphosphate, GTP) on the Gα subunits. Thus, GPCRs act as guanine nucleotide exchange factors (GEF). GTP-bound Gα dissociates from

Table 10.1 Chronological table for the Nobel Prizes that are relevant to lipid or GPCR biology

Year	Event
1964	*The Nobel Prize in Physiology or Medicine (Bloch K and Lynen F)* for their discoveries concerning "the mechanism and regulation of the cholesterol and fatty acid metabolism."
1967	*The Nobel Prize in Physiology or Medicine (Wald G, Granit R, and Hartline H)* for their discoveries concerning "the primary physiological and chemical visual processes in the eye."
1970	*The Nobel Prize in Physiology or Medicine (von Euler U, Katz B, and Axelrod J)* for their discoveries concerning "the humoral transmitters in the nerve terminals and the mechanism for their storage, release and inactivation."
1982	*The Nobel Prize in Physiology or Medicine (Samuelsson BI, Bergstrom SK, and Vane JR)* for their discoveries concerning "prostaglandins and related biologically active substances."
1985	*The Nobel Prize in Physiology or Medicine (Brown MS and Goldstein JL)* for their discoveries concerning "the regulation of cholesterol metabolism."
1994	*The Nobel Prize in Physiology or Medicine (Gilman AG and Rodbell M)* for their "discovery of G-proteins and the role of these proteins in signal transduction in cells."
2004	*The Nobel Prize in Physiology or Medicine (Axel R and Buck LB)* for their "discoveries of odorant receptors and the organization of the olfactory system."
2012	*The Nobel Prize in Chemistry (Lefkowitz RJ and Kobilka B)* for "studies of G-protein-coupled receptors."

Gβγ, resulting in the activation of downstream signaling cascades including Ca^{2+} mobilization, cyclic AMP (cAMP), Rho family of GTPase, mitogen-activated protein kinases (MAPKs), and others [7]. Distinct signaling mediated through arrestins requires receptor phosphorylation at the C-terminal tail by GPCR kinases (GRKs), which promotes receptor internalization. The mechanistic aspects of GPCRs have been thoroughly investigated particularly through a focus on adrenergic receptors or a rhodopsin, resulting in multiple Nobel Prize winners in the 120-years history of Nobel Prize (Table 10.1).

10.2 Tidying-Up of Lipid GPCR Histories

The GPCR history dates back to 1876 when the photosensitive pigment, "rhodopsin," was discovered by *Franz Boll* and isolated by *Wilhelm Kühne*. The requirement of covalent ligand retinal for physico-chemical processes of the "rhodopsin cycle" was discovered by *George Wald* who won the Nobel Prize in Physiology or Medicine (1967) and shared with *Ragnar Granit* and *Haldan Hartline*. In 1973, ATP-dependent cyclic nucleotide phosphodiesterase (PDE) was reported to transduce visual signals through rhodopsin [8], while it was found that the coupling of rhodopsin to PDE is dependent on GTP [9]. The involvement of rhodopsin-coupled GTPase that mediates visual signals from rhodopsin to PDE was proposed in 1977 [10, 11] and confirmed by *Fung B.K.* et al. in 1981 [12]. It is now known as transducin which is a heterotrimeric GTPase composed of α, β, and γ subunits. The complete amino acid sequence of bovine rhodopsin was solved by *Yuri Ovchinnikov* in 1982 and *Paul Hargrave* in 1983, which highlighted the receptor containing 7 transmembrane domains (7TMs) [9]. In 1983, the gene encoding bovine rhodopsin was cloned by *Jeremy Nathans* and *David Hogness* [13]. Until then, only the structural features of rhodopsin had been studied. In 2000, *Krzysztof Palczewski* et al. solved the X-ray crystal structure of dark-adopted bovine rhodopsin as the first GPCR structure [14]. These studies, as well as the adrenergic receptors studied by *Robert Lefkowitz* and *Brain Kobilka* [15], opened up opportunities for understanding GPCR biology and GPCR-targeted drug discovery.

Rhodopsin is a photosensitive receptor requiring a lipophilic retinal in its orthosteric binding site, indicating that lipids could bind to the orthosteric site of other GPCRs. To date, the endoge-

Table 10.2 Chronological table for lipid GPCR identification

Year	Events	Refs
1990	Cloning and identification of **CB$_1$**/*CNR1*	[16]
	Cloning of *EDG-1*	[17]
1991	Cloning and identification of **PAFR**/*PTAFR*	[18]
	Cloning and identification of **TP**/*TBXA2R*	[19]
1992	Cloning and identification of **EP$_3$**/*PTGER3*	[20]
	Discovery of AEA as endogenous ligand for **CB$_1$**	[21]
1993	Cloning and identification of **EP$_2$**/*PTGER2* and **EP$_1$**/*PTGER1*	[22, 23]
	Cloning and identification of **CB$_2$**/*CNR2*	[24]
	Cloning of *EDG-5* (also known as *AGR16* or *H218*)	[25, 26]
1994	Cloning and identification of **FP**/*PTGFR*, **IP**/*PTGIR* and **DP$_1$**/*PTGDR*	[27–32]
	Cloning and identification of **ALX/FPR2**	[33]
1995	Discovery of 2-AG as endogenous ligand for **CB$_{1/2}$**	[34, 35]
	Cloning of *EDG-2*	[36]
1996	Cloning and identification of **LPA$_1$**/*LPAR1* (*EDG-2*, also known as *VZG-1*)	[37]
	Cloning of *EDG-3*	[38]
1997	Identification of *EDG-5* as **S1P$_2$**/*S1PR2* and *EDG-3* as **S1P$_3$**/*S1PR3*	[39]
	Cloning and identification of **BLT$_1$**/*LTB4R*	[40]
1998	Identification of *EDG-1* as **S1P$_1$**/*S1PR1*	[41]
	Cloning and identification of **LPA$_2$**/*LPAR2* (*EDG-4*)	[42]
	Cloning of *EDG-6*	[43]
1999	Cloning and identification of **CysLT$_1$**/*CYSLTR1*	[44]
	Cloning and identification of **LPA$_3$**/*LPAR3* (*EDG-7*)	[45]
	Cloning of *CRTH2*	[46]
2000	Identification of *EDG-6* as **S1P$_4$**/*S1PR4*	[47, 48]
	Cloning and identification of **BLT$_2$**/*LTB4R2*	[49]
	Cloning and identification of **CysLT$_2$**/*CYSLTR2*	[50]
	Cloning and identification of *EDG-8* as **S1P$_5$**/*S1PR5*	[51]
2001	Identification of *CRTH2* as **DP$_2$**/*PTGDR2*	[52]
2002	Cloning and identification of **OXER$_1$**/*OXER1*(GPR170)	[53–55]
	Cloning and identification of **GPBA** (GPR131)	[56, 57]
2003	Identification of *P2RY9/GPR23* as **LPA$_4$**/*LPAR4*	[58]
	Identification of GPR40, GPR43 and GPR41 as **FFA1**/*FFAR1*, **FFA2**/*FFAR2*, **FFA3**/*FFAR3*, respectively.	[59–61]
2005	Discovery of RvE1 as an endogenous ligand for **ERV$_1$**/*CMKLR1*	[62]
	Identification of *GPR120* as **FFA4**/FFAR4	[63]
	Identification of *GPR30* as **GPER**/*GPER*	[64]
2006	Identification of *GPR92* as **LPA$_5$**/*LPAR5*	[65]
	Identification of *GPR34* as **lysoPS$_1$**	[66]
2007	Identification of *GPR55* as **LPI$_1$**/*LPIR1*	[67]
2008	Discovery of 12-HHT as an endogenous ligand for **BLT$_2$**	[68]
	Identification of *P2YR5* as **LPA$_6$**/*LPAR6*	[69, 70]
2010	Discovery of RvD1 as an endogenous ligand for **DRV$_1$**/*GPR32*	[71]
2011	Discovery of oxysterols as an endogenous ligand for **EBI$_2$**/*GPR183*	[72]
	Discovery of 12-HETE as an endogenous ligand for **12-HETER**/*GPR31*	[73]
2012	Identification of *P2YR10* as **lysoPS$_2$**, and *GPR174* as **lysoPS$_3$**	[74]
2013	Identification of *GPR99* as **CysLT$_3$**/*CYSLTR3*	[75]
2015	Discovery of lysophosphatidylglucoside as another endogenous ligand for **LPI$_1$**/*LPIR1*	[76]
	Discovery of RvD2 as an endogenous ligand for **DRV$_2$**/*GPR18*	[77]

nous ligands for ~12% of ~400 non-olfactory GPCRs are lipophilic molecules which include eicosanoids (prostanoids and leukotrienes), lysophospholipids, sphingosine 1-phosphate (S1P), endocannabinoids, platelet-activating factors (PAF), free fatty acids (FFAs), sterols, and others. Tables 10.2 and 10.3 chronologically summarize the history of lipid GPCR identification and lipid GPCR structures, respectively.

Table 10.3 Chronological table for lipid GPCR structures

Year	Events	Refs
2000	Dark-adopted bovine **rhodopsin** structure (*the first GPCR structure*)	[14]
2007	**β2AR** structure in complex with an inverse agonist (carazolol) and a monoclonal antibody (Mab5) against the native third intracellular loop of β2AR, or in complex with carazolol and T4 lysozyme (T4L) fusion in place of the third intracellular loop of β2AR (*the second GPCR structure*)	[78]
2012	**$S1P_1$** structure in complex with an antagonist (ML056) (*the first lipid GPCR structure*) PDB: 3V2W, 3V2Y	[79]
2014	**FFA1** structure in complex with agonists (TAK-875) PDB: 4PHU	[80]
2015	**LPA_1** structure in complex with antagonists (ONO-9780307, ONO-9910539, and ONO-3080573) PDB: 4Z34, 4Z35, 4Z36	[81]
2016 ~ 17	**CB_1** structure in complex with antagonists (AM6538, AM841, AM11542, and taranabant) PDB: 5TGZ, 5XR8, 5XRA, 5 U09	[82–84]
2017	**LPA_6** structure in the absence of any ligand PDB: 5XSZ	[85]
	FFA1 structure in complex with agonists (MK-866 and AgoPAM AP8) PDB: 5TZR, 5TZY	[86]
2018	**PAFR** structure in complex with antagonists (SR 27417 and ABT491) PDB: 5ZKP, 5ZKQ	[87]
	BLT_1 structure in complex with an antagonist (BIIL260) PDB: 5X33	[88]
	TP structure in complex with antagonists (ramatroban and daltroban) PDB: 6IIU, 6IIV	[89]
	DP_2 structure in complex with antagonists (fevipiprant and CAY10471) PDB: 6D26, 6D27	[90]
2019	**EP_3** structure in complex with an endogenous ligand (PGE_2) or an agonist (misoprostol) PDB: 6AK3, 6MQT	[91]
	EP_4 structure in complex with an antagonist (ONO-AE3-208) and Fab PDB: 5YHL, 5YWY, 5YF	[92]
	CB_1 structure in complex with an agonist (MDMB-Fubinaca) and Gi protein PDB: 6N4B	[93]
	CB_2 structure in complex with an agonist (AM10257) PDB: 5ZTY	[94]
	$CysLT_1$ structure in complex with antagonists (pranlukast and zafirlukast) PDB: 6RZ4, 6RZ5	[95]
	CB_1 structure in complex with an agonist (CP55940) and a NAM (ORG27596) PDB: 6KQI	[96]
	$CysLT_2$ structure in complex with antagonists (ONO-2570366, ONO-2770372, and ONO-2080365) PDB: 6RZ6, 6RZ7, 6RZ8, 6RZ9	[97]

10.2.1 Prostanoid Receptors (TP, $EP_{1,2,3,4}$, $DP_{1,2}$, IP, FP)

Prostanoids are a class of eicosanoids including prostaglandins (PGs) and thromboxane (TX) [98]. Prostanoids are derived from arachidonic acid (AA) that is liberated from membrane phospholipids by phospholipases [99–101]. AA is metabolized into an unstable intermediate of PGs, PGH_2, by cyclooxygenase (COX-1/*PTGS1* and COX-2/*PTGS2*) [102]. PGH_2 is subsequently metabolized to PGs and TX by terminal enzymes. Here we introduce a brief history of prostanoid receptor identification [98, 101].

In 1934–1935, *Ulf von Euler* identified a lipophilic substance in semen that induces uterine contraction and named it PG which derives from prostate gland [103]. He also established the role of norepinephrine as a neurotransmitter in 1945 and won the Nobel Prize in Physiology or Medicine in 1970 with *Sir Bernard Katz* and *Julius Axelrod*. Between 1957–1963, *Sune K. Bergstrom* isolated prostaglandins [104], and he won the Nobel Prize in Physiology or Medicine in 1982 shared with *Bengt I. Samuelsson* and *John R. Vane* (Table 10.1). In 1964, *Sune K. Bergstrom* and *David van Dorp* independently demonstrated the conversion of PGE_2 from AA [105, 106]. In 1968, *E.J. Corey's* group succeeded in the total synthesis of PGs [107, 108]. In 1975, *B.I. Samuelsson's group* discovered an unstable intermediate, TXA_2, in the conversion of PGG_2 to TXB_2 [109]. In 1976, *J.R. Vane* and colleague identified an artery-derived unstable PG (prostacyclin/PGI_2) that inhibits platelet aggregation [110], whose structure was also solved in the same year [111]. *Osamu Hayaishi* strongly suggested to Ono Pharmaceutical Co., Ltd., the development of PGs for clinical purposes which resulted in launching the program in 1966 on the basis of information provided by *S.K. Bergstrom, B.I. Samuelsson*, and *E.J. Corey* [112]. In 1974, Ono Pharmaceutical Co., Ltd. (Japan) placed the world's first PG product, PROSTARMON F Injection (dinoprost, a synthetic analogue of the naturally occurring $PGF_{2\alpha}$) on the market as a labor-inducing drug. Since then, Ono Pharmaceutical Co., Ltd. has taken a lead in developing PG-related drugs including Prostarmon E™ (dinoprostone, a synthetic analogue of the naturally occurring PGE_2) as an orally available labor-inducing drug in 1976, Prostandin™ (alprostadil, a synthetic analogue of the naturally occurring PGE_1) for peripheral circulatory disorders in 1979, Preglandin™ (gemeprost, a PGE_1 analog) for therapeutic abortion in mid-pregnancy in 1984, Ronok™ (ornoprostil, a PGE_1 derivative) for gastric ulcers in 1987, Cataclot™ (ozagrel, a TXA_2 synthase inhibitor) for injection for postoperative cerebral vaspospasm after subarachnoid hemorrhage surgery, Opalmon™ (limaprost, a PGE_1 derivative) for thromboangiitis obliterans in 1988, and Vega™ (ozagrel, a TXA_2 synthase inhibitor) for bronchial asthma in 1992. Other prostanoid receptor-targeted drugs were developed by many pharmaceutical companies such as Doner® (beraprost, a synthetic analog of PGI_2; Toray) for arterial occlusive disorders and pulmonary hypertension in 1992, Seratrodast (a TP antagonist; Takeda Pharmaceutical Co. Ltd.) for asthma, Flolan™ (epoprostenol, a synthetic analogue of PGI_2; GlaxoSmithKline plc) for pulmonary hypertension in 1995, and Xalatan® (latanoprost, a $PGF_{2\alpha}$ analog; Pfizer Inc.) for glaucoma in 1996.

In 1989, *Shuh Narumiya*'s group solved partial amino-acid sequences of a human TX receptor (TP/*TBXA2R*) [113] using a TP-selective antagonist S-145 synthesized by *Hitoshi Arita* at Shionogi & Co., Ltd. [114]. Just a few weeks after the PAFR report in 1991 (*see below*), they succeeded in cloning the *TBXA2R* gene by screening a cDNA library of human megakaryocytic leukemia cells using an oligonucleotide probe design based on the amino-acid sequence [19]. PGE receptors had been pharmacologically subdivided into EP_1, EP_2, and EP_3, which were cloned one after another by a Japanese group (*Shuh Narumiya, Atsushi Ichikawa, Manabu Negishi, Yukihiko Sugimoto,* et al.). In 1992, the first PGE receptor, EP_3/*PTGER3*, was cloned from a mouse cDNA library by PCR based on the sequence of human *TBXA2R* [20]. In 1993, EP_2/*PTGER2* and EP_1/*PTGER1* were cloned by hybridization with the mouse EP_3 cDNA and TP cDNA, respectively [22, 23]. Subsequently, the PGF receptor (FP/*PTGFR*) was cloned in 1994 [27]. In the same year, PGI (prostacyclin) receptor (IP/*PTGIR*) was independently cloned by groups in Japan [28, 29] and Merck & Co. [30] using PCR and hybridization methods. PGD receptor (DP_1/*PTGDR*) was cloned by the Japanese group in 1994 [31] and by Merck's group in 1995 [32]. In 2001, a GPCR originally cloned as a type 2 helper T cell (Th2)-specific receptor (CRTH2, chemoattractant receptor-homologous molecule expressed on Th2 cells) was identified as the second PGD receptor (DP_2/*PTGDR2*) [52]. DP_2 has no significant homology in amino acid sequence with DP or all other PG receptors. Structures of TP [89], DP_2 [90], EP_3 [91], and EP_4 [92] were solved in complex with their antagonists.

10.2.2 Leukotriene, HETE, and Oxoeicosanoid Receptors ($BLT_{1,2}$, $CysLT_{1,2,3}$, GPR31, GPR39, $OXER_1$)

Leukotrienes (LTs) are also derived from AA by the action of 5-lipoxygenase (5-LOX/*ALOX5*). Lipoxygenases (12-LOX/*ALOX12*, 15-LOX/*ALOX15*, eLOX3/*ALOXE3*) produce hydroxyeicosatetraenoic acids (HETEs) that are further oxidized to oxoeicosanoids. 5-LOX metabolize AA to produce the intermediate metabolite, 5-hydroepoxy eicosatetoraenoic acid (5-HpETE), that are further metabolized to unstable LTA_4 and 5-HETE. LTA_4 is metabolized to LTB_4 by LTA_4 hydrolase and to cysteinyl LTs (cys-LTs, particularly LTC_4 by LTC_4 synthase). Here, we introduce the identification history of leukotriene and related lipid receptors.

LTB_4 receptors (**$BLT_{1,2}$**) – Leukotrienes ("leuko" = leukocytes, "trienes" = the three conjugated double bonds in the chemical structures) were introduced by *Bengt I. Samuelsson* [115] whose group identified a chemical structure of leukotriene B_4 (LTB_4: 5(S),12(R)-dihydroxy-6Z,8E,10E,14Z-eicosatetranoic acid) using ^{14}C-labeled arachidonic acid in 1979 [116]. After 18 years, the first LTB_4 receptor (BLT_1/*LTB4R*) was cloned by *Takehiko Yokomizo* in *Takao Shimizu*'s group by using a subtraction strategy between retinoic acid-differentiated vs. undifferentiated HL-60 cells [40]. In 2000, they also cloned the second, lower affinity LTB_4 receptor (BLT_2/*LTB4R2*) from the promotor region of *LTB4R* [117]. They also showed that hydrosxyeicosanoids (12-HETE and 15-HETE) could bind with BLT_2 but not BLT_1 [49]. Their continuing efforts result in the identification of a natural, high affinity ligand for BLT_2, which is 12-HHT (12(S)-Hydroxyheptadeca-5Z,8E,10E-trienoic acid), an equimolar product of thromboxane synthase [68]. In 2018, the crystal structure of BLT_1 in complex with the antagonist (BIIL260) was reported [88].

Cys-LT receptors (**$CysLT_{1,2,3}$**) – In 1938, *Wilhelm Feldberg* and *Charles Kellaway* reported a slow and sustained contraction of smooth muscles by substances produced after antigen challenge [118]. In 1959, *K. Frank Austen* and *Walter Brocklehu*rst termed them as a slow-reacting substance of anaphylaxis (SRS-A) that is separated from the activity of histamine [119]. In 1973, Fisons Ltd. (currently Sanofi) developed an SRS-A antagonist (FPL 55712) [120]. In 1979, *Robert Murphy, Sven Hammarström, and Bengt I. Samuelsson* identified the SRS-A as a cysteine-containing derivative of 5-hydroxy-7,9,11,14-icosatetraenoic acid (leukotriene C) [121]. In the following year, *E.J. Corey's* group established a full structure of the SRS-A as 5(S)-hydroxy-6(R)-glutathionyl-7,9-trans-11,14-cis-eicosatetraenoic acid, and named LTC_4 [122]. Finally, three bioactive cys-LTs (LTC_4, LTD_4, and LTE_4) were found to constitute SRS-A [122]. Pharmacological studies of FPL 55712, which showed selectivity against LTD_4 and LTE_4 as compared to LTC_4 [123, 124], predicted at least 2 cys-LT receptors existed in mammalian tissues [125]. SRS-A antagonists, particularly inhibiting the actions of LTD_4, were expected to be potent anti-asthmatic drugs. Before the cloning of the cys-LT receptors, three drugs were clinically applied to treat asthma. Ono Pharmaceutical Co., Ltd. launched a program in 1981 to develop cys-LT receptor antagonists for clinical application. Pranlukast (ONO-1078) was developed from a lead compound in 1985 [126, 127], entered a phase I clinical trial in 1986, and placed on the market as Onon™ in 1995 in Japan for the first time in the world. In 1996, zafirulukast (Accolate™, AstraZeneca) was approved by the Food and Drug Administration (FDA) in the United States (US), followed by montelukast (Singulair®, Merck & Co.) approval in 1998. In 1999, *Jilly F. Evans a*nd colleagues identified the first cys-LT receptor, $CysLT_1$/*CYSLTR1* from an expressed sequence tag (EST) entry (GenBank) based on the identity to BLT_1 [44]. $CysLT_1$ shows a preference for LTD_4 whose action is inhibited by anti-asthmatic drugs. In 2000, they also cloned $CysLT_2$/*CYSLTR2* that binds to cys-LTs ($LTC_4 = LTD_4 >> LTE_4$) [50]. GPR99 was originally described as an orphan GPCR with homology to the P2Y subfamily [128], whose ligand 2-oxoglutarate was reported in 2004 [129]. Although its affinity was in the high micromolar range, a committee for Receptor Nomenclature and Drug Classification (NC-IUPHAR) pairs GPR99 and 2-oxoglutarate and named GPR99 as OXGR1 with its close family GPR91 as succinate receptor, SUCNR1 [130]. However, in 2013, *Yoshihide Kanaoka* and *K. Frank Austen* identified GPR99 as the third cys-LT receptor that shows preference for LTE_4 with an affinity in the low nanomolar range [75]. They also showed that LTE_4-induced vascular leakage was prevented in GPR99-deficient mice, indicating that $CysLT_3$/*CYSLTR3* is an appropriate nomenclature for GPR99. In 2019, the crystal structure of $CysLT_1$ in complex with the antagonists (zafirlukast and pranlukast) [95] and the crystal structure of CysLT2 in complex with the antagonists (ONO-2570366, ONO-2770372, and ONO-2080365) were reported [97].

HETE receptor **(12-HETER/*GPR31*, *GPR39*)** – *Kenneth V. Honn's* group studied how 12-HETE regulates PC3 prostate cancer cell survival, growth, and angiogenesis [131–133]. PC3 cells showed binding with radiolabeled 12-HETE, suggesting its specific receptor existed. In 2011, they screened the binding efficacy of membrane fractions to 16 candidate GPCRs transfected CHO cells, which identified GPR31 as a 12-HETE receptor (12-HETER) [73]. GPR31 showed specific binding with 12(S)-HETE, while its enantiomer, 12(R)-HETE, did not show any binding. In 2018, *Alkayed N.J.* et al. reported that GPR39 is activated by 15-HETE, which is inhibited by 14,15-epoxyeicosatrienoate (14,15-EET) [134]. GPR39 belonging to the ghrelin receptor family was proposed to be activated by a peptide derived from the ghrelin precursor, obestatin [135], and it was controversial [136]. Later, *Holst B.* et al. reported that zinc ions induce GPR39 signaling including IP and cAMP accumulation [137]. *Alkayed N.J.* et al. showed zinc ion potentiates the effect of 15-HETE on Ca^{2+} mobilization. Since no specific binding between 15-HETE vs. GPR39 has been provided, further studies are needed to confirm this binding partner.

Oxoeicosanoid receptor **($OXER_1$/GPR170)** – In 1992, 5-hydroxyeicosanoid dehydrogenase was found to metabolize 5-HETE to the oxidized form, 5-oxo-ETE [138]. Soon after, the potent chemoattractant activity of 5-oxo-ETE was reported in many kinds of leukocytes [139]. Between 2002 and 2003, three independent groups identified the 5-oxo-ETE receptor ($OXER_1$/*OXER1*) [53–55]. *Tatsuya Haga's* group identified 50 GPCR candidate genes in 2002 [140]. Among them, GPCR48 was characterized as a 5-oxo-ETE receptor through screening of ~1000 natural and synthetic compounds in a GTP-γS binding assay using a GPCR48-$G\alpha_{i1}$ fusion protein [55]. A group of Tanabe Seiyaku Co., Ltd. identified the TG1019 gene by querying a BLAST database with a consensus sequence for the peptide ligand GPCR [53]. They screened a natural bioactive compound library and found 5-oxo-ETE as a ligand for TG1019. Novartis cloned the same gene (named R527) and identified the ligand from an ~2000 agonist library using a Ca^{2+} assay in HEK293 cells stably expressing R527 and $G\alpha_{16}$ [54].

10.2.3 Specialized Pro-Resolving Mediator (SPM) Receptors (ALX, $DRV_{1,2}$, ERV_1)

SPMs are a new class of lipid mediators derived from polyunsaturated fatty acids (PUFA) by the action of LOs, COXs, and cytochrome P450 monooxygenase enzymes. SPMs (lipoxins, resolvins, protectins/neuroprotectins, maresins) orchestrate the resolution of inflammation [141–145]. There are many excellent reviews available by *Charles Serhan* [145–151].

Lipoxin receptor **(ALX/FPR2)** – Lipoxins (LXA_4 and LXB_4), discovered by *Charles Serhan*, *Mats Hamberg*, and *Bengt I. Samuelsson* in 1984, are AA metabolites containing three hydroxy residues and four double bonds [147]. In 1994, based on the observation that LXA_4 exhibited a specific binding in human neutrophils and differentiated HL-60 cells, *Charles Serhan's* group cloned a HL-60-derived cDNA encoding a GPCR (*ALX/FPR2*) that displayed a high homology to the N-Formylmethionine-leucyl-phenylalanine (FMLP) receptor (FRP1) [33], which was previously cloned as an FRP1 homolog in 1992 [152–154]. ALX/FRP2 overexpression in CHO cells showed a specific binding to LXA_4, GTP hydrolysis, and AA release [33]. Although contradictory results about LXA_4-ALX/FPR2 pairing have been reported, LXA_4-dependent pro-resolving effects were not observed in *Fpr2/Fpr3* double-knockout mice [139, 155].

Resolvin receptors **(ERV1/CMKLR1, DRV1/*GPR32*, DRV2/GPR18)** – Resolvins (E-, D- and T-series) are metabolites of ω-3 PUFA (EPA, eicosapentaenoic acid and DHA, docosahexaenoic acid) and n-3 DPA (docosapentaenoic acid), respectively [146]. In 2005, *Makoto Arita* in *Charles Serhan*'s group screened 10 GPCRs closely related to ALX by luciferase assay, resulting in the identification of ChemR23 (CMKLR1, chemokine like receptor 1) as a resolvin E1 (RvE1) receptor [62]. ChemR23 was originally identified as a receptor for a chemoattractant, chemerin, in 2003 [156]. *Arita M.* et al. also reported the specific binding of RvE1 to ERV_1/ChemR23 (Kd = ~11 nM) and the inhibitory effects of RvE1 on dendritic cell functions [62]. In 2010, *Charles Serhan*'s group showed a specific binding of resolvin D1 (RvD1) on the membrane fraction of human leukocytes [71]. Screening of 8 GPCRs by luciferase assay, and further validation by β-arrestin assay, identified that RvD1 directly acts at DRV1/GPR32 and ALX. In 2015, *Nan Chiang* in *Charles Serhan*'s group identified GPR18 as resolvin D2 (RvD2) receptors by screening 77 orphan GPCRs with a β-arrestin assay [77]. They showed specific binding of RvD2 to DRV2/GPR18 (Kd = ~10 nM). Furthermore, RvD2 limited infection-induced polymorphonuclear cell infiltration, which was lost in GPR18-deficient mice [77]. N-arachidonylglycine was originally reported as the endogenous ligand for GPR18 in 2006 [157]. Other groups proposed that N-palmitoylglycine or cannabinoids could activate GPR18 [158, 159]. However, β-arrestin or TGF-a shedding assay did not support these results [74, 160], and constitutive activity of GPR18 was reported [161].

10.2.4 Lysophospholipid Receptors ($LPA_{1,2,3,4,5,6}$, $lysoPS_{1,2,2L,3}$, GPR55/LPI_1)

Lysophospholipids, such as lysophosphatidic acid (LPA, 1 or 2-acyl-sn-glycero-3-phosphate), lysophosphatidylserine (lysoPS), lysophosphatidylinositol (LPI), and lysophosphatidylcholine (LPC), are phospholipids in which one acyl chain is replaced with a hydroxy group. Here we provide a brief history of lysophospholipid receptors.

LPA receptors **($LPA_{1,2,3,4,5,6}$)** – Biological activity of LPA was reported in the early 1960s. *Tokumura A.* et al. first isolated LPA from soybeans in 1978 [162]. In 1989, *Wouter H. Moolenaar's* group proposed the involvement of G proteins in LPA-induced cell proliferation [163]. In 1996, *Jerold Chun's* group reported the first LPA receptor (LPA_1/*LPAR1*) [37]. His group cloned cDNAs in neocortical cell lines by using primers for the 2nd and 7th transmembrane domains (TM II and TM VII) of known GPCR sequences. One of the cDNAs was uniquely expressed in the embryonic neuroproliferative layer of the cerebral cortex, which was originally named ventricular zone gene-1 (*VZG-1*). Overexpression of *VZG-1* induced cell rounding in the presence of a heat-stable serum factor – LPA. *VZG-1* was identical to the endothelium differentiation gene (*EDG-2*) cloned in 1995 [36] based on the sequence similarity with *edg-1* cloned by *Timothy Hla* and *Thomas Maciag* in 1990 [17], which is now known as sphingosine 1-phosphate (S1P) receptor 1 ($S1P_1$/*S1PR1*). The binding affinity (K_D) of 1-oleoyl-LPA to LPA_1 was estimated as subnanomolar range [164]. In 1998, *Edward J. Goetzl's* group cloned *EDG-4* from the EST entry (GenBank) based on the identity of *EDG-2/VZG-1* [42] and characterized it as LPA_2/*LPAR2*. In 1999, *Junken Aoki* in *Keizo Inoue's* group cloned *EDG-7* from cDNAs of human Jurkat cells by using primer pairs for TM II and TM VI of *EDG-2* and *EDG-4*, which was characterized as LPA_3/*LPAR3* [45]. In 2003, *Takao Shimizu's* group screened a bioactive lipid library in orphan GPCR-expressing cells, resulting in deorphanization of P2RY5/GPR23 as the fourth LPA receptor (LPA_4/*LPAR4*) that is structurally distinct from the EDG family of LPA receptors [58]. In 2006, *Jerold Chun's* group identified GPR92 as LPA_5/*LPAR5* through an unbiased screen of LPA-dependent cytoskeletal changes in cDNA library transfected cells [65]. In 2008, genetic and mutation screening of hypotrichosis simplex identified homozygous truncating mutations in *P2RY5* which was proposed as a sixth LPA receptor [69, 70]. In the following year, *Keisuke Yanagida* in *Takao Shimizu's* group confirmed *P2RY5* as LPA_6/*LPAR6* that couples to $G\alpha_{12/13}$ with preferential binding to 2-acyl-LPA [165]. In 2015, the structure of LPA_1 in complex with antagonists (ONO-9780307, ONO-9910539, and ONO-3080573) was reported [81]. In 2017, LPA_6 structure was solved in the absence of any ligands [85].

LysoPS receptors **($LysoPS_{1,2,2L,3}$)** – In 2006, researchers of Takeda Pharmaceutical Co., Ltd. reported that GPR34 is a receptor for lysoPS that inhibits forskolin-induced cAMP accumulation and induces degranulation in mast cells [66]. In 2012, *Junken Aoki's* group confirmed that GPR34 is a lysoPS receptor with a ligand preference to 2-acyl-LysoPS [166]. In 2012, *Asuka Inoue* in *Junken Aoki's* group developed a TGF-α shedding assay that identified three additional lysoPS receptors: P2RY10, A630033H20, and GPR174 [74]. They proposed that GPR34, P2RY10, A630033H20, and GPR174 be designated as LPS_1, LPS_2, LPS_{2L}, and LPS_3, respectively [167]. However, to avoid confusion with lipopolysaccharide which is commonly referred to and widely used as "LPS; lipopolysaccharide," we used $lysoPS_1$, $lysoPS_2$, $lysoPS_{2L}$, and $lysoPS_3$ in the IUPHAR review [168].

LPI receptor **(GPR55/LPI_1)** – GPR55 was identified from the EST database and cloned in 1999 [169]. In 2007, GlaxoSmithKline and AstraZeneca independently proposed that GPR55 as the third cannabinoid receptor [170–172]. GlaxoSmithKline showed GPR55 binds to an abnormal cannabidiol (Abn-CBD), a synthetic regioisomer of cannabidiol, using a GTP-γS assay, while GPR55-deficiency in mice did not alter Abn-CBD-induced lowering of mean arterial pressure in wild-type mice [171]. AstraZeneca showed a direct binding of endocannabinoids, Δ^9-THC, and synthetic cannabinoids like CP55940 through a GTP-γS assay [172]. In the same year, *Oka S.* et al. reported that LPI induces ERK phosphorylation and intracellular Ca^{2+} mobilization via GPR55 [67]. They also found that 2-arachidonoyl-LPI had greater activity in ERK phosphorylation than other LPI species [173]. In 2009, an assay that detects β-arrestin recruitment to the ligand-binding GPCR confirmed an agonistic activity of a CB_1 antagonist/inverse agonist, AM251 [160]. A TGF-α shedding assay revealed that LPI signaling through GPR55 preferably coupled with $G\alpha_{13}$ rather than $G\alpha_{12}$ [74]. In 2015, a TGF-α shedding assay revealed lysophosphatidylglucoside as an endogenous ligand for GPR55, which showed a tenfold lower EC_{50} than LPI [76].

10.2.5 Sphingosine 1-Phosphate Receptors ($S1P_{1,2,3,4,5}$)

In 1991, *Sarah Spiegel*'s group first reported that S1P induces cellular proliferation and intracellular Ca^{2+} mobilization [174]. In 1990, *Timothy Hla* and *Thomas Maciag* cloned a gene, *EDG-1*, from human endothelial cells by differential hybridization between quiescent vs. differentiated endothelium [17, 175], which contains seven transmembrane domains with structural similarities to GPCRs. In 1993–94, two groups independently cloned a gene encoding GPCR (AGR16 or H218), later termed *EDG-5* [25, 26], and currently known as $S1P_2$/*S1PR2*. In 1996, *EDG-3* was cloned by PCR with degenerate primers designed from CB_1 sequences [38], which is now designated as $S1P_3$/*S1PR3*. In 1997, *Edward J. Goetzl's* group revealed that S1P induces intracellular signaling via *EDG-5* and *EDG-3* [39]. In 1998, *Timothy Hla's* group identified that a heat-stable serum-borne lipid, S1P, induced cell-cell aggregates in *EDG-1* overexpressing cells, bound to EDG-1 with high affinity (Kd = ~8 nM), and induced receptor internalization [41]. *EDG-1* was designated as $S1P_1$/*S1PR1*. In the same year,

EDG-6 was cloned from *in vitro* differentiated dendritic cells by PCR [43]. In 2000, two groups including Kirin Brewery Co., Ltd. and *Sarah Spiegel* independently reported EDG-6 as the fourth S1P receptor ($S1P_4$/*S1PR4*) [47, 48]. Also, *Kevin R. Lynch's* group cloned *EDG-8* and characterized it as S1P receptor ($S1P_5$/*S1PR5*) by a binding assay and cAMP accumulation [51]. In 2012, $S1P_1$ structure in complex with an antagonist (ML056) was reported as the first lipid GPCR structure [79]. One of the most important drug discovery contributions from the sphingolipid biology field was the development of fingolimod (Gilenya®, Novartis International AG) for relapsing-remitting multiple sclerosis (MS). We have reviewed S1P receptor modulators including fingolimod and siponimod (Mayzent®, Novartis International AG) [176–179].

10.2.6 Cannabinoid Receptors ($CB_{1,2}$)

Cannabis or marijuana (*Cannabis sativa, Cannabis indica,* and *Cannabis ruderalis*), belonging to the *Cannabaceae* family of the *Plantae* Kingdom, is one of the oldest cultivated plants in Asia and it was grown for industrial (as hemp), medicinal, recreational, and ritual purposes since the third millennium BCE (Before the Common Era). The Greek historian *Herodotus* recorded cannabis usage in his book, "*The Histories*" in the 5th BCE, which is the oldest reported record regarding the practice of smoking or inhaling cannabis fumes [180]. A recent study scientifically verified evidence for psychoactive compound-containing cannabis smoking in western China beginning at least 2500 years ago [181]. The world's oldest Chinese Pharmacopeia "*Shennong Ben Ts'ao King*" described the use of cannabis achenes (seeds) as a laxative for over 1800 years [182]. In 1899, *Thomas Wood* and colleagues first isolated cannabinol (CBN), a metabolite of tetrahydrocannabinol (THC). In 1940, the full structure of CBN and its total synthesis was reported by two independent groups in United Kingdom (*Lord Alexander R. Todd*, who won the Nobel Prize in Chemistry in 1957) [183], and in USA (*Roger Adams*) [184]. In 1961, to combat drug abuse (cannabis, heroin, opium, cocaine, etc.), the United Nations drew up a treaty, "Single Convention on Narcotic Drugs," [185] making these drugs illegal. Several countries (Netherlands, Spain, Canada, and several states in the USA) have more recently legalized the recreational use of marijuana. In 1964, *Yechiel Gaoni* and *Raphael Mechoulam* isolated and defined the full structure of Δ^9-THC from hashish (a resin of cannabis) [186]. Since then, the pharmacological actions of cannabis components have been investigated thoroughly [187]. In 1990, a research group in the National Institute of Mental Health, USA, cloned a first cannabinoid receptor (CB_1) from a rat cerebral cortex cDNA library using an oligonucleotide probe derived from the sequence of bovine substance-K receptor [16]. The major psychoactive cannabinoid in marijuana (Δ^9-THC) and its synthetic analogue (CP 55940) inhibited forskolin-stimulated cAMP accumulation in CHO-K1 cells stably expressing rat CB_1. However, endogenous ligands remained unknown until 1992 when *William Devane* and *Raphael Mechoulam* discovered the anandamide (N-arachidonoyl-ethanolamine, AEA) as a lipophilic endogenous ligand (endocannabinoid) for CB_1/*CNR1* [21], although cannabinoids are highly hydrophilic molecules. *Munro S.* et al. identified the peripheral cannabinoid receptor, CB_2/*CNR2*, using a cDNA library of human leukemic cell line HL60 in 1993, which showed a displacement of the radioactive chemical compound by Δ^9-THC as well as AEA [24]. Another endocannabinoid, 2-arachidonoyl-glycerol (2-AG), was independently reported by *Mechoulam R.* et al. and *Sugiura T.* et al. in 1995 [34, 35]. CB_1 structure was solved in complex with antagonists (AM6538 [82], AM841, AM11542 [83], and taranabant [84] in 2016–2017, MDMB-Fubinaca with $G\alpha_i$ protein in 2019 [93]), and CB_2 in complex with an antagonist (AM10257) in 2019 [94]. Most recently, the CB_1 structure in complex with a negative allosteric modulator, ORG27569, and an agonist, CP55940, was reported [96].

10.2.7 Platelet-Activating Factor (PAF) Receptor (PAFR)

PAF was first proposed in 1972 by *Jacques Benveniste*, which induced histamine release from platelets [188]. In 1979, the chemical structure of PAF was determined as 1-O-alkyl-2-acetyl-sn-glyceryl-3-phosphorylcholine by *Donald Hanahan* [189]. PAF is produced from LPC by the action of acetyltransferase which was cloned by *Hideo Shindou* in *Takao Shimizu*'s group in 2007 [190]. PAF-dependent phospholipase C activation through G proteins implied the presence of a receptor for PAF [191]. In 1991, *Takao Shimizu*'s group discovered PAFR/*PTAFR* through functional cloning using electrophysiological detection in *Xenopus laevis* oocytes that expressed a guinea-pig lung cDNA library [18]. PAFR was identified as a GPCR with 7TM. Given the fact that no endogenous ligands were known when CB_1 was cloned, PAFR is among the first to be a cloned lipid GPCR. A dual histamine H1 receptor and PAFR antagonist, rupatadine (Rupall™, Pediapharm Inc.), has been on the market in Canada since 2017. The crystal structure of human PAFR in complex with antagonists (SR 27417 and ABT491) was reported in 2018 [87].

10.2.8 Free Fatty Acid (FFA) Receptors (FFA1,2,3,4)

Sawzdargo M. et al. identified a group of intronless genes (FFA1/*GPR40*, FFA3/*GPR41*, *GPR42* and FFA2/*GPR43*) located in tandem at human chromosome 19q13.1 [192]. In 2003, Takeda Pharmaceutical Co., Ltd. and GlaxoSmithKline independently reported that FFAs are the endogenous ligands for an orphan GPR40 [59, 60]. Takeda Pharmaceutical Co., Ltd. screened over 1000 chemical compounds on CHO cells transiently expressing GPR40 in intracellular Ca^{2+} mobilization, resulting in the identification of long-chain FFAs [60]. They also showed that FFAs amplify glucose-stimulated insulin secretion (GSIS) from pancreatic β cells through FFA1/GPR40. Takeda Pharmaceutical Co., Ltd. synthesized a GPR40-selective ago-allosteric agonist, TAK-875 (fasiglifam), as a novel glucose-dependent insulin secretagogues [193], which acts as an ago-allosteric modulator of GPR40 [194]. Clinical trials of fasiglifam in type 2 diabetes showed positive results on reducing glycated hemoglobin HbA1c with minimum risk of hypoglycemia [195], while its development was terminated due to concerns about liver safety [196]. GlaxoSmithKline applied the same strategy in HEK293 cells with an agonist library of ~1500 known and putative natural GPCRs [59]. They also identified GPR41 and GPR43 as additional FFA receptors [61]. A pheromone pathway-responsive FUS1-lacZ reporter assay in yeast (*Saccharomyces cerevisiae*) expressing GPR43 was used to screen the compound library, which identified several compounds containing acetate counterions activated by GPR43. Further investigation demonstrated that short chain carboxylic acids activate not only GPR43 but also GPR41. Although GPR42 lacked responses against carboxylate ligands, a single amino acid mutation (W174R) in GPR42 restored responses to propionate. GPR42 still remains as Class A orphan GPCR. In 2003, a database search in the human genome revealed seven evolutionarily conserved Class A GPCRs including GPR120 [197]. In 2005, *Gozoh Tsujimoto's* group screened over 1000 chemical compounds in HEK293 cells stably expressing FFA4/GPR120-enhanced green fluorescent protein (EGFP) and assessed receptor internalization with a high content screening system [63]. They identified that long-chain FFAs induced intracellular Ca^{2+} signals, ERK phosphorylation in FFA4/GPR120-expressing HEK293 cells, and glucagon-like peptide-1 (GLP-1) secretion in intestinal endocrine cells. FFA receptor-targeted drug discovery was summarized in a recent review [198]. FFA1 structure was solved in complex with agonists (TAK-875, MK-8666, and AgoPAM AP8), which revealed two distinct allosteric binding sites [80, 86].

10.2.9 Sterol Lipid GPCRs (GPER, GPBA, MRGPRX4, EBI2)

To date, three receptors in Class A GPCRs have been proposed as sterol lipid receptors as follows:

Estrogen receptor **(GPER/GPR30)** –The biological functions of estrogen have been well-studied, particularly in the female reproductive system. Many actions of estrogen are mediated via the nuclear estrogen receptors (ER) from which the first ERα was cloned [199, 200] after 30 years following the first evidence of ER presented by *Elwood V. Jensen* in the 1950s [201]. Subsequently, the second nuclear ER (ERβ) was cloned in 1985. In the late 1960s, *Clara M. Szego* showed that acute estrogen exposure increased uterine cAMP production [202], which was inhibited by glucocorticoids and β blockers [203, 204]. She also reported estrogen receptors in plasma membrane fractions and partial purification of the receptor [205]. In 2005 GPR30, that was cloned from an orphan GPCR in 1997, was identified as the third and the first transmembrane ER by *Eric R. Prossnitz* [64]. They showed that GPR30 was uniquely localized to the endoplasmic reticulum. 17β estradiol binds to the receptor (K_i = 6.6 nM) and transduces intracellular Ca^{2+} (EC_{50} = 0.5 nM) and phosphatidylinositol 3,4,5-trisphosphate (PIP_3) signaling via GPR30. Data supporting the pairing between estrogen vs. GPR30 led to its designation as a G protein-coupled estrogen receptor (GPER) by the IUPHAR in 2007 [206].

Bile acid receptor **(GPBA/GPR131, MRGPRX4)** – Bile acids are mixtures of amphipathic steroid acids that are produced by the liver, stored in the gallbladder, and their homeostasis is controlled by enterohepatic circulation [207]. Bile acids facilitate digestion and absorption of lipids as digestive surfactants. In 1999, the farnesoid X receptor (FXR) was identified as a nuclear bile acid receptor that controls bile acid synthesis [208]. In 2002–2003, Banyu Pharmaceutical Co., Ltd. and Takeda Chemical Industries, Ltd., independently reported that bile acids increase intracellular cAMP levels via a GPCR, BG37 [56] or TGR5 [57], both now known as G protein-coupled bile acid receptor (GPBA). GPBA plays roles in the control of glucose homeostasis, inflammation, etc. [209]. In 2019, an additional bile acid receptor was reported by two independent groups [210, 211]. Both groups identified that bile acids elicited Ca^{2+} signals in neurons via MRGPRX4 (EC_{50} = ~5 μM). They proposed that bile acids-induced itch in humans (cholestatic itch) is independent from histamine and GPBA, and dependent on MRGPRX4. These suggest that MRGPRX4 seems to be the second bile acid receptor. In addition to bile acids, the same group also reported that MRGPRX4 is a receptor for a heme metabolite, bilirubin, that is another major component of bile, which showed an EC_{50} value of 61.9 μM as determined by a Ca^{2+} assay [212].

Oxysterol receptor **(EBI2/GPR183)**– In 2009, *Jason G. Cyster's* group reported that Epstein-Barr virus-induced gene 2 (EBI2, also known as GPR183) mediates B cell segregation between the outer and center follicle [213], while its endogenous ligand has not been determined. EBI2 was originally cloned from Burkitt's lymphoma cells, which are mainly expressed in B cells [214]. In 2011, EBI2 was identified as a receptor for oxysterol (7α,25-dihydroxycholesterol, 7α,25-OHC) [72]. Oxysterols induce intracellular Ca^{2+} mobilization via EBI2 with EC_{50} values in a range of 2 nM to 3 μM depending on the oxysterol species [72].

10.2.10 Perplexing Lipid GPCRs (GPR3, GPR6, GPR12, GPR35, GPR63, GPR84, GPR87, GPR119)

About ~25% of the Class A Rhodopsin family of GPCRs remain as orphan GPCRs whose endogenous ligands have not been determined. Lipid ligands for some of these receptors have been proposed, but additional studies are needed to confirm or reject the proposed ligand-receptor pairs.

GPR3, GPR6, GPR12, GPR63 – In 1993, GPR3 was cloned as GPCR21 by *Saeki Y.* et al. from a mouse brain cDNA library [215]. They also isolated GPR12 (GPCR01) [215] whose rat homolog was cloned in 1991 [216]. Human GPR3, GPR6, and GPR12 were isolated between 1994 and 1995 by three independent groups [217–219]. In 2002, *Evi Kostenis's* group reported that GPR3, GPR6, and GPR12 showed constitutive activation of adenylate cyclase in the presence of serum, while the constitutive activity was reduced by removing lipids by charcoal stripping [220]. They screened 200 bioactive lipids which resulted in the identification of S1P and dihydrosphingosine 1-phosphate (DHS1P) as ligands for these receptors. In the same year, *H. Chica Schaller's* group identified sphingosylphosphorylcholine (SPC) as a high-affinity ligand for GPR12 by examining Ca^{2+} signaling and GIRK (G-protein-gated inwardly rectifying K^+ channel)-mediated inward currents in *Xenopus oocytes* [221]. They also reported S1P to be a high-affinity ligand for GPR6 [222]. However, other groups were not able to confirm these observations, which were not reproduced by β-arrestin assay [160]. In 2007, cannabidiol (CBD) was reported as an inverse agonist for GPR3 and GPR6 [223, 224]. In 2003, *Evi Kostenis's* group reported that S1P and dioleoylphosphatidic acid are low affinity ligands for GPR63 [225], but again the β-arrestin assay did not support this observation [160].

GPR35 – In 2006, Kynurenic acid, a metabolite of L-tryptophan, was first reported as a ligand for GPR35 by screening of ~300 biochemical intermediates using an aequorin assay [226]. In 2010, *Oka S. et.al.* reported that 2-acyl-LPA activates intracellular Ca^{2+} mobilization via GPR35 [227]. In 2012, a group in Corning Inc. reported the agonist activity of tyrosine metabolites [228]. Several surrogate ligands for GPR35 like zaprinast have been reported [229]. Controversy still remains about the endogenous ligand for GPR35.

GPR84, GPR87 – In 2001, GPR84, along with GPR86, GPR87, GPR90, and GPR91, was discovered by EST data mining [230]. Amgen Inc. reported medium chain fatty acids as ligands for GPR84 in 2006 [231]. Independent studies also supported the results by β-arrestin, cAMP, and Ca^{2+} assays [229] or TGF-α shedding assay [74]. However, hydroxy fatty acids and 6-*n*-octylaminouracil were proposed to activate GPR84 potently as compared to fatty acids [232]. In 2007, GPR87 was proposed as an LPA receptor [233], where the GPR87-$G\alpha_{16}$ fusion protein in CHO cells was used to determine the Ca^{2+} mobilization (EC_{50} = ~36 nM). Another group demonstrated the responses of GPR87 to 100 μM LPA by β-arrestin assay [234].

GPR119 – GPR119 was identified as one of ~60 novel GPCRs in 2002 [140]. In 2005, Yamanouchi Pharmaceutical Co. Ltd. (currently Astellas Pharma Inc.) reported GPR119 as a novel endogenous receptor for LPC, LPE, and LPI that induces insulin secretion from pancreatic β-cells [235]. In 2006, based on the homology between cannabinoid receptors vs. GPR119, Prosidion Limited/OSI Pharmaceuticals tested a series of cannabinoids and fatty-acid ethanolamides using a yeast fluorescent reporter assay, resulting in the identification of oleoylethanolamide as a ligand for GPR119 [236]. This was reproduced by a TGF-α shedding assay [74]. Then, in 2010, Arena Pharmaceuticals Inc. proposed that N-oleoyldopamine enhances glucose homeostasis via activation of GPR119-$G\alpha_s$ signaling [237]. Another group reported that 2-oleoyl glycerol and other 2-monoacylglycerols induces cAMP accumulation in GPR119-expressing COS-7 cells [238]. Many pharmaceutical companies have tried to deorphanize GPR119 since receptor signaling is an attractive target for the treatment of type 2 diabetes and obesity [239]. However, all the identified ligands showed lower potency than other lipid-GPCR pairs, implying the existence of a real, natural ligand for GPR119.

10.2.11 Lipid GPCRs Belonging to the Class B2 Adhesion Family (ADGRB1, ADGRF1)

Among 33 Adhesion family GPCRs [240], two receptors have been proposed to bind with lipids as follows:

ADGRB1/BAI1 – ADGRB1/BAI1 was originally cloned as a p53-regulated gene in brain in 1997 [241]. In 2007, ADGRB1/BAI1 was discovered to be interacting with ELMO1 (engulfment and cell motility 1) [242] which promotes internalization of apoptotic cell corpses [243, 244]. Because thrombospondin type 1 repeats on the N-terminal of ADGRB1/BAI1 that binds with phosphatidylserine [242], it is proposed to be one of the receptors for the eat me signal exposed on apoptotic cells like Tim-4, Stabilin-2, CD300f, and RAGE [245].

ADGRF1/GPR110 – GPR110, and 7 other closely related GPCRs, were identified from the human genome database by hidden Markov models [246]. All contained a GPCR proteolytic site (GPS) domain in their N-termini [240]. In 2016, *Lee J.W.* et al. reported that ADGRF1/GPR110 is the receptor for N-docosahexaenoylethanolamine (they call it synaptamide) that binds to an extracellular N-terminus of ADGRF1/GPR110 [247]. The synaptamide-induced bioactivity via the receptor works in a cAMP-dependent manner [247]. Since only one group has reported this ligand-receptor interaction, further studies are needed to confirm this pairing.

10.3 Tidying-Up of Drugs Targeting Lipid GPCRs

GPCRs are attractive targets for drug discovery. According to a review by *Hauser A.S.* et al. in 2007 [248], the FDA approved 475 drugs (~34% of all FDA-approved drugs) targeting 108 unique GPCRs (~27% of non-olfactory GPCRs), which account for ~27% of the global market share of therapeutic drugs [248]. Most of the approved GPCR drugs (~77%) target opioid and aminogenic GPCRs (dopamine, acetylcholine, 5-hydroxytryptamine, histamine, adrenaline) [248]. Approved drugs targeting lipid GPCRs (Table 10.4) are curated from the GPCRdb [4, 5], while some drugs were omitted because they seem to be mistakenly registered in the DrugBank database [249]. Several lipid GPCRs have been crystalized with approved drugs, including EP_3 with misoprostol, TP with ramatroban, and $CysLT_1$ with pranlukast (Fig. 10.2). There are 24 approved clinical drugs targeting pharmacologically well-characterized lipid GPCRs, of which 16 drugs are agonists for prostanoid or cannabinoid receptors (Table 10.4). The remaining 8 drugs are antagonists for TP, $CysLT_1$, PAFR, and $S1P_1$. For example, alprostadil is the pharmaceutical name of PGE_1 that was derived from dihomo-γ-linolenic acid. Alprostadil (Caverject™, Pfizer, approved in 1981; Muse™, Meda Pharmaceuticals, approved in 1996) is used to treat erectile dysfunction in adult males. The mechanism of action (MOA) of alprostadil is similar but different from phosphodiesterase 5 (PDE5) inhibitors (sildenafil, Viagra®; avanafil, Stendra®; tadalafil, Cialis®; vardenafil, Levitra® or Staxyn®) that increase cGMP levels in smooth muscle. On the other hand, activation of the PGE_1 receptor by alprostadil increases cAMP in smooth muscle cells, resulting in relaxation and penile erection formation. Considering the rank order of affinity of PGE_1 for the EPs ($EP_3 = EP_4 > EP_2 > EP_1 = IP$ [252]) and the receptor expression pattern, this efficacy might be mediated by EP_4.

The most recent (as of 2019) FDA-approved lipid GPCR drug is siponimod (Mayzent®, Novartis International AG) for the treatment of secondary progressive multiple sclerosis (SPMS) [253]. Siponimod, as well as fingolimod, functionally antagonize $S1P_1$ on lymphocytes, resulting in the sequestration of pathogenic lymphocytes in the peripheral lymphoid organs [254]. A major difference between siponimod vs.

Table 10.4 Approved lipid GPCR drugs

Drugs	Types	Targets	MOA	Indications
Alprostadil (PGE_1)	Small molecule	**EP_1, EP_2, EP_3, EP_4**	Agonist	Sexual dysfunction
Misoprostol (PGE_1 analogue)	Small molecule	**EP_3, EP_2[a]**	Agonist	Anti-ulcer
Dinoprostone (PGE_2)	Small molecule	**EP_1, EP_2, EP_3, EP_4**	Agonist	Abortifacient
Grapiprant	Small molecule	**EP_4**	Antagonist	Osteoarthritis (*for dogs*)
Bimatoprost ($PGF_{2\alpha}$ analogue)	Prodrug	**FP**	Agonist	Antiglaucomic
Carboprost ($PGF_{2\alpha}$ analogue)	Small molecule	**FP**	Agonist	Abortifacient
Latanoprost ($PGF_{2\alpha}$ analogue)	Small molecule	**FP**	Agonist	Antiglaucomic
Tafluprost ($PGF_{2\alpha}$ analogue)	Small molecule	**FP**	Agonist	Reduce intraocular pressure
Travoprost ($PGF_{2\alpha}$ analogue)	Small molecule	**FP**	Agonist	Antiglaucomic
Seratrodast	Small molecule	**TP**	Antagonist	Anti-asthmatic
Ramatroban	Small molecule	**TP, DP_2**	Antagonist	Anti-allergic
Epoprostenol (PGI_2)	Small molecule	**IP**	Agonist	Antihypertensive, platelet aggregation inhibitor
Iloprost (PGI_2 analogue)	Small molecule	**IP**	Agonist	Antihypertensive
Treprostinil (PGI_2 analogue)	Small molecule	**IP**	Agonist	Antihypertensive
Selexipag	Small molecule	**IP**	Agonist	Pulmonary arterial hypertension
Montelukast	Small molecule	**$CysLT_1$**	Antagonist	Anti-asthmatic
Pranlukast	Small molecule	**$CysLT_1$**	Antagonist	Anti-asthmatic
Zafirlukast	Small molecule	**$CysLT_1$**	Antagonist	Anti-asthmatic
Fingolimod (Sphingosine analogue)	Prodrug	**$S1P_1$, $S1P_3$[a], $S1P_4$[a], $S1P_5$[a]**	Functional antagonist	Relapsing remitting multiple sclerosis
Siponimod	Small molecule	**$S1P_1$, $S1P_5$[a]**	Functional antagonist	Secondary progressive multiple sclerosis
Dronabinol (Tetrahydrocannabinol)	Small molecule	**CB_1, CB_2[a]**	Agonist	Analgesics, appetite stimulant, antiemetics
Nabilone	Small molecule	**CB_1, CB_2**	Agonist	Antiemetics
Cannabidiol	Plant extract Small molecule	**CB_1, CB_2**	Agonist	Seizures associated with Lennox-Gastaut syndrome or Dravet syndrome
Rupatadine	Small molecule	**PAFR**	Antagonist	Anti-allergic
Icosapent	Small molecule	**FFA1[a]**	Agonist	Hypertriglyceridemia

[a]Secondary target

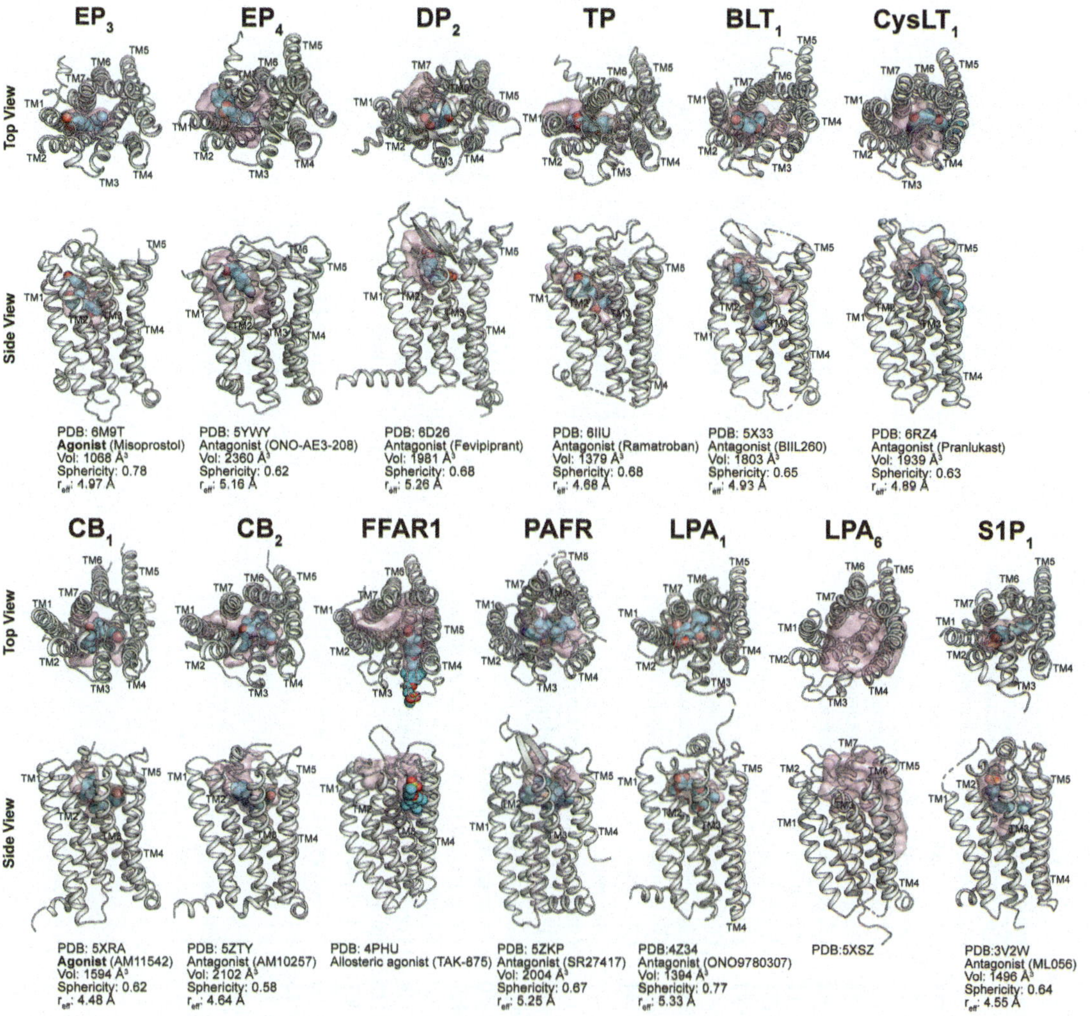

Fig. 10.2 Lipid GPCR structures. Receptors are shown as cartoon representations and ligands are represented as spherical molecules. Surface representations in pink are binding pockets that were estimated using the HOLLOW program [250] with a spherical constraint, followed by manual extraction based on the existence of chemical compounds. The volume (Å^3), sphericity (ψ: a measure of how much the volume resembles a sphere; $\psi = 1.0$ is an exact sphere), and effective radius (r_{eff}: a radius of a sphere with the same surface area to volume ratio) of these binding pockets are also provided, which were analyzed using the ^{3}V program [251]

fingolimod is a receptor selectivity that siponimod selectively binds to $S1P_1$ and $S1P_5$ [255], while fingolimod binds to $S1P_1$, $S1P_3$, $S1P_4$, and $S1P_5$. Moreover, fingolimod is a pro-drug that requires *in vivo* phosphorylation by sphingosine kinases to bind with S1P receptors. There are many excellent reviews written by front-line researchers available [176, 177, 179, 256–259].

Lipid GPCR drugs currently tested in clinical trials are curated mainly from the GPCRdb [4, 5]. Among them, drugs deposited in clinicaltrials.gov are summarized in Table 10.5. Although the list may be inexhaustive, we found more than 100 trials testing 44 drugs targeting 19 lipid GPCRs. Many drugs are in Phase 3 clinical trials including DP_2 antagonists for asthma, IP agonists for pulmonary arterial hypertension (PAH), FP and EP_2 agonists for glaucoma and ocular hypertension, and S1P receptor modulators for MS, Cohn's disease, ulcerative colitis, etc. Also, 11 approved drugs are being tested for new or other indications such as alprostadil for cardiovascular

Table 10.5 Lipid GPCR drugs in clinical trials as of October 2019

Target	Drugs	MOA	NCT number	Indications	Phase	Sponsor
EP_1	***Alprostadil***	Agonist		Congenital heart disease, acute kidney injury, respiratory distress syndrome, etc.	1 ~ 4	*Multiple sponsors*
	ONO-8539	Antagonist	NCT01705275	Glaucoma/ocular hypertension	1^{c}	Ono Pharmaceutical
			NCT00876421	Overactive bladder	2^{c}	Ono Pharmaceutical
EP_2	DE-117	Agonist	NCT03697811	Glaucoma/ocular hypertension	3^{r}	Santen
			NCT03691649	Glaucoma/ocular hypertension	3^{r}	Santen
			NCT03691662	Glaucoma/ocular hypertension	3^{r}	Santen
EP_4	Grapiprant (ARY-007)	Antagonist	NCT03696212	Non-small cell lung cancer	$1 \sim 2^{r}$	Arrys Therapeutics, etc.
			NCT03658772	Microsatellite stable colorectal cancer	1^{r}	Arrys Therapeutics, etc
FP	PDC31	Antagonist	NCT01250587	Dysmenorrhea	1^{c}	PDC Biotech GmbH
	Latanoprostene BUNOD (BOL-303259-X, NCX116)	Agonist	NCT01749904	Glaucoma/ocular hypertension	3^{c}	Bausch & Lomb Incorporated
			NCT01749930	Glaucoma/ocular hypertension	3^{c}	Bausch & Lomb Incorporated
			NCT01895972	Glaucoma/ocular hypertension	3^{c}	Mount Sinai
			NCT03949244	Glaucoma	4^{r}	*Multiple sponsors*
FP/EP_3	***Latanoprost***	Agonist		Glaucoma	2 ~ 4	Allergan
	Bimatoprost	Agonist		Glaucoma/ocular hypertension	2 ~ 4	*Multiple sponsors*
	ONO-9054	Agonist	NCT02083289	Glaucoma/ocular hypertension	2^{c}	Ono Pharmaceutical
TP	Ifetroban (BMS180291)	Antagonist	NCT02682511	Scleroderma	2^{r}	Cumberland Pharmaceuticals
			NCT03028350	Asthma, aspirin-induced	2^{r}	Cumberland Pharmaceuticals
			NCT02216357	Aspirin exacerbated respiratory disease (AERD)	2^{c}	Cumberland Pharmaceuticals
			NCT01436500	Hepatorenal syndrome	2^{c}	Cumberland Pharmaceuticals
			NCT02802228	Portal hypertension/liver cirrhosis	2^{c}	Cumberland Pharmaceuticals
			NCT03694249	Breast cancer	2^{r}	Cumberland Pharmaceuticals
			NCT03326063	Nasal polyps, asthma	2^{r}	Brigham and Women's Hospital
IP	Ralinepag (APD811)	Agonist	NCT02279160	Pulmonary arterial hypertension (PAH)	2^{c}	Arena Pharmaceuticals
			NCT02279745	PAH	2^{a}	United Therapeutics
			NCT03683186	PAH	3^{a}	United Therapeutics
			NCT03626688	PAH	3^{r}	United Therapeutics
	Iloprost	Agonist		Pulmonary hypertension	1 ~ 4	*Multiple sponsors*
	Selexipag	Agonist		Pulmonary hypertension	1 ~ 4	Actelion

DP_2	Fevipiprant	Antagonist	NCT03681093	Asthma	3^{r}	Novartis
	(QAW039)		NCT03629249	Asthma	3^{r}	Novartis
	Setipiprant	Antagonist	NCT02781311	Alopecia	2^{c}	Allergan
	(ACT-129968)		NCT01225315	Asthma	2^{c}	Idorsia Pharm.
			NCT01484119	Allergic rhinitis	3^{c}	Idorsia Pharm.
	ADC3680	Antagonist	NCT01730027	Asthma	2^{c}	Pulmagen Therapeutics
	OC459	Antagonist	NCT02660489	Asthma/rhinovirus	2^{u}	Chiesi Farmaceutici S.p.A.
	(OC000459)		NCT02002208	Eosinophilic asthma	2^{c}	Chiesi Farmaceutici S.p.A.
			NCT02560610	Atopic dermatitis	2^{c}	Chiesi Farmaceutici S.p.A.
	ACT-774312	Antagonist	NCT03688555	Nasal polyposis	2^{r}	Idorsia Pharm.
	ATX2417	Antagonist	NCT02316912	Asthma	1^{c}	Chiesi Farmaceutici S.p.A.
	ARRY-502	Antagonist	NCT01561690	Asthma	2^{c}	Array BioPharm
$CysLT_1$	***Montelukast***	Antagonist	NCT03402503	Alzheimer's disease	2^{r}	IntelGenx
			NCT03991988	Alzheimer's disease	2^{r}	Emory Univ.
			NCT03545997	Sleep apnea	2^{r}	Universitaire de Grenoble
			NCT03109288	Multiple sclerosis	$1 \sim 2^{r}$	NIAID
LPA_1	`SAR100842	Antagonist	NCT01651143	Systemic sclerosis	2^{c}	Sanofi
	BMS-986020	Antagonist	NCT01766817	Idiopathic pulmonary fibrosis	2^{c}	Bristol-Myers Squibb
			NCT02588625	Scleroderma	2^{c}	Bristol-Myers Squibb
	^{18}F-BMS-986327	PET ligand	NCT04069143	Idiopathic pulmonary fibrosis	1^{n}	Bristol-Myers Squibb
$S1P_1$	***Fingolimod***			Multiple sclerosis	$3 \sim 4$	*Multiple sponsors*
	(FTY720)		NCT03943498	Chemotherapy-induced peripheral	1^{r}	Mayo Clinic and NCI
			NCT04088630	neuropathy, etc.	1^{r}	Wake Forest University
			NCT03941743	Intracerebral hemorrhage (IHC)	1^{r}	Mayo Clinic and NCI
	Siponimod		NCT03623243	Breast carcinoma	3^{r}	Novartis
	(BAF312)		NCT03338998	Multiple sclerosis	2^{r}	Novartis
	Ozanimod		NCT02576717	Hemorrhagic stroke, IHC	3^{a}	Celgene
	(RPC1063)		NCT03440385	Multiple sclerosis	3^{r}	Celgene
			NCT02435922	Crohn's disease	3^{r}	Celgene
	Ponesimod		NCT02907177	Ulcerative colitis	3^{r}	Actelion
	(ACT-128800)		NCT03232073	Multiple sclerosis	3^{a}	Actelion
			NCT01093326	Multiple sclerosis	2^{a}	Actelion
			NCT03882255	Multiple sclerosis	1^{r}	Janssen Pharmaceuticals
	Cenerimod		NCT03742037	In patients receiving Propranolol	2^{r}	Idorsia Pharmaceuticals
	Etrasimod		NCT03945188	Systemic lupus erythematosus	3^{r}	Arena Pharmaceuticals
	(APD334)		NCT03996369	Ulcerative colitis	3^{n}	Arena Pharmaceuticals
			NCT03950232	Ulcerative colitis	3^{n}	Arena Pharmaceuticals
	Amiselimod		NCT02389790	Ulcerative colitis	2^{c}	Mitsubishi Tanabe Pharma
	(MT1303)		NCT01890655	Crohn's disease	2^{c}	Mitsubishi Tanabe Pharma
			NCT01987843	Multiple sclerosis	2^{c}	Mitsubishi Tanabe Pharma
				Plaque psoriasis		

(continued)

Table 10.5 (continued)

Target	Drugs	MOA	NCT number	Indications	Phase	Sponsor
CB_1/ CB_2	***Cannabidiol***	Agonist		Pain, epilepsy, Rett syndrome, anxiety	3^{nra}	*Multiple sponsors*
	Dronabinol	Agonist		Pain, trichotillomania, PTSD, schizophrenia	3^{nra}	*Multiple sponsors*
	Cannabidivarin (GWP420006)	Agonist	NCT03202303	Autism spectrum disorder (ASD)	2^{r}	GW Research Ltd
			NCT03848481	Prader-Willi syndrome	2^{n}	Montefiore Medical Center
	Nabilone	Agonist	NCT03251326	PTSD	$1 \sim 2^{c}$	NIDA
			NCT03773796	Parkinson's disease	3^{r}	Medical University Innsbruck
	RYI-018	MAb	NCT03261739	Nonalcoholic fatty liver disease	1^{c}	Bird Rock Bio
CB_2	Lenabasum (JBT-101)	Agonist	NCT03398837	Diffuse cutaneous systemic sclerosis	3^{a}	Corbus Pharmaceuticals
			NCT03813160	Dermatomyositis	3^{r}	Corbus Pharmaceuticals
			NCT03451045	Cystic fibrosis	2^{r}	Corbus Pharmaceuticals
PAFR	***Rupatadine***	Antagonist	NCT03770923	Rheumatoid arthritis	3^{r}	Sherief Abd-Elsalam, etc
ALX?	BLXA4-ME (lipoxin analog)	Agonist?	NCT02342691	Gingival inflammation	$1 \sim 2^{a}$	NIDCR
ERV1? BLT_1?	RX-10045 (Resolvin E_1 analog)	Agonist?	NCT01675570	Dry eye syndrome	2^{c}	C.T. Development America
			NCT01639846	Allergic conjunctivitis	2^{c}	C.T. Development America
			NCT02329743	Inflammation/pain//cataract	2^{c}	A.T. Resolve SARL
			NCT00799552	Dry eye syndrome	2^{c}	Resolvyx Pharmaceuticals
DRV1?	Resolvin D_1	Agonist?	NCT03483311	Psoriasis		Cairo Univ.
GPR84	GLGP1205	Antagonist	NCT02337608	Ulcerative colitis	2^{c}	Galapagos NV
			NCT03725852	Idiopathic pulmonary fibrosis	2^{r}	Galapagos NV
GPR84/ FFA1	PBI-4050	Antagonist/ Agonist	NCT03184584	Alström syndrome	2^{a}	ProMetic Biosciences
			NCT02538536	Idiopathic pulmonary fibrosis	2^{c}	ProMetic Biosciences
			NCT02562573	Type 2 diabetes, metabolic syndrome	2^{c}	ProMetic Biosciences
			NCT02955888	Cystic fibrosis	2^{t}	ProMetic Biosciences
	PBI-4547	Antagonist/ Agonist	NCT04068259		1^{r}	ProMetic Biosciences

Bold italic drugs are approved drugs. [a]active, [c]completed, [n]not yet recruiting, [r]recruiting, [t]terminated diseases, montelukast for Alzheimer's disease and MS, fingolimod for neuropathy, intracerebral hemorrhage (IHC), and carcinoma, and rupatadine for arthritis (Table 10.5). Grapiprant is an orally bioavailable EP_4 antagonist [260], which was approved by the FDA's Center for Veterinary Medicine for control of pain and inflammation in canines associated with osteoarthritis (Galliprant®, Aratana Therapeutics). Since PGE_2-EP_2/EP_4 signaling suppresses cytotoxic T lymphocyte survival and function [260, 261], a combination therapy of the anti-PD-1 (programmed cell death protein 1) antibody and an EP_4 antagonist was proposed as a treatment for cancer and chronic viral infections. Thus, the clinical trials for galliprant aim to treat patients with non-small cell lung cancer in combination with anti-PD-1 antibodies.

10.4 Lipid GPCR Drugs that Spark Joy!

In this section, some key concepts of GPCR pharmacology (allosteric modulation, biased agonism, and functional antagonism) are introduced as the next generation modes of drug action for lipid GPCR drug discovery. Also, monoclonal antibodies are introduced as promising biological drugs that could modulate lipid GPCR functions with higher selectivity than small molecules.

10.4.1 Allosteric Modulators

According to the definition by IUPHAR [262], allosterism is a modification of the orthosteric ligand-induced protein's properties by another ligand that binds to the spatially distinct site (allosteric site). Allosteric modulators that increase or decrease the affinity and/or efficacy of orthosteric ligands are defined as positive or negative allosteric modulators (PAMs or NAMs), respectively. An allosteric ligand that does not alter the action of the orthosteric ligand is called a neutral allosteric ligand (NAL) or silent allosteric modulator (SAM). Allosteric agonists increase the protein activity independently of orthosteric ligand binding, whereas allosteric inverse agonists reduce it.

As of 2019, 34,371 allosteric drugs targeting GPCRs, which is 42% of total modulator entries, have been deposited in the Allosteric Database (ASD) [263]. About 1% (369 drugs) are under clinical development. ASD displayed 4 approved allosteric modulators targeting GPCRs including cinacalcet (PAM for calcium-sensing receptor), ticagrelor (reversible allosteric antagonist for P2Y12), and niclosamide (PAM for neuropeptide Y4 receptor). Ivermectin is listed in the ASD as an approved PAM for the $GABA_B$ receptor. However, because it is reported as a modulator for the $GABA_A$ receptor [264], it might be mistakenly deposited in the database. ASD listed 18 drugs targeting lipid GPCRs including $CysLT_1$, CB_1, CB_2, and FFA1.

The Protein Data Bank (PDB) is the archive of structural data of biological macromolecules including crystal structures of GPCRs in complex with allosteric ligands [265]. Allosteric sites for Class A GPCRs are widely distributed. For example, the M2 muscarinic acetylcholine receptor showed a binding of its allosteric modulator, LY2119629, on the extracellular vestibule just above the orthosteric agonist [266]. On the other hand, a NAM (Compound-15) for β2 adrenergic receptor (β2AR) binds to a pocket located on the cytoplasmic side, which is similar to the allosteric site of CC chemokine receptors (CCR2 and CCR9) [267]. In 2018, the BLT_1 structure was solved in complex with an antagonist BIIL260 [88], which revealed that benzamidine moiety within the BIIL260 mimics the entire sodium ion-centered water cluster. Benzamidine serves as a NAM for BLT_1 by interacting with highly conserved amino acids ($D66^{2.50}$, $S106^{3.39}$, and $S277^{7.46}$) in Class A GPCRs (Fig. 10.3). The NAM activity of benzamidine was also proved in the β_1 adrenergic receptor, which provides a rational for developing novel inverse agonists specific for each GPCR.

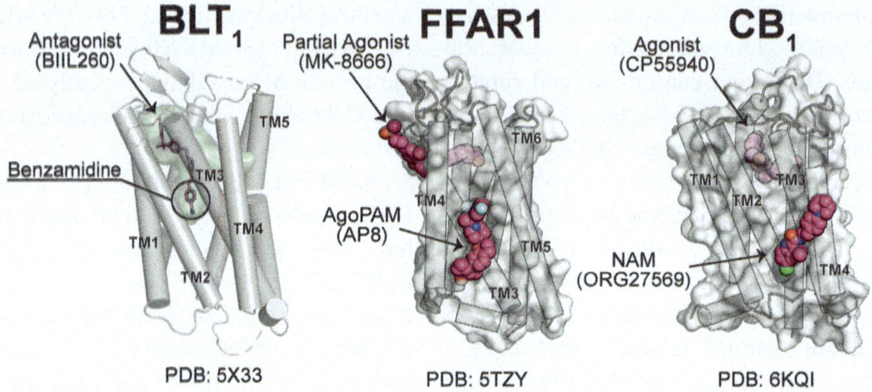

Fig. 10.3 Allosteric sites of lipid GPCRs. Receptors are shown in cartoon and/or surface representation. Ligands are represented as stick or spherical molecules. Surface representation of BLT_1 in green is a binding pocket

Crystal structures of FFA1 and CB_1 in combination with allosteric modulators showed extra-helical allosteric sites in the inner leaflet of the plasma membrane. TAK-875 (also known as fasiglifam) was developed for the treatment of type 2 diabetes [198]. TAK-875 activates $G\alpha_q$ signaling via FFA1/GPR40 in pancreatic β-cells, which enhances insulin secretion in a glucose-dependent manner and improves glucose homeostasis [198]. It was first reported as a selective partial agonist of FFA1 [268], while further studies characterized TAK-875 as an ago-allosteric modulator [194]. Crystal structure revealed that TAK-875 binds to FFA1 between TM3 and TM4, and the extracellular loop 2 (ECL2) forms the roof of the binding cavity. Another allosteric modulator, MK-8666, also binds to the same binding cavity. An additional allosteric site for AgoPAM AP8 was identified on the TM3/4/5 interface (Fig. 10.3), where there is a binding site for allosteric modulators for complement 5a receptor (C5AR), avacopan, and NDT9513727. On the other hand, a NAM of CB_1, ORG27569, binds at the TM2/3/4 interface (Fig. 10.3). ORG27569 enhances CB_1 agonist binding and inhibits $G\alpha_i$ signaling. Allosteric modulation of CB_1 signaling is a novel approach to manipulate the endocannabinoid system for therapeutic benefits including neuropsychiatric, behavioral, neuroendocrine, and immunological disorders [269].

10.4.2 Biased Agonism and Functional Antagonism

Ligand selectively that activates one pathway over another is referred to as "biased agonism." GPCR biased agonism is separated into G protein-biased and β-arrestin-biased agonists. For example, among 16 clinically relevant β-adrenergic receptor antagonists, carvedilol is uniquely reported as a β-arrestin-biased agonist because it shows inverse efficacy for stimulating $G\alpha_s$ signaling and activation of β-arrestin-mediated receptor internalization and MAPK phosphorylation [270]. GPCR drug discovery focusing on biased agonism is well documented in several reviews [271–274].

Biased agonists for some lipid GPCRs have been reported. For example, ONO Pharmaceutical Co. Ltd. reported EP_2-selective G protein-biased agonists, which activate cAMP signaling without affecting β-arrestin recruitment [275, 276]. Takeda Pharmaceutical Co. Ltd. reported five $G\alpha_q$-biased NAMs for LPA_1, LQ1-5 [277], that did not show any antagonistic activity for $G\alpha_i$ signaling and β-arrestin recruitment, while LPA-induced Ca^{2+} mobilization was blocked by these compounds. In 2015, *Hla's* group proposed that high density lipoprotein (HDL)-bound S1P acts as a biased agonist for $S1P_1$ [278]. About 65% of plasma S1P is carried by an apolipoprotein M (ApoM) positive subfraction of HDL and the remaining ~35% of plasma S1P binds to albumin [279]. $ApoM^+HDL$-bound S1P attenuates TNF-α-induced endothelial inflammatory responses via selective recruitment of β-arrestin 2 on the cell surface. On the other hand, albumin-bound S1P triggers greater $G\alpha_i$ activation and induces more receptor endocytosis than $ApoM^+HDL$-bound S1P, suggesting that distinct bound forms of S1P induce opposite biological effects. Biased agonists are promising drugs that selectively target pathogenic or therapeutic signaling pathways while avoiding side effects.

Functional antagonism is classically considered as physiological antagonism or indirect antagonism [280]. An example of physiological antagonism is the action of glucagon (increased blood glucose levels) that physiologically antagonize insulin-induced hypoglycemia via a Class B GPCR, glucagon receptor (GCGR). Indirect antagonism is defined as an inhibition of agonist actions by blockade of intermediate signaling molecules (e.g., protein kinase A inhibitors block agonist-induced β adrenergic receptor). The FDA-approved MS drug, fingolimod (FTY720), is also proposed as a functional antagonist [176]. However, because it shares the $S1P_1$ orthosteric site with S1P, this "functional antagonism" is distinct from physiological/indirect antagonism. Phosphorylated fingolimod (fingolimod-p) initially activates $S1P_1$ as an agonist, followed by inducing sustained receptor internalization, which results in desensitization to the endogenous agonist, S1P. This mechanism of action may have merits and demerits as compared to a pure antagonist like NIBR-0213 [281]. Fingolimod causes bradycardia through G protein-coupled inwardly rectifying potassium channel (GIRK) on cardiomyocytes, while NIBR-0213 has no effect on GIRK but induces pulmonary vascular leakage. The discovery of fingolimod as a functional antagonist may signal a new trend in lipid GPCR drug discovery.

10.4.3 GPCR-Directed Monoclonal Antibodies (MAbs)

The pharmaceutical market for therapeutic MAbs has been exponentially growing since the CD3-specific MAb, muromonab (Orthoclone OKT3, Janssen-Cilag) was approved by the FDA in 1986 [282]. In 2018, the FDA approved the first GPCR-targeted MAb drug, erenumab (Aimovig®, Amgen) that inhibits the calcitonin gene-related peptide (CGRP) receptor, for the treatment of migraine [283]. CGRP is a neuropeptide that modulates nociceptive signals within the trigeminovascular system and plays a key role in migraine. The MOA of erenumab blocks the binding of CGRP to its receptor. In the same year, MAbs that neutralize circulating CGRP peptides were also approved by the FDA for migraine prevention (fremanezumab, Ajoby®, Teva Pharmaceutical Industries Ltd.; and galcanezumab, Emgality®, Eli Lilly and Company).

In 2018, mogamulizumab (Libtayo®, Kyowa Hakko Kirin) was also approved by the FDA for the treatment of cutaneous T cell lymphoma [284]. The proposed MOA of mogamulizumab is that it binds to CC-chemokine receptor 4 (CCR4) on cancer cells, which induces antibody-dependent cellular cytotoxicity. Other GPCR-targeted antibodies have been developed and tested in clinical trials including the CCR5 antibody (leronlimab or PRO 140, CytoDyn Inc.) for acquired immune deficiency syndrome/human immunodeficiency virus (AIDS/HIV) and solid tumors, and the CX3C chemokine receptor 1 (CX3CR1) nanobody (BI 655088, Ablynx/Boehringer Ingelheim) for renal failure.

Among lipid GPCRs, CB_1 is an attractive and validated target for treating nonalcoholic steatohepatitis (NASH) and metabolic disorders [285]. Because small-molecule CB_1 inhibitors cause psychotropic effects, CB_1-directed MAbs that do not pass through the blood-brain barrier have superior specificity and selectivity over small molecules. The CB_1-directed MAb, RYI-018 (Nimacimab, Bird Rock Bio), is currently being developed for the treatment of fibrotic, metabolic, and inflammatory diseases.

The EP_4 crystal structure was solved with antagonist ONO-AE3-208 and an antibody Fab001 [92]. The antibody developed by NB Health Laboratory Co. Ltd. inhibits PGE_2-dependent cAMP production by binding to the extracellular loops (ECLs) of EP_4, indicating the NAM activity of the antibody. Considering the problems with small molecules targeting EP_4 that include selectivity and off-target effects, the antibody seems to be promising for the treatment of chronic inflammation such as colitis, asthma, and chronic obstructive pulmonary disease (COPD).

10.5 Conclusions

In summary, this review provides an overview of lipid GPCR histories, lipid GPCR drugs that were approved or are in clinical trials, and key concepts driving the next generation of GPCR pharmacology. In the nearly 30 years of lipid GPCR history from the cloning of the first lipid GPCRs, 46 lipid GPCRs have been identified and characterized, in which 14 receptors have been crystalized. Only 5% of the 475 FDA-approved GPCR drugs target lipid GPCRs, while 44 drugs targeting 19 lipid GPCRs have been tested in more than 100 clinical trials. Accumulating knowledge about the physiological and pathophysiological functions of bioactive lipids will enable development of more lipid GPCR drugs for diseases including neurological, neuropsychiatric, metabolic, cardiovascular, allergic, immune diseases, and many others.

Acknowledgements The authors would like to thank Christian Munk for supplying the GPCR phylogenetic data in Fig. 10.1, Ms. Danielle Jones (SBP) for editorial assistance. This work was supported by NINDS of the National Institutes of Health under award number R01NS103940 (Y.K.). The content is solely the responsibility of the authors and does not necessarily represent the official views of the National Institutes of Health.

Conflict of Interest H.M. is an employee of ONO Pharmaceutical Co., Ltd. Y.K. declares no competing financial interests.

References

1. Alexander SP, Christopoulos A, Davenport AP, Kelly E, Marrion NV, Peters JA, Faccenda E, Harding SD, Pawson AJ, Sharman JL, Southan C, Davies JA, Collaborators C (2017) THE CONCISE GUIDE TO PHARMACOLOGY 2017/18: G protein-coupled receptors. Br J Pharmacol 174(Suppl 1):S17–S129
2. Niimura Y (2009) Evolutionary dynamics of olfactory receptor genes in chordates: interaction between environments and genomic contents. Hum Genomics 4:107–118
3. Alexander SPH, Christopoulos A, Davenport AP, Kelly E, Mathie A, Peters JA, Veale EL, Armstrong JF, Faccenda E, Harding SD, Pawson AJ, Sharman JL, Southan C, Davies JA, Collaborators C (2019) THE CONCISE GUIDE TO PHARMACOLOGY 2019/20: G protein-coupled receptors. Br J Pharmacol 176(Suppl 1):S21–S141
4. Munk C, Isberg V, Mordalski S, Harpsoe K, Rataj K, Hauser AS, Kolb P, Bojarski AJ, Vriend G, Gloriam DE (2016) GPCRdb: the G protein-coupled receptor database – an introduction. Br J Pharmacol 173:2195–2207
5. Pandy-Szekeres G, Munk C, Tsonkov TM, Mordalski S, Harpsoe K, Hauser AS, Bojarski AJ, Gloriam DE (2018) GPCRdb in 2018: adding GPCR structure models and ligands. Nucleic Acids Res 46:D440–D446
6. Kobilka BK (2007) G protein coupled receptor structure and activation. Biochim Biophys Acta 1768:794–807
7. Weis WI, Kobilka BK (2018) The molecular basis of G protein-coupled receptor activation. Annu Rev Biochem 87:897–919
8. Miki N, Keirns JJ, Marcus FR, Freeman J, Bitensky MW (1973) Regulation of cyclic nucleotide concentrations in photoreceptors: an ATP-dependent stimulation of cyclic nucleotide phosphodiesterase by light. Proc Natl Acad Sci U S A 70:3820–3824
9. Costanzi S, Siegel J, Tikhonova IG, Jacobson KA (2009) Rhodopsin and the others: a historical perspective on structural studies of G protein-coupled receptors. Curr Pharm Des 15:3994–4002
10. Wheeler GL, Bitensky MW (1977) A light-activated GTPase in vertebrate photoreceptors: regulation of light-activated cyclic GMP phosphodiesterase. Proc Natl Acad Sci U S A 74:4238–4242
11. Wheeler GL, Matuto Y, Bitensky MW (1977) Light-activated GTPase in vertebrate photoreceptors. Nature 269:822–824
12. Fung BK, Hurley JB, Stryer L (1981) Flow of information in the light-triggered cyclic nucleotide cascade of vision. Proc Natl Acad Sci U S A 78:152–156
13. Nathans J, Hogness DS (1983) Isolation, sequence analysis, and intron-exon arrangement of the gene encoding bovine rhodopsin. Cell 34:807–814
14. Palczewski K, Kumasaka T, Hori T, Behnke CA, Motoshima H, Fox BA, Le Trong I, Teller DC,

Okada T, Stenkamp RE, Yamamoto M, Miyano M (2000) Crystal structure of rhodopsin: a G protein-coupled receptor. Science 289:739–745
15. Lefkowitz RJ (2013) A brief history of G-protein coupled receptors (Nobel Lecture). Angew Chem Int Ed Engl 52:6366–6378
16. Matsuda LA, Lolait SJ, Brownstein MJ, Young AC, Bonner TI (1990) Structure of a cannabinoid receptor and functional expression of the cloned cDNA. Nature 346:561–564
17. Hla T, Maciag T (1990) An abundant transcript induced in differentiating human endothelial cells encodes a polypeptide with structural similarities to G-protein-coupled receptors. J Biol Chem 265:9308–9313
18. Honda Z, Nakamura M, Miki I, Minami M, Watanabe T, Seyama Y, Okado H, Toh H, Ito K, Miyamoto T et al (1991) Cloning by functional expression of platelet-activating factor receptor from guinea-pig lung. Nature 349:342–346
19. Hirata M, Hayashi Y, Ushikubi F, Yokota Y, Kageyama R, Nakanishi S, Narumiya S (1991) Cloning and expression of cDNA for a human thromboxane A2 receptor. Nature 349:617–620
20. Sugimoto Y, Namba T, Honda A, Hayashi Y, Negishi M, Ichikawa A, Narumiya S (1992) Cloning and expression of a cDNA for mouse prostaglandin E receptor EP3 subtype. J Biol Chem 267:6463–6466
21. Devane WA, Hanus L, Breuer A, Pertwee RG, Stevenson LA, Griffin G, Gibson D, Mandelbaum A, Etinger A, Mechoulam R (1992) Isolation and structure of a brain constituent that binds to the cannabinoid receptor. Science 258:1946–1949
22. Honda A, Sugimoto Y, Namba T, Watabe A, Irie A, Negishi M, Narumiya S, Ichikawa A (1993) Cloning and expression of a cDNA for mouse prostaglandin E receptor EP2 subtype. J Biol Chem 268:7759–7762
23. Watabe A, Sugimoto Y, Honda A, Irie A, Namba T, Negishi M, Ito S, Narumiya S, Ichikawa A (1993) Cloning and expression of cDNA for a mouse EP1 subtype of prostaglandin E receptor. J Biol Chem 268:20175–20178
24. Munro S, Thomas KL, Abu-Shaar M (1993) Molecular characterization of a peripheral receptor for cannabinoids. Nature 365:61–65
25. MacLennan AJ, Browe CS, Gaskin AA, Lado DC, Shaw G (1994) Cloning and characterization of a putative G-protein coupled receptor potentially involved in development. Mol Cell Neurosci 5:201–209
26. Okazaki H, Ishizaka N, Sakurai T, Kurokawa K, Goto K, Kumada M, Takuwa Y (1993) Molecular cloning of a novel putative G protein-coupled receptor expressed in the cardiovascular system. Biochem Biophys Res Commun 190:1104–1109
27. Sugimoto Y, Hasumoto K, Namba T, Irie A, Katsuyama M, Negishi M, Kakizuka A, Narumiya S, Ichikawa A (1994) Cloning and expression of a cDNA for mouse prostaglandin F receptor. J Biol Chem 269:1356–1360
28. Katsuyama M, Sugimoto Y, Namba T, Irie A, Negishi M, Narumiya S, Ichikawa A (1994) Cloning and expression of a cDNA for the human prostacyclin receptor. FEBS Lett 344:74–78
29. Namba T, Oida H, Sugimoto Y, Kakizuka A, Negishi M, Ichikawa A, Narumiya S (1994) cDNA cloning of a mouse prostacyclin receptor. Multiple signaling pathways and expression in thymic medulla. J Biol Chem 269:9986–9992
30. Boie Y, Rushmore TH, Darmon-Goodwin A, Grygorczyk R, Slipetz DM, Metters KM, Abramovitz M (1994) Cloning and expression of a cDNA for the human prostanoid IP receptor. J Biol Chem 269:12173–12178
31. Hirata M, Kakizuka A, Aizawa M, Ushikubi F, Narumiya S (1994) Molecular characterization of a mouse prostaglandin D receptor and functional expression of the cloned gene. Proc Natl Acad Sci U S A 91:11192–11196
32. Boie Y, Sawyer N, Slipetz DM, Metters KM, Abramovitz M (1995) Molecular cloning and characterization of the human prostanoid DP receptor. J Biol Chem 270:18910–18916
33. Fiore S, Maddox JF, Perez HD, Serhan CN (1994) Identification of a human cDNA encoding a functional high affinity lipoxin A4 receptor. J Exp Med 180:253–260
34. Mechoulam R, Ben-Shabat S, Hanus L, Ligumsky M, Kaminski NE, Schatz AR, Gopher A, Almog S, Martin BR, Compton DR et al (1995) Identification of an endogenous 2-monoglyceride, present in canine gut, that binds to cannabinoid receptors. Biochem Pharmacol 50:83–90
35. Sugiura T, Kondo S, Sukagawa A, Nakane S, Shinoda A, Itoh K, Yamashita A, Waku K (1995) 2-Arachidonoylglycerol: a possible endogenous cannabinoid receptor ligand in brain. Biochem Biophys Res Commun 215:89–97
36. Masana MI, Brown RC, Pu H, Gurney ME, Dubocovich ML (1995) Cloning and characterization of a new member of the G-protein coupled receptor EDG family. Receptors Channels 3:255–262
37. Hecht JH, Weiner JA, Post SR, Chun J (1996) Ventricular zone gene-1 (vzg-1) encodes a lysophosphatidic acid receptor expressed in neurogenic regions of the developing cerebral cortex. J Cell Biol 135:1071–1083
38. Yamaguchi F, Tokuda M, Hatase O, Brenner S (1996) Molecular cloning of the novel human G protein-coupled receptor (GPCR) gene mapped on chromosome 9. Biochem Biophys Res Commun 227:608–614
39. An S, Bleu T, Huang W, Hallmark OG, Coughlin SR, Goetzl EJ (1997) Identification of cDNAs encoding two G protein-coupled receptors for lysosphingolipids. FEBS Lett 417:279–282
40. Yokomizo T, Izumi T, Chang K, Takuwa Y, Shimizu T (1997) A G-protein-coupled receptor for leukotriene B4 that mediates chemotaxis. Nature 387:620–624

41. Lee MJ, Van Brocklyn JR, Thangada S, Liu CH, Hand AR, Menzeleev R, Spiegel S, Hla T (1998) Sphingosine-1-phosphate as a ligand for the G protein-coupled receptor EDG-1. Science 279:1552–1555
42. An S, Bleu T, Hallmark OG, Goetzl EJ (1998) Characterization of a novel subtype of human G protein-coupled receptor for lysophosphatidic acid. J Biol Chem 273:7906–7910
43. Graler MH, Bernhardt G, Lipp M (1998) EDG6, a novel G-protein-coupled receptor related to receptors for bioactive lysophospholipids, is specifically expressed in lymphoid tissue. Genomics 53:164–169
44. Lynch KR, O'Neill GP, Liu Q, Im DS, Sawyer N, Metters KM, Coulombe N, Abramovitz M, Figueroa DJ, Zeng Z, Connolly BM, Bai C, Austin CP, Chateauneuf A, Stocco R, Greig GM, Kargman S, Hooks SB, Hosfield E, Williams DL Jr, Ford-Hutchinson AW, Caskey CT, Evans JF (1999) Characterization of the human cysteinyl leukotriene CysLT1 receptor. Nature 399:789–793
45. Bandoh K, Aoki J, Hosono H, Kobayashi S, Kobayashi T, Murakami-Murofushi K, Tsujimoto M, Arai H, Inoue K (1999) Molecular cloning and characterization of a novel human G-protein-coupled receptor, EDG7, for lysophosphatidic acid. J Biol Chem 274:27776–27785
46. Abe H, Takeshita T, Nagata K, Arita T, Endo Y, Fujita T, Takayama H, Kubo M, Sugamura K (1999) Molecular cloning, chromosome mapping and characterization of the mouse CRTH2 gene, a putative member of the leukocyte chemoattractant receptor family. Gene 227:71–77
47. Van Brocklyn JR, Graler MH, Bernhardt G, Hobson JP, Lipp M, Spiegel S (2000) Sphingosine-1-phosphate is a ligand for the G protein-coupled receptor EDG-6. Blood 95:2624–2629
48. Yamazaki Y, Kon J, Sato K, Tomura H, Sato M, Yoneya T, Okazaki H, Okajima F, Ohta H (2000) Edg-6 as a putative sphingosine 1-phosphate receptor coupling to Ca(2+) signaling pathway. Biochem Biophys Res Commun 268:583–589
49. Yokomizo T, Kato K, Hagiya H, Izumi T, Shimizu T (2001) Hydroxyeicosanoids bind to and activate the low affinity leukotriene B4 receptor, BLT2. J Biol Chem 276:12454–12459
50. Heise CE, O'Dowd BF, Figueroa DJ, Sawyer N, Nguyen T, Im DS, Stocco R, Bellefeuille JN, Abramovitz M, Cheng R, Williams DL Jr, Zeng Z, Liu Q, Ma L, Clements MK, Coulombe N, Liu Y, Austin CP, George SR, O'Neill GP, Metters KM, Lynch KR, Evans JF (2000) Characterization of the human cysteinyl leukotriene 2 receptor. J Biol Chem 275:30531–30536
51. Im DS, Heise CE, Ancellin N, O'Dowd BF, Shei GJ, Heavens RP, Rigby MR, Hla T, Mandala S, McAllister G, George SR, Lynch KR (2000) Characterization of a novel sphingosine 1-phosphate receptor, Edg-8. J Biol Chem 275:14281–14286
52. Hirai H, Tanaka K, Yoshie O, Ogawa K, Kenmotsu K, Takamori Y, Ichimasa M, Sugamura K, Nakamura M, Takano S, Nagata K (2001) Prostaglandin D2 selectively induces chemotaxis in T helper type 2 cells, eosinophils, and basophils via seven-transmembrane receptor CRTH2. J Exp Med 193:255–261
53. Hosoi T, Koguchi Y, Sugikawa E, Chikada A, Ogawa K, Tsuda N, Suto N, Tsunoda S, Taniguchi T, Ohnuki T (2002) Identification of a novel human eicosanoid receptor coupled to G(i/o). J Biol Chem 277:31459–31465
54. Jones CE, Holden S, Tenaillon L, Bhatia U, Seuwen K, Tranter P, Turner J, Kettle R, Bouhelal R, Charlton S, Nirmala NR, Jarai G, Finan P (2003) Expression and characterization of a 5-oxo-6E,8Z,11Z,14Z-eicosatetraenoic acid receptor highly expressed on human eosinophils and neutrophils. Mol Pharmacol 63:471–477
55. Takeda S, Yamamoto A, Haga T (2002) Identification of a G protein-coupled receptor for 5-oxo-eicosatetraenoic acid. Biomed Res 23:101–108
56. Maruyama T, Miyamoto Y, Nakamura T, Tamai Y, Okada H, Sugiyama E, Nakamura T, Itadani H, Tanaka K (2002) Identification of membrane-type receptor for bile acids (M-BAR). Biochem Biophys Res Commun 298:714–719
57. Kawamata Y, Fujii R, Hosoya M, Harada M, Yoshida H, Miwa M, Fukusumi S, Habata Y, Itoh T, Shintani Y, Hinuma S, Fujisawa Y, Fujino M (2003) A G protein-coupled receptor responsive to bile acids. J Biol Chem 278:9435–9440
58. Noguchi K, Ishii S, Shimizu T (2003) Identification of p2y9/GPR23 as a novel G protein-coupled receptor for lysophosphatidic acid, structurally distant from the Edg family. J Biol Chem 278:25600–25606
59. Briscoe CP, Tadayyon M, Andrews JL, Benson WG, Chambers JK, Eilert MM, Ellis C, Elshourbagy NA, Goetz AS, Minnick DT, Murdock PR, Sauls HR Jr, Shabon U, Spinage LD, Strum JC, Szekeres PG, Tan KB, Way JM, Ignar DM, Wilson S, Muir AI (2003) The orphan G protein-coupled receptor GPR40 is activated by medium and long chain fatty acids. J Biol Chem 278:11303–11311
60. Itoh Y, Kawamata Y, Harada M, Kobayashi M, Fujii R, Fukusumi S, Ogi K, Hosoya M, Tanaka Y, Uejima H, Tanaka H, Maruyama M, Satoh R, Okubo S, Kizawa H, Komatsu H, Matsumura F, Noguchi Y, Shinohara T, Hinuma S, Fujisawa Y, Fujino M (2003) Free fatty acids regulate insulin secretion from pancreatic beta cells through GPR40. Nature 422:173–176
61. Brown AJ, Goldsworthy SM, Barnes AA, Eilert MM, Tcheang L, Daniels D, Muir AI, Wigglesworth MJ, Kinghorn I, Fraser NJ, Pike NB, Strum JC, Steplewski KM, Murdock PR, Holder JC, Marshall FH, Szekeres PG, Wilson S, Ignar DM, Foord SM, Wise A, Dowell SJ (2003) The Orphan G protein-coupled receptors GPR41 and GPR43 are activated

by propionate and other short chain carboxylic acids. J Biol Chem 278:11312–11319

62. Arita M, Bianchini F, Aliberti J, Sher A, Chiang N, Hong S, Yang R, Petasis NA, Serhan CN (2005) Stereochemical assignment, antiinflammatory properties, and receptor for the omega-3 lipid mediator resolvin E1. J Exp Med 201:713–722
63. Hirasawa A, Tsumaya K, Awaji T, Katsuma S, Adachi T, Yamada M, Sugimoto Y, Miyazaki S, Tsujimoto G (2005) Free fatty acids regulate gut incretin glucagon-like peptide-1 secretion through GPR120. Nat Med 11:90–94
64. Revankar CM, Cimino DF, Sklar LA, Arterburn JB, Prossnitz ER (2005) A transmembrane intracellular estrogen receptor mediates rapid cell signaling. Science 307:1625–1630
65. Lee CW, Rivera R, Gardell S, Dubin AE, Chun J (2006) GPR92 as a new G12/13- and Gq-coupled lysophosphatidic acid receptor that increases cAMP, LPA5. J Biol Chem 281:23589–23597
66. Sugo T, Tachimoto H, Chikatsu T, Murakami Y, Kikukawa Y, Sato S, Kikuchi K, Nagi T, Harada M, Ogi K, Ebisawa M, Mori M (2006) Identification of a lysophosphatidylserine receptor on mast cells. Biochem Biophys Res Commun 341:1078–1087
67. Oka S, Nakajima K, Yamashita A, Kishimoto S, Sugiura T (2007) Identification of GPR55 as a lysophosphatidylinositol receptor. Biochem Biophys Res Commun 362:928–934
68. Okuno T, Iizuka Y, Okazaki H, Yokomizo T, Taguchi R, Shimizu T (2008) 12(S)-Hydroxyheptadeca-5Z, 8E, 10E-trienoic acid is a natural ligand for leukotriene B4 receptor 2. J Exp Med 205:759–766
69. Pasternack SM, von Kugelgen I, Al Aboud K, Lee YA, Ruschendorf F, Voss K, Hillmer AM, Molderings GJ, Franz T, Ramirez A, Nurnberg P, Nothen MM, Betz RC (2008) G protein-coupled receptor P2Y5 and its ligand LPA are involved in maintenance of human hair growth. Nat Genet 40:329–334
70. Shimomura Y, Wajid M, Ishii Y, Shapiro L, Petukhova L, Gordon D, Christiano AM (2008) Disruption of P2RY5, an orphan G protein-coupled receptor, underlies autosomal recessive woolly hair. Nat Genet 40:335–339
71. Krishnamoorthy S, Recchiuti A, Chiang N, Yacoubian S, Lee CH, Yang R, Petasis NA, Serhan CN (2010) Resolvin D1 binds human phagocytes with evidence for proresolving receptors. Proc Natl Acad Sci U S A 107:1660–1665
72. Hannedouche S, Zhang J, Yi T, Shen W, Nguyen D, Pereira JP, Guerini D, Baumgarten BU, Roggo S, Wen B, Knochenmuss R, Noel S, Gessier F, Kelly LM, Vanek M, Laurent S, Preuss I, Miault C, Christen I, Karuna R, Li W, Koo DI, Suply T, Schmedt C, Peters EC, Falchetto R, Katopodis A, Spanka C, Roy MO, Detheux M, Chen YA, Schultz PG, Cho CY, Seuwen K, Cyster JG, Sailer AW (2011) Oxysterols direct immune cell migration via EBI2. Nature 475:524–527
73. Guo Y, Zhang W, Giroux C, Cai Y, Ekambaram P, Dilly AK, Hsu A, Zhou S, Maddipati KR, Liu J, Joshi S, Tucker SC, Lee MJ, Honn KV (2011) Identification of the orphan G protein-coupled receptor GPR31 as a receptor for 12-(S)-hydroxyeicosatetraenoic acid. J Biol Chem 286:33832–33840
74. Inoue A, Ishiguro J, Kitamura H, Arima N, Okutani M, Shuto A, Higashiyama S, Ohwada T, Arai H, Makide K, Aoki J (2012) TGFalpha shedding assay: an accurate and versatile method for detecting GPCR activation. Nat Methods 9:1021–1029
75. Kanaoka Y, Maekawa A, Austen KF (2013) Identification of GPR99 protein as a potential third cysteinyl leukotriene receptor with a preference for leukotriene E4 ligand. J Biol Chem 288:10967–10972
76. Guy AT, Nagatsuka Y, Ooashi N, Inoue M, Nakata A, Greimel P, Inoue A, Nabetani T, Murayama A, Ohta K, Ito Y, Aoki J, Hirabayashi Y, Kamiguchi H (2015) Neuronal development. Glycerophospholipid regulation of modality-specific sensory axon guidance in the spinal cord. Science 349:974–977
77. Chiang N, Dalli J, Colas RA, Serhan CN (2015) Identification of resolvin D2 receptor mediating resolution of infections and organ protection. J Exp Med 212:1203–1217
78. Rasmussen SG, Choi HJ, Rosenbaum DM, Kobilka TS, Thian FS, Edwards PC, Burghammer M, Ratnala VR, Sanishvili R, Fischetti RF, Schertler GF, Weis WI, Kobilka BK (2007) Crystal structure of the human beta2 adrenergic G-protein-coupled receptor. Nature 450:383–387
79. Hanson MA, Roth CB, Jo E, Griffith MT, Scott FL, Reinhart G, Desale H, Clemons B, Cahalan SM, Schuerer SC, Sanna MG, Han GW, Kuhn P, Rosen H, Stevens RC (2012) Crystal structure of a lipid G protein-coupled receptor. Science 335:851–855
80. Srivastava A, Yano J, Hirozane Y, Kefala G, Gruswitz F, Snell G, Lane W, Ivetac A, Aertgeerts K, Nguyen J, Jennings A, Okada K (2014) High-resolution structure of the human GPR40 receptor bound to allosteric agonist TAK-875. Nature 513:124–127
81. Chrencik JE, Roth CB, Terakado M, Kurata H, Omi R, Kihara Y, Warshaviak D, Nakade S, Asmar-Rovira G, Mileni M, Mizuno H, Griffith MT, Rodgers C, Han GW, Velasquez J, Chun J, Stevens RC, Hanson MA (2015) Crystal structure of antagonist bound human lysophosphatidic acid receptor 1. Cell 161:1633–1643
82. Hua T, Vemuri K, Pu M, Qu L, Han GW, Wu Y, Zhao S, Shui W, Li S, Korde A, Laprairie RB, Stahl EL, Ho JH, Zvonok N, Zhou H, Kufareva I, Wu B, Zhao Q, Hanson MA, Bohn LM, Makriyannis A, Stevens RC, Liu ZJ (2016) Crystal structure of the human cannabinoid receptor CB1. Cell 167(750–762):e714
83. Hua T, Vemuri K, Nikas SP, Laprairie RB, Wu Y, Qu L, Pu M, Korde A, Jiang S, Ho JH, Han GW, Ding K, Li X, Liu H, Hanson MA, Zhao S, Bohn LM, Makriyannis A, Stevens RC, Liu ZJ (2017) Crystal

structures of agonist-bound human cannabinoid receptor CB1. Nature 547:468–471
84. Shao Z, Yin J, Chapman K, Grzemska M, Clark L, Wang J, Rosenbaum DM (2016) High-resolution crystal structure of the human CB1 cannabinoid receptor. Nature 540:602–606
85. Taniguchi R, Inoue A, Sayama M, Uwamizu A, Yamashita K, Hirata K, Yoshida M, Tanaka Y, Kato HE, Nakada-Nakura Y, Otani Y, Nishizawa T, Doi T, Ohwada T, Ishitani R, Aoki J, Nureki O (2017) Structural insights into ligand recognition by the lysophosphatidic acid receptor LPA6. Nature 548:356–360
86. Lu J, Byrne N, Wang J, Bricogne G, Brown FK, Chobanian HR, Colletti SL, Di Salvo J, Thomas-Fowlkes B, Guo Y, Hall DL, Hadix J, Hastings NB, Hermes JD, Ho T, Howard AD, Josien H, Kornienko M, Lumb KJ, Miller MW, Patel SB, Pio B, Plummer CW, Sherborne BS, Sheth P, Souza S, Tummala S, Vonrhein C, Webb M, Allen SJ, Johnston JM, Weinglass AB, Sharma S, Soisson SM (2017) Structural basis for the cooperative allosteric activation of the free fatty acid receptor GPR40. Nat Struct Mol Biol 24:570–577
87. Cao C, Tan Q, Xu C, He L, Yang L, Zhou Y, Zhou Y, Qiao A, Lu M, Yi C, Han GW, Wang X, Li X, Yang H, Rao Z, Jiang H, Zhao Y, Liu J, Stevens RC, Zhao Q, Zhang XC, Wu B (2018) Structural basis for signal recognition and transduction by platelet-activating-factor receptor. Nat Struct Mol Biol 25:488–495
88. Hori T, Okuno T, Hirata K, Yamashita K, Kawano Y, Yamamoto M, Hato M, Nakamura M, Shimizu T, Yokomizo T, Miyano M, Yokoyama S (2018) Na(+)-mimicking ligands stabilize the inactive state of leukotriene B4 receptor BLT1. Nat Chem Biol 14:262–269
89. Fan H, Chen S, Yuan X, Han S, Zhang H, Xia W, Xu Y, Zhao Q, Wu B (2019) Structural basis for ligand recognition of the human thromboxane A2 receptor. Nat Chem Biol 15:27–33
90. Wang L, Yao D, Deepak R, Liu H, Xiao Q, Fan H, Gong W, Wei Z, Zhang C (2018) Structures of the human PGD2 receptor CRTH2 reveal novel mechanisms for ligand recognition. Mol Cell 72(48–59):e44
91. Morimoto K, Suno R, Hotta Y, Yamashita K, Hirata K, Yamamoto M, Narumiya S, Iwata S, Kobayashi T (2019) Crystal structure of the endogenous agonist-bound prostanoid receptor EP3. Nat Chem Biol 15:8–10
92. Toyoda Y, Morimoto K, Suno R, Horita S, Yamashita K, Hirata K, Sekiguchi Y, Yasuda S, Shiroishi M, Shimizu T, Urushibata Y, Kajiwara Y, Inazumi T, Hotta Y, Asada H, Nakane T, Shiimura Y, Nakagita T, Tsuge K, Yoshida S, Kuribara T, Hosoya T, Sugimoto Y, Nomura N, Sato M, Hirokawa T, Kinoshita M, Murata T, Takayama K, Yamamoto M, Narumiya S, Iwata S, Kobayashi T (2019) Ligand binding to human prostaglandin E receptor EP4 at the lipid-bilayer interface. Nat Chem Biol 15:18–26
93. Krishna Kumar K, Shalev-Benami M, Robertson MJ, Hu H, Banister SD, Hollingsworth SA, Latorraca NR, Kato HE, Hilger D, Maeda S, Weis WI, Farrens DL, Dror RO, Malhotra SV, Kobilka BK, Skiniotis G (2019) Structure of a signaling cannabinoid receptor 1-G protein complex. Cell 176(448–458):e412
94. Li X, Hua T, Vemuri K, Ho JH, Wu Y, Wu L, Popov P, Benchama O, Zvonok N, Locke K, Qu L, Han GW, Iyer MR, Cinar R, Coffey NJ, Wang J, Wu M, Katritch V, Zhao S, Kunos G, Bohn LM, Makriyannis A, Stevens RC, Liu ZJ (2019) Crystal structure of the human cannabinoid receptor CB2. Cell 176(459–467):e413
95. Luginina A, Gusach A, Marin E, Mishin A, Brouillette R, Popov P, Shiriaeva A, Besserer-Offroy E, Longpre JM, Lyapina E, Ishchenko A, Patel N, Polovinkin V, Safronova N, Bogorodskiy A, Edelweiss E, Hu H, Weierstall U, Liu W, Batyuk A, Gordeliy V, Han GW, Sarret P, Katritch V, Borshchevskiy V, Cherezov V (2019) Structure-based mechanism of cysteinyl leukotriene receptor inhibition by antiasthmatic drugs. Sci Adv 5:eaax2518
96. Shao Z, Yan W, Chapman K, Ramesh K, Ferrell AJ, Yin J, Wang X, Xu Q, Rosenbaum DM (2019) Structure of an allosteric modulator bound to the CB1 cannabinoid receptor. Nat Chem Biol 15:1199–1205
97. Gusach A, Luginina A, Marin E, Brouillette RL, Besserer-Offroy E, Longpre JM, Ishchenko A, Popov P, Patel N, Fujimoto T, Maruyama T, Stauch B, Ergasheva M, Romanovskaia D, Stepko A, Kovalev K, Shevtsov M, Gordeliy V, Han GW, Katritch V, Borshchevskiy V, Sarret P, Mishin A, Cherezov V (2019) Structural basis of ligand selectivity and disease mutations in cysteinyl leukotriene receptors. Nat Commun 10:5573
98. Funk CD (2001) Prostaglandins and leukotrienes: advances in eicosanoid biology. Science 294:1871–1875
99. Burke JE, Dennis EA (2009) Phospholipase A2 structure/function, mechanism, and signaling. J Lipid Res 50(Suppl):S237–S242
100. Murakami M, Miki Y, Sato H, Murase R, Taketomi Y, Yamamoto K (2019) Group IID, IIE, IIF and III secreted phospholipase A2s. Biochim Biophys Acta Mol Cell Biol Lipids 1864:803–818
101. Shimizu T (2009) Lipid mediators in health and disease: enzymes and receptors as therapeutic targets for the regulation of immunity and inflammation. Annu Rev Pharmacol Toxicol 49:123–150
102. Smith WL, DeWitt DL, Garavito RM (2000) Cyclooxygenases: structural, cellular, and molecular biology. Annu Rev Biochem 69:145–182
103. Marcus AJ (1978) The role of lipids in platelet function: with particular reference to the arachidonic acid pathway. J Lipid Res 19:793–826
104. Bergström S, Sjövall J (1957) The isolation of prostaglandin. Acta Chem Scand 11:1086–1086

105. Bergström S, Danielsson H, Samuelsson B (1964) The enzymatic formation of prostaglandin E2 from arachidonic acid prostaglandins and related factors 32. Biochim Biophys Acta 90:207–210
106. Van D, Beerthuis RK, Nugteren DH, Vonkeman H (1964) The biosynthesis of prostaglandins. Biochim Biophys Acta 90:204–207
107. Corey EJ, Andersen NH, Carlson RM, Paust J, Vedejs E, Vlattas I, Winter RE (1968) Total synthesis of prostaglandins. Synthesis of the pure dl-E1, -F1-alpha-F1-beta, -A1, and -B1 hormones. J Am Chem Soc 90:3245–3247
108. Corey EJ, Vlattas I, Andersen NH, Harding K (1968) A new total synthesis of prostaglandins of the E1 and F1 series including 11-epiprostaglandins. J Am Chem Soc 90:3247–3248
109. Hamberg M, Svensson J, Samuelsson B (1975) Thromboxanes: a new group of biologically active compounds derived from prostaglandin endoperoxides. Proc Natl Acad Sci U S A 72:2994–2998
110. Moncada S, Gryglewski R, Bunting S, Vane JR (1976) An enzyme isolated from arteries transforms prostaglandin endoperoxides to an unstable substance that inhibits platelet aggregation. Nature 263:663–665
111. Whittaker N, Bunting S, Salmon J, Moncada S, Vane JR, Johnson RA, Morton DR, Kinner JH, Gorman RR, McGuire JC, Sun FF (1976) The chemical structure of prostaglandin X (prostacyclin). Prostaglandins 12:915–928
112. Narumiya S, Shimizu T, Yamamoto S (2016) In Memoriam: Osamu Hayaishi (1920–2015). J Lipid Res 57:517–522
113. Ushikubi F, Nakajima M, Hirata M, Okuma M, Fujiwara M, Narumiya S (1989) Purification of the thromboxane A2/prostaglandin H2 receptor from human blood platelets. J Biol Chem 264:16496–16501
114. Hanasaki K, Arita H (1988) Characterization of a new compound, S-145, as a specific TXA2 receptor antagonist in platelets. Thromb Res 50:365–376
115. Samuelsson B (1983) Leukotrienes: mediators of immediate hypersensitivity reactions and inflammation. Science 220:568–575
116. Borgeat P, Samuelsson B (1979) Transformation of arachidonic acid by rabbit polymorphonuclear leukocytes. Formation of a novel dihydroxyeicosatetraenoic acid. J Biol Chem 254:2643–2646
117. Yokomizo T, Kato K, Terawaki K, Izumi T, Shimizu T (2000) A second leukotriene B(4) receptor, BLT2. A new therapeutic target in inflammation and immunological disorders. J Exp Med 192:421–432
118. Feldberg W, Kellaway CH (1938) Liberation of histamine and formation of lysocithin-like substances by cobra venom. J Physiol 94:187–226
119. Brocklehurst WE (1960) The release of histamine and formation of a slow-reacting substance (SRS-A) during anaphylactic shock. J Physiol 151:416–435
120. Augstein J, Farmer JB, Lee TB, Sheard P, Tattersall ML (1973) Selective inhibitor of slow reacting substance of anaphylaxis. Nat New Biol 245:215–217
121. Murphy RC, Hammarstrom S, Samuelsson B (1979) Leukotriene C: a slow-reacting substance from murine mastocytoma cells. Proc Natl Acad Sci U S A 76:4275–4279
122. Lewis RA, Drazen JM, Austen KF, Clark DA, Corey EJ (1980) Identification of the C(6)-S-conjugate of leukotriene A with cysteine as a naturally occurring slow reacting substance of anaphylaxis (SRS-A). Importance of the 11-cis-geometry for biological activity. Biochem Biophys Res Commun 96:271–277
123. Drazen JM, Austen KF, Lewis RA, Clark DA, Goto G, Marfat A, Corey EJ (1980) Comparative airway and vascular activities of leukotrienes C-1 and D in vivo and in vitro. Proc Natl Acad Sci U S A 77:4354–4358
124. Krell RD, Osborn R, Vickery L, Falcone K, O'Donnell M, Gleason J, Kinzig C, Bryan D (1981) Contraction of isolated airway smooth muscle by synthetic leukotrienes C4 and D4. Prostaglandins 22:387–409
125. Lee TH, Austen KF, Corey EJ, Drazen JM (1984) Leukotriene E4-induced airway hyperresponsiveness of guinea pig tracheal smooth muscle to histamine and evidence for three separate sulfidopeptide leukotriene receptors. Proc Natl Acad Sci U S A 81:4922–4925
126. Nakagawa N, Obata T, Kobayashi T, Okada Y, Nambu F, Terawaki T, Aishita H (1992) In vivo pharmacologic profile of ONO-1078: a potent, selective and orally active peptide leukotriene (LT) antagonist. Jpn J Pharmacol 60:217–225
127. Obata T, Nambu F, Kitagawa T, Terashima H, Toda M, Okegawa T, Kawasaki A (1987) ONO-1078: an antagonist of leukotrienes. Adv Prostaglandin Thromboxane Leukot Res 17A:540–543
128. Wittenberger T, Hellebrand S, Munck A, Kreienkamp HJ, Schaller HC, Hampe W (2002) GPR99, a new G protein-coupled receptor with homology to a new subgroup of nucleotide receptors. BMC Genomics 3:17
129. He W, Miao FJ, Lin DC, Schwandner RT, Wang Z, Gao J, Chen JL, Tian H, Ling L (2004) Citric acid cycle intermediates as ligands for orphan G-protein-coupled receptors. Nature 429:188–193
130. Davenport AP, Alexander SP, Sharman JL, Pawson AJ, Benson HE, Monaghan AE, Liew WC, Mpamhanga CP, Bonner TI, Neubig RR, Pin JP, Spedding M, Harmar AJ (2013) International Union of Basic and Clinical Pharmacology. LXXXVIII. G protein-coupled receptor list: recommendations for new pairings with cognate ligands. Pharmacol Rev 65:967–986
131. Nie D, Hillman GG, Geddes T, Tang K, Pierson C, Grignon DJ, Honn KV (1998) Platelet-type 12-lipoxygenase in a human prostate carcinoma

stimulates angiogenesis and tumor growth. Cancer Res 58:4047–4051
132. Nie D, Krishnamoorthy S, Jin R, Tang K, Chen Y, Qiao Y, Zacharek A, Guo Y, Milanini J, Pages G, Honn KV (2006) Mechanisms regulating tumor angiogenesis by 12-lipoxygenase in prostate cancer cells. J Biol Chem 281:18601–18609
133. Pidgeon GP, Tang K, Cai YL, Piasentin E, Honn KV (2003) Overexpression of platelet-type 12-lipoxygenase promotes tumor cell survival by enhancing alpha(v)beta(3) and alpha(v)beta(5) integrin expression. Cancer Res 63:4258–4267
134. Alkayed NJ, Cao Z, Qian ZY, Nagarajan S, Liu X, Nelson J, Xie F, Li B, Fan W, Liu L, Grafe MR, Xiao X, Barnes AP, Kaul S (2018) Bidirectional control of coronary vascular resistance by Eicosanoids via a Novel GPCR. bioRxiv
135. Zhang JV, Ren PG, Avsian-Kretchmer O, Luo CW, Rauch R, Klein C, Hsueh AJ (2005) Obestatin, a peptide encoded by the ghrelin gene, opposes ghrelin's effects on food intake. Science 310:996–999
136. Chartrel N, Alvear-Perez R, Leprince J, Iturrioz X, Reaux-Le Goazigo A, Audinot V, Chomarat P, Coge F, Nosjean O, Rodriguez M, Galizzi JP, Boutin JA, Vaudry H, Llorens-Cortes C (2007) Comment on "Obestatin, a peptide encoded by the ghrelin gene, opposes ghrelin's effects on food intake". Science 315:766; author reply 766
137. Holst B, Egerod KL, Schild E, Vickers SP, Cheetham S, Gerlach LO, Storjohann L, Stidsen CE, Jones R, Beck-Sickinger AG, Schwartz TW (2007) GPR39 signaling is stimulated by zinc ions but not by obestatin. Endocrinology 148:13–20
138. Powell WS, Gravelle F, Gravel S (1992) Metabolism of 5(S)-hydroxy-6,8,11,14-eicosatetraenoic acid and other 5(S)-hydroxyeicosanoids by a specific dehydrogenase in human polymorphonuclear leukocytes. J Biol Chem 267:19233–19241
139. Back M, Powell WS, Dahlen SE, Drazen JM, Evans JF, Serhan CN, Shimizu T, Yokomizo T, Rovati GE (2014) Update on leukotriene, lipoxin and oxoeicosanoid receptors: IUPHAR review 7. Br J Pharmacol 171:3551–3574
140. Takeda S, Kadowaki S, Haga T, Takaesu H, Mitaku S (2002) Identification of G protein-coupled receptor genes from the human genome sequence. FEBS Lett 520:97–101
141. Buckley CD, Gilroy DW, Serhan CN (2014) Proresolving lipid mediators and mechanisms in the resolution of acute inflammation. Immunity 40:315–327
142. Chiang N, Serhan CN (2017) Structural elucidation and physiologic functions of specialized proresolving mediators and their receptors. Mol Asp Med 58:114–129
143. Levy BD, Serhan CN (2014) Resolution of acute inflammation in the lung. Annu Rev Physiol 76:467–492
144. Norris PC, Serhan CN (2018) Metabololipidomic profiling of functional immunoresolvent clusters and eicosanoids in mammalian tissues. Biochem Biophys Res Commun 504:553–561
145. Serhan CN (2010) Novel lipid mediators and resolution mechanisms in acute inflammation: to resolve or not? Am J Pathol 177:1576–1591
146. Serhan CN, Levy BD (2018) Resolvins in inflammation: emergence of the pro-resolving superfamily of mediators. J Clin Invest 128:2657–2669
147. Serhan CN, Hamberg M, Samuelsson B (1984) Trihydroxytetraenes: a novel series of compounds formed from arachidonic acid in human leukocytes. Biochem Biophys Res Commun 118:943–949
148. Serhan CN, Dalli J, Colas RA, Winkler JW, Chiang N (2015) Protectins and maresins: new pro-resolving families of mediators in acute inflammation and resolution bioactive metabolome. Biochim Biophys Acta 1851:397–413
149. Serhan CN, Chiang N, Dalli J (2018) New pro-resolving n-3 mediators bridge resolution of infectious inflammation to tissue regeneration. Mol Asp Med 64:1–17
150. Serhan CN (2017) Discovery of specialized pro-resolving mediators marks the dawn of resolution physiology and pharmacology. Mol Asp Med 58:1–11
151. Dalli J, Serhan CN (2017) Pro-resolving mediators in regulating and conferring macrophage function. Front Immunol 8:1400
152. Bao L, Gerard NP, Eddy RL Jr, Shows TB, Gerard C (1992) Mapping of genes for the human C5a receptor (C5AR), human FMLP receptor (FPR), and two FMLP receptor homologue orphan receptors (FPRH1, FPRH2) to chromosome 19. Genomics 13:437–440
153. Murphy PM, Ozcelik T, Kenney RT, Tiffany HL, McDermott D, Francke U (1992) A structural homologue of the N-formyl peptide receptor. Characterization and chromosome mapping of a peptide chemoattractant receptor family. J Biol Chem 267:7637–7643
154. Ye RD, Cavanagh SL, Quehenberger O, Prossnitz ER, Cochrane CG (1992) Isolation of a cDNA that encodes a novel granulocyte N-formyl peptide receptor. Biochem Biophys Res Commun 184:582–589
155. Dufton N, Hannon R, Brancaleone V, Dalli J, Patel HB, Gray M, D'Acquisto F, Buckingham JC, Perretti M, Flower RJ (2010) Anti-inflammatory role of the murine formyl-peptide receptor 2: ligand-specific effects on leukocyte responses and experimental inflammation. J Immunol 184:2611–2619
156. Wittamer V, Franssen JD, Vulcano M, Mirjolet JF, Le Poul E, Migeotte I, Brezillon S, Tyldesley R, Blanpain C, Detheux M, Mantovani A, Sozzani S, Vassart G, Parmentier M, Communi D (2003) Specific recruitment of antigen-presenting cells by chemerin, a novel processed ligand from human inflammatory fluids. J Exp Med 198:977–985
157. Kohno M, Hasegawa H, Inoue A, Muraoka M, Miyazaki T, Oka K, Yasukawa M (2006) Identification of N-arachidonylglycine as the

endogenous ligand for orphan G-protein-coupled receptor GPR18. Biochem Biophys Res Commun 347:827–832
158. Console-Bram L, Brailoiu E, Brailoiu GC, Sharir H, Abood ME (2014) Activation of GPR18 by cannabinoid compounds: a tale of biased agonism. Br J Pharmacol 171:3908–3917
159. Rimmerman N, Bradshaw HB, Hughes HV, Chen JS, Hu SS, McHugh D, Vefring E, Jahnsen JA, Thompson EL, Masuda K, Cravatt BF, Burstein S, Vasko MR, Prieto AL, O'Dell DK, Walker JM (2008) N-palmitoyl glycine, a novel endogenous lipid that acts as a modulator of calcium influx and nitric oxide production in sensory neurons. Mol Pharmacol 74:213–224
160. Yin H, Chu A, Li W, Wang B, Shelton F, Otero F, Nguyen DG, Caldwell JS, Chen YA (2009) Lipid G protein-coupled receptor ligand identification using beta-arrestin PathHunter assay. J Biol Chem 284:12328–12338
161. Qin Y, Verdegaal EM, Siderius M, Bebelman JP, Smit MJ, Leurs R, Willemze R, Tensen CP, Osanto S (2011) Quantitative expression profiling of G-protein-coupled receptors (GPCRs) in metastatic melanoma: the constitutively active orphan GPCR GPR18 as novel drug target. Pigment Cell Melanoma Res 24:207–218
162. Tokumura A, Fukuzawa K, Akamatsu Y, Yamada S, Suzuki T, Tsukatani H (1978) Identification of vasopressor phospholipid in crude soybean lecithin. Lipids 13:468–472
163. van Corven EJ, Groenink A, Jalink K, Eichholtz T, Moolenaar WH (1989) Lysophosphatidate-induced cell proliferation: identification and dissection of signaling pathways mediated by G proteins. Cell 59:45–54
164. Mizuno H, Kihara Y, Kussrow A, Chen A, Ray M, Rivera R, Bornhop DJ, Chun J (2019) Lysophospholipid G protein-coupled receptor binding parameters as determined by backscattering interferometry. J Lipid Res 60:212–217
165. Yanagida K, Masago K, Nakanishi H, Kihara Y, Hamano F, Tajima Y, Taguchi R, Shimizu T, Ishii S (2009) Identification and characterization of a novel lysophosphatidic acid receptor, p2y5/LPA6. J Biol Chem 284:17731–17741
166. Kitamura H, Makide K, Shuto A, Ikubo M, Inoue A, Suzuki K, Sato Y, Nakamura S, Otani Y, Ohwada T, Aoki J (2012) GPR34 is a receptor for lysophosphatidylserine with a fatty acid at the sn-2 position. J Biochem 151:511–518
167. Makide K, Uwamizu A, Shinjo Y, Ishiguro J, Okutani M, Inoue A, Aoki J (2014) Novel lysophosphoplipid receptors: their structure and function. J Lipid Res 55:1986–1995
168. Kihara Y, Maceyka M, Spiegel S, Chun J (2014) Lysophospholipid receptor nomenclature review: IUPHAR review 8. Br J Pharmacol 171:3575–3594
169. Sawzdargo M, Nguyen T, Lee DK, Lynch KR, Cheng R, Heng HH, George SR, O'Dowd BF (1999) Identification and cloning of three novel human G protein-coupled receptor genes GPR52, PsiGPR53 and GPR55: GPR55 is extensively expressed in human brain. Brain Res Mol Brain Res 64:193–198
170. Baker D, Pryce G, Davies WL, Hiley CR (2006) In silico patent searching reveals a new cannabinoid receptor. Trends Pharmacol Sci 27:1–4
171. Johns DG, Behm DJ, Walker DJ, Ao Z, Shapland EM, Daniels DA, Riddick M, Dowell S, Staton PC, Green P, Shabon U, Bao W, Aiyar N, Yue TL, Brown AJ, Morrison AD, Douglas SA (2007) The novel endocannabinoid receptor GPR55 is activated by atypical cannabinoids but does not mediate their vasodilator effects. Br J Pharmacol 152:825–831
172. Ryberg E, Larsson N, Sjogren S, Hjorth S, Hermansson NO, Leonova J, Elebring T, Nilsson K, Drmota T, Greasley PJ (2007) The orphan receptor GPR55 is a novel cannabinoid receptor. Br J Pharmacol 152:1092–1101
173. Oka S, Toshida T, Maruyama K, Nakajima K, Yamashita A, Sugiura T (2009) 2-Arachidonoyl-sn-glycero-3-phosphoinositol: a possible natural ligand for GPR55. J Biochem 145:13–20
174. Zhang H, Desai NN, Olivera A, Seki T, Brooker G, Spiegel S (1991) Sphingosine-1-phosphate, a novel lipid, involved in cellular proliferation. J Cell Biol 114:155–167
175. Hla T, Lee MJ, Ancellin N, Thangada S, Liu CH, Kluk M, Chae SS, Wu MT (2000) Sphingosine-1-phosphate signaling via the EDG-1 family of G-protein-coupled receptors. Ann N Y Acad Sci 905:16–24
176. Chun J, Kihara Y, Jonnalagadda D, Blaho VA (2019) Fingolimod: lessons learned and new opportunities for treating multiple sclerosis and other disorders. Annu Rev Pharmacol Toxicol 59:149–170
177. Groves A, Kihara Y, Chun J (2013) Fingolimod: direct CNS effects of sphingosine 1-phosphate (S1P) receptor modulation and implications in multiple sclerosis therapy. J Neurol Sci 328:9–18
178. Kihara Y (2019) Systematic understanding of bioactive lipids in neuro-immune interactions: lessons from an animal model of multiple sclerosis. Adv Exp Med Biol 1161:133–148
179. Kihara Y, Mizuno H, Chun J (2015) Lysophospholipid receptors in drug discovery. Exp Cell Res 333:171–177
180. Abel EL (1980) Marihuana. The first twelve thousand years. Springer, Boston
181. Ren M, Tang Z, Wu X, Spengler R, Jiang H, Yang Y, Boivin N (2019) The origins of cannabis smoking: chemical residue evidence from the first millennium BCE in the Pamirs. Sci Adv 5:eaaw1391
182. Brand EJ, Zhao Z (2017) Cannabis in Chinese medicine: are some traditional indications referenced in ancient literature related to cannabinoids? Front Pharmacol 8:108
183. Jacob A, Todd AR (1940) Cannabis indica. Part II. Isolation of cannabidiol from Egyptian hashish.

Observations on the structure of cannabinol. J Chem Soc:649–653
184. Adams R, Baker BR, Wearn RB (1940) Structure of Cannabinol. III. Synthesis of Cannabinol, l-Hydroxy-3-n-amyl6^b-trimethyl-6-dibenzopyran. J Am Chem Soc 62:2204–2207
185. Single convention on narcotic drugs, Gt. Brit (1965) Single convention on narcotic drugs, 1961; New York, March 30–August 1. In: Presented to parliament by the Secretary of State for Foreign Affairs by Command of Her Majesty June 1965. H.M. Stationery Off., London
186. Gaoni Y, Mechoulam R (1964) Isolation, structure, and partial synthesis of an active constituent of hashish. J Am Chem Soc 86:1646–1647
187. Di Marzo V (2006) A brief history of cannabinoid and endocannabinoid pharmacology as inspired by the work of British scientists. Trends Pharmacol Sci 27:134–140
188. Benveniste J, Henson PM, Cochrane CG (1972) Leukocyte-dependent histamine release from rabbit platelets. The role of IgE, basophils, and a platelet-activating factor. J Exp Med 136:1356–1377
189. Demopoulos CA, Pinckard RN, Hanahan DJ (1979) Platelet-activating factor. Evidence for 1-O-alkyl-2-acetyl-sn-glyceryl-3-phosphorylcholine as the active component (a new class of lipid chemical mediators). J Biol Chem 254:9355–9358
190. Shindou H, Hishikawa D, Nakanishi H, Harayama T, Ishii S, Taguchi R, Shimizu T (2007) A single enzyme catalyzes both platelet-activating factor production and membrane biogenesis of inflammatory cells. Cloning and characterization of acetyl-CoA:LYSO-PAF acetyltransferase. J Biol Chem 282:6532–6539
191. Morrison WJ, Shukla SD (1988) Desensitization of receptor-coupled activation of phosphoinositide-specific phospholipase C in platelets: evidence for distinct mechanisms for platelet-activating factor and thrombin. Mol Pharmacol 33:58–63
192. Sawzdargo M, George SR, Nguyen T, Xu S, Kolakowski LF, O'Dowd BF (1997) A cluster of four novel human G protein-coupled receptor genes occurring in close proximity to CD22 gene on chromosome 19q13.1. Biochem Biophys Res Commun 239:543–547
193. Sasaki S, Kitamura S, Negoro N, Suzuki M, Tsujihata Y, Suzuki N, Santou T, Kanzaki N, Harada M, Tanaka Y, Kobayashi M, Tada N, Funami M, Tanaka T, Yamamoto Y, Fukatsu K, Yasuma T, Momose Y (2011) Design, synthesis, and biological activity of potent and orally available G protein-coupled receptor 40 agonists. J Med Chem 54:1365–1378
194. Yabuki C, Komatsu H, Tsujihata Y, Maeda R, Ito R, Matsuda-Nagasumi K, Sakuma K, Miyawaki K, Kikuchi N, Takeuchi K, Habata Y, Mori M (2013) A novel antidiabetic drug, fasiglifam/TAK-875, acts as an ago-allosteric modulator of FFAR1. PLoS One 8:e76280
195. Burant CF, Viswanathan P, Marcinak J, Cao C, Vakilynejad M, Xie B, Leifke E (2012) TAK-875 versus placebo or glimepiride in type 2 diabetes mellitus: a phase 2, randomised, double-blind, placebo-controlled trial. Lancet 379:1403–1411
196. Kaku K, Enya K, Nakaya R, Ohira T, Matsuno R (2016) Long-term safety and efficacy of fasiglifam (TAK-875), a G-protein-coupled receptor 40 agonist, as monotherapy and combination therapy in Japanese patients with type 2 diabetes: a 52-week open-label phase III study. Diabetes Obes Metab 18:925–929
197. Fredriksson R, Hoglund PJ, Gloriam DE, Lagerstrom MC, Schioth HB (2003) Seven evolutionarily conserved human rhodopsin G protein-coupled receptors lacking close relatives. FEBS Lett 554:381–388
198. Gendaszewska-Darmach E, Drzazga A, Koziolkiewicz M (2019) Targeting GPCRs activated by fatty acid-derived lipids in type 2 diabetes. Trends Mol Med 25(10):915–929
199. Walter P, Green S, Greene G, Krust A, Bornert JM, Jeltsch JM, Staub A, Jensen E, Scrace G, Waterfield M et al (1985) Cloning of the human estrogen receptor cDNA. Proc Natl Acad Sci U S A 82:7889–7893
200. Greene GL, Gilna P, Waterfield M, Baker A, Hort Y, Shine J (1986) Sequence and expression of human estrogen receptor complementary DNA. Science 231:1150–1154
201. O'Malley BW, Khan S (2013) Elwood V. Jensen (1920–2012): father of the nuclear receptors. Proc Natl Acad Sci U S A 110:3707–3708
202. Szego CM, Davis JS (1967) Adenosine 3′,5′-monophosphate in rat uterus: acute elevation by estrogen. Proc Natl Acad Sci U S A 58:1711–1718
203. Szego CM, Davis JS (1969) Inhibition of estrogen-induced elevation of cyclic 3′,5′-adenosine monophosphate in rat uterus. I. By beta-adrenergic receptor-blocking drugs. Mol Pharmacol 5:470–480
204. Szego CM, Davis JS (1969) Inhibition of estrogen-induced cyclic AMP elevation in rat uterus. II: By glucocorticoids. Life Sci 8:1109–1116
205. Soloff MS, Szego CM (1969) Purification of estradiol receptor from rat uterus and blockade of its estrogen-binding function by specific antibody. Biochem Biophys Res Commun 34:141–147
206. Prossnitz ER, Arterburn JB (2015) International Union of Basic and Clinical Pharmacology. XCVII. G protein-coupled estrogen receptor and its pharmacologic modulators. Pharmacol Rev 67:505–540
207. Staels B, Fonseca VA (2009) Bile acids and metabolic regulation: mechanisms and clinical responses to bile acid sequestration. Diabetes Care 32(Suppl 2):S237–S245
208. Parks DJ, Blanchard SG, Bledsoe RK, Chandra G, Consler TG, Kliewer SA, Stimmel JB, Willson TM, Zavacki AM, Moore DD, Lehmann JM (1999) Bile acids: natural ligands for an orphan nuclear receptor. Science 284:1365–1368

209. Lieu T, Jayaweera G, Bunnett NW (2014) GPBA: a GPCR for bile acids and an emerging therapeutic target for disorders of digestion and sensation. Br J Pharmacol 171:1156–1166
210. Yu H, Zhao T, Liu S, Wu Q, Johnson O, Wu Z, Zhuang Z, Shi Y, Peng L, He R, Yang Y, Sun J, Wang X, Xu H, Zeng Z, Zou P, Lei X, Luo W, Li Y (2019) MRGPRX4 is a bile acid receptor for human cholestatic itch. elife 8
211. Meixiong J, Vasavda C, Snyder SH, Dong X (2019) MRGPRX4 is a G protein-coupled receptor activated by bile acids that may contribute to cholestatic pruritus. Proc Natl Acad Sci U S A 116:10525–10530
212. Meixiong J, Vasavda C, Green D, Zheng Q, Qi L, Kwatra SG, Hamilton JP, Snyder SH, Dong X (2019) Identification of a bilirubin receptor that may mediate a component of cholestatic itch. elife 8:e44116
213. Pereira JP, Kelly LM, Xu Y, Cyster JG (2009) EBI2 mediates B cell segregation between the outer and centre follicle. Nature 460:1122–1126
214. Birkenbach M, Josefsen K, Yalamanchili R, Lenoir G, Kieff E (1993) Epstein-Barr virus-induced genes: first lymphocyte-specific G protein-coupled peptide receptors. J Virol 67:2209–2220
215. Saeki Y, Ueno S, Mizuno R, Nishimura T, Fujimura H, Nagai Y, Yanagihara T (1993) Molecular cloning of a novel putative G protein-coupled receptor (GPCR21) which is expressed predominantly in mouse central nervous system. FEBS Lett 336:317–322
216. Eidne KA, Zabavnik J, Peters T, Yoshida S, Anderson L, Taylor PL (1991) Cloning, sequencing and tissue distribution of a candidate G protein-coupled receptor from rat pituitary gland. FEBS Lett 292:243–248
217. Heiber M, Docherty JM, Shah G, Nguyen T, Cheng R, Heng HH, Marchese A, Tsui LC, Shi X, George SR et al (1995) Isolation of three novel human genes encoding G protein-coupled receptors. DNA Cell Biol 14:25–35
218. Marchese A, Docherty JM, Nguyen T, Heiber M, Cheng R, Heng HH, Tsui LC, Shi X, George SR, O'Dowd BF (1994) Cloning of human genes encoding novel G protein-coupled receptors. Genomics 23:609–618
219. Song ZH, Modi W, Bonner TI (1995) Molecular cloning and chromosomal localization of human genes encoding three closely related G protein-coupled receptors. Genomics 28:347–349
220. Uhlenbrock K, Gassenhuber H, Kostenis E (2002) Sphingosine 1-phosphate is a ligand of the human gpr3, gpr6 and gpr12 family of constitutively active G protein-coupled receptors. Cell Signal 14:941–953
221. Ignatov A, Lintzel J, Hermans-Borgmeyer I, Kreienkamp HJ, Joost P, Thomsen S, Methner A, Schaller HC (2003) Role of the G-protein-coupled receptor GPR12 as high-affinity receptor for sphingosylphosphorylcholine and its expression and function in brain development. J Neurosci 23:907–914
222. Ignatov A, Lintzel J, Kreienkamp HJ, Schaller HC (2003) Sphingosine-1-phosphate is a high-affinity ligand for the G protein-coupled receptor GPR6 from mouse and induces intracellular Ca2+ release by activating the sphingosine-kinase pathway. Biochem Biophys Res Commun 311:329–336
223. Laun AS, Shrader SH, Song ZH (2018) Novel inverse agonists for the orphan G protein-coupled receptor 6. Heliyon 4:e00933
224. Laun AS, Song ZH (2017) GPR3 and GPR6, novel molecular targets for cannabidiol. Biochem Biophys Res Commun 490:17–21
225. Niedernberg A, Tunaru S, Blaukat A, Ardati A, Kostenis E (2003) Sphingosine 1-phosphate and dioleoylphosphatidic acid are low affinity agonists for the orphan receptor GPR63. Cell Signal 15:435–446
226. Wang J, Simonavicius N, Wu X, Swaminath G, Reagan J, Tian H, Ling L (2006) Kynurenic acid as a ligand for orphan G protein-coupled receptor GPR35. J Biol Chem 281:22021–22028
227. Oka S, Ota R, Shima M, Yamashita A, Sugiura T (2010) GPR35 is a novel lysophosphatidic acid receptor. Biochem Biophys Res Commun 395:232–237
228. Deng H, Hu H, Fang Y (2012) Multiple tyrosine metabolites are GPR35 agonists. Sci Rep 2:373
229. Southern C, Cook JM, Neetoo-Isseljee Z, Taylor DL, Kettleborough CA, Merritt A, Bassoni DL, Raab WJ, Quinn E, Wehrman TS, Davenport AP, Brown AJ, Green A, Wigglesworth MJ, Rees S (2013) Screening beta-arrestin recruitment for the identification of natural ligands for orphan G-protein-coupled receptors. J Biomol Screen 18:599–609
230. Wittenberger T, Schaller HC, Hellebrand S (2001) An expressed sequence tag (EST) data mining strategy succeeding in the discovery of new G-protein coupled receptors. J Mol Biol 307:799–813
231. Wang J, Wu X, Simonavicius N, Tian H, Ling L (2006) Medium-chain fatty acids as ligands for orphan G protein-coupled receptor GPR84. J Biol Chem 281:34457–34464
232. Suzuki M, Takaishi S, Nagasaki M, Onozawa Y, Iino I, Maeda H, Komai T, Oda T (2013) Medium-chain fatty acid-sensing receptor, GPR84, is a proinflammatory receptor. J Biol Chem 288:10684–10691
233. Tabata K, Baba K, Shiraishi A, Ito M, Fujita N (2007) The orphan GPCR GPR87 was deorphanized and shown to be a lysophosphatidic acid receptor. Biochem Biophys Res Commun 363:861–866
234. Wetter JA, Revankar C, Hanson BJ (2009) Utilization of the tango beta-arrestin recruitment technology for cell-based EDG receptor assay development and interrogation. J Biomol Screen 14:1134–1141
235. Soga T, Ohishi T, Matsui T, Saito T, Matsumoto M, Takasaki J, Matsumoto S, Kamohara M, Hiyama H, Yoshida S, Momose K, Ueda Y, Matsushime H, Kobori M, Furuichi K (2005) Lysophosphatidylcholine enhances glucose-dependent insulin secretion via an orphan G-protein-coupled receptor. Biochem Biophys Res Commun 326:744–751

236. Overton HA, Babbs AJ, Doel SM, Fyfe MC, Gardner LS, Griffin G, Jackson HC, Procter MJ, Rasamison CM, Tang-Christensen M, Widdowson PS, Williams GM, Reynet C (2006) Deorphanization of a G protein-coupled receptor for oleoylethanolamide and its use in the discovery of small-molecule hypophagic agents. Cell Metab 3:167–175
237. Chu ZL, Carroll C, Chen R, Alfonso J, Gutierrez V, He H, Lucman A, Xing C, Sebring K, Zhou J, Wagner B, Unett D, Jones RM, Behan DP, Leonard J (2010) N-oleoyldopamine enhances glucose homeostasis through the activation of GPR119. Mol Endocrinol 24:161–170
238. Hansen KB, Rosenkilde MM, Knop FK, Wellner N, Diep TA, Rehfeld JF, Andersen UB, Holst JJ, Hansen HS (2011) 2-Oleoyl glycerol is a GPR119 agonist and signals GLP-1 release in humans. J Clin Endocrinol Metab 96:E1409–E1417
239. Overton HA, Fyfe MC, Reynet C (2008) GPR119, a novel G protein-coupled receptor target for the treatment of type 2 diabetes and obesity. Br J Pharmacol 153(Suppl 1):S76–S81
240. Hamann J, Aust G, Arac D, Engel FB, Formstone C, Fredriksson R, Hall RA, Harty BL, Kirchhoff C, Knapp B, Krishnan A, Liebscher I, Lin HH, Martinelli DC, Monk KR, Peeters MC, Piao X, Promel S, Schoneberg T, Schwartz TW, Singer K, Stacey M, Ushkaryov YA, Vallon M, Wolfrum U, Wright MW, Xu L, Langenhan T, Schioth HB (2015) International Union of Basic and Clinical Pharmacology. XCIV. Adhesion G protein-coupled receptors. Pharmacol Rev 67:338–367
241. Nishimori H, Shiratsuchi T, Urano T, Kimura Y, Kiyono K, Tatsumi K, Yoshida S, Ono M, Kuwano M, Nakamura Y, Tokino T (1997) A novel brain-specific p53-target gene, BAI1, containing thrombospondin type 1 repeats inhibits experimental angiogenesis. Oncogene 15:2145–2150
242. Park D, Tosello-Trampont AC, Elliott MR, Lu M, Haney LB, Ma Z, Klibanov AL, Mandell JW, Ravichandran KS (2007) BAI1 is an engulfment receptor for apoptotic cells upstream of the ELMO/Dock180/Rac module. Nature 450:430–434
243. Brugnera E, Haney L, Grimsley C, Lu M, Walk SF, Tosello-Trampont AC, Macara IG, Madhani H, Fink GR, Ravichandran KS (2002) Unconventional Rac-GEF activity is mediated through the Dock180-ELMO complex. Nat Cell Biol 4:574–582
244. Gumienny TL, Brugnera E, Tosello-Trampont AC, Kinchen JM, Haney LB, Nishiwaki K, Walk SF, Nemergut ME, Macara IG, Francis R, Schedl T, Qin Y, Van Aelst L, Hengartner MO, Ravichandran KS (2001) CED-12/ELMO, a novel member of the CrkII/Dock180/Rac pathway, is required for phagocytosis and cell migration. Cell 107:27–41
245. Park SY, Kim IS (2017) Engulfment signals and the phagocytic machinery for apoptotic cell clearance. Exp Mol Med 49:e331
246. Fredriksson R, Lagerstrom MC, Hoglund PJ, Schioth HB (2002) Novel human G protein-coupled receptors with long N-terminals containing GPS domains and Ser/Thr-rich regions. FEBS Lett 531:407–414
247. Lee JW, Huang BX, Kwon H, Rashid MA, Kharebava G, Desai A, Patnaik S, Marugan J, Kim HY (2016) Orphan GPR110 (ADGRF1) targeted by N-docosahexaenoylethanolamine in development of neurons and cognitive function. Nat Commun 7:13123
248. Hauser AS, Attwood MM, Rask-Andersen M, Schioth HB, Gloriam DE (2017) Trends in GPCR drug discovery: new agents, targets and indications. Nat Rev Drug Discov 16:829–842
249. Wishart DS, Feunang YD, Guo AC, Lo EJ, Marcu A, Grant JR, Sajed T, Johnson D, Li C, Sayeeda Z, Assempour N, Iynkkaran I, Liu Y, Maciejewski A, Gale N, Wilson A, Chin L, Cummings R, Le D, Pon A, Knox C, Wilson M (2018) DrugBank 5.0: a major update to the DrugBank database for 2018. Nucleic Acids Res 46:D1074–D1082
250. Ho BK, Gruswitz F (2008) HOLLOW: generating accurate representations of channel and interior surfaces in molecular structures. BMC Struct Biol 8:49
251. Voss NR, Gerstein M (2010) 3V: cavity, channel and cleft volume calculator and extractor. Nucleic Acids Res 38:W555–W562
252. Kiriyama M, Ushikubi F, Kobayashi T, Hirata M, Sugimoto Y, Narumiya S (1997) Ligand binding specificities of the eight types and subtypes of the mouse prostanoid receptors expressed in Chinese hamster ovary cells. Br J Pharmacol 122:217–224
253. Al-Salama ZT (2019) Siponimod: first global approval. Drugs 79:1009–1015
254. Gergely P, Nuesslein-Hildesheim B, Guerini D, Brinkmann V, Traebert M, Bruns C, Pan S, Gray NS, Hinterding K, Cooke NG, Groenewegen A, Vitaliti A, Sing T, Luttringer O, Yang J, Gardin A, Wang N, Crumb WJ Jr, Saltzman M, Rosenberg M, Wallstrom E (2012) The selective sphingosine 1-phosphate receptor modulator BAF312 redirects lymphocyte distribution and has species-specific effects on heart rate. Br J Pharmacol 167:1035–1047
255. Pan S, Gray NS, Gao W, Mi Y, Fan Y, Wang X, Tuntland T, Che J, Lefebvre S, Chen Y, Chu A, Hinterding K, Gardin A, End P, Heining P, Bruns C, Cooke NG, Nuesslein-Hildesheim B (2013) Discovery of BAF312 (Siponimod), a potent and selective S1P receptor modulator. ACS Med Chem Lett 4:333–337
256. Brinkmann V, Cyster JG, Hla T (2004) FTY720: sphingosine 1-phosphate receptor-1 in the control of lymphocyte egress and endothelial barrier function. Am J Transplant 4:1019–1025
257. Chiba K, Adachi K (2012) Discovery of fingolimod, the sphingosine 1-phosphate receptor modulator and its application for the therapy of multiple sclerosis. Future Med Chem 4:771–781
258. Cyster JG, Schwab SR (2012) Sphingosine-1-phosphate and lymphocyte egress from lymphoid organs. Annu Rev Immunol 30:69–94

259. Proia RL, Hla T (2015) Emerging biology of sphingosine-1-phosphate: its role in pathogenesis and therapy. J Clin Invest 125:1379–1387
260. Kirkby Shaw K, Rausch-Derra LC, Rhodes L (2016) Grapiprant: an EP4 prostaglandin receptor antagonist and novel therapy for pain and inflammation. Vet Med Sci 2:3–9
261. Miao J, Lu X, Hu Y, Piao C, Wu X, Liu X, Huang C, Wang Y, Li D, Liu J (2017) Prostaglandin E2 and PD-1 mediated inhibition of antitumor CTL responses in the human tumor microenvironment. Oncotarget 8:89802–89810
262. Christopoulos A, Changeux JP, Catterall WA, Fabbro D, Burris TP, Cidlowski JA, Olsen RW, Peters JA, Neubig RR, Pin JP, Sexton PM, Kenakin TP, Ehlert FJ, Spedding M, Langmead CJ (2014) International Union of Basic and Clinical Pharmacology. XC. multisite pharmacology: recommendations for the nomenclature of receptor allosterism and allosteric ligands. Pharmacol Rev 66:918–947
263. Shen Q, Wang G, Li S, Liu X, Lu S, Chen Z, Song K, Yan J, Geng L, Huang Z, Huang W, Chen G, Zhang J (2016) ASD v3.0: unraveling allosteric regulation with structural mechanisms and biological networks. Nucleic Acids Res 44:D527–D535
264. Ayan M, Essiz S (2018) The neural gamma2alpha-1beta2alpha1beta2 gamma amino butyric acid ion channel receptor: structural analysis of the effects of the ivermectin molecule and disulfide bridges. J Mol Model 24:206
265. Berman HM, Westbrook J, Feng Z, Gilliland G, Bhat TN, Weissig H, Shindyalov IN, Bourne PE (2000) The protein data bank. Nucleic Acids Res 28:235–242
266. Kruse AC, Ring AM, Manglik A, Hu J, Hu K, Eitel K, Hubner H, Pardon E, Valant C, Sexton PM, Christopoulos A, Felder CC, Gmeiner P, Steyaert J, Weis WI, Garcia KC, Wess J, Kobilka BK (2013) Activation and allosteric modulation of a muscarinic acetylcholine receptor. Nature 504:101–106
267. Liu X, Ahn S, Kahsai AW, Meng KC, Latorraca NR, Pani B, Venkatakrishnan AJ, Masoudi A, Weis WI, Dror RO, Chen X, Lefkowitz RJ, Kobilka BK (2017) Mechanism of intracellular allosteric beta2AR antagonist revealed by X-ray crystal structure. Nature 548:480–484
268. Negoro N, Sasaki S, Mikami S, Ito M, Suzuki M, Tsujihata Y, Ito R, Harada A, Takeuchi K, Suzuki N, Miyazaki J, Santou T, Odani T, Kanzaki N, Funami M, Tanaka T, Kogame A, Matsunaga S, Yasuma T, Momose Y (2010) Discovery of TAK-875: a potent, selective, and orally bioavailable GPR40 agonist. ACS Med Chem Lett 1:290–294
269. Badal S, Smith KN, Rajnarayanan R (2017) Analysis of natural product regulation of cannabinoid receptors in the treatment of human disease. Pharmacol Ther 180:24–48
270. Wisler JW, DeWire SM, Whalen EJ, Violin JD, Drake MT, Ahn S, Shenoy SK, Lefkowitz RJ (2007) A unique mechanism of beta-blocker action: carvedilol stimulates beta-arrestin signaling. Proc Natl Acad Sci U S A 104:16657–16662
271. Rankovic Z, Brust TF, Bohn LM (2016) Biased agonism: an emerging paradigm in GPCR drug discovery. Bioorg Med Chem Lett 26:241–250
272. Komatsu H, Fukuchi M, Habata Y (2019) Potential utility of biased GPCR signaling for treatment of psychiatric disorders. Int J Mol Sci 20:3207
273. Gundry J, Glenn R, Alagesan P, Rajagopal S (2017) A practical guide to approaching biased agonism at G protein coupled receptors. Front Neurosci 11:17
274. Tan L, Yan W, McCorvy JD, Cheng J (2018) Biased ligands of G protein-coupled receptors (GPCRs): structure-functional selectivity relationships (SFSRs) and therapeutic potential. J Med Chem 61:9841–9878
275. Ogawa S, Watanabe T, Sugimoto I, Moriyuki K, Goto Y, Yamane S, Watanabe A, Tsuboi K, Kinoshita A, Kigoshi H, Tani K, Maruyama T (2016) Discovery of G protein-biased EP2 receptor agonists. ACS Med Chem Lett 7:306–311
276. Ogawa S, Watanabe T, Moriyuki K, Goto Y, Yamane S, Watanabe A, Tsuboi K, Kinoshita A, Okada T, Takeda H, Tani K, Maruyama T (2016) Structural optimization and structure-functional selectivity relationship studies of G protein-biased EP2 receptor agonists. Bioorg Med Chem Lett 26:2446–2449
277. Shimizu Y, Nakayama M (2017) Discovery of novel Gq-biased LPA1 negative allosteric modulators. SLAS Discov 22:859–866
278. Galvani S, Sanson M, Blaho VA, Swendeman SL, Obinata H, Conger H, Dahlback B, Kono M, Proia RL, Smith JD, Hla T (2015) HDL-bound sphingosine 1-phosphate acts as a biased agonist for the endothelial cell receptor S1P1 to limit vascular inflammation. Sci Signal 8:ra79
279. Christoffersen C, Obinata H, Kumaraswamy SB, Galvani S, Ahnstrom J, Sevvana M, Egerer-Sieber C, Muller YA, Hla T, Nielsen LB, Dahlback B (2011) Endothelium-protective sphingosine-1-phosphate provided by HDL-associated apolipoprotein M. Proc Natl Acad Sci U S A 108:9613–9618
280. Neubig RR, Spedding M, Kenakin T, Christopoulos A, International Union of Pharmacology Committee on Receptor, and C. Drug (2003) International Union of Pharmacology Committee on receptor nomenclature and drug classification. XXXVIII. Update on terms and symbols in quantitative pharmacology. Pharmacol Rev 55:597–606
281. Quancard J, Bollbuck B, Janser P, Angst D, Berst F, Buehlmayer P, Streiff M, Beerli C, Brinkmann V, Guerini D, Smith PA, Seabrook TJ, Traebert M, Seuwen K, Hersperger R, Bruns C, Bassilana F, Bigaud M (2012) A potent and selective S1P(1) antagonist with efficacy in experimental autoimmune encephalomyelitis. Chem Biol 19:1142–1151
282. Leavy O (2010) Therapeutic antibodies: past, present and future. Nat Rev Immunol 10:297

283. Dolgin E (2018) First GPCR-directed antibody passes approval milestone. Nat Rev Drug Discov 17:457–459
284. Ureshino H, Kamachi K, Kimura S (2019) Mogamulizumab for the treatment of adult T-cell leukemia/lymphoma. Clin Lymphoma Myeloma Leuk 19:326–331
285. Dibba P, Li AA, Cholankeril G, Ali Khan M, Kim D, Ahmed A (2019) Potential mechanisms influencing the inverse relationship between cannabis and nonalcoholic fatty liver disease: a commentary. Nutr Metab Insights 12:1178638819847480

Index

Y. Kihara (ed.), *Druggable Lipid Signaling Pathways*, Advances in Experimental Medicine and Biology 1274, https://doi.org/10.1007/978-3-030-50621-6

The manufacturer's authorised representative in the EU is Springer Nature Customer Service Centre GmbH, Europaplatz 3, 69115 Heidelberg, Germany. If you have any concerns regarding our products, please contact ProductSafety@springernature.com

Printed and bound by CPI Group (UK) Ltd, Croydon, CR0 4YY
28/09/2026
02224085-0019